Methods in Molecular Biology

Series Editor
John M. Walker
School of Life and Medical Sciences
University of Hertfordshire
Hatfield, Hertfordshire, AL10 9AB, UK

For further volumes:
http://www.springer.com/series/7651

Chemotaxis

Methods and Protocols

Edited by

Tian Jin

Laboratory of Immunogenetics, National Institute of Allergy and Infectious Diseases, NIH, Rockville, MD, USA

Dale Hereld

National Institute Alcohol Abuse and Alcoholism, National Institutes of Health (NIH), Rockville, MD, USA

Editors
Tian Jin
Laboratory of Immunogenetics
National Institute of Allergy and Infectious Diseases, NIH
Rockville, MD, USA

Dale Hereld
National Institutes Alcohol Abuse and Alcoholism
National Institutes of Health (NIH)
Rockville, MD, USA

ISSN 1064-3745 ISSN 1940-6029 (electronics)
Methods in Molecular Biology
ISBN 978-1-4939-3478-2 ISBN 978-1-4939-3480-5 (eBook)
DOI 10.1007/978-1-4939-3480-5

Library of Congress Control Number: 2016937336

Printed on acid-free paper

This Humana Press imprint is published by Springer Nature
The registered company is Springer Science+Business Media LLC New York

Preface

Cell movement is fundamental to the development and other vital functions of organisms. Many motile cells can detect shallow gradients of specific chemical signals in their environments and migrate accordingly. This directed cell movement is called chemotaxis and is essential for various cell types to carry out their biological functions. This book is the second edition of *Chemotaxis: Methods and Protocols* in the Springer Protocols series *Methods in Molecular Biology*. This edition includes new chapters describing methods for studying cell movement, molecular components involved in chemotaxis, spatiotemporal dynamics of signaling components, and quantitative modeling, as well as several updated chapters from the first edition.

Various methods to investigate directional cell growth and movement are presented in Chapters 1–20. These chapters contain experimental procedures to visualize and measure migration behaviors of different kinds of organisms, including chemotropism in the budding yeast, cell growth and migration of *D. discoideum*, border cell migration in *Drosophila*, chemotaxis of mouse and human neutrophils, and HIV-induced T cell chemotactic response. The volume also contains microscopy procedures for studying breast cancer cell migration, tumor cell invasion in vivo, and axon guidance. These methods allow quantitative measurement and description of cell migration behaviors.

Fluorescence microscopy permits us to directly monitor the dynamics of many signaling events in single cells in real time. Chapters 21–25 describe methods that measure spatiotemporal dynamics of signaling components involved in chemotaxis. Several chapters introduce cutting-edge imaging techniques, such as TIRF, BRET, FRET, and single-molecule microscopy. These techniques allow us to reveal the dynamics of signaling components in live cells and to track signaling events in single cells in space and time.

Quantitative models can address the complexity of the signaling network that underlies chemotaxis. To this end, Chapter 26 presents mathematical models of experimentally generated chemoattractant gradients. Finally, Chapter 27 introduces a computational modeling approach to understanding the chemotactic signaling network.

We are grateful to all the authors for contributing their expertise and believe that this book will provide the reader with an overview of and practical guidance on the diverse methodologies that are propelling chemotaxis research forward.

Rockville, MD, USA

Dale Hereld
Tian Jin

Contents

Contributors

SARAH J. ANNESLEY • *Discipline of Microbiology, La Trobe University, Bundoora, VIC, Australia*
ROBERT A. ARKOWITZ • *Institute of Biology Valrose, CNRS/INSERM/Université Nice-Sophia Antipolis, Nice, France*
KARL J. AUFDERHEIDE • *Department of Biology, Texas A&M University, College Station, TX, USA*
ABDUL BASIT • *Department of Pharmacology, Vascular Biology and Therapeutic Program, Yale University School of Medicine, New Haven, CT, USA*
CARSTEN BETA • *Institute of Physics and Astronomy, University of Potsdam, Potsdam, Germany*
SAYAK BHATTACHARYA • *Department of Electrical and Computer Engineering, The John Hopkins University, Baltimore, MD, USA*
DENNIS BREITSPRECHER • *NanoTemper Technologies GmbH, Munich, Germany*
TILL BRETSCHNEIDER • *Warwick Systems Biology Centre, University of Warwick, Coventry, UK*
STEFAN BRÜHMANN • *Hannover Medical School, Institute for Biophysical Chemistry, Hannover, Germany*
JOSEPH BRZOSTOWSKI • *Imaging Core Facility, Laboratory of Immunogenetics, National Institute of Allergy and Infectious Diseases, NIH, Rockville, MD, USA*
XIUMEI CAO • *Shanghai Institute of Immunology, Shanghai Jiao Tong University School of Medicine, Shanghai, China*
PASCALE G. CHAREST • *Department of Chemistry and Biochemistry, University of Arizona, Tucson, AZ, USA*
MARCO COCOROCCHIO • *Centre for Biomedical Sciences, School of Biological Sciences, Royal Holloway University of London, Surrey, UK*
JUSTIN COOPER-WHITE • *Australian Institute for Bioengineering and Nanotechnology and School of Chemical Engineering, The University of Queensland, St. Lucia, QLD, Australia*
WEI DAI • *Molecular, Cellular, and Developmental Biology Department, University of California, Santa Barbara, CA, USA*
BRUCE W. DRAPER • *Department of Molecular and Cellular Biology, School of Medicine, University of California-Davis, Sacramento, CA, USA*
MARY ECKE • *Max Planck Institute of Biochemistry, Martinsried, Germany*
JAN FAIX • *Institute for Biophysical Chemistry, Hannover Medical School, Hannover, Germany*
PAUL R. FISHER • *Discipline of Microbiology, La Trobe University, Bundoora, VIC, Australia*
CHRISTOF FRANKE • *Hannover Medical School, Institute for Biophysical Chemistry, Hannover, Germany*
MATTHIAS GERHARDT • *Institute of Physics and Astronomy, University of Potsdam, Potsdam, Germany*

GÜNTHER GERISCH • *Max Planck Institute of Biochemistry, Martinsried, Germany*
NICK GLASS • *Australian Institute for Bioengineering and Nanotechnology and School of Chemical Engineering, The University of Queensland, St. Lucia, QLD, Australia*
GEOFFREY J. GOODHILL • *Queensland Brain Institute and School of Mathematics and Physics, The University of Queensland, St. Lucia, QLD, Australia*
PETER J.M. VAN HAASTERT • *Department of Cell Biochemistry, University of Groningen, Groningen, The Netherlands*
IL-YOUNG HWANG • *Laboratory of Immunoregulation, National Institute of Allergy and Infectious Diseases, NIH, Bethesda, MD, USA*
PABLO A. IGLESIAS • *Department of Electrical and Computer Engineering, The John Hopkins University, Baltimore, MD, USA; Department of Cell Biology, The John Hopkins University, Baltimore, MD, USA*
ROBERT H. INSALL • *CR-UK Beatson Institute for Cancer Research, Glasgow, UK*
A.F.M. TARIQUL ISLAM • *Department of Chemistry and Biochemistry, University of Arizona, Tucson, AZ, USA*
CHRIS JANETOPOULOS • *Department of Biological Sciences, University of the Sciences, Philadelphia, PA, USA*
TIAN JIN • *Laboratory of Immunogenetics, Chemotaxis Signaling Section, National Institute of Allergy and Infectious Diseases, NIH, Rockville, MD, USA*
SAM A. JOHNSON • *Light Microscopy Core Facility, Duke University, Durham, NC, USA*
ROBERT R. KAY • *MRC Laboratory of Molecular Biology, Cambridge, UK*
JOHN H. KEHRL • *Laboratory of Immunoregulation, National Institute of Allergy and Infectious Diseases, NIH, Bethesda, MD, USA*
NANCY D. KIM • *Division of Rheumatology, Allergy, and Immunology, Center for Immunology and Inflammatory Diseases, Harvard Medical School, Massachusetts General Hospital, Boston, MA, USA*
ALAN R. KIMMEL • *Laboratory of Cellular and Developmental Biology, National Institute of Diabetes and Digestive and Kidney Diseases, NIH, Bethesda, MD, USA*
SPENCER KUHL • *W.M. Keck Dynamic Image Analysis Facility, Department of Biological Sciences, University of Iowa, Iowa City, IA, USA*
SUI LAY • *Discipline of Microbiology, La Trobe University, Bundoora, VIC, Australia*
DANIEL J. LEW • *Department of Pharmacology and Cancer Biology, Duke University School of Medicine, Durham, NC, USA*
DANIEL F. LUSCHE • *W.M. Keck Dynamic Image Analysis Facility, Department of Biological Science, University of Iowa, Iowa City, IA, USA*
ANDREW D. LUSTER • *Division of Rheumatology, Allergy and Immunology, Center for Immunology and Inflammatory Diseases, Harvard Medical School, Massachusetts General Hospital, Boston, MA, USA*
LAURA MARTÍNEZ-MUÑOZ • *Department of Immunology and Oncology, Centro Nacional de Biotecnología/CSIC, Madrid, Spain*
SATOMI MATSUOKA • *Laboratory for Cell Signaling Dynamics, RIKEN Quantitative Biology Center, Osaka, Japan*
ALLISON W. MCCLURE • *Department of Pharmacology and Cancer Biology, Duke University School of Medicine, Durham, NC, USA*
NETRA PAL MEENA • *Laboratory of Cellular and Developmental Biology, National Institute of Diabetes and Digestive and Kidney Diseases, NIH, Bethesda, MD, USA*

MARIO MELLADO • *Department of Immunology and Oncology, Centro Nacional de Biotecnología/CSIC, Madrid, Spain*
CHIE MIYABE • *Division of Rheumatology, Allergy and Immunology, Center for Immunology and Inflammatory Diseases, Harvard Medical School, Massachusetts General Hospital, Boston, MA, USA*
YOSHISHIGE MIYABE • *Division of Rheumatology, Allergy and Immunology, Center for Immunology and Inflammatory Diseases, Harvard Medical School, Massachusetts General Hospital, Boston, MA, USA*
YUKIHIRO MIYANAGA • *Laboratory of Single Molecule Biology, Department of Biological Sciences, Graduate School of Science, Osaka University, Osaka, Japan*
ALEX MOGILNER • *Courant Institute and Department of Biology, New York University, New York, NY, USA*
DENISE J. MONTELL • *Molecular, Cellular, and Developmental Biology Department, University of California, Santa Barbara, CA, USA*
ANDREW J. MUINONEN-MARTIN • *Melanoma Clinic, Leeds Cancer Centre, St. James's University Hospital, Leeds, UK*
LAURA MUNOZ • *Centre for Biomedical Sciences, School of Biological Sciences, Royal Holloway University of London, Surrey, UK*
AKIHIKO NAKAJIMA • *Graduate School of Arts and Sciences, University of Tokyo, Tokyo, Japan*
HUYEN NGUYEN • *Queensland Brain Institute and School of Mathematics and Physics, The University of Queensland, St. Lucia, QLD, Australia*
MAX NOBIS • *CR-UK Beatson Institute for Cancer Research, Glasgow, UK*
GRANT P. OTTO • *Centre for Biomedical Sciences, School of Biological Sciences, Royal Holloway University of London, Surrey, UK*
CHUNG PARK • *Laboratory of Immunoregulation, National Institute of Allergy and Infectious Diseases, NIH, Bethesda, MD, USA*
MARTEN POSTMA • *Swammerdam Institute for Life Sciences, University of Amsterdam, Amsterdam, The Netherlands*
ZAC PUJIC • *Queensland Brain Institute, The University of Queensland, St. Lucia, QLD, Australia*
WEI QUAN • *Laboratory of Immunogenetics, Chemotaxis Signaling Section, National Institute of Allergy and Infectious Diseases, NIH, Rockville, MD, USA*
JOSÉ MIGUEL RODRÍGUEZ-FRADE • *Department of Immunology and Oncology, Centro Nacional de Biotecnología/CSIC, Madrid, Spain*
OANA SANISLAV • *Discipline of Microbiology, La Trobe University, Bundoora, VIC, Australia*
SATOSHI SAWAI • *Graduate School of Arts and Sciences, University of Tokyo, Tokyo, Japan*
AMANDA SCHERER • *W.M. Keck Dynamic Image Analysis Facility, Department of Biological Sciences, University of Iowa, Iowa City, IA, USA*
DAVID R. SOLL • *Department of Biology, The University of Iowa, Iowa City, IA, USA*
BRANDEN M. STEPANSKI • *Department of Chemistry and Biochemistry, University of Arizona, Tucson, AZ, USA*
DAVID E. STONE • *Department of Biological Sciences, University of Illinois at Chicago, Chicago, IL, USA*
YAO-HUI SUN • *Department of Dermatology, School of Medicine, University of California-Davis, Sacramento, CA, USA*

YUXIN SUN • *Department of Dermatology, School of Medicine, University of California-Davis, Sacramento, CA, USA*
OLIVIA SUSANTO • *CR-UK Beatson Institute for Cancer Research, Glasgow, UK*
WENWEN TANG • *Department of Pharmacology, Vascular Biology and Therapeutic Program, Yale University School of Medicine, New Haven, CT, USA*
RICHARD A. TYSON • *Warwick Systems Biology Centre, University of Warwick, Coventry, UK*
MASAHIRO UEDA • *Laboratory of Single Molecule Biology, Department of Biological Sciences, Graduate School of Science, Osaka University, Osaka, Japan*
EDWARD VOSS • *W.M. Keck Dynamic Image Analysis Facility, Department of Biological Sciences, University of Iowa, Iowa City, IA, USA*
MICHAEL WALZ • *Institute of Physics and Astronomy, University of Potsdam, Potsdam, Germany*
Q. JANE WANG • *Department of Pharmacology and Chemical Biology, University of Pittsburgh School of Medicine, Pittsburgh, PA, USA*
XI WEN • *Laboratory of Immunogenetics, Chemotaxis Signaling Section, National Institute of Allergy and Infectious Diseases, NIH, Rockville, MD, USA*
DEBORAH WESSELS • *W.M. Keck Dynamic Image Analysis Facility, Department of Biological Sciences, University of Iowa, Iowa City, IA, USA*
ROBIN S.B. WILLIAMS • *Centre for Biomedical Sciences, School of Biological Sciences, Royal Holloway University of London, Surrey, UK*
MORITZ WINTERHOFF • *Institute for Biophysical Chemistry, Hannover Medical School, Hannover, Germany*
CHI-FANG WU • *Department of Pharmacology and Cancer Biology, Duke University School of Medicine, Durham, NC, USA*
DIANQING WU • *Department of Pharmacology, Vascular Biology and Therapeutic Program, Yale University School of Medicine, New Haven, CT, USA*
YUNTAO WU • *National Center for Biodefense and Infectious Diseases, School of System Biology, George Mason University, Manassas, VA, USA*
XUEHUA XU • *Laboratory of Immunogenetics, Chemotaxis Signaling Section, National Institute of Allergy and Infectious Diseases, NIH, Rockville, MD, USA*
JIANSHE YAN • *Shanghai Institute of Immunology, Shanghai Jiao Tong University School of Medicine, Shanghai, China*
MICHELLE YUN • *Laboratory of Immunogenetics, Chemotaxis Signaling Section, National Institute of Allergy and Infectious Diseases, NIH, Rockville, MD, USA*
EVGENY ZATULOVSKIY • *Department of Biology, Stanford University, Stanford, CA, USA*
MIN ZHAO • *Departments of Dermatology and Ophthalmology, School of Medicine, University of California-Davis, Sacramento, CA, USA*
KAN ZHU • *Department of Dermatology, School of Medicine, University of California-Davis, Sacramento, CA, USA*

Chapter 1

In Situ Assays of Chemotropism During Yeast Mating

David E. Stone and Robert A. Arkowitz

Abstract

Virtually all eukaryotic cells can grow in a polarized fashion in response to external signals. Cells can respond to gradients of chemoattractants or chemorepellents by directional growth, a process referred to as chemotropism. The budding yeast *Saccharomyces cerevisiae* undergoes chemotropic growth during mating, in which two haploid cells of opposite mating type grow towards one another. Mating pheromone gradients are essential for efficient mating in yeast and different yeast mutants are defective in chemotropism. Two methods of assessing the ability of yeast strains to respond to pheromone gradients are presented here.

Key words Chemotropism, Mating pheromone, Bud scar, Gradient, Zygote

1 Introduction

In the budding yeast *Saccharomyces cerevisiae*, haploid cells respond to mating pheromone peptides via G-protein-coupled receptors and their associated heterotrimeric G-proteins. Activation of these receptors results in cell cycle arrest, transcriptional activation via a MAP-kinase cascade, and polarized growth toward the mating partner [1–3]. While mating pheromone itself is sufficient to induce the cell cycle arrest, transcriptional activation, and cell shape changes (to form a pear-shaped cell called a shmoo) that occur during mating, the mating pheromone gradient that is generated during the mating process is necessary for the oriented growth toward the mating partner [4–7]; this chemotropic process is similar to that required for axonal guidance [1].

Yeast chemotropism comprises two distinct phenomena: (1) *Orientation*, defined as the initial alignment of actin cables along the axis of polarity, clustering of polarity markers, and the observable emergence of the mating projection toward the strongest pheromone source. Orientation requires molecular mechanisms we collectively term *gradient sensing*; (2) *Reorientation*, defined as the observable bending of a mating projection or initiation of a

Tian Jin and Dale Hereld (eds.), *Chemotaxis: Methods and Protocols*, Methods in Molecular Biology, vol. 1407,
DOI 10.1007/978-1-4939-3480-5_1,

second projection in response to a change in the position of the strongest pheromone source. Reorientation requires molecular mechanisms we collectively term *gradient tracking*. Although orientation has been much studied and modeled, reorientation has received relatively little attention, despite its broad relevance to processes such as axonal guidance and angiogenesis.

By definition, chemotropism is the directed growth aligned by a chemical gradient. This phenomenon depends on two experimentally separable capabilities: environmental-cue mediated positioning of the polarity site, and polarized growth continually reassessed by gradient tracking. Historically, genes and mechanisms have been implicated in yeast chemotropism on the basis of indirect, endpoint analyses such as the partner discrimination [8, 9] and pheromone confusion assays [4]. Because these methods begin with high-density mating mixtures, in which cells are touching potential mating partners, they require little to no chemotropic growth and continuous tracking of dynamic gradients is minimal. With the advent of microfluidic devices capable of forming stable, linear pheromone concentration gradients, direct observation of chemotropic shmooing, first established by Segal [6] has become common in recent years. Numerous studies now use microfluidic devices of various designs [10–18]. The ostensible advantage of such devices is that the gradient parameters (pheromone concentration and slope) can be easily manipulated. Moreover, time-lapse imaging and the use of fluorescent reporters are easily incorporated in the assays. However, there are also a number of limitations in the use of gradient-forming microfluidic devices.

In order to elicit a robust directional response, artificial gradients must be much steeper than those thought to be effective in situ. Artificial gradients are also relatively static in contrast to the highly dynamic and complex gradients that cells encounter in mating mixtures. Rather than the single source of pheromone that cells are exposed to in microfluidic chambers, mating cells must interpret the relative strength of pheromone signals coming from all directions. In addition, the myriad pheromone sources decrease over time, as partners form zygotes and turn off mating functions. Even the strength (concentration) and shape (slope) of a single pheromone gradient is likely to change as the responding cell grows towards its source, in part due to the secreted pheromone protease Bar1. With one exception [10], artificial gradient makers cannot rotate the direction of the gradient in increments less than 180°, which is required to study reorientation. Finally, there is evidence that Bar1 secretion has an important role in shaping local pheromone gradients [11, 14, 17, 19]. As a result, the artificial gradients formed in microfluidic devices fall far short of mimicking the complexities of pheromone gradients in a mating mixture.

The advantage of being able to form a range of predetermined gradients in vitro might well outweigh the limitations enumerated above if the chambers were able to elicit a more robust chemotropic response. Unfortunately, yeast cells simply do not orient effectively in artificial pheromone gradients produced by the microfluidic devices published thus far: The average angle of orientation is about 60° across all studies [10–18], as compared to 5–10° measured in situ using method 1 described below. Artificial gradients are not particularly effective, and studies using them are therefore likely to miss important aspects of the yeast chemotropic response.

In situ assays are needed to study yeast chemotropism. Here we describe two simple, accurate, and reproducible methods to assay chemotropism in mating mixtures. These methods can be used to study both orientation and reorientation.

2 Materials

2.1 Cell Growth and Labeling

1. YEPD rich medium (*see* **Note 1**) can either be purchased ready-made (Sigma-Aldrich, St. Louis, MO; Clontech, Mountain View, CA; or MP Biomedicals, Solon, OH) or made from individual components. YEPD medium comprises 11 g yeast extract, 22 g bactopeptone, 55 mg adenine sulfate, and 20 g glucose per liter (*see* **Notes 2** and **3**). YEPD plates contain, in addition, 22 g agar per liter (*see* **Note 4**).
2. Synthetic dropout medium can either be purchased ready-made (Sigma-Aldrich, Clontech or MP Biomedicals) or made from individual components (*see* **Note 5**). Synthetic dropout medium comprises 8 g yeast nitrogen base, 55 mg adenine sulfate, 55 mg tyrosine, 55 mg uracil (for Ura– strains), and 20 g glucose per liter. Synthetic dropout plates contain, in addition, 22 g agar per liter. A 100× concentrated amino acid dropout stock contains: 2 g arginine, 1 g histidine, 6 g isoleucine, 6 g leucine, 4 g lysine, 1 g methionine, 6 g phenylalanine, 5 g threonine, and 4 g tryptophan per liter. Amino acids necessary for selection are omitted. Synthetic complete (SC) medium has no amino acids omitted from the dropout.
3. Calcofluor white Fluorescent Brightener 28 (M2R) (Sigma-Aldrich or MP Biomedicals) is made up as a stock solution of 10 mg/mL in water, and diluted solutions of 1 mg/mL (in water) are prepared from this stock and used for daily experiments (*see* **Note 6**).
4. Concanavalin A-Alexa Fluor conjugates, 488, 594, or 633 (Molecular Probes) are made up as a stock solution of 5 mg/mL in 100 mM sodium bicarbonate pH 8.3 containing 1 mM $CaCl_2$ and 1 mM $MnCl_2$ and stored frozen. Prior to use, this stock solution is centrifuged in a microfuge at maximum speed

(e.g., Eppendorf microfuge, 16,100 × *g*) for 1 min to remove particulate matter and diluted solutions of 10 μg/mL (in water) are prepared fresh.

5. YEPD agar pads for microscopy experiments, are generated by melting YEPD agar in a microwave and letting it cool in a heat block to 50–60 °C. A roughly 2.5 × 2 cm^2 piece of 0.8–1 mm thick non-reinforced silicone sheet or sealing gasket from a protein gel dryer (*see* **Note 7**) in which a ~7 mm diameter hole has been punched out (using for example a cork borer) is also cleaned with water and then ethanol, dried and then pressed with a gloved hand onto a microscope slide (also cleaned with water and then ethanol) such that it adheres. Roughly 10–20 μl of the molten media is then rapidly spotted into the hole of the silicone on the slide such onto a slide heated on a 50–60 °C heat block, such that there is excess liquid (held in by surface tension) and no air bubbles. A cleaned coverslip is gently placed on the liquid, pressing gently with a gloved finger and then removed from the heat block to cool. After cooling the slide is kept in a humidified chamber (a Tupperware box equilibrated with water) for 1–3 days at 30 °C prior to use (*see* **Note 8**).

3 Methods

We describe here two methods that can be used to assess the relative ability of yeast cells to sense and track pheromone gradients in situ: (1) measurement of the fusion angle upon zygote formation, and (2) quantification of partner discrimination in low-density mating mixtures. The first method is based on the observation that the angle created when two mating cells fuse is a function of how precisely they orient towards one another. When orientation is optimal, the two cells of a mating pair grow directly toward one another, and consequently, their angle of fusion is ~0°. In contrast, large fusion angles are indicative of poor orientation (Fig. 1) [20]. The second method is based on the partner discrimination assay (PDA) [8, 9] and takes advantage of the fact that cells able to sense pheromone gradients can distinguish between potential mating partners that secrete pheromone and those that do not, preferentially mating with the former. Cells that are fully functional gradient sensors will mate only with pheromone secretors, even when non-secretors outnumber them 20:1. In the original PDA, however, mating was performed on filter disks at high cell density, which minimized the need for the formation of chemotropic mating projections that depend on continuous gradient tracking, as well as initial orientation, to find their targets. In the modified assay, the mating mixture is spread on a plate at lower cell density. Partner discrimination requires chemotropic shmooing, and therefore reflects the full range of chemotropic functions. The modified assay is especially well suited to study reorientation.

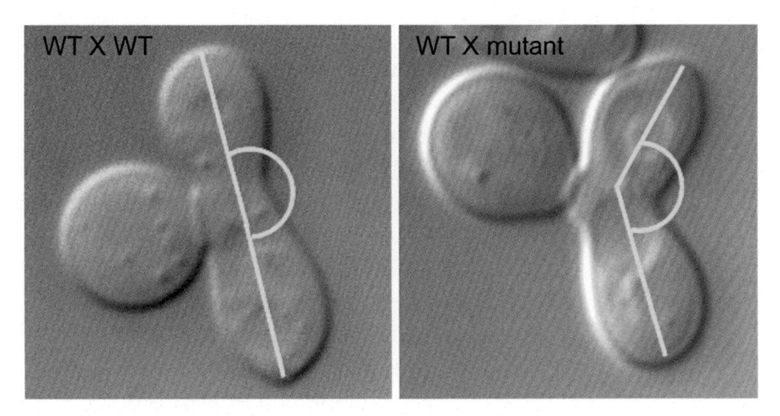

Fig. 1 The zygote formation assay can detect subtle defects in gradient sensing. Representative straight and angled zygotes from a wild type (WT) mating mixture (*left*), and a mixture of WT cells with chemotropism-defective cells. The mean angle of orientation (*yellow arc*) ± sem was 8.68 ± 0.65° for the WT cross (n = 114), and 18.27 ± 1.80° for the unilateral mutant cross (n = 92; p = 0.0001) after 4 h of mating

3.1 Zygote Formation Assay

1. Wild-type or mutant yeast *MAT***a** strains and a wild-type *MAT*α tester strain are grown overnight at in selective medium and back diluted into 5 mL of rich media (YEPD) to an OD_{600} of 0.1 in the morning. Cells are then grown to an OD_{600} of 0.6–0.8 (at least two doublings).
2. Volumes corresponding to 0.25 OD_{600} of each mating type are transferred to individual sterile microcentrifuge tubes. The cells are then centrifuged at maximum speed in an Eppendorf centrifuge at room temperature for 10–20 s, the supernatants removed and the *MAT***a** cells resuspended in 0.5 mL conditioned media, i.e., supernatant media from respectively grown cultures and the *MAT*α resuspended in 0.5 mL fresh media. The wild-type *MAT*α mating partner cells are stained with 10 μg/mL ConA-Alexa Fluor conjugates (488, 594, or 633 nm from Molecular Probes, Eugene, Oregon) for 10–20 min and then washed 3× with YEP media and resuspended in an equal volume of conditioned media (*see* **Notes 9** and **10**). The entire volumes of both cell mating types are combined in an eppendorf tube, which is mixed by vortexing briefly. The mating mixture is then sonicated in a bath sonicator four to six times for 1 s (with 1 s pause between pulses) at roughly 20–40 % power (*see* **Note 11**).
3. The coverslip on the agar pad slide is carefully removed by sliding it across the surface. A fresh coverslip is used to remove, by scraping with the edge, the excess agar on the silicone or sealing gasket surface, such that the circular agar surface is the only agar that rises slightly above the silicone or sealing gasket surface.

4. Cells (4–8 μL) are carefully spotted on the agar pad, avoiding in particular damaging the agar surface and air bubbles. This works out to ~20,000–40,000 cells per pad. However this density can by modulated depending on the strain. The liquid on the agar pad is air-dried (5–10 min at room temperature), rotating the slide in three dimensions as would a nutator to ensure even distribution of liquid while drying (*see* **Note 12**). Once the liquid has dried, a cleaned square coverslip is gently placed on the pad, making sure not to trap air bubbles. Four small drops of clear nail polish are then put, one at each corner, to fix the coverslip to the slide. Wait for sufficient time (5–10 min) for the nail polish to dry.
5. The slide is transferred to a humidified (H_2O saturated) microscope incubation chamber (such as Solent and Oko-lab; http://www.solentsci.com/products.htm and http://oko-lab.com/H201-Enclosure.page). Zygotes (dumbbell-shaped cells) or potential mating pairs (one fluorescent cell and one non-fluorescent cell) with no cells between them (typically with 1–2 cell diameters between cells) are identified. The *X*/*Y* coordinates of five to ten mating pairs are marked (either noted or saved in microscope point visiting software), and a transmission or DIC (the latter is preferable) image is acquired together with fluorescent image to identify mating partner.
6. Transmission or DIC images of each mating pair are acquired every hour (*see* **Notes 13** and **14**) for 3–6 h, depending on when zygotes form (Fig. 2).
7. Fusion angles are determined when zygotes first appear. The ImageJ angle tool is used to draw a straight line from the center of the back of one mating partner to the fusion zone, and then to the center of the back of the other mating partner (*see* **Note 15**). Angles of growth of two cells of opposite mating type growing towards one another can also be measured at different time points by drawing a straight line from the center of back of the cell to the shmoo tip for both cells.
8. Typically angles from between 50 and 150 cells are determined (Fig. 1).

Fig. 2 (continued) *Arrowheads* indicate localization of the receptor to the default polarity site; *Arrows* indicate redistribution of the receptor to the chemotropic site. (**b**) The α-factor receptor polarizes towards distant mating partners prior to morphogenesis. *MAT***a** cells expressing the endogenously tagged STE2-GFP reporter were mixed with *MAT*α cells pre-stained with ConA-Alexa Fluor 594 (labeled α) and incubated at 30 °C for 30 min. *Arrows* indicate Ste2-GFP crescents. Bar, 5 μm. From Suchkov et al. [23]. Reprinted with permission. (**c**) Dynamic localization of the pheromone receptor as shmooing cells reorient towards mating partners. *MAT***a** cells expressing the endogenously tagged STE2-GFP reporter were induced to shmoo in isotropic α-factor, then mixed with *MAT*α cells and incubated at 30 °C for the indicated times. *Arrowheads* indicate the redistribution of the receptor from the initial shmoo site to the chemotropic site

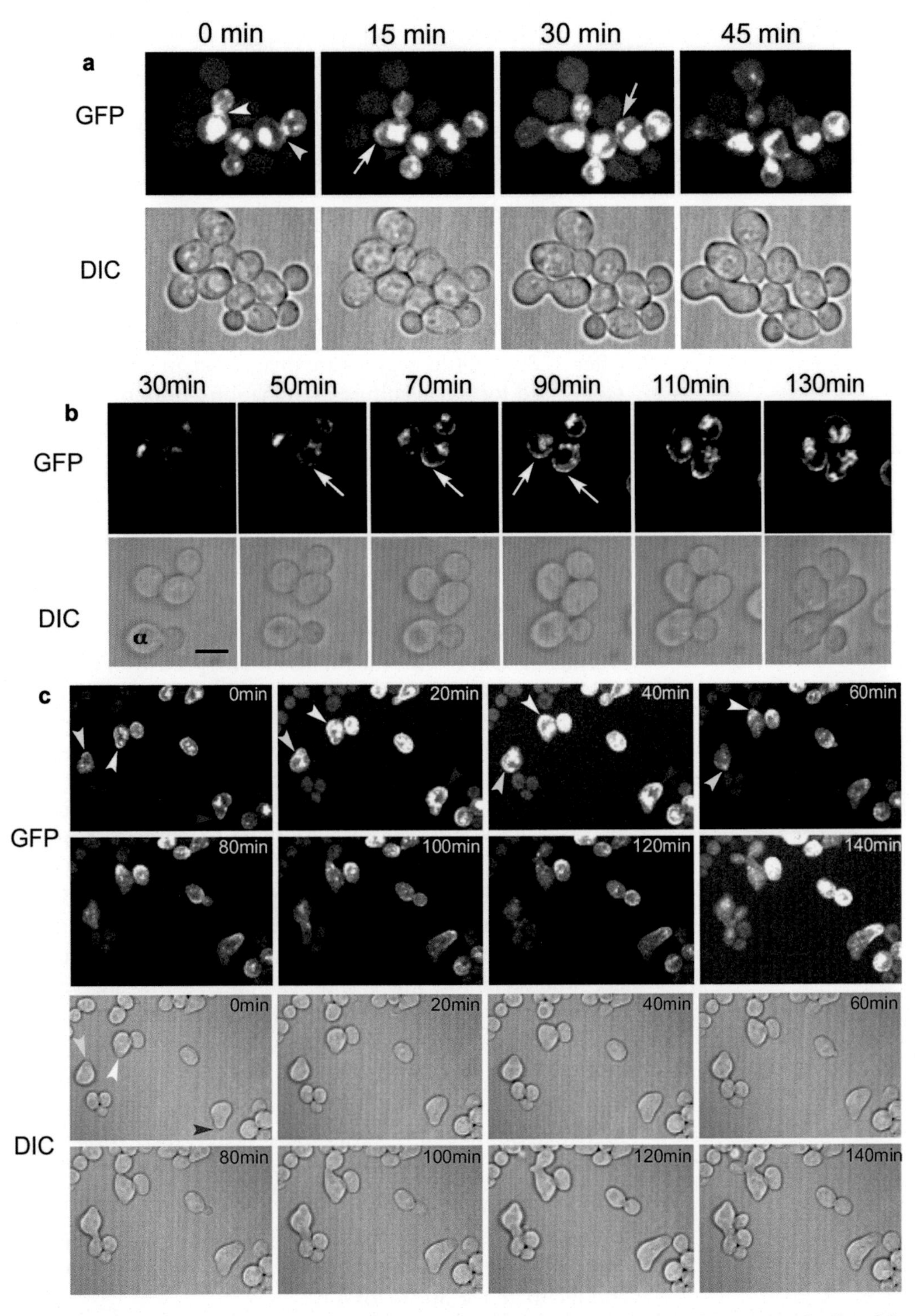

Fig. 2 Examples of time-lapse zygote formation assay experiments on agar pads. (**a**) Dynamic localization of the pheromone receptor as cells orient toward mating partners. *MAT***a** cells expressing an endogenously tagged α-factor receptor, STE2-GFP, were mixed with *MAT*α cells and incubated at 30 °C for the indicated times.

3.2 Variation on Agar Plates

1. Cells grown similarly as above except after centrifugation, cells are resuspended each in 0.1 mL of fresh media.
2. In **step 2** above, cells are resuspended in 100 μL of conditioned media, combined, vortexed, sonicated, and then plated on YEPD rich media plates. For plating, the mixtures are spotted onto the plates and 10–15 sterilized glass beads added to each plate. After spotting and adding glass beads to several plates, the plates are stacked and slowly shaken back and forth, turning every ~10 s. After shaking for 2–3 min and ensuring the plates are visibly dry, the beads are poured into a waste container. Plates are incubated face up at 30 °C.
3. After 2, 3, and 4 h, zygotes are imaged on an inverted microscope used for tissue culture examination (400× magnification is optimal).

3.3 Low-Density Partner Discrimination Assay

1. Wild-type or mutant yeast *MAT***a** strains, the *MAT*α pheromone-secreting strain, and the *MAT*α non-secreting strain (*see* **Note 16**) are grown overnight at 30 °C in rich media (YEPD) and back diluted to an OD_{600} of 0.1 in the morning. Cells are then grown to an OD_{600} of 0.6–0.8 (at least two doublings).
2. At least 0.5 mL of the pheromone-secreting *MAT*α strain culture is centrifuged in a sterile tube at maximum speed in an Eppendorf centrifuge at room temperature for 10–20 s. The supernatant is carefully removed without taking cells and kept in a separate tube (referred to as conditioned media). Next, volumes corresponding to 2.27 OD_{600} of non-secreting *MAT*α strain and 0.227 OD_{600} of the pheromone-secreting *MAT*α strain are similarly centrifuged (non-secretor–secretor ratio of 10:1, *see* **Note 17**) and 2.38 OD_{600} of non-secreting *MAT*α strain and 0.119 OD_{600} of pheromone-secreting *MAT*α strain are also centrifuged (non-secretor–secretor ratio of 20:1, *see* **Note 17**). This will require two to four centrifugations of a full 1.5 mL eppendorf tube for the non-secreting strain and all supernatants are carefully removed. The conditioned media (100 μL for each tube) is then used to resuspend each cell pellet.
3. The 10:1 and 20:1 *MAT*α cells are combined, each in a sterile microfuge tube. Next, 5 μL of an 0.1 OD_{600} dilution of the *MAT***a** cells is added to each tube, which is then vortexed (*see* **Note 18**).
4. The mixtures are sonicated for 3 s in 0.5 s bursts (*see* **Note 19**).
5. Each mating mixture is then spread (either with glass beads or a spreader) onto two standard-sized (100 mm) YEPD plates and incubated at 30 °C for 4 h.
6. One of the two mating mixtures is replica-plated to medium that selects for diploids formed with *MAT*α secretors, and the other to medium that selects for diploids formed with *MAT*α non-secretors (*see* **Note 20**).

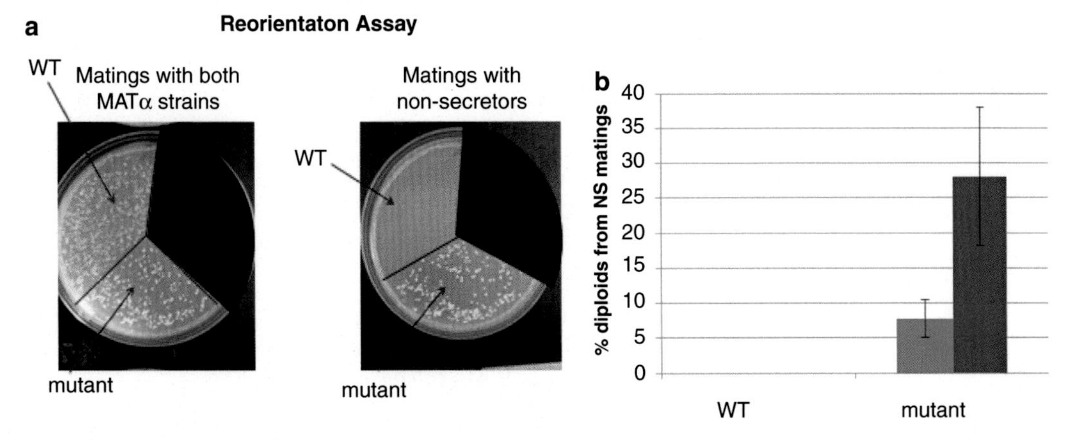

Fig. 3 The low-density partner discrimination assay can be used to screen for defects in orientation and reorientation. Approximately 500 *MAT***a** cells of each mutant strain to be tested are spread on YEPD plates, grown for 2 days, then either replica-plated directly to mating mixtures (to assay orientation), or replicated first to plates containing pheromone for 3 h and then to the mating mixtures (to assay reorientation). The mating mixtures contain a total of 2.5×10^7 *MAT***a** cells in a 20:1 ratio of pheromone non-secretors to secretors. (**a**) Diploid selection plates showing the results of a reorientation assay. (**b**) The mutant strain exhibits a defect in orientation and a more pronounced defect in reorientation. *Bars* indicate the mean % ± sem of *MAT***a** cells that mated with non-secretors in four trials of the orientation (*blue*) and reorientation (*red*) assays. WT *MAT***a** cells did not mate with non-secretors

7. The selective plates are incubated at 30 °C for 2–3 days.
8. The number of colonies on each plate are counted and the percent diploids of each type are determined out of the maximum number that could form per plate (~5000, the number of *MAT*a cells in the mating mixtures; Fig. 3).

4 Notes

1. All solutions should be prepared in water that has a resistivity of 18 MΩ cm. This standard is referred to as "water" in this text.
2. YEPD medium is the same as YPD medium but also contains 50 mg/L adenine sulfate.
3. YEP (or YEP agar) is autoclaved and a one-tenth volume of a 20 % (w/v) glucose solution (0.2 μm filter-sterilized) is subsequently added.
4. Sterilized yeast medium is stored in 90-mL volumes without a carbon source (glucose) or amino acid supplements.
5. Yeast medium amino acid supplements can be stored for up to 1 year at 4 °C in 10 mL aliquots in 15-mL Falcon tubes.
6. Calcofluor white solutions should be stored in a tube wrapped in aluminum foil (to protect from light) at 4 °C. Stock solutions

should be made fresh each week as it has a tendency to precipitate if kept longer.

7. Non-reinforced silicone sheeting can be purchased from for example Bentec Medical, Summit Medical or Silicone Medical Products or Bio-Rad model 583 Transparent sealing gasket can be used.
8. Slides kept for more than 3 days tend to dry out or become contaminated.
9. Instead of using ConA labeled mating partner, a GFP labeled partner (e.g., Spa2-GFP or cells expressing a plasma membrane targeted GFP or soluble GFP) can be used.
10. Similarly, the mating partner can be stained with calcofluor white. The calcofluor white concentration may be varied to obtain the best staining. Too high a concentration can lead to excessive fluorescence making it difficult to definitively identify the originally labeled (*MATa*) cells.
11. The mating mixtures are sonicated to separate cells that may be adhering to one another.
12. The distribution of yeast cells can be easily visualized by microscope, prior to putting a coverslip on, using an inverted microscope for tissue culture examination.
13. Image acquisition can also be much more frequent, with acquisition every 10–15 min possible (*see* **Note 14**).
14. A modification of this method is to follow the localization of a reporter, such as a reporter for active Cdc42 or the site of polarized growth (e.g., Spa2-GFP). This provides additional information on the position of the growth site and can be particularly informative with image acquisition every 10–15 min (Fig. 2).
15. To facilitate identification of the shmoo tip and mating partner, a GFP fusion such as Spa2-GFP can be expressed in the strain. This makes it possible to follow the position of the growth sites using the Spa2-GFP signal.
16. We use the *MATα* secretor and non-secretor strains published by Schrick et al. [21], 8906-1-4b (*MATα cry1 ade6 leu2-3,112 lys2o trp1a can1 ura3-52 SUP4-3ts*) and 8907-4-1b (*MATα cry1 ade6 his4-580a leu2-3,112 tyr1a ura3-52 can1 cyh2 SUP4-3ts mfb1::ura3FOA mfb2::LEU2*). Transformation of strain 8906-1-4b with the integrative *URA3* plasmid, YIplac111 [22], provides a convenient way to distinguish the secretor cells (URA^+ leu^-) from the non-secretor cells (ura^- LEU^+). The *MAT***a** cells should be *TRP1* and *URA3*, so that diploids with secretors *MATα* strains will be TRP^+/URA^+ and diploids with non-secretors *MATα* strains will be URA^+/LEU^+.

17. Similar to the original PDA [8, 9], this assay challenges the *MAT***a** experimental cells to locate pheromone-secreting *MAT*α cells amongst a large excess of non-secretors. The mating mixtures contain a total of 2.5×10^7 *MAT*α cells (OD_{600} of $1.0 = 10^7$ cells/mL) in non-secretor–secretor ratios of 10:1 and 20:1.
18. To assay *reorientation* rather than *orientation*, pretreat the log-phase *MAT***a** cells with α-factor until all the cells have begun to shmoo, wash the cells twice with ice-cold sterile water, then proceed to **step 4**.
19. To sonicate, either a microtip sonicator, set at 35 % and repeating duty cycle = 0.5 is used or a sonicator bath as above.
20. Replica plating is carried out with a replica-plating tool and black velveteen squares (available from VWR and Fisher Scientific, for example). Details of the replica plating technique can be found at http://en.wikipedia.org/wiki/Replica_plating. Assuming the *MAT***a** strain is *URA3* and *TRP1*, then diploids from the secretor strain can be selected for on –TRP/–URA and diploids from the non-secretor strain can be selected on –URA/–LEU (*see* **Note 16**).

Acknowledgments

The authors would like to thank Edward Draper and Madhushalini Sukumar for images and for helpful discussion. This work was supported by the Centre National de la Recherche Scientifique (RAA), the Association pour la Recherche sur le Cancer (SFI20121205755) (RAA), the SIGNALIFE LabEX (RAA), and National Science Foundation (MCB1024718 and MCB-1415589) (DES).

References

1. Arkowitz RA (2009) Chemical gradients and chemotropism in yeast. Cold Spring Harb Perspect Biol 1:a001958
2. Bi E, Park HO (2012) Cell polarization and cytokinesis in budding yeast. Genetics 191:347–387
3. Dohlman HG, Thorner JW (2001) Regulation of G protein-initiated signal transduction in yeast: paradigms and principles. Annu Rev Biochem 70:703–754
4. Dorer R, Pryciak PM, Hartwell LH (1995) *Saccharomyces cerevisiae* cells execute a default pathway to select a mate in the absence of pheromone gradients. J Cell Biol 131:845–861
5. Nern A, Arkowitz RA (1998) A GTP-exchange factor required for cell orientation. Nature 391:195–198
6. Segall JE (1993) Polarization of yeast cells in spatial gradients of α mating factor. Proc Natl Acad Sci U S A 90:8332–8336
7. Valtz N, Peter M, Herskowitz I (1995) FAR1 is required for oriented polarization of yeast cells in response to mating pheromones. J Cell Biol 131:863–873
8. Jackson CL, Hartwell LH (1990) Courtship in *S. cerevisiae*: both cell types choose mating partners by responding to the strongest pheromone signal. Cell 63:1039–1051

9. Jackson CL, Hartwell LH (1990) Courtship in *Saccharomyces cerevisiae*: an early cell-cell interaction during mating. Mol Cell Biol 10:2202–2213
10. Brett ME, DeFlorio R, Stone DE, Eddington DT (2012) A microfluidic device that forms and redirects pheromone gradients to study chemotropism in yeast. Lab Chip 12:3127–3134
11. Diener C, Schreiber G, Giese W et al (2014) Yeast mating and image-based quantification of spatial pattern formation. PLoS Comput Biol 10:e1003690
12. Dyer JM, Savage NS, Jin M et al (2013) Tracking shallow chemical gradients by actin-driven wandering of the polarization site. Curr Biol 23:32–41
13. Hao N, Nayak S, Behar M et al (2008) Regulation of cell signaling dynamics by the protein kinase-scaffold Ste5. Mol Cell 30:649–656
14. Jin M, Errede B, Behar M et al (2011) Yeast dynamically modify their environment to achieve better mating efficiency. Sci Signal 4:ra54
15. Kelley JB, Dixit G, Sheetz JB et al (2015) RGS proteins and septins cooperate to promote chemotropism by regulating polar cap mobility. Curr Biol 25:275–285
16. Lee SS, Horvath P, Pelet S et al (2012) Quantitative and dynamic assay of single cell chemotaxis. Integr Biol (Camb) 4:381–390
17. Moore TI, Chou CS, Nie Q et al (2008) Robust spatial sensing of mating pheromone gradients by yeast cells. PLoS One 3:e3865
18. Moore TI, Tanaka H, Kim HJ et al (2013) Yeast G-proteins mediate directional sensing and polarization behaviors in response to changes in pheromone gradient direction. Mol Biol Cell 24:521–534
19. Barkai N, Rose MD, Wingreen NS (1998) Protease helps yeast find mating partners. Nature 396:422–423
20. Deflorio R, Brett ME, Waszczak N et al (2013) Phosphorylation of Gβ is crucial for efficient chemotropism in yeast. J Cell Sci 126:2997–3009
21. Schrick K, Garvik B, Hartwell LH (1997) Mating in *Saccharomyces cerevisiae*: the role of the pheromone signal transduction pathway in the chemotropic response to pheromone. Genetics 147:19–32
22. Gietz RD, Sugino A (1988) New yeast-*Escherichia coli* shuttle vectors constructed with in vitro mutagenized yeast genes lacking six-base pair restriction sites. Gene 74:527–534
23. Suchkov DV, DeFlorio R, Draper E et al (2010) Polarization of the yeast pheromone receptor requires its internalization but not actin-dependent secretion. Mol Biol Cell 21:1737–1752

Chapter 2

Imaging Polarization in Budding Yeast

Allison W. McClure, Chi-Fang Wu, Sam A. Johnson, and Daniel J. Lew

Abstract

We describe methods for live-cell imaging of yeast cells that we have exploited to image yeast polarity establishment. As a rare event occurring on a fast time-scale, imaging polarization involves a trade-off between spatiotemporal resolution and long-term imaging without excessive phototoxicity. By synchronizing cells in a way that increases resistance to photodamage, we discovered unexpected aspects of polarization including transient intermediates with more than one polarity cluster, oscillatory clustering of polarity factors, and mobile "wandering" polarity sites.

Key words Microscopy, Imaging, Cell polarity, Yeast, Polarity establishment, Cdc42, GFP

1 Introduction

The budding yeast *Saccharomyces cerevisiae* has served as an extraordinarily tractable model system in which to develop a molecular understanding of cell biological phenomena. Despite its small cell size (diameter ~4–6 μm, volume ~40–80 fL depending on whether the cell is haploid or diploid [1]), live-cell imaging approaches have revealed critical features of many cell biological processes in yeast cells. Combining live-cell imaging with genetic perturbations provides a powerful strategy for elucidating the molecular design principles underlying the phenomena under investigation. Here, we provide detailed protocols for imaging polarity establishment in yeast.

Polarity establishment is central to cell migration and to the function of many differentiated cell types. Yeast cells polarize to form a bud during vegetative growth and to form a mating projection during conjugation. Polarity establishment occurs in the G1 phase of the cell cycle and is triggered by activation of G1 cyclin/cyclin-dependent kinase complexes [2]. The location of polarity establishment is influenced by inherited landmark proteins and often occurs near to the preceding cytokinesis site [2]. Because many of the same proteins concentrate at both the cytokinesis site and the polarity site, it can be difficult to clearly distinguish polarity

Tian Jin and Dale Hereld (eds.), *Chemotaxis: Methods and Protocols*, Methods in Molecular Biology, vol. 1407,
DOI 10.1007/978-1-4939-3480-5_2,

establishment from the completion of the prior cytokinesis. To circumvent this issue, we often image cells in which *RSR1*, encoding a key transducer of spatial information from the landmarks to the polarity machinery, has been deleted. In cells lacking Rsr1, polarity establishment occurs at more-or-less random locations.

A central challenge for imaging polarity establishment concerns phototoxicity [3]. Because polarity establishment occurs at low frequency (only once per cell cycle or approximately once per 100 min in rapid growth conditions), cells must be imaged for prolonged periods in order to detect significant numbers of events, and imaging must be performed at high temporal resolution because the process itself is rapid, occurring in just 1–2 min. Because the location of polarity establishment cannot be accurately predicted, cells must also be imaged at high spatial resolution, with multiple *z*-plane acquisition so that polarity sites can be detected wherever they occur over the cell surface. In combination, the requirements for prolonged imaging at high spatiotemporal resolution translate to high fluorescence excitation exposure, which can easily lead to unacceptable phototoxicity. To make matters worse, it appears that cells are more sensitive to phototoxic stress in G1 than they are later in the cell cycle, and imaging conditions that are normally considered to be "low-light" can delay or block polarity establishment [4].

To lower the amount of light needed, we use widefield fluorescence or spinning disk confocal microscopes equipped with sensitive EM-CCD cameras. In addition, we image yeast cells on agarose slabs because these cells are less photosensitive than those imaged in more stressful conditions (e.g., in microfluidic devices). The total imaging time can be reduced by synchronizing the cells so that many undergo polarity establishment within a short time interval. We tested several synchrony protocols and fortuitously discovered that hydroxyurea arrest-release had the unanticipated side-benefit of making cells more photoresistant. We speculate that this treatment (which strongly induces the DNA damage response [5]) prepares cells to mitigate the damaging effects of light.

If yeast cells in G1 are exposed to sufficient amounts of mating pheromone, they arrest with low G1 cyclin/CDK activity, but they nevertheless polarize to form a mating projection. The polarity site in mating cells is not static, but rather wanders around the cortex [6]. Wandering is suppressed by high doses of pheromone, and this process probably contributes to the successful tracking of pheromone gradients during chemotropism [6]. To characterize polarity patch wandering behavior in the absence of suppression by added pheromone, we exploit the fact that pheromone-induced MAPK activation is sufficient to trigger cell-cycle arrest and polarization but does not constrain wandering (pheromone constrains wandering via a separate pathway). Thus, artificial activation of the MAPK using an artificial scaffold protein, Ste5-CTM [7], leads to a uniform population of polarized cells in which wandering of the polarity site is prominent. We induce expression of Ste5-CTM using an orthologous promoter system [8] to avoid affecting cell metabolism by changing the media.

In this chapter, we provide detailed protocols for imaging and quantification of polarity establishment and polarity site wandering in budding and mating yeast.

2 Materials

2.1 Yeast Strains and Fluorescent Probes

1. The following protocols have been developed using two yeast strains: YEF-473 [9] and BF264-15Du [10]. We expect the protocols to apply to other lab yeast strains, though some small changes especially to imaging parameters (e.g., exposure time) might need to be applied.
2. While Cdc42 is considered the master regulatory of polarity, we do not usually use Cdc42 fluorescent probes because they are not fully functional [4]. Instead, we use Bem1-GFP and Spa2-mCherry as our primary markers for the polarity patch. These probes are very functional and bright, making them ideal for live cell microscopy.

2.2 Medium

For live cell imaging, grow cells and prepare samples with complete synthetic medium (CSM). YEP medium is not recommended because it is auto-fluorescent, leading to high background fluorescence intensity.

1. Complete synthetic medium (CSM): Add 950 ml deionized water to a 1 l glass bottle. Add 6.7 g yeast nitrogen base without amino acids (Difco™, Becton Dickinson and Company), 0.74 g CSM (MP Biomedicals), and 0.1 g adenine (*see* **Note 1**) to the bottle and mix them together. Autoclave the liquid at 121 °C for 20 min.
2. 40 % dextrose solution: Dissolve 400 g dextrose (Macron Fine Chemicals™) in 1 l deionized water. Sterilize the solution by filtration using membrane filters with pore size 0.2 μm.

2.3 Reagents for Synchronizing/Arresting Cells

1. 2 M Hydroxyurea (HU) solution: Add 1.52 g HU (Sigma-Aldrich) in about 5 ml deionized water. Make up to 10 ml with water and vortex vigorously to dissolve HU. Aliquot 0.5 ml HU solution in eppendorf tubes for individual uses. Store at −20 °C.
2. α-factor solution: Dilute 10 mg of α-factor (Genesee Scientific) in 5.938 ml water to make 1 mM stock solution and aliquots. For *bar1* cells, make up 25 μM stock solution aliquots. Store at −20 °C, and only freeze-thaw once.

3. β-estradiol solution: Dilute 13.6 mg of β-estradiol (Sigma-Aldrich) in 5 ml ethanol to make 10 mM stock solution. Make up 50 μM stock solution and store at −20 °C.

2.4 Mounting Slab Components

1. CSM containing 2 % dextrose (*see* Subheading 2.2).
2. Agarose (Denville Scientific Inc.).
3. 75 × 25 mm Gold Seal Micro Slides (Gold Seal Products).
4. 18 × 18 mm No. 1 (0.14 mm thickness) microscope cover glass (Globe Scientific Inc.).
5. Vaseline for sealing the slab: Scoop Vaseline in a 50 ml conical tube and place in a boiling water bath. Once Vaseline is melted, pour into a 10 ml syringe with a 16 G needle.
6. 0.2 μm Tetraspeck beads (Life Technologies), diluted 1:5 in water.

2.5 Microscopes and Imaging Settings

We have optimized our protocols using two microscopes that we describe here. There are many parameters to consider when imaging live yeast cells, and we discuss the trade-offs between parameters to help when using other microscopes, fluorescent probes, conditions, etc.

1. Widefield fluorescence microscope: Axio Observer Z1 (Carl Zeiss, Thornwood, NY) with an X-CITE 120XL metal halide fluorescence light source, a 100×/1.46 Plan Apochromat oil immersion objective, and a QuantEM backthinned EM-CCD camera (Photometrics, Tucson, AZ). The microscope is equipped with a XL S1 incubator and heating system (PeCon GmbH). The GFP and RFP filter cubes have the following filter sets (excitation/dichroic/emission): 470/495/525 nm and 560/585/630 nm, respectively.
2. Spinning disk confocal microscope: Andor Revolution XD (Olympus) with a CSU-X1 spinning disk unit (Yokogawa Electric Corporation), a 100×/1.4 UPlanSApo oil immersion objective, and an iXon3 897 EM-CCD camera (Andor Technology). The microscope is equipped with a PZ-2300FT automated stage, which contains a temperature-containing chamber (Applied Scientific Instrumentation Inc.).
3. Temperature control. Our live cell imaging experiments require temperatures ranging from 24 to 37 °C. Microscopes using small heaters (e.g., Andor Spinning Disk) require ~10 min to heat, and microscopes that are fully enclosed (e.g., Axio Observer Widefield) requires >30 min to heat (*see* **Note 2**).
4. Choosing imaging fields. The number of fields used and the number of cells per field depends on the experimental question, the length of the experiment, the cell size, among other considerations. We have found that when imaging multiple fields in an experiment, the cells from fields not currently being imaged are still exposed to indirect light. Therefore, for certain experiments we only use one field. When finding a suitable

field with the desired density of cells, we have found that the cell density is often low in the middle of the slab where the cells have been placed, but the cell density is higher near the edge.

5. Light exposure. The phototoxicity caused by excitation light has been measured previously [3]. The maximum photon flux that did not cause a delay in cell division was 3.0×10^{11} photons/μm² per 3D image. Since the light source of different microscopes varies (e.g., laser power or % fluorescence excitation power), we quantified the light intensity from our imaging conditions. The excitation light intensity at different laser power settings (spinning disk microscope) or percent of fluorescence excitation power settings (widefield fluorescence microscope) is shown in Fig. 1. The range of light used in our experiments is marked in blue boxes. Based on our imaging settings, the range of the photon flux shone on cells is listed in Table 1. In the calculations, we assume that the entire area is evenly illuminated by the excitation light; however, in reality,

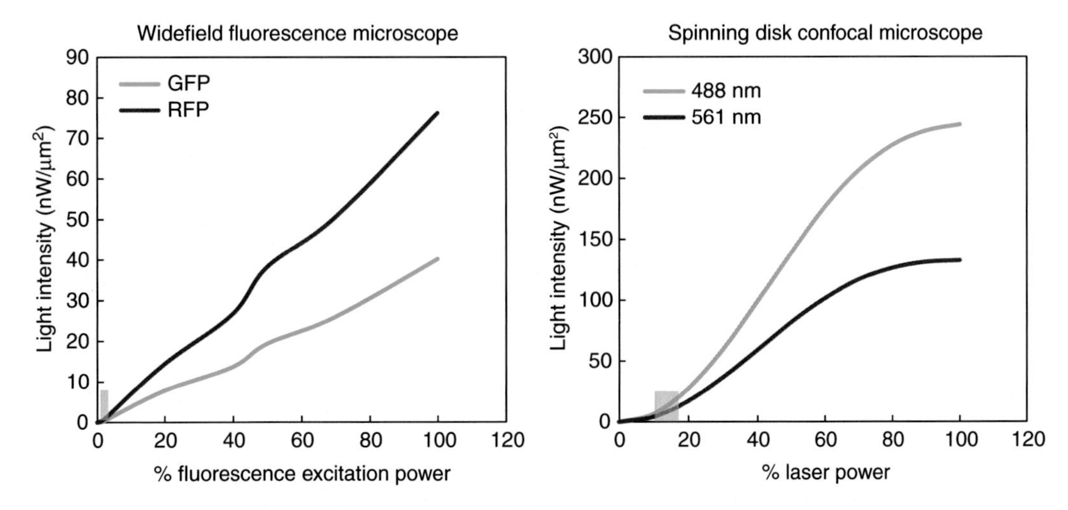

Fig. 1 Measured light intensity by microscope. Light intensity was measured using an X-cite power meter (Excelitas Technologies) over different settings of transmitted light for the widefield fluorescence microscope and laser power for the spinning disk confocal microscope. The *blue boxes* represent the range of settings used during live-cell yeast imaging

Table 1
Estimated photon flux by microscope

Microscope	Light intensity (nW/μm²)	Photon flux (photons/μm² per 3D image)
Widefield	GFP: 0.15–0.4 RFP: 0.27–0.75	2.7×10^{9}–1.1×10^{10}
Spinning disk	488: 6.7–17.0 561: 4.2–10.0	8.0×10^{10}–2.5×10^{11}

the illumination is uneven with higher intensity in the center than at the edge. Thus, the field we imaged may have a higher photon flux than that shown in the table, but the actual value should be below the 3.0×10^{11} photons/μm^2 limit.

6. Z-stack. Cell structures, including the polarity site, are not always in the medial plane of the cell. Therefore, it is usually necessary to image cells with multiple z-plane acquisition. We used the Nyquist rate calculator (http://www.svi.nl/NyquistCalculator) to determine the minimal sampling density that is required to capture all information of the objects. Based on the microscopic parameters we used, the critical sampling distance (or Nyquist rate) for the widefield fluorescence microscope is 235 nm. With this distance, we took 30 *z*-plane images to cover the entire cell (HU-treated cells are larger, with diameter ~6–7 μm). The Nyquist rate for spinning disk microscope is 131 nm, which requires 54 *z*-planes to cover the same depth, but the increased light exposure would lead to more phototoxicity. Therefore, we use 30 *z*-planes (0.24 μm apart) for live cell imaging on both the widefield and confocal microscopes. When imaging multiple fields and multiple fluorescent channels, especially with small time intervals, sometimes the acquisition is reduced to 15 *z*-planes (0.5 μm apart) to reduce phototoxicity and reduce the imaging time. Additionally, the z-stack for each fluorescent channel is acquired separately rather than alternating between the fluorescent channels at each *z*-plane to reduce motor movements and decrease time of acquisition (*see* **Note 3**).
7. Time interval between image acquisitions. We usually use 45 s or 1.5 min intervals for live cell imaging. Shorter time intervals result in more phototoxicity, and it is important to keep in mind the trade-off between temporal resolution, spatial resolution, and phototoxicity.

3 Methods

3.1 Preparing Agarose Slabs

1. Combine 0.5 g agarose with 2.5 ml CSM + 2 % dextrose in a 15 ml conical tube. Add α-factor or β-estradiol if needed.
2. Loosen the cap of the tube and microwave (900 W) for 35–45 s. Forcefully invert or vortex every time the media boils to the top of the tube (about four times). Make sure the agarose completely dissolves. The final solution will be very viscous (*see* **Note 4**).
3. Using a cut pipette tip, quickly pipette 90–180 μl of the agarose solution onto a glass slide and immediately drop a coverslip on the agarose droplet. Repeat this procedure to make several slabs in order to ensure one is level with the glass. The slabs will solidify within 1 min.

4. Store the prepared slabs in a box (e.g., empty pipette tip box) with some water to prevent the slab from drying out. Use the slabs within a few hours of preparing.

3.2 Mounting Cells onto Agarose Slabs

1. Centrifuge sample (1 ml) at 16,000 × *g* for 1 min (*see* **Note 5**). Aspirate most of the media, leaving ~10–20 μl media to resuspend cells.
2. Pipette ~0.7 μl cells into tip and set pipette aside.
3. If the agarose slab extends beyond the coverslip, remove with a razor blade prior to loading cells.
4. Gently lift a corner of the coverslip away from the slab using a razor blade, and hold the coverslip in one hand.
5. Use other hand to pick up the pipette and drop cells onto slab.
6. Replace coverslip on slab.
7. Gently press sides of coverslip to ensure proper contact and reduce air pockets.
8. Seal slabs with Vaseline. Heat Vaseline in needle over flame prior to sealing each edge of coverslip.

3.3 Synchronizing Cells with HU Treatment Prior to Imaging

1. Grow cells overnight at 30 °C (or 24 °C if the strain is temperature sensitive) in CSM + 2 % dextrose to a log phase culture (OD_{600} = 0.1–0.7) (*see* **Note 6**).
2. The next morning, dilute the cell culture OD_{600} = 0.15. Mix 4.5 ml of the diluted cell culture and 0.5 ml 2 M HU solution (final working concentration 200 mM) in a 50 ml flask or glass test tube. Agitate the sample using a rotary shaker or roller drum for 3 h at 30 °C or 4 h at 24 °C to arrest cells in S phase (*see* **Note 7**).
3. Centrifuge 5 ml treated cells in eppendorf tubes at 16,000 × *g* (or in conical tubes at 2000 × *g*) and discard supernatant. Wash the cells twice with 1 ml CSM + 2 % dextrose to remove any remaining HU. Resuspend the cells in 5 ml CSM + 2 % dextrose in a 50 ml flask or a glass test tube. Agitate the sample for 1 h at 30 °C or 2 h at 24 °C to release the cells from HU arrest. At 30 °C, most strains take 75–90 min after releasing from HU arrest to enter G1 of the next cell cycle.
4. Mount cells onto an agarose slab and find a good field for imaging (*see* Subheading 2.5). We recommend using a field with 10–15 large-budded cells, which are not touching each other to prevent overlapping signal from neighboring cells during quantification. Also, most polarity markers localize to the bud neck during cytokinesis, so choose a field with cells that show signal at the bud neck. These cells are finishing their cell cycle and will polarize in the following G1 phase soon.

Table 2
Settings for imaging polarity establishment

Widefield	Channel	Transmitted light	EM gain	Exposure (ms)
	DIC	100 %	150	50
	GFP (470 nm)	2 %	750	250
	RFP (560 nm)	2 %	750	250
Spinning disk	Channel	Laser power	EM gain	Exposure (ms)
	DIC	N/A	200	100
	488 nm	10 %	200	200
	561 nm	10 %	200	200

If using one fluorescence channel, use a 30 *z*-plane stack with images 0.24 μm apart (*see* **Note 10**). If using two channels, use a 15 *z*-plane stack with images 0.5 μm apart. Image at 45 s intervals for 45–90 min total (*see* Subheading 2.5 for more details)

3.4 Imaging Polarity Establishment

1. Ensure that the temperature control chamber has reached a stable temperature at least 30 min prior to imaging.
2. Image cells (*see* Subheading 2.5) using settings listed in Table 2.

3.5 Image Deconvolution and Analysis of Polarity Establishment

1. Deconvolve the z-stack images taken by a widefield fluorescence microscope using deconvolution software (e.g., Huygens Essential) to reduce the blur from out-of-focus light. The classic maximum-likelihood estimation and predicted point spread function method with a signal-to-noise ratio of 10 (*see* **Note 8**) is used with a constant background across all images from the same channel on the same day.
2. Images taken by a spinning disk confocal microscope are denoised by the Hybrid 3D Median Filter plugin in ImageJ (http://rsb.info.nih.gov/ij/plugins/hybrid3dmedian.html). Alternatively, the raw images can be further enhanced by deconvolution with a signal-to-noise ratio of 3 (*see* **Note 8**).
3. Using image quantification software (e.g., Volocity), set a pixel intensity (fluorescence) threshold that selects the pixels corresponding to the polarity patch but not the rest of the cell. If the software is unable to import 3D information for quantification, use the sum projection of the z-stack.
4. The sum intensity of the polarity patch is recorded from its first detection (we generally focus on early times before bud emergence). The intensity at each time point is normalized to the peak intensity value within the imaging period, and plotted against time.

Table 3
Settings for imaging polarity patch movement

Spinning disk	Channel	Laser power	EM gain	Exposure (ms)
	DIC	N/A	200	100
	488 nm	15–18 %	200	200
	561 nm	15 %	200	200

Use a 30 *z*-plane stack with images 0.24 μm apart. Image at 1.5 min intervals for 30–45 min (*see* Subheading 2.5 for more details)

3.6 Arresting Cells with Ste5-CTM Fusion and Imaging Polarity Patch Wandering

1. Grow cells overnight at 30 °C in CSM + 2 % dextrose to a log phase culture (OD_{600} = 0.1–0.7) (*see* **Note 6**).
2. Dilute cells to 1.5 ml of OD_{600} = 0.1, and treat with 20 nM β-estradiol for 4 h (*see* **Note 9**).
3. Centrifuge 1 ml of culture at 16,000 × *g* for 1 min, and aspirate most of the liquid. Add 1 μl of Tetraspeck beads (diluted 1:5 in water), and resuspend pelleted cells.
4. Mount approximately 0.7 μl onto agarose slabs and seal with Vaseline. If using α-factor slabs, let cells sit on slabs for 20 min prior to imaging.
5. Image cells (*see* Subheading 2.5) using settings listed in Table 3.

3.7 Tracking Polarity Patch Wandering and Calculating Mean Squared Displacement (MSD)

1. Using image processing software (e.g., Volocity), set a pixel intensity (fluorescence) threshold that selects the pixels that correspond to the polarity patch but not the rest of the cell. Use the same threshold for all the samples being compared.
2. Remove any objects detected with an area <0.1 μm^2.
3. Determine the centroid of a Tetraspeck bead in the field of view for each movie. This will serve to control for stage drift during imaging.
4. Determine the location of the centroid of the polarity patch (in three dimensions) for each cell at each time point.
5. If there are multiple polarity patches for a given time point for a given cell, remove all centroid measurements for that cell at that time point.
6. From each polarity patch centroid location, subtract the Tetraspeck bead centroid location.
7. Combine centroids from continuous trajectories: patches that have been continuously measured in a given cell.
8. Calculate the MSD for each trajectory. For all pairs of time points ti,tj and corresponding centroid positions pi,pj, calculate the squared distance between pi and pj. Then, average all squared distances for each time interval $T = tj - ti$ per cell.

9. For each cell, average the MSD of all trajectories. We consider each cell an observation unit for statistical analysis, and we typically measure approximately 50 cells per condition in an experiment.

4 Notes

1. One of the yeast strains (BF264-15D) used in our lab carries a mutation (*ade1*) that can lead to the accumulation of a fluorescent red metabolite under low-adenine conditions, which interferes with imaging. To prevent this, 0.5 mM adenine is added to the CSM. For strains with a wild-type adenine synthesis pathway, extra adenine is not needed.
2. Sometimes the temperature recorded by the microscope is not accurate. We recommend determining the temperature that the sample is exposed to rather than relying on the microscope's thermometer.
3. When comparing the appearance or movement of two fluorescent probes, note that the second fluorescent channel will be imaged with a few seconds delay compared to the first channel.
4. The 15 ml conical tube can be kept in a 95 °C heat block to prevent the agarose solution from solidifying after microwaving.
5. Certain plastics are better at pelleting dilute cell samples. If the cells are not well-pelleted after a 1 min centrifuge spin at 16,000 × *g*, aspirate the media, scrape the sides of the tube using a pipette tip, and spin again.
6. We have found that growing multiple dilutions of starter cultures overnight is useful for ensuring one culture will be the appropriate density.
7. Some strains take longer to respond to HU so longer treatment is needed to obtain a synchronized population. Check cells under microscope after 3 h HU treatment to ensure good synchrony (>70 % large-budded cells).
8. The signal-to-noise ratio listed here is not the ratio on the raw images; instead, it is the ratio by which the raw images will be enhanced after deconvolution. The signal-to-noise ratio of raw images can be influenced by excitation light intensity, the brightness of fluorescence probes, the sensitivity of camera, etc., so the enhancing ratio we suggest here may not apply to images generated from other microscopes. We suggest choosing a signal-to-noise ratio by deconvolving the same image with different ratios and choosing the one that effectively reduces the blur without introducing erroneous signals not seen on the raw image.

9. Because the cells are asynchronous when we initially treat them with β-estradiol, the cells arrest at different times. 3–4 h treatment is usually sufficient for >95 % cells to arrest. As the arrested cells continue to grow, larger size yields higher MSD, so it is essential to maintain the same β-estradiol treatment time for all samples.
10. Bud emergence can occur outside the current focal plane. If the timing of bud emergence is important for analysis, image using a z-stack in the DIC channel.

Acknowledgments

We thank Audrey Howell and Jayme Dyer for their role in developing the protocols discussed here. A.W.M. and C.-F.W. contributed equally to this work. Work in the Lew lab was supported by NIH/NIGMS grants GM62300 and GM103870 to D.J.L.

References

1. Klis FM, de Koster CG, Brul S (2014) Cell wall-related bionumbers and bioestimates of *Saccharomyces cerevisiae* and *Candida albicans*. Eukaryot Cell 13:2–9
2. Howell AS, Lew DJ (2012) Morphogenesis and the cell cycle. Genetics 190:51–77
3. Carlton PM, Boulanger J, Kervrann C et al (2010) Fast live simultaneous multiwavelength four-dimensional optical microscopy. Proc Natl Acad Sci U S A 107:16016–16022
4. Howell AS, Jin M, Wu CF et al (2012) Negative feedback enhances robustness in the yeast polarity establishment circuit. Cell 149:322–333
5. Harper JW, Elledge SJ (2007) The DNA damage response: ten years after. Mol Cell 28:739–745
6. Dyer JM, Savage NS, Jin M et al (2013) Tracking shallow chemical gradients by actin-driven wandering of the polarization site. Curr Biol 23:32–41
7. Pryciak PM, Huntress FA (1998) Membrane recruitment of the kinase cascade scaffold protein Ste5 by the Gβγ complex underlies activation of the yeast pheromone response pathway. Genes Dev 12:2684–2697
8. Louvion JF, Havaux-Copf B, Picard D (1993) Fusion of GAL4-VP16 to a steroid-binding domain provides a tool for gratuitous induction of galactose-responsive genes in yeast. Gene 131:129–134
9. Bi E, Pringle JR (1996) ZDS1 and ZDS2, genes whose products may regulate Cdc42p in *Saccharomyces cerevisiae*. Mol Cell Biol 16:5264–5275
10. Richardson HE, Wittenberg C, Cross F, Reed SI (1989) An essential G1 function for cyclin-like proteins in yeast. Cell 59:1127–1133

Chapter 3

Migration of *Dictyostelium discoideum* to the Chemoattractant Folic Acid

Karl J. Aufderheide and Chris Janetopoulos

Abstract

Dictyostelium discoideum can be grown axenically in a cultured media or in the presence of a natural food source, such as the bacterium *Klebsiella aerogenes* (KA). Here we describe the advantages and methods for growing *D. discoideum* on a bacterial lawn for several processes studied using this model system. When grown on a bacterial lawn, *D. discoideum* show positive chemotaxis towards folic acid (FA). While these vegetative cells are highly unpolarized, it has been shown that the signaling and cytoskeletal molecules regulating the directed migration of these cells are homologous to those seen in the motility of polarized cells in response to the chemoattractant cyclic adenosine monophosphate (cAMP). Growing *D. discoideum* on KA stimulates chemotactic responsiveness to FA. A major advantage of performing FA-mediated chemotaxis is that it does not require expression of the cAMP developmental program and therefore has the potential to identify mutants that are purely unresponsive to chemoattractant gradients. The cAMP-mediated chemotaxis can appear to fail when cells are developmentally delayed or do not up-regulate genes needed for cAMP-mediated migration. In addition to providing robust chemotaxis to FA, cells grown on bacterial lawns are highly resistant to light damage during fluorescence microscopy. This resistance to light damage could be exploited to better understand other biological processes such as phagocytosis or cytokinesis. The cell cycle is also shortened when cells are grown in the presence of KA, so the chances of seeing a mitotic event increases.

Key words Folate, Vegetative, Signal transduction, Motility, Chemotaxis, Cytokinesis, Phagocytosis

1 Introduction

The discovery of *D. discoideum* by K.B. Raper in 1935 [1] proved critical to studies on a number of cellular processes that occur in eukaryotic cells. This social amoeba was soon found to be a genetically and biochemically tractable model system [2, 3]. Raper was able to do many of his experiments because he found he was able to grow *D. discoideum* on lawns of bacteria [4]. Years later, others would add to the ease by which these amoeba are grown by isolating strains that could be cultured in axenic liquid broth media [5, 6]. The original NC4 strains isolated by Raper can grow only by phagocytosis of bacteria; some labs still find them favorable for

Tian Jin and Dale Hereld (eds.), *Chemotaxis: Methods and Protocols*, Methods in Molecular Biology, vol. 1407,
DOI 10.1007/978-1-4939-3480-5_3, © Springer Science+Business Media New York 2016

their studies [7]. Two axenic strains were isolated from the original wild isolate. AX1 (which gave rise to AX2) and AX3 (which later gave rise to AX4), were independently isolated by two different laboratories [2]. These axenic strains can also grow in the presence of bacteria. This feature can provide certain advantages and will largely be the focus of the protocols described in this chapter.

1.1 Chemotaxis

Depending upon their physiological state, *D. discoideum* cells can exhibit chemotaxis, defined as the directed migration of cells towards or away from gradients of signaling molecules, e.g., to either of the chemoattractants FA or cAMP [8]. In general, chemotaxis is implicated in a myriad of physiological activities in single-celled and multicellular species, including inflammation, lymphocyte homing, axon guidance, angiogenesis, and numerous cell movements during early development [8–11]. Defects in chemotaxis contribute to pathological conditions including infectious and allergic diseases, atherosclerosis, and tumor metastasis [12–15]. Vegetative *D. discoideum* cells feed on bacteria and other microbes by sensing and migrating toward FA and likely other potential chemical signals [16, 17]. These molecules activate seven-transmembrane receptor signal pathways in a manner similar to that seen for the chemokines responsible for leukocyte chemotaxis [18]. When nutrients are limiting, *D. discoideum*, cells enter a cAMP-dependent developmental cycle that results in the formation of multicellular fruiting bodies [19–22]. During aggregation, *D. discoideum* cells polarize, forming a distinct front and rear. Cells are highly chemotactic at this stage of development.

Compromises in the cAMP signaling pathway can readily be identified since cells will remain as smooth monolayers when grown on a non-nutrient surface [23, 24]. It is often useful to grow cells on a bacterial lawn, since cells can be seeded at low density. Clonal populations will form plaques over the course of several days and enter development as they clear out the bacteria. Developmental and migratory mutants can be readily identified by the size and shape of the plaque and the ability to form fruiting bodies with viable spores. Work derived from understanding this developmental process has provided much insight into the role of many molecules critical for cell signaling and motility.

1.2 Expression of Developmentally Regulated Genes

Altered gene expression in starving cells makes them more sensitive to cAMP [25–27]. Both the cAMP receptor (cAR1) and the heterotrimeric G protein alpha subunit, Gα2, increase in expression as do a plethora of other developmentally regulated genes [28–30]. This can pose a problem, since mutants that appear to be defective in aggregation may just be developmentally delayed or may not develop at all. Many cell lines where this is suspected should also be assayed for their ability to show chemotaxis to FA, which does not involve the cAMP developmental program. This will rule out a developmental defect that results in the cells not being competent to respond to cAMP. Many downstream signal transduction molecules that are used during FA chemotaxis are also critical for the signal relay necessary for early cAMP development and cell polarization.

1.3 FA and cAMP Chemotaxis Use Similar Signaling Pathways

The morphological characteristics of *D. discoideum* cells migrating towards FA are distinct from starved cells chemotaxing to cAMP. Vegetative cells are amoeboid-shaped and unpolarized, with multiple pseudopods and are nevertheless capable of migrating directionally in a FA gradient [17, 31–38]. While the leading edge can sometimes extend more than one pseudopod, polarized cells typically lack lateral pseudopods as they migrate towards a cAMP source [39–44]. The underlying sensing mechanism regulating directional motility largely functions in a similar manner regardless of cell shape or the chemoattractant used [45]. Activation of the small G protein Ras, the phosphoinositide 3-kinase 2, PI3K2, one of five PI3Ks containing a Ras binding domain in *D. discoideum* [46–48], and the synthesis of the plasma membrane phosphoinositides $PI(3,4)P_2$ and $PI(3,4,5)P_3$ (as assayed by the Pleckstrin homology (PH) domain of the Cytosolic Regulator of Adenylyl Cyclase) all occurred in response to uniform treatments of FA (Fig. 1), as has been previously shown for cAMP [49, 50]. The cytoskeletal markers dynacortin and coronin translocated to the plasma membrane, while cortexillin and myosin II moved in a reciprocal manner, relocalizing from the plasma membrane to the cytosol, and mirrored the movements of the tumor suppressor and phosphatase PTEN (Fig. 1, right). In a gradient of FA, activated Ras and PI3K2 were at the pseudopodial projections on the high side of the gradient (Fig. 2) as has been similarly shown at the leading edge of chemotaxing cells in cAMP gradients [51, 52]. Conversely, the tumor suppressor and phosphatase PTEN was distributed on the lateral and trailing edge of the plasma membrane of the cell [47, 53], as previously shown for cAMP. Myosin II and cortexillin localized in similar manner to PTEN in response to a FA gradient. This reciprocal spatial regulation of PI3K2 and PTEN confines $PI(3,4,5)P_3$ to the leading edge, while also likely maintaining higher levels of $PI(4,5)P_2$ at the rear, which helps specify the back of the cell [54, 55].

The receptors and heterotrimeric G proteins specific for cAMP are well characterized [56–65]. Cells lacking Gβ do not chemotax or respond to uniform stimulation of either FA or cAMP [66–68]. Since there appears to be only one functional β and γ subunit in the *D. discoideum* genome, it should not be surprising that similar effector molecules would be activated by βγ subunits, regardless of the chemoattractant. PKBA (human protein kinase B (PKB) homolog) is recruited to the plasma membrane and binds to $PI(3,4,5)P_3$ through its Pleckstrin homology (PH) domain and is

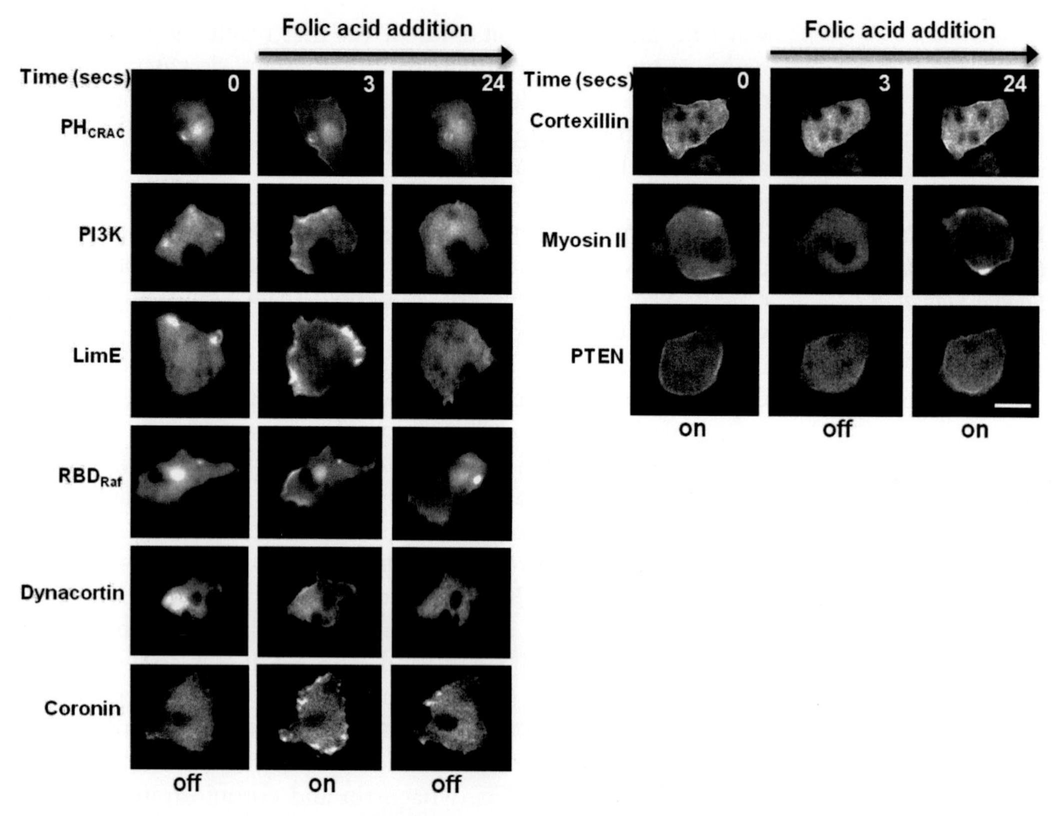

Fig. 1 Membrane redistribution of signaling and cytoskeletal proteins by uniform folic acid stimulation. Fluorescent images of indicated GFP markers during uniform folic acid stimulation (100 μM). *Left.* PH_{Crac}-GFP, PI3K2-GFP, LimE-RFP, RBD_{Raf}-GFP GFP-dynacortin, and coronin-GFP translocated to the plasma membrane upon FA addition (prior to frame 2). The cells adapt and the molecules return to the cytosol (frame 3). *Right.* GFP-cortexillin-1, myosin II-GFP, and PTEN-GFP proteins are on the plasma membrane prior to stimulus and re-localize to the cytosol. Time between frames designated in seconds. Scale bar: 10 μm. Reproduced from Srinivasan et al. 2013 [45] with permission from the Company of Biologists

activated by phosphorylation of its hydrophobic motif and activation loop [69–76]. PKBA/PKBR1 activation and substrate phosphorylation have been examined in response to either FA or cAMP and are largely similar, although several unique targets for each chemoattractant having been identified [45, 51, 69]. While these two kinases are critical for cAMP mediated responses, cells lacking both are still able to chemotax to FA when grown on KA (Fig. 3).

By eliminating the role of polarity and phenotypes due to developmental delays in cAMP-mediated migration, one can determine the core regulators of the gradient sensing mechanism. Indeed, screens have been applied to isolate pure chemotaxis mutants using FA-mediated migration [77]. The feasibility of FA-mediated chemotaxis has been shown by testing mutants lacking functional Ras C and G [45]. In previous studies, Ras C/G double nulls were shown to be completely blind in a cAMP gradient and were thought

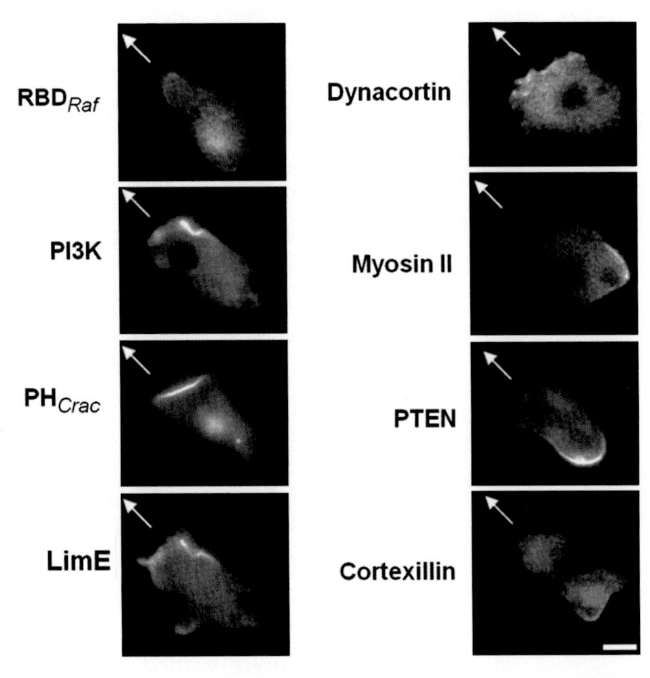

Fig. 2 Localization of signaling proteins during folic acid gradient. Representative images of indicated fluorescently tagged proteins migrating towards 10 μM FA-filled micropipette. *Arrow* indicates the position of the micropipette. GFP tags shown include RBD$_{Raf}$, PI3K-GFP, PH$_{Crac}$-GFP, LimE-RFP, and dynacortin-GFP, which localized on pseudopodial projections in the direction of the micropipette. Myosin II-GFP, PTEN-GFP, and GFP-cortexillin-1 de-localized from the pseudopods during migration towards FA. Reproduced from Srinivasan et al. 2013 [45] with permission from the Company of Biologists

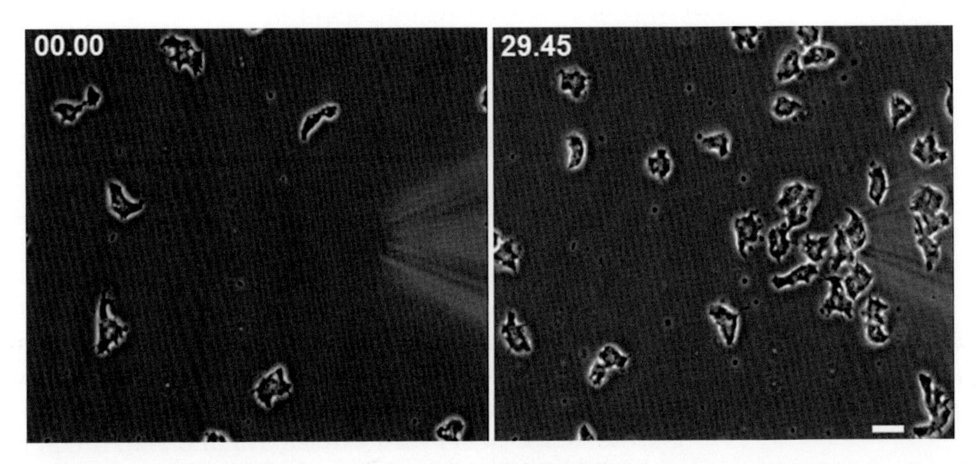

Fig. 3 Chemotaxis of PKBA/PKBR1 null cells. Frames of PKBA/PKBR1 null cells at indicated time points in a FA gradient. Cells migrate up the concentration gradient and towards the FA-loaded (10 μM) micropipette. Scale bar: ~10 μm. Reproduced from Srinivasan et al. 2013 [45] with permission from the Company of Biologists

to be required for directed migration [78]. In these studies, cAR1 expression was significantly reduced and delayed during early development in Ras G null cells and undetectable in Ras C/G double null cells. This new study found there was loss of PI3K activity in Ras mutants, but Ras C, G and CG double nulls all migrated quite well towards FA. When Ras CG nulls were carefully pulsed after starving, polarized cells responded to cAMP [45]. While Ras proteins are clearly activated during chemotaxis, compensatory or redundant mechanisms may compensate for the loss of these proteins. For instance, Ras D and Ras B have been shown to have increased expression levels in Ras G nulls, and Ras D has also been shown to go up in CG nulls [45]. Interestingly, in at least one case, Ras mutants were tested for FA chemotaxis and did not respond [79]. It is likely the differences in FA chemotaxis results were because cells grown on KA are highly chemotactic to FA while cells grown axenically in HL-5 medium are less so. In our hands, cells grown in HL-5 medium respond poorly to FA. We recommend always performing FA chemotaxis on *D. discoideum* cells grown on KA lawns.

Cells migrate with a biased random walk in response to FA, whereas polarized cells responding to cAMP are more persistent and do not typically make lateral pseudopods. Underdeveloped cells (cells starved 3–4 h) that were unpolarized, responded to cAMP by migrating in a manner similar to vegetative cells grown on KA and migrating towards FA [45]. In both situations where the cells were unpolarized, random pseudopods were likely generated by an oscillatory mechanism that was independent of the heterotrimeric G proteins, but that can be biased by the chemotactic signal transduction system. This may be a useful characteristic so that feeding *D. discoideum* can rapidly reorient to the correct direction of a moving bacterium. Unpolarized cells can also generate phagosomes along the entire periphery of the cell. Thus, the major signal transduction proteins, lipids, and cytoskeletal elements function similarly in both types of directed migration.

2 Materials

2.1 General Media (Recipes Adapted from Dictybase and [80])

1. HL5 growth medium: We typically use Formedium HL5 without glucose: Add 22 g of premixed HL-5 and 10 g of dextrose per liter. If making HL5 from scratch [81]: 0.5 % (w/v) proteose peptone, 0.5 % (w/v) Thiotone E peptone, 55.5 mM glucose, 0.5 % (w/v) yeast extract, 1.3 mM $Na_2HPO_4 \cdot 7H_2O$, and 2.57 mM KH_2PO_4. Bring to a volume of 1 l. Adjust pH with HCl to pH 6.4–6.6. Autoclave to sterilize.
2. SM Agar Plate: Formedium SM Agar. Suspend 41.7 g in 1 l of distilled or deionized H_2O. Autoclave and let cool to 55 °C and then pour plates. Plates can be stored inverted at 4 °C for weeks.

3. Developmental buffer (DB): 5 mM Na_2HPO_4, 5 mM KH_2PO_4, 1 mM $CaCl_2$, and 2 mM $MgCl_2$. Prepare the phosphate solution as 25 mM (5×), adjust the pH to 6.5 and autoclave. Make 10× $CaCl_2$ (10 mM) and $MgCl_2$ (20 mM) solutions each separately and autoclave. To make 1 l of DB, mix 600 ml of distilled, autoclaved water with 200 ml of 5× phosphate solution and 100 ml of 10× $CaCl_2$ and 10× $MgCl_2$ solution, respectively.
4. Luria broth (LB): Nutrient-rich media commonly used to culture bacteria. 10 g tryptone, 5 g yeast extract, and 10 g NaCl. Add to 1 l of deionized water and autoclave.
5. H-50 transformation buffer: 20 mM HEPES, 50 mM KCl, 10 mM NaCl, 1 mM $MgSO_4$, 5 mM $NaHCO_3$, and 1.3 mM NaH_2PO_4 in 1 l of distilled water. Adjust pH to 7.0 with HCl or NaOH as appropriate. Autoclave and store cold or frozen.
6. G418 and hygromycin: sulfate (Invitrogen) used at 20 μg/ml in the HL5 growth medium of the transformants. Hygromycin (Sigma-Aldrich), used at 20 μg/ml. We purchase the liquid stocks.
7. Folic acid (Fischer Scientific; *see* Subheading 3 for preparation).
8. *Dictyostelium* and Bacterial Cells: AX2, AX3, and NC4 wild-type *Dictyostelium* strains as well as *Klebsiella aerogenes* obtained from the DictyBase Stock Center.

2.2 Equipment

1. Incubator set at 22 °C. This can be a shaking incubator for growth in suspension or a stationary incubator for growth in plastic dishes. Typically, cultures can be maintained at room temperature and grown on laboratory benches, provided that the temperature in the laboratory is constant and below 25 °C.
2. Sterile 100 mm petri dishes.
3. Sterile 15 and 50 ml conical tubes.
4. Hemocytometer.
5. Eppendorf Injectman micromanipulator and Femtojet pump.
6. Bacteria loop.
7. Inverted light microscope (preferably with phase-contrast optics) with low magnification dry lenses.
8. Stereoscope.
9. Fluorescence microscope with oil immersion and GFP and RFP filter sets.
10. Micropipette (P1000, P200, P20).
11. 100 mm sterile petri dishes.
12. 35 mm sterile petri dishes.
13. Sterile 5 and 10 ml pipettes.

14. Nunc Lab-Tek 1-well and 8-well chambers with cover glass bottoms.
15. Glass spreader.
16. Ethanol.
17. Bunsen burner.
18. Sterile 1.5 ml Eppendorf tubes.
19. Autoclave.

3 Methods

3.1 Preparing a KA Plate

1. Centrifuge approximately $1–5 \times 10^5$ *D. discoideum* cells at $420 \times g$ for ~2 min in a 15 or 50 ml conical tube and obtain a pellet.
2. Decant HL5 and resuspend cells in ~500 μl HL5 (no antibiotics; *see* **Note 1**).
3. Obtain a freshly grown KA plate and a sterile SM plate (*see* **Notes 2–5**).
4. Use a heat-sterilized bacterial loop and scoop a loopful of KA from the bacterial plate. If using KA from an overnight liquid culture, use approximately 0.2 ml.
5. Place bacteria into the conical tube containing cells and ~500 μl HL5 and pipette up and down to break up any clumps.
6. Resuspend cell/KA mixture and place on SM plate (*see* **Notes 6** and **7**).
7. Use sterile cell spreader to evenly spread KA/*D. discoideum* cell culture.
8. Allow to dry for a few minutes next to a flame from a Bunsen burner and then flip plate onto its top and incubate at 22 °C overnight.
9. The following morning the plates should have a confluent KA lawn that looks like creamy white film. One must use this plate before the *D. discoideum* clear the lawn. The bacteria will initially grow faster, but once they deplete the nutrients in the SM plate, the *D. discoideum* will catch up and continue to consume them (*see* **Notes 8** and **9**).

3.2 Preparing Cells for Microscopy

1. Obtain cells/bacteria from SM plate. This can be done in a number of ways. Either scoop out some cells on a sterile loop, or add DB to plate and pipette up and down. Take mixture and add to conical tube, wash with DB and spin at $420 \times g$ for 3 min. Bacteria will not pellet at low speeds. Repeat until most bacteria are gone.

2. Dilute cells appropriately for your microscopy experiments and add to your chambered slide (*see* **Note 10**). If we are adding a uniform stimulus, we typically use the 8-well slides. For uniform stimulus and chemotaxis assays, we use the 1-well chamber that can be used with our Eppendorf micropipette system (*see* **Note 11**).

3.3 Folic Acid Preparation

1. Folic acid must always be made fresh. We have tried freezing and it does not work.
2. The key for preparing folic acid is to add just enough, but not too much, NaOH so that the FA dissolves.
3. Add 5.5 mg FA to 12.5 ml dH_2O, and then add 13.5 of μl 2 N NaOH to make a 1.25 mM stock solution.
4. Dilute to 10 μM for FA chemotaxis.

   Note: spin down your FA in a microcentrifuge at high speed after making it and only use the supernatant for micropipette loading. This will get rid of any undissolved FA or particulates that could potentially clog the micropipette. We typically load 5–10 μl of 10 μM FA for our chemotaxis assays. If we are doing uniform stimulations using the "clean" function on the micropipette, we typically use 100 μM FA.

3.4 Imaging Cell Division and Phagocytosis

Cells taken directly out of HL5 media and imaged using fluorescence microscopy will round up and die rapidly. Given the wealth of information that can be obtained by imaging dynamic cellular events that occur over short periods of time, this presents a problem. We have found that cells grown on bacteria are highly resistant to phototoxicity when grown in the presence of KA. We also know that when cells are starved for several hours, they are also resistant to phototoxicity. We have exploited these findings for imaging cell division and phagocytosis (Fig. 4).

3.5 Cell Preparation

1. Cells can be grown overnight in HL-5 media (*see* **Note 12**) with the appropriate drug selection markers and then washed twice in DB, centrifuged, and seeded in an 8-well chamber for 120 min (*see* **Note 13**).
2. A low density of KA can be added just prior to imaging to feed the *D. discoideum* cells that have been starving for 90 min (*see* **Note 14**). The developmental process at this time is still reversible, and cells will begin feeding on the bacteria (*see* **Note 15**). This has two affects: (1) the cells will reenter the cell cycle and rapidly divide, which is beneficial if you wish to catch mitotic events and cytokinesis (Fig. 4). (2) The cells are resistant to phototoxicity and can be imaged with relatively short intervals between acquisitions (*see* **Note 16**). We have imaged continuously for several minutes with 2 s intervals using spinning disk

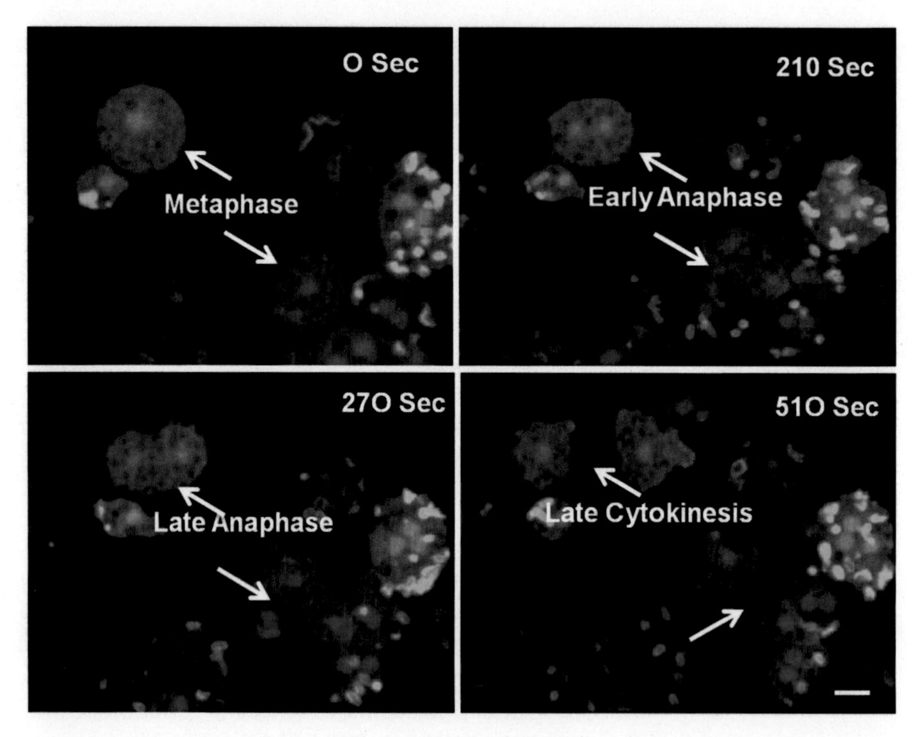

Fig. 4 Cytokinesis and phagocytosis of cells imaged with *Klebsiella aerogenes*. Cells expressing PH-GFP and LimE-RFP were imaged every 2 s. Shown are projected images from a 0.5 μm z stack using 63 × 1.4 NA lens on a Zeiss Axio Observer outfitted with a spinning disk confocal head (Intelligent Imaging and Innovations). Times are indicated. Two cells were at metaphase (labeled) at time zero and both divide. The rest of the cells were continuously consuming KA, as depicted by the numerous small fluorescent phagocytic cups evident in all the interphase cells. Cells were completely healthy during the movie, which was over 10 min in duration. Scale bar: ~10 μm

confocal microscopy (Fig. 4) and continuously for several minutes using lattice light-sheet microscopy (data not shown) [82].

3. The feeding procedure can be repeated if the cells consume all the bacteria (*see* **Notes 17–19**).

4 Notes

1. To insure consistent growth rates, use medium and plates warmed to 22 °C.
2. Pour thick SM plates; 1 l should yield around thirty 100 mm petri dish plates. We also use smaller petri dishes (Nunc Cat. #153066) to save on media and to make many more plates.
3. Dry plates overnight at room temperature and then store at 4 °C. If the plates are too dry, the bacteria will not grow well.
4. Use deionized (or distilled) water to prepare all media and solutions.

5. We reuse plastic petri dishes for DB agar plates since the plates do not need to sterile. Agar can be pulled out with a pipette tip and disposed of in the waste.
6. Different cell lines and mutants have varying capacity to clear KA lawns and grow in the presence of KA. You may need to add more *D. discoideum* for cells with phagocytosis defects, for instance.
7. The inoculum of *D. discoideum* cells is important; if too many cells are used, they will consume the bacteria too quickly and enter the developmental cycle.
8. The doubling time of *D. discoideum* grown in the presence of KA is about 4 h.
9. Once every few months, we find that the *D. discoideum* cells fail to grow or those that appear seem dead. When this happens, we go back to our frozen KA stock and start a new culture.
10. We reuse the Lab-Tek 8- and 1-well chambers. Wash right away with deionized water and use a cotton swab to scrub away any cell debris. Caution should be taken when scrubbing these. One needs to hold the chamber from the side so that you do not separate the coverslip from the plastic wells, as it is just held together by silicon.
11. For chemotaxis assays, we melt away the side of the 1-well chamber so that the micropipette does not hit the sides. We leave a few millimeters on the side so that the 1 well will hold about 2.5 ml of DB.
12. The HL5 medium should be stored at 4 °C, but bottles can be kept at room temperature for regular use.
13. When we analyze mutants, we mix the WT and mutant cells on the same bacterial plate, keeping the total cell density the same as would be done for one cell line. One cell line needs to be expressing a GFP fusion protein near 100 % to differentiate the cell lines in the microscope (or cells can express different GFP variants for identification). This allows you to have an internal control when doing chemotaxis experiments. A similar control could be performed by developing cells together for cAMP-mediated chemotaxis experiments. If chemotaxis fails for both lines, the data can be discarded without compromising overall findings.
14. When preparing cells for microscopy, instead of washing cells in DB and centrifuging, one can take an inoculation loop and obtain a small amount of the bacterial/*Dictyostelium* cell mixture and add directly to a Lab-Tek well containing DB. The *Dictyostelium* cells will sink, while the bacteria will not and one can wash out the bacteria by adding and removing the DB.

15. Cells incubated at room temperature should be carefully monitored and any variations in temperature noted. This may be critical when you are growing KA lawns with *D. discoideum*, as fluctuations in temperature may be beneficial for growth of one organism, and not the other.
16. When adding KA to cells under microscopic observation, be careful to not add too much. This has two deleterious effects: Oxygen levels may drop, killing the *D. discoideum*, and KA bacteria are also autofluorescent, so this will lower the signal to noise for fluorescent proteins expressed in the amoebae.
17. KA is considered non-pathogenic. For further conversation on the taxonomy and pathology of the genus *Klebsiella see* ref. 83.
18. There is G418 resistant KA that can be obtained from Dictybase, although we are not aware of KA strains resistant to the other drugs used for selection of expression vectors.
19. For further discussion on using other bacterial strains or for troubleshooting problems growing *D. discoideum* on KA lawns see: http://www.dictybase.org/ListServ_archive/listserv_archive_bacteria.html

Acknowledgement

We thank Dictybase for providing cell lines and recipes for protocols. KJA was supported by a Faculty Development Leave from Texas A&M University. Funding support for CJ was provided by NIH grant R01-GM080370.

References

1. Raper KB (1935) *Dictyostelium discoideum*, a new species of slime mold from decaying forest leaves. J Agric Res 50:135–147
2. Romeralo M, Baldauf S, Escalante R (eds) (2013) Dictyostelids: evolution, genomics and cell biology. Springer, New York
3. Kessin RH (2001) *Dictyostelium*: evolution, cell biology, and the development of multicellularity. Cambridge University Press, Cambridge, UK
4. Raper KB (1937) Growth and development of *Dictyostelium discoideum* with different bacterial associates. J Agric Res 55:289–316
5. Sussman R, Sussman M (1967) Cultivation of *Dictyostelium discoideum* in axenic medium. Biochem Biophys Res Commun 29:53–55
6. Loomis WF Jr (1971) Sensitivity of *Dictyostelium discoideum* to nucleic acid analogues. Exp Cell Res 64:484–486
7. Veltman DM, Lemieux MG, Knecht DA, Insall RH (2014) PIP_3-dependent macropinocytosis is incompatible with chemotaxis. J Cell Biol 204:497–505
8. Devreotes PN, Zigmond SH (1988) Chemotaxis in eukaryotic cells: a focus on leukocytes and *Dictyostelium*. Annu Rev Cell Biol 4:649–686
9. Niggli V (2003) Microtubule-disruption-induced and chemotactic-peptide-induced migration of human neutrophils: implications for differential sets of signalling pathways. J Cell Sci 116:813–822
10. Park HT, Wu J, Rao Y (2002) Molecular control of neuronal migration. Bioessays 24:821–827
11. Rickert P, Weiner OD, Wang F et al (2000) Leukocytes navigate by compass: roles of PI3Kγ and its lipid products. Trends Cell Biol 10:466–473

12. Patel DD, Haynes BF (2001) Leukocyte homing to synovium. Curr Dir Autoimmun 3:133–167
13. Libby P (2002) Inflammation in atherosclerosis. Nature 420:868–874
14. Moore MA (2001) The role of chemoattraction in cancer metastases. Bioessays 23:674–676
15. Lazennec G, Richmond A (2010) Chemokines and chemokine receptors: new insights into cancer-related inflammation. Trends Mol Med 16:133–144
16. Maeda Y, Mayanagi T, Amagai A (2009) Folic acid is a potent chemoattractant of free-living amoebae in a new and amazing species of protist, Vahlkampfia sp. Zoolog Sci 26:179–186
17. Pan P, Hall EM, Bonner JT (1972) Folic acid as second chemotactic substance in the cellular slime moulds. Nat New Biol 237:181–182
18. Artemenko Y, Lampert TJ, Devreotes PN (2014) Moving towards a paradigm: common mechanisms of chemotactic signaling in *Dictyostelium* and mammalian leukocytes. Cell Mol Life Sci 71:3711–3747
19. Bonner J (1978) The life cycle of the cellular slime molds. Nat Hist 87:70–79
20. Bonner JT (1971) Aggregation and differentiation in the cellular slime molds. Annu Rev Microbiol 25:75–92
21. Katoh M, Chen G, Roberge E et al (2007) Developmental commitment in *Dictyostelium discoideum*. Eukaryot Cell 6:2038–2045
22. Loomis WF (1979) Biochemistry of aggregation in *Dictyostelium*. A review. Dev Biol 70:1–12
23. Pitt GS, Milona N, Borleis J et al (1992) Structurally distinct and stage-specific adenylyl cyclase genes play different roles in *Dictyostelium* development. Cell 69:305–315
24. Garcia GL, Rericha EC, Heger CD et al (2009) The group migration of *Dictyostelium* cells is regulated by extracellular chemoattractant degradation. Mol Biol Cell 20:3295–3304
25. Williams HP, Harwood AJ (2003) Cell polarity and *Dictyostelium* development. Curr Opin Microbiol 6:621–627
26. Zhang M, Goswami M, Sawai S et al (2007) Regulation of G protein-coupled cAMP receptor activation by a hydrophobic residue in transmembrane helix 3. Mol Microbiol 65:508–520
27. Manahan CL, Iglesias PA, Long Y, Devreotes PN (2004) Chemoattractant signaling in *Dictyostelium discoideum*. Annu Rev Cell Dev Biol 20:223–253
28. Abe F, Maeda Y (1994) Precise expression of the cAMP receptor gene, CAR1, during transition from growth to differentiation in *Dictyostelium discoideum*. FEBS Lett 342:239–241
29. Parent CA, Devreotes PN (1996) Constitutively active adenylyl cyclase mutant requires neither G proteins nor cytosolic regulators. J Biol Chem 271:18333–18336
30. Verkerke-Van Wijk I, Kim JY, Brandt R et al (1998) Functional promiscuity of gene regulation by serpentine receptors in *Dictyostelium discoideum*. Mol Cell Biol 18:5744–5749
31. Jowhar D, Wright G, Samson PC et al (2010) Open access microfluidic device for the study of cell migration during chemotaxis. Integr Biol (Camb) 2:648–658
32. De Wit RJW, Rinke De Wit TF (1986) Developmental regulation of the folic acid chemosensory system in *Dictyostelium discoideum*. Dev Biol 118:385–391
33. Devreotes PN (1983) The effect of folic acid on cAMP-elicited cAMP production in *Dictyostelium discoideum*. Dev Biol 95:154–162
34. Hadwiger JA, Srinivasan J (1999) Folic acid stimulation of the Gα4 G protein-mediated signal transduction pathway inhibits anterior prestalk cell development in *Dictyostelium*. Differentiation 64:195–204
35. Kesbeke F, van Haastert PJM, De Wit RJW, Snaar-Jagalska BE (1990) Chemotaxis to cyclic AMP and folic acid is mediated by different G-proteins in *Dictyostelium discoideum*. J Cell Sci 96:669–673
36. Maeda M, Firtel RA (1997) Activation of the mitogen-activated protein kinase ERK2 by the chemoattractant folic acid in *Dictyostelium*. J Biol Chem 272:23690–23695
37. van Haastert PJ, De Wit RJ, Konijn TM (1982) Antagonists of chemoattractants reveal separate receptors for cAMP, folic acid and pterin in *Dictyostelium*. Exp Cell Res 140:453–456
38. Kortholt A, Kataria R, Keizer-Gunnink I et al (2011) *Dictyostelium* chemotaxis: essential Ras activation and accessory signalling pathways for amplification. EMBO Rep 12:1273–1279
39. Andrew N, Insall RH (2007) Chemotaxis in shallow gradients is mediated independently of PtdIns 3-kinase by biased choices between random protrusions. Nat Cell Biol 9:193–200
40. Chubb JR, Wilkins A, Wessels DJ et al (2002) Pseudopodium dynamics and rapid cell movement in *Dictyostelium* Ras pathway mutants. Cell Motil Cytoskeleton 53:150–162
41. Devreotes P, Janetopoulos C (2003) Eukaryotic chemotaxis: distinctions between directional sensing and polarization. J Biol Chem 278:20445–20448
42. Insall R, Andrew N (2007) Chemotaxis in *Dictyostelium*: how to walk straight using parallel pathways. Curr Opin Microbiol 10:578–581

43. van Haastert PJ, Bosgraaf L (2009) The local cell curvature guides pseudopodia towards chemoattractants. HFSP J 3:282–286
44. van Haastert PJ, Postma M (2007) Biased random walk by stochastic fluctuations of chemoattractant-receptor interactions at the lower limit of detection. Biophys J 93:1787–1796
45. Srinivasan K, Wright GA, Hames N et al (2013) Delineating the core regulatory elements crucial for directed cell migration by examining folic-acid-mediated responses. J Cell Sci 126:221–233
46. Janetopoulos C, Borleis J, Vazquez F et al (2005) Temporal and spatial regulation of phosphoinositide signaling mediates cytokinesis. Dev Cell 8:467–477
47. Funamoto S, Meili R, Lee S et al (2002) Spatial and temporal regulation of 3-phosphoinositides by PI 3-kinase and PTEN mediates chemotaxis. Cell 109:611–623
48. Kae H, Lim CJ, Spiegelman GB, Weeks G (2004) Chemoattractant-induced Ras activation during *Dictyostelium* aggregation. EMBO Rep 5:602–606
49. Parent CA, Blacklock BJ, Froehlich WM et al (1998) G protein signaling events are activated at the leading edge of chemotactic cells. Cell 95:81–91
50. Dormann D, Weijer G, Dowler S, Weijer CJ (2004) In vivo analysis of 3-phosphoinositide dynamics during *Dictyostelium* phagocytosis and chemotaxis. J Cell Sci 117:6497–6509
51. Kamimura Y, Xiong Y, Iglesias PA et al (2008) PIP_3-independent activation of TorC2 and PKB at the cell's leading edge mediates chemotaxis. Curr Biol 18:1034–1043
52. Sasaki AT, Firtel RA (2009) Spatiotemporal regulation of Ras-GTPases during chemotaxis. Methods Mol Biol 571:333–348
53. Iijima M, Devreotes P (2002) Tumor suppressor PTEN mediates sensing of chemoattractant gradients. Cell 109:599–610
54. Janetopoulos C, Ma L, Devreotes PN, Iglesias PA (2004) Chemoattractant-induced phosphatidylinositol 3,4,5-trisphosphate accumulation is spatially amplified and adapts, independent of the actin cytoskeleton. Proc Natl Acad Sci U S A 101:8951–8956
55. Ma L, Janetopoulos C, Yang L et al (2004) Two complementary, local excitation, global inhibition mechanisms acting in parallel can explain the chemoattractant-induced regulation of $PI(3,4,5)P_3$ response in *Dictyostelium* cells. Biophys J 87:3764–3774
56. Milne JL, Caterina MJ, Devreotes PN (1997) Random mutagenesis of the cAMP chemoattractant receptor, cAR1, of *Dictyostelium*. Evidence for multiple states of activation. J Biol Chem 272:2069–2076
57. Prabhu Y, Eichinger L (2006) The *Dictyostelium* repertoire of seven transmembrane domain receptors. Eur J Cell Biol 85:937–946
58. Dormann D, Kim JY, Devreotes PN, Weijer CJ (2001) cAMP receptor affinity controls wave dynamics, geometry and morphogenesis in *Dictyostelium*. J Cell Sci 114:2513–2523
59. Kim JY, Borleis JA, Devreotes PN (1998) Switching of chemoattractant receptors programs development and morphogenesis in *Dictyostelium*: receptor subtypes activate common responses at different agonist concentrations. Dev Biol 197:117–128
60. Pupillo M, Kumagai A, Pitt GS et al (1989) Multiple α subunits of guanine nucleotide-binding proteins in *Dictyostelium*. Proc Natl Acad Sci U S A 86:4892–4896
61. Sonnemann J, Aichem A, Schlatterer C (1998) Dissection of the cAMP induced cytosolic calcium response in *Dictyostelium discoideum*: the role of cAMP receptor subtypes and G protein subunits. FEBS Lett 436:271–276
62. Kortholt A, van Haastert PJ (2008) Highlighting the role of Ras and Rap during *Dictyostelium* chemotaxis. Cell Signal 20:1415–1422
63. Chen MY, Insall RH, Devreotes PN (1996) Signaling through chemoattractant receptors in *Dictyostelium*. Trends Genet 12:52–57
64. Janetopoulos C, Jin T, Devreotes P (2001) Receptor-mediated activation of heterotrimeric G-proteins in living cells. Science 291:2408–2411
65. Kumagai A, Hadwiger JA, Pupillo M, Firtel RA (1991) Molecular genetic analysis of two Gα protein subunits in *Dictyostelium*. J Biol Chem 266:1220–1228
66. Kumagai A, Pupillo M, Gundersen R et al (1989) Regulation and function of Gα protein subunits in *Dictyostelium*. Cell 57:265–275
67. Wu L, Valkema R, van Haastert PJ, Devreotes PN (1995) The G protein β subunit is essential for multiple responses to chemoattractants in *Dictyostelium*. J Cell Biol 129:1667–1675
68. Hadwiger JA (2007) Developmental morphology and chemotactic responses are dependent on Gα subunit specificity in *Dictyostelium*. Dev Biol 312:1–12
69. Liao XH, Buggey J, Kimmel AR (2010) Chemotactic activation of *Dictyostelium* AGC-family kinases AKT and PKBR1 requires separate but coordinated functions of PDK1 and TORC2. J Cell Sci 123:983–992
70. Chung CY, Potikyan G, Firtel RA (2001) Control of cell polarity and chemotaxis by

Akt/PKB and PI3 kinase through the regulation of PAKa. Mol Cell 7:937–947

71. Kamimura Y, Devreotes PN (2010) Phosphoinositide-dependent protein kinase (PDK) activity regulates phosphatidylinositol 3,4,5-trisphosphate-dependent and -independent protein kinase B activation and chemotaxis. J Biol Chem 285:7938–7946
72. Sarbassov DD, Guertin DA, Ali SM, Sabatini DM (2005) Phosphorylation and regulation of Akt/PKB by the rictor-mTOR complex. Science 307:1098–1101
73. Guertin DA, Stevens DM, Thoreen CC et al (2006) Ablation in mice of the mTORC components raptor, rictor, or mLST8 reveals that mTORC2 is required for signaling to Akt-FOXO and PKCα, but not S6K1. Dev Cell 11:859–871
74. Bozulic L, Hemmings BA (2009) PIKKing on PKB: regulation of PKB activity by phosphorylation. Curr Opin Cell Biol 21:256–261
75. DiNitto JP, Lambright DG (2006) Membrane and juxtamembrane targeting by PH and PTB domains. Biochim Biophys Acta 1761:850–867
76. Feng J, Park J, Cron P et al (2004) Identification of a PKB/Akt hydrophobic motif Ser-473 kinase as DNA-dependent protein kinase. J Biol Chem 279:41189–41196
77. Segall JE, Fisher PR, Gerisch G (1987) Selection of chemotaxis mutants of *Dictyostelium discoideum*. J Cell Biol 104:151–161
78. Bolourani P, Spiegelman GB, Weeks G (2006) Delineation of the roles played by RasG and RasC in cAMP-dependent signal transduction during the early development of *Dictyostelium discoideum*. Mol Biol Cell 17:4543–4550
79. Bolourani P, Spiegelman G, Weeks G (2010) Ras proteins have multiple functions in vegetative cells of *Dictyostelium*. Eukaryot Cell 9:1728–1733
80. Fey P, Kowal AS, Gaudet P et al (2007) Protocols for growth and development of *Dictyostelium discoideum*. Nat Protoc 2:1307–1316
81. Watts DJ, Ashworth JM (1970) Growth of myxameobae of the cellular slime mould *Dictyostelium discoideum* in axenic culture. Biochem J 119:171–174
82. Chen BC, Legant WR, Wang K et al (2014) Lattice light-sheet microscopy: imaging molecules to embryos at high spatiotemporal resolution. Science 346:1257998
83. Brisse S, Grimont F, Grimont P (2006) The genus *Klebsiella*. Prokaryotes 6:159–196

Chapter 4

Mitochondrial Stress Tests Using Seahorse Respirometry on Intact *Dictyostelium discoideum* Cells

Sui Lay, Oana Sanislav, Sarah J. Annesley, and Paul R. Fisher

Abstract

Mitochondria not only play a critical and central role in providing metabolic energy to the cell but are also integral to the other cellular processes such as modulation of various signaling pathways. These pathways affect many aspects of cell physiology, including cell movement, growth, division, differentiation, and death. Mitochondrial dysfunction which affects mitochondrial bioenergetics and causes oxidative phosphorylation defects can thus lead to altered cellular physiology and manifest in disease. The assessment of the mitochondrial bioenergetics can thus provide valuable insights into the physiological state, and the alterations to the state of the cells. Here, we describe a method to successfully use the Seahorse XF^{e}24 Extracellular Flux Analyzer to assess the mitochondrial respirometry of the cellular slime mold *Dictyostelium discoideum*.

Key words *Dictyostelium discoideum*, Mitochondrial respirometry, Seahorse extracellular flux analyzer, XF^{e}24, AOX pathway, Matrigel® basement membrane matrix, Oxygen consumption rate, OCR, Mitochondrial inhibitors

1 Introduction

Chemotactic motility and the actomyosin cytoskeleton play major roles in normal cellular growth and division [1], in embryogenesis and differentiation [2–4], immune response [5], but can also be involved in a series of pathological conditions such as metastasis [6] or vascular disease [7]. In unicellular organisms it controls the motile behavior. Chemotactic motility not only costs energy in the form of mitochondrially generated ATP, but it is regulated by signaling pathways that interact with those that sense and regulate the cell's energy status.

Amongst the different cell types that have been used for the investigation of the chemotactic signaling, *Dictyostelium discoideum* has proven to be a very useful model for the discovery of many of the conserved chemotaxis pathways of eukaryotic cells [8–10]. Numerous signaling pathways are involved in the regulation of cell

Tian Jin and Dale Hereld (eds.), *Chemotaxis: Methods and Protocols*, Methods in Molecular Biology, vol. 1407,
DOI 10.1007/978-1-4939-3480-5_4, © Springer Science+Business Media New York 2016

polarity and actin cytoskeleton rearrangements, including the PTEN, Ras, PI3K [11, 12], PLC [13], PLA2 pathways [12, 14], ROS signaling cascade [15], and the TOR signaling pathways. The TOR pathways involve two multiprotein complexes, TOR Complex 1 (TORC1) and TOR Complex 2 (TORC2) [16–19] that cross-talk with one another and together sense chemoattractants as well as the cell's oxidative, energy, and nutritional status. In response to these inputs they regulate chemotaxis, endocytosis, autophagy, cell growth, and mitochondrial biogenesis. Both complexes have different subunit compositions, function, and regulation, but have in common, a Target of Rapamycin (TOR) subunit with kinase activity.

TORC2 is a key regulator of cytoskeletal reorganization and chemotaxis [20]. In *Dictyostelium* the chemoattractant cAMP binds to the surface receptor cAR1 and activates a series of heterotrimeric G proteins (RasC, RasG) that recruit and activate TORC2. This in turn leads to phosphorylation of PKB/Akt substrates (PKBA and PKBR1) and thence further activation of additional effectors that are directly responsible for the spatial rearrangements of the cytoskeleton and adenylyl cyclase activity [21, 22]. The roles of TORC2 and PKB/Akt signaling are also conserved in other organisms. For example, in *Saccharomyces cerevisiae*, TORC2 regulates actin polarization via a PKB/Akt homologue [23, 24]. In various cancer cells Akt kinases modulate motility, invasion, and metastasis [25]. The major effector of TORC2 in mammalian cells is PKC. It is involved in the regulation of cAMP production as well as the polarization of F-actin and myosin II [20].

The major role of TORC1 is the regulation of mRNA translation, cell growth, and cell-cycle progression as well as autophagy (reviewed in [26]). One of its downstream targets is the protein kinase S6K1 which is activated by TORC1 and in turn inhibits TORC2. Conversely, TORC1 is an indirect downstream target for TORC2 and PKB/Akt which inhibit the tumor suppressor complex, TSC1-TSC2, thereby activating Rheb and TORC1 [19]. The interaction between TORC1 and the pathways controlling cell motility and the cytoskeleton is not only through the feedback loops that connect Akt, TORC1, S6K1, and TORC2 signaling. Thus TORC1 has been shown to play a role in enhancing microtubule assembly and stability by phosphorylating CLIP-170, a microtubule associated protein [27, 28]. Furthermore, several studies have shown that both S6K1 and 4E-BP1, direct targets of TORC1, are able to regulate cell motility by controlling actin organization or the expression of various key proteins of the cytoskeletal architectural signaling pathway [20, 29–31].

All these processes are great consumers of cellular energy and therefore, not surprisingly, activation of the TORC1 pathway also leads to increased mitochondrial biogenesis in order to maintain homeostatically the ATP supply of the cell. TORC1 can directly regulate the expression and activity of mitochondrial proteins [32–34] and the turnover of defective mitochondria by mitophagy [35] as well as the uptake and utilization of carbohydrates by the cell [36]. Mitochondrial activity in its turn can modulate the activity of the TORC1 pathway. Insufficient mitochondrial respiratory activity leads to the accumulation of AMP and decreased ATP levels within the cell [37]. This activates AMPK, a major energy-sensing protein kinase that stimulates energy-producing, catabolic processes and switches off many energy-consuming pathways including TORC1 targets. The inhibition of TORC1 by AMPK can occur directly through phosphorylation of a regulatory component, Raptor, within TORC1, or indirectly, by activation of TSC1-TSC2 [38]. The TSC1-TSC2 complex associates with TORC2 to stimulate its activity [39]. The extensive cross-talk between these various signaling pathways allows the TOR complexes to function as key check-point sensors for cellular metabolism, integrating signals from energy and nutrient uptake with cell growth and motility (chemotaxis) whilst also playing a role in controlling mitochondrial biogenesis and function.

Indeed, mitochondria also directly play a role in affecting the chemotactic response. The mitochondrial protein Tortoise (TorA) for example, has been shown to be essential for directional movement of *Dictyostelium* cells in a cAMP gradient [40]. The chemotactic response to the cAMP waves of *torA* null mutants is attenuated, whereby more waves are required for aggregation and the resulting aggregates are smaller than the wild type. That is, the mutant cells are no longer capable of responding to a spatial gradient with direct movement, and instead form lateral pseudopods at higher rates than wild-type cells. Furthermore, the knockout of MidA, a mitochondrial protein required for Complex I assembly [41], and RnlA, a mitochondrial ribosomal protein [42] as well as knockdown of the nuclear-encoded chaperonin 60, a molecular chaperone required for correct folding of mitochondrial proteins [43], have all been shown to result in a series of well-defined mitochondrial disease phenotypes, in which phototactic and thermotactic signaling and/or chemotactic responses are affected [44].

The ability of cells to respond to stress under conditions of increased energy demand, such as occurs during chemotactic motility, is in large part influenced by the bioenergetic capacity of mitochondria. This bioenergetic capacity is determined by several factors, including the ability of the cell to deliver substrate to mitochondria and the functional capacity of enzymes involved in electron transport. Numerous methods, many of which are biochemical assays, have been developed to assess the activity and function of mitochondria. These methods include cytochrome c oxidase or succinate dehydrogenase activity assays, mitochondrial membrane potential, the production of reactive oxygen species (ROS) and ATP content assays, as well as Western blot and Blue native polyacrylamide gel electrophoresis of mitochondrial proteins and protein complexes [45–47]. Major disadvantages of all of these methods are that they cannot be performed in real time on intact cells and they measure steady state levels rather than physiological fluxes.

We have developed a method for applying to *Dictyostelium* the multi-well, microplate-based Seahorse XF^{e}24 Extracellular Flux Analyzer (Seahorse Bioscience) to assay the physiological function of mitochondrial respiratory complexes in intact cells. This allows the testing of mutant cells and the effects of various compounds on the mitochondrial function of live intact cells or isolated mitochondria [48–50]. The test utilizes four injection ports to directly introduce the test compounds into a 24 well plate in parallel, and by using two sensor probes, one oxygen-sensing fluorophore and a pH sensor, is able to simultaneously measure the oxygen consumption rates (OCR) and the extracellular acidification rates (ECAR) of a monolayer of cells within a small volume of medium. The technology allows for many aspects of mitochondrial function and respiratory control to be measured in a single experiment, including basal respiration, ATP turnover, proton leak, coupling efficiency, maximum respiration rate, spare respiratory capacity, and nonmitochondrial respiration as well as the contributions of each of the respiratory complexes and the alternative oxidase (Figs. 1 and 2).

In this chapter, we describe a protocol for successful measurement of mitochondrial function in intact axenically grown *Dictyostelium discoideum* cells using the Seahorse XF^{e}24 extracellular Flux Analyzer. We detail the optimized and validated *Dictyostelium*-specific mitochondrial stress test drug regimen that allows the activities in respiration of the various mitochondrial components to be determined, the assay medium and substrates to be used. We describe the critical steps for plate and instrument setup required for consistent and reproducible results.

2 Materials

2.1 Equipment and Consumables

1. XF^{e}24 Extracellular Flux Analyzer (Seahorse Bioscience, MA, USA).
2. Seahorse XF^{e}24 FluxPak with PS Cell Culture Microplates (Seahorse Bioscience) (*see* **Note 1**).

2.2 Cell Culture Media, Buffers, and Experimental Compounds

1. HL5 medium: For 1 L, add 14 g proteose peptone, 7 g yeast extract, 0.5 g KH_2PO_4, 0.5 g Na_2HPO_4 in 500 mL distilled water and adjust pH to 6.4–6.6. Add 10 g glucose in 500 mL distilled water. Combine both solutions after autoclaving (*see* **Note 2**).
2. SIH medium (ForMedium™, Norfolk, UK). Dissolve 20.3 g powdered medium in 1 L distilled water. Sterilize by autoclaving. The full composition can be found in Table 1 (*see* **Note 3**).

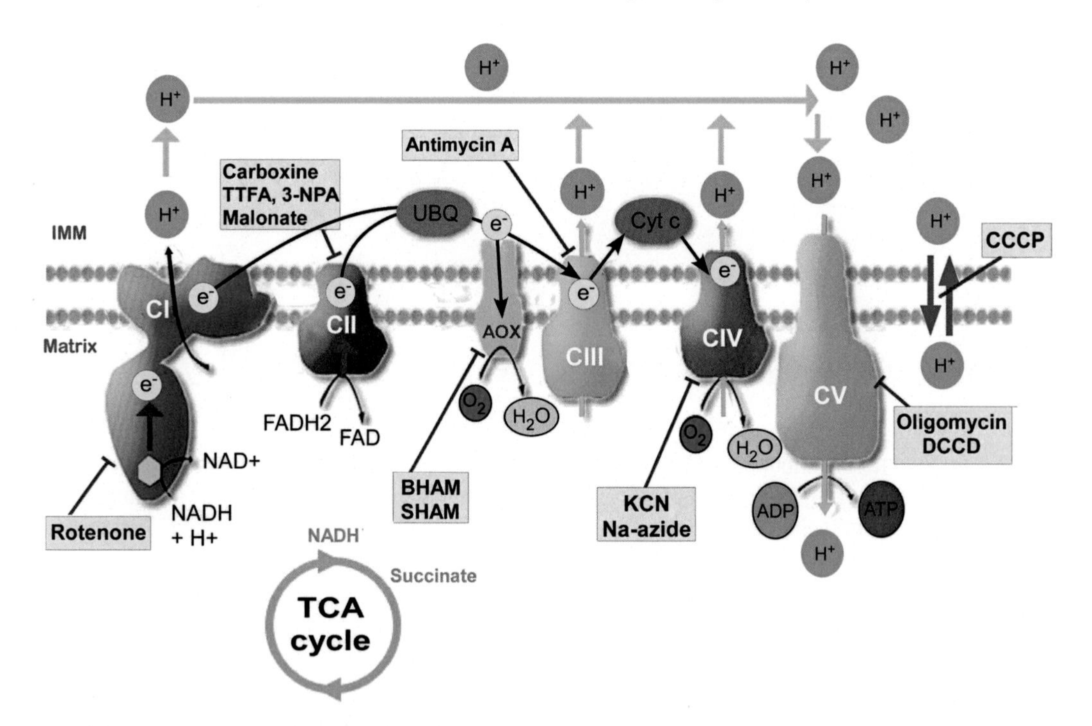

Fig. 1 The mitochondrial OXPHOS pathway and various mitochondrial inhibitors. At the mitochondrial inner membrane, oxidative phosphorylation occurs when electrons from reduction of NADH (by Complex I, CI) or succinate (by Complex II, CII, via $FADH_2$) pass through a series of additional enzyme complexes (Complex III, CIII; Complex IV, CIV; alternative oxidase, AOX) to a terminal electron acceptor (O_2) via a series of redox reactions. These reactions are coupled to the creation of a proton gradient across the mitochondrial inner membrane which is used either to make ATP via ATP synthase (Complex V) or to drive other mitochondrial membrane potential-dependent processes, such as mitochondrial protein import. Many organisms also have an alternative oxidase (AOX) enzyme that forms part of the electron transport chain (ETC). It provides an alternative pathway for electrons passing through the ETC and bypasses proton-pumping steps at CIII and CIV to reduce oxygen and consequently reduces ATP yield. The expression of AOX is influenced by stresses such as cold, reactive oxygen species and infection by pathogens, as well as other factors that reduce electron flow through the ETC. In *Dictyostelium*, AOX is differentially regulated during development [51]. In assessing cellular energy metabolism, the measurement of the activities of the various ETC complexes may be measured by the addition of complex-specific inhibitors (as shown)

3. L (−) malic acid (L-hydroxybutanedioic acid).
4. Sodium pyruvate.
5. SIH assay medium: Supplement SIH medium with 20 mM sodium pyruvate and 5 mM malic acid. Add 110 mg sodium pyruvate and 33.5 mg of malic acid to 50 mL (sufficient for one assay) of SIH medium. Adjust pH to 7.4 with 2 M NaOH. Prepared fresh on the day of use (*see* **Note 4**).
6. Matrigel® growth factor reduced (GFR) basement membrane matrix, phenol-red free, LDEV-free, 10 mL (Corning, MA, USA) (*see* **Note 5**).

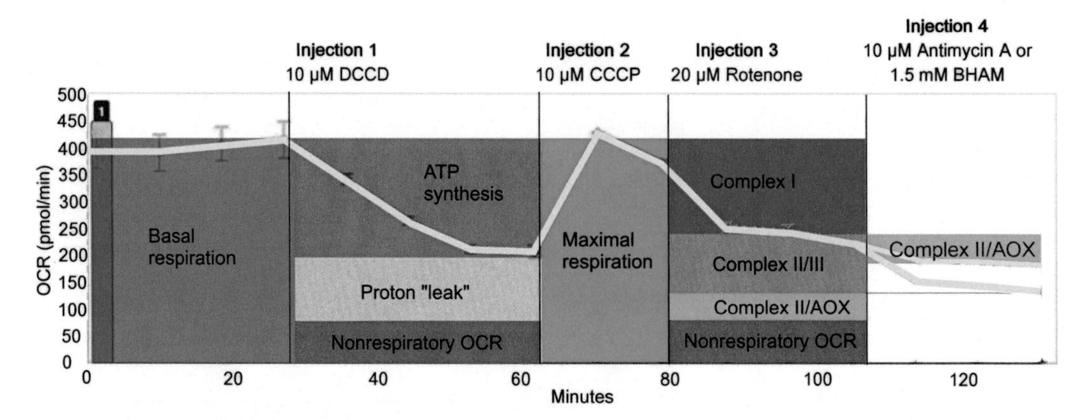

Fig. 2 Components of mitochondrial function revealed by the Seahorse assay. In a typical Seahorse mitochondrial assay run, the contributions of various mitochondrial complexes and measurement of key parameters of mitochondrial function in cellular bioenergetics can be determined. The Seahorse analyzer uses two optical sensors to simultaneously make real-time measurements of the extracellular acidification rate (ECAR) and oxygen consumption rate (OCR) levels in a transiently isolated volume of medium above a monolayer of cells within a microplate. Through measurement of the OCR, an indicator of mitochondrial respiration, valuable insights into the physiological state of the cells, and the alterations to the state of the cells is gained. This is achieved by treatment of the cells with subsequent injections of four drugs; (injection 1) DCCD (ATP synthase inhibitor), (injection 2) CCCP (ETC uncoupler), (injection 3) rotenone (Complex I inhibitor), and (injection 4) antimycin A (Complex III inhibitor) or BHAM (AOX inhibitor) and monitoring the resultant levels. The OCR following either of the fourth injected drugs is measured in parallel, in two sets of replicate wells, to allow the relative contribution of oxygen consumption by either the cytochrome c (ETC) or alternative oxidase (AOX) pathways to be determined

2.3 Mitochondrial Drugs

The mitochondrial drugs listed below are prepared as 1000× stock solutions in 100 % DMSO. To avoid multiple freeze-thaws that could affect their efficacy, 25 μL single-use aliquots should be prepared and stored at −20 °C.

1. 10 mM *N,N′*-dicyclohexylcarbodiimide (DCCD, Sigma). Add 20.6 mg in 10 mL DMSO (*see* **Note 6**).
2. 10 mM carbonyl cyanide 3-chlorophenol hydrazone (CCCP, Sigma-Aldrich). Add 20.5 mg in 10 mL DMSO.
3. 20 mM rotenone (Sigma-Aldrich). Add 78.9 mg in 10 mL DMSO.
4. 10 mM antimycin A (Sigma-Aldrich). Add 54.9 mg in 10 mL DMSO.
5. 1.5 M benzohydroxamic acid (BHAM, Sigma-Aldrich, cat. no. 412260). Add 1.028 g in 5 mL DMSO.

Table 1
Chemical composition of ForMedium® SIH Medium

Formula	mg/L	Formula	mg/L
Amino acids		Vitamins	
Arg	700	Biotin	0.02
Asp	300	Cyanocobalamin	0.01
Asp A	150	Folic acid	0.2
Cys	300	Lipoic acid	0.4
Glu A	545	Riboflavin	0.5
Gly	900	Thiamine	0.6
His	300	Micro elements	
Ile	600	$Na_2EDTA \cdot 2H_2O$	4.84
Leu	900	$ZnSO_4$	2.3
Lys	1250	H_3BO_3	1.11
Met	350	$MnCl_2 \cdot 4H_2O$	0.51
Phe	550	$CoCl_2 \cdot 6H_2O$	0.17
Pro	800	$CuSO_4 \cdot 5H_2O$	0.15
Thr	500	$(NH_4)_6Mo_7O_{24} \cdot 4H_2O$	0.1
Trp	350	Carbon source	
Val	700	Glucose	10,000
Minerals			
NH_4Cl	53.5		
$CaCl_2 \cdot 2H_2O$	2.94		
$FeCl_3$	16.2		
$MgCl_2 \cdot 6H_2O$	81.32		
KH_2PO_4	870		

3 Methods

The protocol described below has been adapted to allow for the assessment of mitochondrial respiration in axenically grown *Dictyostelium discoideum* cells using the Seahorse XF^{e}24 Extracellular Flux Analyzer. While the Seahorse XF^{e}24 Analyzer is not optimally designed for testing this model system, by implementing some key adjustments to the plating of cells, assay medium, compounds utilized, and the instrument setup, it is possible to perform mitochondrial assays successfully in *Dictyostelium*.

3.1 Tasks for the Day before the Assay

1. Open one XF24 FluxPak that contains four parts; a lid (top), an assay sensor cartridge (second), a hydration booster (third), and a 24-well microplate (base). Remove the top three sections of the FluxPak, and add 1 mL of XF calibrant solution into each of the 24 wells. Replace the layers onto microplate and incubate at 21 °C for at least 16 h before use (*see* **Note 7**).
2. *Dictyostelium* cells are grown in HL5 medium. Check cells and adjust cell concentrations (if necessary) to ensure that they will be in exponential phase ($0.8–3.0 \times 10^6$ cells/mL) at the time of the assay (*see* **Note 8**).
   On the Day of the Assay:

3.2 Setup of the Seahorse Analyzer

1. Turn on Seahorse instrument and open the XF Wave application on the computer.
2. Set instrument target temperature to 23 °C, turn heater off and set temperature tolerance ±3.5 °C (*see* **Note 9**).
3. Set up assay design in the XF Wave software using the Assay Wizard.
   (a) Define groups and conditions, including: (1) Injections, (2) Pretreatments, (3) Assay medium, and (4) Cell types.
   (b) Generate groups (automatically or manually). Groups are the sets of wells, data from which will be pooled automatically, analyzed, and displayed by the Seahorse XF^{e}24 Analyzer software [52].
   (c) Distribute groups to plate map (background wells are already selected) (*see* **Note 10**).
   (d) Define assay protocol (*see* Table 2).
   (e) Review experimental details and include any additional notes.
   (f) Save design. The instrument is set up and ready for use.

3.3 Coating the Cell Culture Plate with Matrigel® Basement Membrane Matrix

The coating of the cell culture microplate with Matrigel® is a crucial step in the setup of the assay to ensure cell adherence during the assay, without which it is not possible to generate reliable results (*see* **Note 11**).

1. Prepare Matrigel® plate coating suspension by mixing a 45 μL aliquot of thawed Matrigel® with 90 μL (1:3 dilution) of cold SIH medium (*see* **Note 12**).
2. Apply 4.5 μL of the diluted Matrigel®-SIH solution onto the center of each well of a cold cell culture microplate, then using a swirling motion with the pipette tip gently distribute it onto the entire surface of the well.

Table 2
Typical Seahorse XF^e^24 Analyzer protocol setup for assaying *Dictyostelium* cells

Command	Time (min)	Repeats
Calibrate	Fixed	1
Equilibrate	Fixed	1
Mix	3	3
Wait	2	
Measure	3	
Inject Port A		
Mix	3	4
Wait	2	
Measure	3	
Inject Port B		
Mix	3	3
Wait	2	
Measure	3	
Inject Port C		
Mix	3	3
Wait	2	
Measure	3	
Inject Port D		
Mix	2	3
Wait	2	
Measure	3	

3. When all wells have been coated with Matrigel® suspension, completely air-dry the plate (lid off) for ~20–45 min at room temperature (*see* **Note 13**).
4. The plate is now ready to be used (*see* **Note 14**).

3.4 Seeding of Cells onto Cell Culture Plate

1. Harvest ~3×10^6 cells (*see* **Note 15**) by centrifugation at $3000 \times g$ for 20 s and remove the supernatant.
2. Wash the cell pellet with 3 mL SIH assay medium and centrifuge as above. Repeat once.
3. Resuspend the cells in ~3 mL SIH assay medium and adjust cell density to 1×10^6/mL.

4. Seed 1×10^5 cells/well (*see* **Notes 16** and **17**) into the desired wells in a two-step process (*see* **Note 17**): (1) add 100 μL of cells to the wells, allow them to attach to the plate surface for 10–20 min; (2) top up wells with assay medium to a total volume of 500 μL and settle for another 10 min before use in assay.
5. Background correction wells (by default), A1, B4, C3, D6 should be filled with 500 μL of SIH assay medium only (i.e., no cells).

3.5 Preparation of XF FluxPak Plate

1. Remove one aliquot of 1000× stocks of DCCD, CCCP, rotenone, antimycin-A, and BHAM from −20 °C freezer and thaw at room temperature.
2. Prepare 2 mL (for injection ports A, B, C) and 1 mL (for injection port D) of desired compound by diluting (*see* **Note 18**) aliquots 100-fold to 10× of the final working concentration in SIH assay medium (*see* **Note 19**).
3. Remove the hydration booster (third) layer from the XF FluxPak. It is no longer needed. Replace it with the sensor cartridge which is put back onto the cell culture microplate.
4. Load the indicated volumes (56–76 μL) of diluted compounds into the desired injection ports, A–D in the sensor cartridge according to your plate map/groupings (*see* Table 3) (*see* **Note 19**) (*see* Fig. 3).

Table 3
Assay cartridge injection port setup

Port	Compound (function)	Injected concentration (μM) [10×]	Injection volume (μL)	Final well concentration (μM) [1×]
Port A (Groups A and B of each set)	DCCD (ATP synthase inhibitor)	100	56	10
Port B (Groups A and B of each set)	CCCP (ETC uncoupler)	100	62	10
Port C (Groups A and B of each set)	Rotenone (Complex I inhibitor)	200	69	20
Port D (Group A of each set)	Antimycin A (Complex III inhibitor)	100	76	10
Port D (Group B of each set)	BHAM (AOX inhibitor)	15,000	76	1500

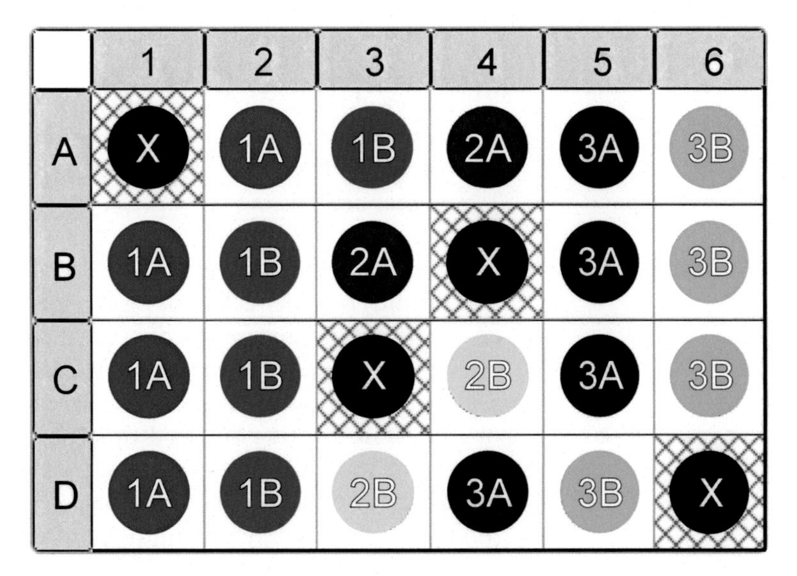

Fig. 3 Typical plate setup for mitochondrial stress test of *Dictyostelium* cells. This plate configuration allows for the parallel analysis of two test strains or conditions (two sets of four wells per strain) and one control parental strain (two sets of two wells). Each strain is tested with a combination of five drugs in the order; (1) 10 μM DCCD, (2) 10 μM CCCP, (3) 20 μM rotenone, (4) 10 μM antimycin A or 1.5 mM BHAM. Effectively, drug responses for the first three drugs injected are obtained for eight replicate wells per test strain, or four for the parental control strain. The fourth injection then tests four or two wells respectively with either antimycin A or with BHAM. This approach allows the contributions of the cytochrome c (Complex III/IV) and alternative oxidase pathways in oxygen consumption to be determined within the same assay. Legend: *1*, *2*, *3* represent the strain where *A*, antimycin-A used in last injection, and *B*, BHAM used in the last injection. The first three drug injections are the same in all wells—DCCD, CCCP, and rotenone, respectively. *X* indicates the baseline control (blank) wells

3.6 Running the Assay

1. Open the assay design created earlier on the Seahorse Wave software, as described in Subheading 3.2, press "Start" and follow instructions as prompted.
2. Remove the lid and insert the XF assay sensor cartridge. Check for correct orientation of the cartridge and place onto the instrument tray.
3. When the ~20 min calibration step is finished, the sensor cartridge is kept inside the machine, while the calibrant-containing utility plate is ejected, to be removed from the tray by the experimenter. At this point, the prepared cell culture plate should be loaded, by placing it onto the instrument tray, replacing the calibrant-containing utility plate.
4. Press "Continue" and allow the assay to run until complete (*see* **Note 20**).
5. After the XF^{e}24 Extracellular Flux Analyzer run is finished, carefully remove the assayed cell culture microplate and press

"Continue" to end the program and proceed to data analysis. An analysis (.asy) file is generated automatically which can be analyzed using the Wave software. Any necessary background corrections and other normalizations can also be made in the software. The data can also be exported in excel format for further analysis.

3.7 Data Analysis

The results of the assay can be reported as absolute values (pmol O_2 per min) on a per cell basis or on a per unit of protein basis. For energetic profiling experiments that measure the responses to the various drugs utilized, the oxygen consumption rate (OCR) can also be analyzed as a percentage relative to the resting (basal) OCR or the OCR at other selected time points within the assay (e.g., relative to the OCR immediately after CCCP injection). Based on the drug injection regime described in this chapter, a number of fundamental parameters of mitochondrial respiratory function can be determined. These are listed below. It is worth noting that by changing the order of the drug injections and also by using different compounds and/or substrates as well as combinations of all these, various other aspects of the cellular bioenergetics can be measured (*see* Figs. 2 and 4).

1. Basal OCR—steady state oxygen consumption of cells.
2. ATP turnover—represents the OCR devoted to the synthesis of ATP, and can be measured by the decline in OCR after the addition of DCCD (*see* **Note 21**).
3. Maximum OCR—can be reached after introduction of the proton ionophore CCCP and is dependent on the ETC activity and substrate delivery.
4. Contribution of Complex I to the maximum OCR—is represented by the drop in OCR after the addition of Complex I inhibitor rotenone (in the presence of CCCP).
5. Contribution of Complex II/III/IV to the maximum OCR—can be inferred from the drop in OCR after the addition of Complex III inhibitor antimycin A (in the presence of rotenone and CCCP).
6. Contribution of Complex II/AOX to the maximum OCR—can be inferred from the drop in OCR after the addition of AOX inhibitor BHAM (in the presence of rotenone and CCCP).
7. Contribution of Complex II to the maximum OCR—can be calculated from the sum of **steps 5** and **6** above.
8. "Spare capacity"—is revealed by the difference in maximum OCR and the basal levels.
9. Nonrespiratory OCR—is the residual OCR after addition of all respiratory inhibitors and includes contributions from all oxidases

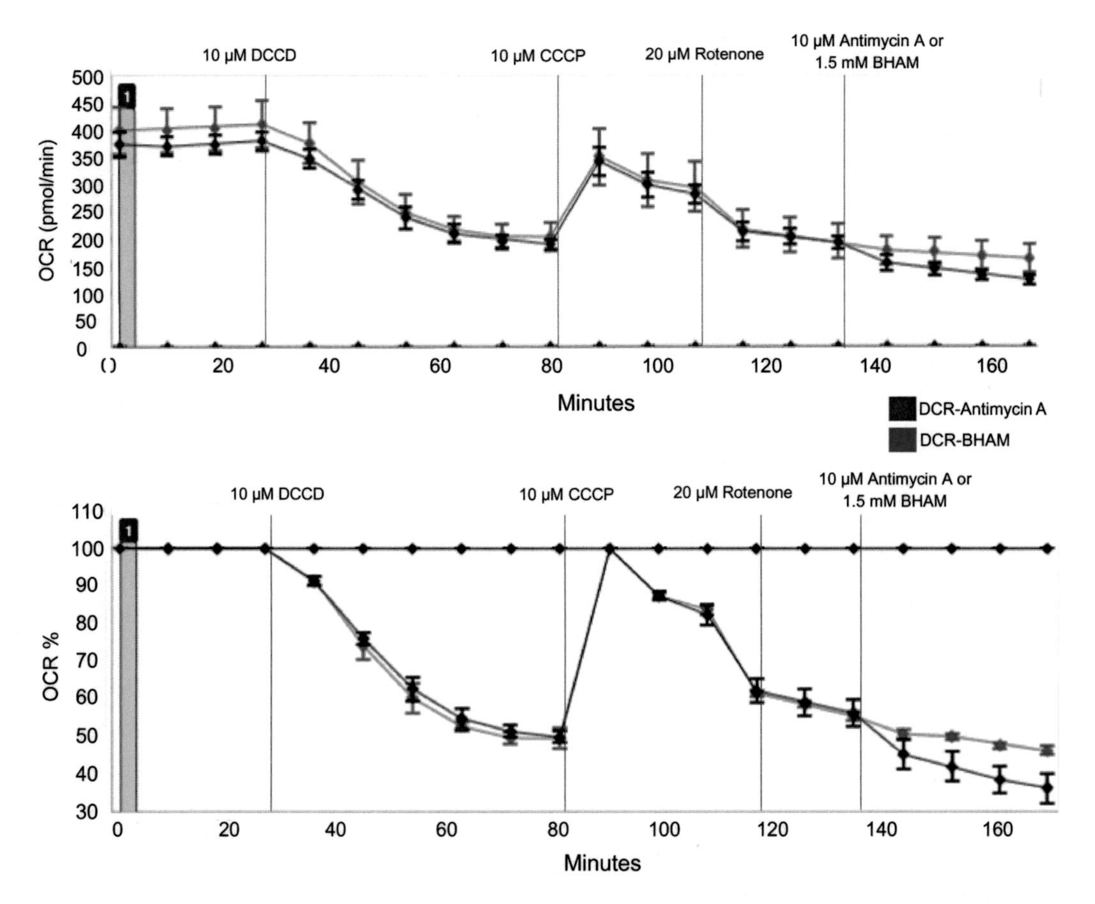

Fig. 4 A typical XF^{e}24 Seahorse *Dictyostelium* mitochondrial assay. The Seahorse assay allows the contributions of various mitochondrial complexes and measurement of key parameters of mitochondrial function (*refer to* Subheading 3.7) to be determined. (**a**) In this run, the same strain is tested with a combination of five mitochondrial drugs. The first three drugs are the same for both set of wells while the fourth drug is different. This approach provides a pair-wise comparison of the contributions to respiration of the cytochrome c pathway (complex II/complex III) using antimycin A (DCR-antimycin A), or the alternative oxidase pathway, using BHAM (DCR-BHAM). (**b**) Data from the experiment in Panel (**a**) has been normalized relative to the basal respiration levels (up to the time of the CCCP injection) and the maximum OCR rates (after the CCP injection)

and oxygenases that are not involved in the electron transport chain (but could be mitochondrial or nonmitochondrial).

10. Proton "leak"—respiratory OCR devoted to mitochondrial functions other than ATP synthesis, calculated from the difference between the OCR after DCCD addition and the residual OCR.

4 Notes

1. The FluxPak contains 18 XF24 extracellular flux assay kits (with each assay kit containing one sensor cartridge, one utility plate, one hydration booster plate and one lid), 20 XF24 polystyrene

cell culture microplates and one bottle of 500 mL XF calibrant solution. These items can also be purchased separately.

2. Autoclaving the glucose and other components separately prevents the variable caramelization of the glucose that might otherwise affect the physiological state of the cells.
3. SIH medium is a specific minimal medium that can support the growth of *Dictyostelium* cells. Seahorse Bioscience's XF medium could not be used for the assay as it lacked key microelements required for *Dictyostelium* growth. SIH medium will darken over extended periods of storage. Prepare only small batches for short-term use. However the color does not seem to affect the assay.
4. The correct pH of the assay medium is critical and pH differences will cause apparent OCR differences.
5. It is best to prepare 45 μL aliquots of Matrigel® and store them at −20 °C for long term storage, while keeping aliquots for short term use at 4 °C to have ready-to-use thawed aliquots.
6. DCCD is light-sensitive (keep in the dark) and unstable in solution, even in storage at −20 °C. We find it works best to prepare fresh lots once a month and keep single use aliquots wrapped in aluminum foil.
7. If the sensor cartridge is not used within 2 days of hydrating, seal the plate to avoid liquid evaporation from wells and store at 4 °C until required. Allow the cartridge to equilibrate to room temperature before use.
8. Cells should be growing in the exponential phase at the time of assay, ~$0.8–3.0 \times 10^6$ cells/mL. These cells are washed and resuspended in SIH assay medium and then seeded onto the cell culture microplate for testing soon after. We do not include an overnight incubation on the microplates because of difficulties in achieving similar cell densities for different strains with different growth rates.
9. The Seahorse Flux Analyzer has no cooling function and therefore depending on the ambient room temperature it may be difficult to achieve and/or maintain a constant temperature of ~23 °C inside the instrument. We usually set our room air conditioning to 18 °C to achieve instrument assay temperatures within an acceptable range of ~23–27 °C. For further detailed information about the instrument setup and result analyses, please refer to the XF^e Wave user manual [52].
10. We normally run two sets of four wells for each test strain—effectively, we will obtain eight replicate wells per strain with drug responses for the first three drugs injected. The fourth injection tests four wells each with antimycin A or with BHAM. This allows us to calculate the contributions of Complex III/IV and AOX which provide alternative pathways for electrons after

Complex I and II. We normally also include two wells of AX2 with antimycin A and two wells of AX2 with BHAM (as internal parental or wild type controls) (*see* Fig. 3). It is noteworthy that additional variations or combinations of injection solutions are possible depending on the experimental questions being addressed. The order of drug injections can also be altered, depending on the experimental questions, but it should not be assumed that effects of two inhibitors are additive regardless of injection order. For example, the effects of rotenone and either BHAM or Antimycin A are additive (e.g., rotenone effect + BHAM effect = effect of rotenone/BHAM in combination) only if the rotenone is added first.

11. Many of our earlier mitochondrial stress test assays produced inconsistent and apparently erratic results. However, observations of the wells following the assay's completion revealed that the main problem was caused by changes in the distribution of the cells during the assay. Initially the cells were uniformly distributed at the same density across the entire well and the same in all wells, but they became detached during the experiment, and accumulated into different sections (to the edges or center) of wells depending on the row of wells in which they were located. The cells would consistently cluster around the edges of the wells in the top rows (A and B) or accumulate in the center of the wells in the bottom rows (C and D) of the plate (*see* Figs. 5 and 6). This was caused by a number of factors. Firstly, the sequential addition of mitochondrial drugs caused the cells to detach. In contrast, *Dictyostelium* cells in untreated (no drug) wells continued to adhere to the plate surface and yielded consistent results throughout the experiment. Secondly, the mechanical/hydrodynamic forces exerted during injection and mixing in the wells appear to vary across the plate, such that once detached, even small variations across the plate would produce large differences in cell distribution. To improve cell adhesion onto the well surface, several reagents including gelatine, poly-l-lysine, agarose overlays, and Corning® Cell-Tak™ cell and tissue adhesive were tested without success. Of the reagents tested, only Matrigel® basement membrane matrix (Corning) was able to provide the sufficient cell adhesive properties to maintain and remedy the cell detachment issues. This plate-coating treatment provides for very reproducible results with very similar outcomes across all wells (*see* Fig. 5).
12. Matrigel® solution will solidify at temperatures above 16 °C and will revert to a liquid state (melts) at lower temperatures. Pipette tips, SIH assay medium and cell culture microplates should therefore be prechilled at 4 °C. Ensure the Matrigel® is completely melted at 4 °C before use. The suspension should have

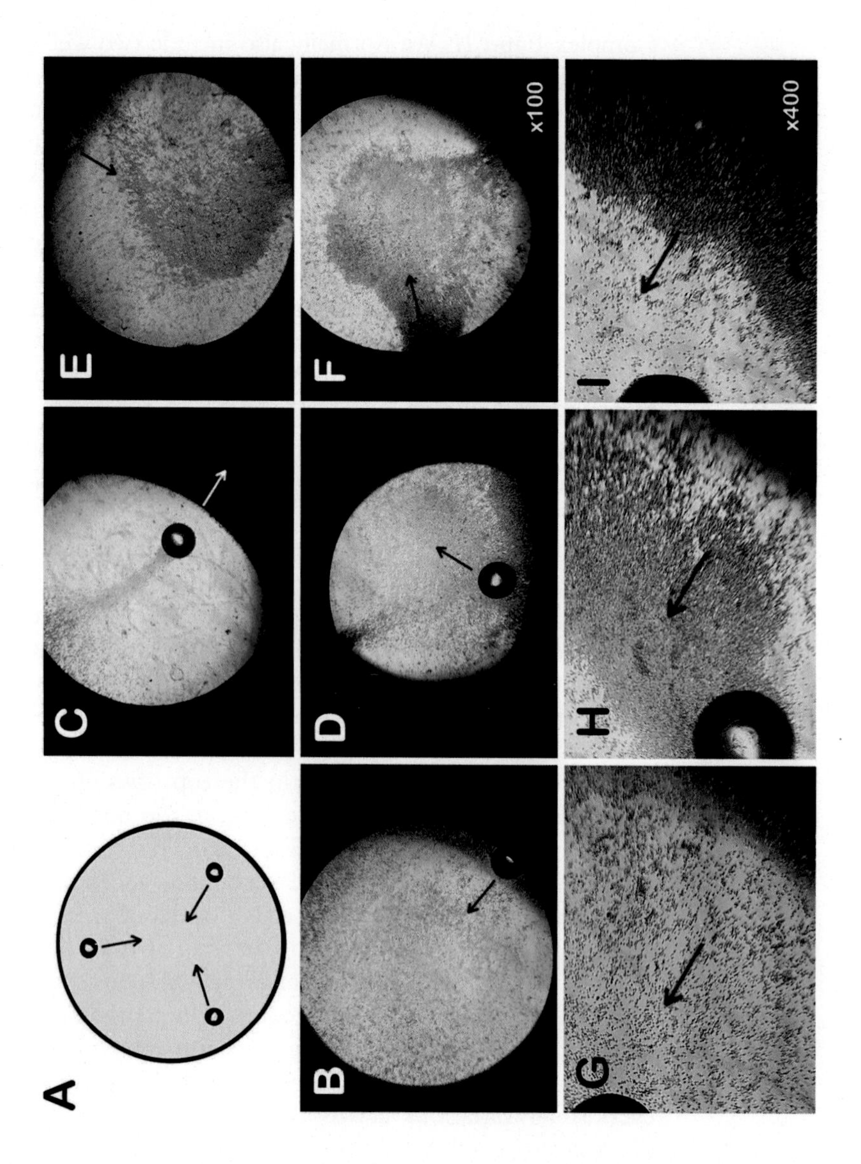

Fig. 5 Cell detachment variability during Seahorse assays. A schematic of the well surface area upon which cells are seeded (**a**) depicts the three round protrusions which may be used for reference to orientate within the well for figures shown in panels (**b–f**). *Arrows* are shown for additional guidance to indicate direction to the center of the wells. Panels (**c–f**) show representative row to row variation of cell detachment and cell redistribution in non-treated wells compared with a Matrigel® coated well (**b**) at the end of an assay. Because of the central location of the oxygen and pH probe sensors, cells clumping at the edges of wells in (**c** and **d**) provide underestimated OCR readings while the cells accumulating in the center of the wells (**e** and **f**) will give overestimated OCR readings. Panels (**g–i**) show various degrees of cell clumping in the wells; from an even monolayer, to progressively clumped cells at the edge of a well. Precoating of the wells with Matrigel® alleviates the cell detachment and redistribution issues resulting in even cell distribution and attachment throughout an assay (b and i)

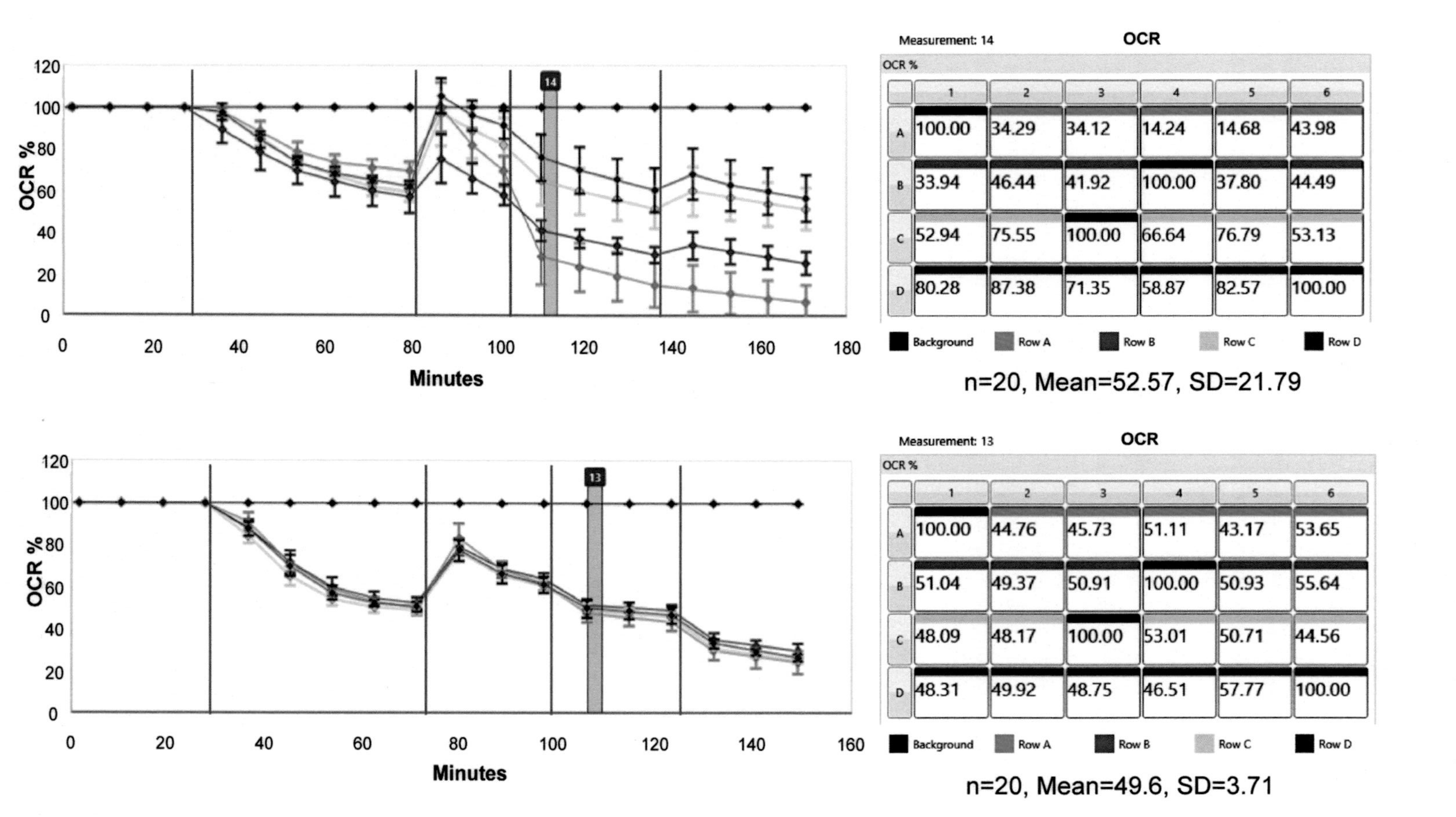

Fig. 6 Row to row variability in Seahorse assays. The plotted data for a non-treated plate (**a**) and a Matrigel® coated plate (b), clearly reveals the row to row variability in the non-treated wells during a Seahorse mitochondrial stress test assay. In each plot, 20 replicate wells have been seeded with *Dictyostelium* AX2 cells and their mitochondrial function was assessed following an identical treatment regimen with four mitochondrial drugs; 10 μM DCCD, 10 μM CCCP, 20 μM rotenone, and 10 μM Antimycin A. For the purposes of illustration, the replicates have been grouped by rows on the plate. Clearly, the untreated wells (**a**) showed significant row to row differences. This is due to the consistent yet uneven cell detachment and redistribution on a plate during the assay (*refer to* Fig. 4). In contrast, the OCR data from Matrigel® coated wells (**b**) showed reproducible results with dramatically reduced variability amongst wells. The mean value and standard deviation for all 20 replicate wells in both assays following the third injection is shown. These differences were consistently observed in repeat assays

a homogenous liquid texture without clumps for optimal even coating of wells and subsequent consistent cell adherence.

13. Once the Matrigel® has fully dried, the well surface will appear hazy.
14. A Matrigel® coated XF24 cell culture microplate can be prepared on the day of use or can be prepared in advance and stored at room temperature for at least a week before use.
15. A relevant internal control (e.g., parental strain) should always be included in each assay plate.
16. It is recommended to use cell seeding densities that produce basal OCR levels that fall in the linear detection range (~100–400 pmol/min) of the Seahorse Extracellular Flux Analyzer. Cell seeding densities should be optimized when testing strains of different parental backgrounds to fall within this desired range, but care is needed to distinguish elevated or reduced OCR levels that are consequence of "alterations" in the cellular bioenergetics in the cells. We find 1×10^5 cells per well would typically give basal OCR levels of ~200–400 pmol/min depending on the strain.
17. Carefully add the cells to the center of the wells, and do not scratch the Matrigel® coating. The two-step seeding is used to allow for a more even distribution of cells over the well surface and decreases the time required for cell attachment. Alternatively, for quicker cell seeding in a single step, a plate spinner with swing-out rotor can be used. Simply transfer cells in the final 500 μL of assay medium and allow cells to adhere to the well surface by spinning the plate at 680 × g for 2 min with no brake at room temperature.
18. The concentrated stock compounds are dissolved in 100 % (v/v) DMSO. It is important to maintain the DMSO concentration of the 10× working stocks at 1 % (v/v) to keep the drugs in solution when they are diluted during the assay. High levels of DMSO (i.e., >1 % (v/v)) in the final cell suspension can adversely affect cellular respiration rates. While the background wells don't have cells, we usually also inject the same compounds as per the test wells.
19. Check port orientation before loading the drugs. Make sure the correct ports are used for each drug. When injecting solutions, even if some wells will not be used or will be used as background, be sure to add the same volume of either drug solution or assay medium to each well to establish correct volumes for baseline measurements. Hold the tips at a 45° angle. Place the tips halfway into the injection ports with the bevel of the tip against the opposite wall of the injection port. Do not insert the tips completely to the bottom of the injection ports as this may cause compound leakage through the port. Gently and com-

pletely load the compounds into the ports via a single stream. It is important to avoid air bubbles. Do not tap any portion of the cartridge in an attempt to alleviate air bubbles as it may cause compound leakage through the injection port. The volume of the injected compounds is based on a starting well volume of 500 μL. Each subsequent injection results in a tenfold dilution of the mitochondrial drug.

20. At the completion of assay, it is worthwhile to observe the cells in each well and to note any discrepancies in cell distribution and cell counts. This can be caused by pipetting errors or inaccurate cell counts. To quantitate the protein content of each well, and to allow for normalization of the data, carefully aspirate the assay medium from each well and add 100 μL of cell lysis buffer for the protein quantitation method of your choice. Discard all used plates appropriately—cartridge sensor, cell culture plate, utility (calibrant) plate.
21. *Dictyostelium* cells are sensitive to the ATP synthase inhibitor, DCCD. The compound however does not seem to work immediately upon injection. We find it generally reaches maximal inhibition after the fourth reading. In order to test that it was not a problem of inhibitor or substrates not being permeable into the cell, we tested the effects of adding various detergents such as digitonin, saponin and streptolysin O. Oligomycin A, the most commonly used ATPase inhibitor in mammalian systems, was ineffective at concentrations up to 20 μM whether or not the cells were permeabilized with any of these reagents, suggesting that *Dictyostelium* ATP synthase is insensitive to the compound.

Acknowledgements

S. Lay and O. Sanislav contributed equally to this work, which was supported by the Australian Research Council Discovery Project grant DP140104276.

References

1. Mayr-Wohlfart U, Waltenberger J, Hausser H et al (2002) Vascular endothelial growth factor stimulates chemotactic migration of primary human osteoblasts. Bone 30:472–477
2. Biber K, Zuurman MW, Dijkstra IM, Boddeke HW (2002) Chemokines in the brain: neuroimmunology and beyond. Curr Opin Pharmacol 2:63–68
3. Fernandis AZ, Ganju RK (2001) Slit: a roadblock for chemotaxis. Sci STKE 2001:pe1
4. Rubel EW, Cramer KS (2002) Choosing axonal real estate: location, location, location. J Comp Neurol 448:1–5
5. Schneider L, Cammer M, Lehman J et al (2010) Directional cell migration and chemotaxis in wound healing response to PDGF-AA are coordinated by the primary cilium in fibroblasts. Cell Physiol Biochem 25:279–292
6. Condeelis JS, Wyckoff JB, Bailly M et al (2001) Lamellipodia in invasion. Semin Cancer Biol 11:119–128

7. Worthley SG, Osende JI, Helft G et al (2001) Coronary artery disease: pathogenesis and acute coronary syndromes. Mt Sinai J Med 68: 167–181
8. Chung CY, Funamoto S, Firtel RA (2001) Signaling pathways controlling cell polarity and chemotaxis. Trends Biochem Sci 26:557–566
9. Iijima M, Huang YE, Devreotes P (2002) Temporal and spatial regulation of chemotaxis. Dev Cell 3:469–478
10. Franca-Koh J, Kamimura Y, Devreotes P (2006) Navigating signaling networks: chemotaxis in *Dictyostelium discoideum*. Curr Opin Genet Dev 16:333–338
11. Funamoto S, Meili R, Lee S et al (2002) Spatial and temporal regulation of 3-phosphoinositides by PI 3-kinase and PTEN mediates chemotaxis. Cell 109:611–623
12. van Haastert PJ, Keizer-Gunnink I, Kortholt A (2007) Essential role of PI3-kinase and phospholipase A2 in *Dictyostelium discoideum* chemotaxis. J Cell Biol 177:809–816
13. Keizer-Gunnink I, Kortholt A, Van Haastert PJ (2007) Chemoattractants and chemorepellents act by inducing opposite polarity in phospholipase C and PI3-kinase signaling. J Cell Biol 177:579–585
14. Chen L, Iijima M, Tang M et al (2007) PLA2 and PI3K/PTEN pathways act in parallel to mediate chemotaxis. Dev Cell 12:603–614
15. Niles BJ, Powers T (2014) TOR complex 2-Ypk1 signaling regulates actin polarization via reactive oxygen species. Mol Biol Cell 25:3962–3972
16. Lee S, Comer FI, Sasaki A et al (2005) TOR complex 2 integrates cell movement during chemotaxis and signal relay in *Dictyostelium*. Mol Biol Cell 16:4572–4583
17. Stephens L, Milne L, Hawkins P (2008) Moving towards a better understanding of chemotaxis. Curr Biol 18:R485–R494
18. Jacinto E, Loewith R, Schmidt A et al (2004) Mammalian TOR complex 2 controls the actin cytoskeleton and is rapamycin insensitive. Nat Cell Biol 6:1122–1128
19. Zoncu R, Efeyan A, Sabatini DM (2011) mTOR: from growth signal integration to cancer, diabetes and ageing. Nat Rev Mol Cell Biol 12:21–35
20. Liu L, Das S, Losert W, Parent CA (2010) mTORC2 regulates neutrophil chemotaxis in a cAMP- and RhoA-dependent fashion. Dev Cell 19:845–857
21. Cai H, Das S, Kamimura Y et al (2010) Ras-mediated activation of the TORC2-PKB pathway is critical for chemotaxis. J Cell Biol 190:233–245
22. Charest PG, Shen Z, Lakoduk A et al (2010) A Ras signaling complex controls the RasC-TORC2 pathway and directed cell migration. Dev Cell 18:737–749
23. Schmidt A, Kunz J, Hall MN (1996) TOR2 is required for organization of the actin cytoskeleton in yeast. Proc Natl Acad Sci U S A 93: 13780–13785
24. Kamada Y, Fujioka Y, Suzuki NN et al (2005) Tor2 directly phosphorylates the AGC kinase Ypk2 to regulate actin polarization. Mol Cell Biol 25:7239–7248
25. Stambolic V, Woodgett JR (2006) Functional distinctions of protein kinase B/Akt isoforms defined by their influence on cell migration. Trends Cell Biol 16:461–466
26. Wullschleger S, Loewith R, Hall MN (2006) TOR signaling in growth and metabolism. Cell 124:471–484
27. Choi JH, Adames NR, Chan TF et al (2000) TOR signaling regulates microtubule structure and function. Curr Biol 10:861–864
28. Jiang X, Yeung RS (2006) Regulation of microtubule-dependent protein transport by the TSC2/mammalian target of rapamycin pathway. Cancer Res 66:5258–5269
29. Berven LA, Willard FS, Crouch MF (2004) Role of the p70(S6K) pathway in regulating the actin cytoskeleton and cell migration. Exp Cell Res 296:183–195
30. Liu L, Chen L, Chung J, Huang S (2008) Rapamycin inhibits F-actin reorganization and phosphorylation of focal adhesion proteins. Oncogene 27:4998–5010
31. Murooka TT, Rahbar R, Platanias LC, Fish EN (2008) CCL5-mediated T-cell chemotaxis involves the initiation of mRNA translation through mTOR/4E-BP1. Blood 111: 4892–4901
32. Cunningham JT, Rodgers JT, Arlow DH et al (2007) mTOR controls mitochondrial oxidative function through a YY1-PGC-1α transcriptional complex. Nature 450:736–740
33. Risson V, Mazelin L, Roceri M et al (2009) Muscle inactivation of mTOR causes metabolic and dystrophin defects leading to severe myopathy. J Cell Biol 187:859–874
34. Ramanathan A, Schreiber SL (2009) Direct control of mitochondrial function by mTOR. Proc Natl Acad Sci U S A 106:22229–22232
35. Gilkerson RW, De Vries RL, Lebot P et al (2012) Mitochondrial autophagy in cells with mtDNA mutations results from synergistic loss of transmembrane potential and mTORC1 inhibition. Hum Mol Genet 21:978–990

36. Edinger AL, Thompson CB (2002) Akt maintains cell size and survival by increasing mTOR-dependent nutrient uptake. Mol Biol Cell 13:2276–2288
37. Hardie DG (2007) AMP-activated/SNF1 protein kinases: conserved guardians of cellular energy. Nat Rev Mol Cell Biol 8:774–785
38. Gwinn DM, Shackelford DB, Egan DF et al (2008) AMPK phosphorylation of raptor mediates a metabolic checkpoint. Mol Cell 30:214–226
39. Huang J, Manning BD (2008) The TSC1-TSC2 complex: a molecular switchboard controlling cell growth. Biochem J 412:179–190
40. van Es S, Wessels D, Soll DR et al (2001) Tortoise, a novel mitochondrial protein, is required for directional responses of *Dictyostelium* in chemotactic gradients. J Cell Biol 152:621–632
41. Torija P, Vicente JJ, Rodrigues TB et al (2006) Functional genomics in *Dictyostelium*: MidA, a new conserved protein, is required for mitochondrial function and development. J Cell Sci 119:1154–1164
42. Wilczynska Z, Barth C, Fisher PR (1997) Mitochondrial mutations impair signal transduction in *Dictyostelium discoideum* slugs. Biochem Biophys Res Commun 234:39–43
43. Kotsifas M, Barth C, de Lozanne A et al (2002) Chaperonin 60 and mitochondrial disease in *Dictyostelium*. J Muscle Res Cell Motil 23: 839–852
44. Bokko PB, Francione L, Bandala-Sanchez E et al (2007) Diverse cytopathologies in mitochondrial disease are caused by AMP-activated protein kinase signaling. Mol Biol Cell 18:1874–1886
45. Miró O, Casademont J, Grau JM et al (1998) Histological and biochemical assessment of mitochondrial function in dermatomyositis. Br J Rheumatol 37:1047–1053
46. Hiona A, Sanz A, Kujoth GC et al (2010) Mitochondrial DNA mutations induce mitochondrial dysfunction, apoptosis and sarcopenia in skeletal muscle of mitochondrial DNA mutator mice. PLoS One 5:e11468
47. Chance B, Williams GR (1955) Respiratory enzymes in oxidative phosphorylation. I Kinetics of oxygen utilization. J Biol Chem 217:383–393
48. Watanabe M, Houten SM, Mataki C et al (2006) Bile acids induce energy expenditure by promoting intracellular thyroid hormone activation. Nature 439:484–489
49. Wang X, Moraes CT (2011) Increases in mitochondrial biogenesis impair carcinogenesis at multiple levels. Mol Oncol 5:399–409
50. Choi SW, Gerencser AA, Nicholls DG (2009) Bioenergetic analysis of isolated cerebrocortical nerve terminals on a microgram scale: spare respiratory capacity and stochastic mitochondrial failure. J Neurochem 109: 1179–1191
51. Kimura K, Kuwayama H, Amagai A, Maeda Y (2010) Developmental significance of cyanide-resistant respiration under stressed conditions: experiments in *Dictyostelium* cells. Dev Growth Differ 52:645–656
52. Seahorse Bioscience (2014) XFe Wave user guide. Retrieved 5 May 2014 from http://www.seahorsebio.com/resources/pdfs/user-guide-xfe-wave.pdf

Chapter 5

Studying Chemoattractant Signal Transduction Dynamics in *Dictyostelium* by BRET

A.F.M. Tariqul Islam, Branden M. Stepanski, and Pascale G. Charest

Abstract

Understanding the dynamics of chemoattractant signaling is key to our understanding of the mechanisms underlying the directed migration of cells, including that of neutrophils to sites of infections and of cancer cells during metastasis. A model frequently used for deciphering chemoattractant signal transduction is the social amoeba *Dictyostelium discoideum*. However, the methods available to quantitatively measure chemotactic signaling are limited. Here, we describe a protocol to quantitatively study chemoattractant signal transduction in *Dictyostelium* by monitoring protein–protein interactions and conformational changes using Bioluminescence Resonance Energy Transfer (BRET).

Key words Bioluminescence resonance energy transfer, *Dictyostelium discoideum*, Chemotaxis, Directed cell migration, GPCR, Heterotrimeric G proteins, Protein–protein interaction, Signaling dynamics

1 Introduction

The mechanisms of chemoattractant signal transduction controlling the directional migration of cells are not understood. Much of our current understanding of chemotactic signaling comes from studies performed with the model organism *Dictyostelium discoideum*. *Dictyostelium* is a particularly powerful experimental model for studying chemotaxis due, in part, to its genome simplicity and ease of genetic manipulations, which allows straightforward assessment of individual protein function [1]. Upon starvation, *Dictyostelium* cells enter a developmental program and, in the early phase, acquire the competency to perform chemotaxis to cAMP, which they themselves secrete. Chemotaxis to cAMP drives the aggregation of millions of cells that ultimately form a fruiting body containing spores, which allows *Dictyostelium* to survive starvation [2]. In the same way neutrophils and other types of mammalian cells detect various chemoattractants, *Dictyostelium* cells detect and transduce the cAMP chemoattractant signal through G protein-coupled receptors (GPCRs) [3].

Tian Jin and Dale Hereld (eds.), *Chemotaxis: Methods and Protocols*, Methods in Molecular Biology, vol. 1407,
DOI 10.1007/978-1-4939-3480-5_5, © Springer Science+Business Media New York 2016

The quantitative study of chemoattractant signaling dynamics in both *Dictyostelium* and neutrophils has mostly been performed with fluorescent reporters and Förster (or Fluorescence) Resonance Energy Transfer (FRET) methods, using sophisticated imaging systems [4–9]. The ability of monitoring the interaction between signaling proteins in live cells using a proximity-based method such as FRET, in which two proteins of interest are respectively fused to a fluorescent energy donor and fluorescent acceptor, is particularly powerful in analyzing spatiotemporal chemoattractant signaling dynamics [10]. However, quantitative FRET is challenging and requires correction of a large number of potential artifacts such as the non-ratiometric expression of the two proteins under study, cross talks of the energy donor and acceptor emissions into the FRET channel, autofluorescence, and photobleaching [11]. The Bioluminescence Resonance Energy Transfer (BRET) technology is related to FRET but uses a bioluminescent enzyme (luciferase) as energy donor and, therefore, offers the advantages of FRET without the problems related to fluorescence excitation [12]. Hence, BRET is very sensitive and quantitative, and its analysis is extremely simple. Consequently, although BRET does not generate enough light to be imaged at the subcellular level with currently available optics, it is one of the best methods available to assess and quantitatively study the dynamics of protein–protein interactions, as well as protein conformational changes, in vivo.

There are three main BRET methods that have been developed, which are characterized by the use of different luciferase substrates and fluorescent protein (FP) acceptors, resulting in distinct spectral properties (Fig. 1). In all BRET methods, the energy donor used is the luciferase from the sea pansy *Renilla reniformis* (Rluc), or its improved variants Rluc2 (C124A/M185V) and Rluc8 (A55T/C124A/S130A/K136R/A143M/M185V/M253L/S287L), which display increased bioluminescence [13]. Upon oxidation of an appropriate coelenterazine substrate, Rluc produces light whose energy can be transferred to a nearby fluorescent protein. In the first generation of BRET, $BRET^1$, coelenterazine h or its protected form EnduRen™, which has increased signal stability, are used in combination with YFP variants [14, 15]; in $BRET^2$, coelenterazine 400a (also known as DeepBlueC) or the newly developed Prolume Purple coelenterazine (methoxy-e-coelenterazine) are used in combination with GFP^2 or GFP10 [16, 17]; and in $BRET^3$, coelenterazine h or EnduRen™ are used in combination with mOrange, a variant of *Discosoma* spp. RFP (DsRed) [18]. $BRET^2$ offers the largest spectral resolution of all three BRET methods and, especially since the development of the improved Rluc2 and Rluc8 used together with the new Prolume Purple coelenterazine that result in high signal intensities comparable to those of $BRET^1$ and $BRET^3$, $BRET^2$ is currently the preferred BRET method. However, the availability of three BRET

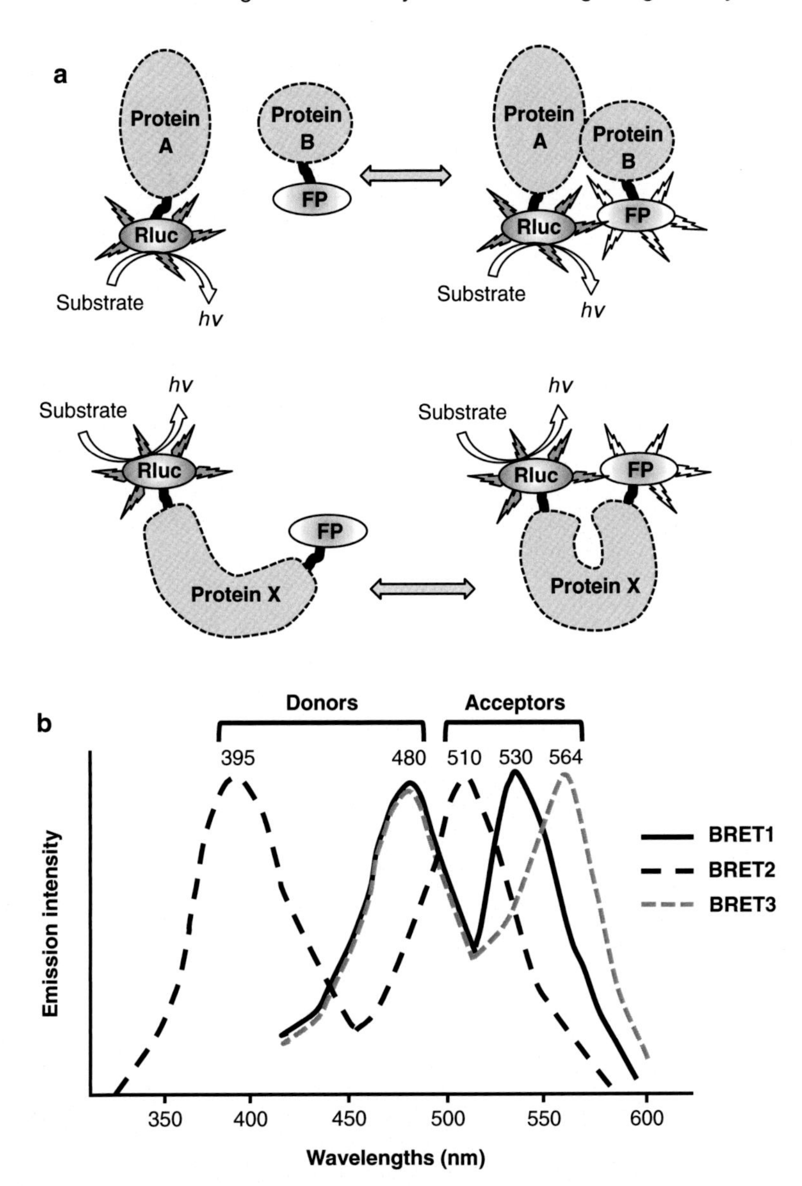

Fig. 1 Detection of molecular proximity and conformational changes in proteins by BRET. (**a**) To assess and monitor the interaction between two proteins of interest, these are independently fused to the BRET donor Rluc and to a fluorescent protein (FP) BRET acceptor. Since the BRET signal is extremely sensitive to the distance and relative orientation between Rluc and the FP acceptor, it can also be used to assess and monitor protein conformational changes. In that case, the same protein is fused to Rluc at one extremity and to the FP acceptor at the other extremity. The Rluc-mediated oxidation of its coelenterazine substrate produces light (luminescence; $h\nu$) that, if in close proximity (1–10 nm) to the FP acceptor, can lead to the non-radiative transfer of energy to the FP, which then emits light at a different wavelength (fluorescence). (**b**) The three major BRET methods have different spectral properties. $BRET^1$ and $BRET^3$ use the same coelenterazine substrates (h or EnduRen™) and produce luminescence that peaks at 480 nm. These two BRET methods differ by the FP acceptor used: YFP variants for $BRET^1$ that emit fluorescence peaking at 530 nm, and mOrange for $BRET^3$ that emits fluorescence peaking at 564 nm. The use of coelenterazine 400a or Prolum Purple in $BRET^2$ produces a blue-shifted luminescence [with peaks at 395 nm for 400a (depicted) and 405 for Prolum Purple (not shown)] that is spectrally compatible with GFP^2 and GFP10, which emit fluorescence peaking at 510 nm

methods is particularly useful for the study of protein complexes, where multiple BRET methods could be used simultaneously.

BRET is particularly widely used to study the oligomerization of GPCRs and their interactions with different proteins in mammalian cells, and several protocols for such studies have been recently described [19–23]. Here, we describe a BRET^2 protocol adapted to the study of chemoattractant signal transduction in *Dictyostelium*, in which we provide an example study using the Gα2 and Gβ subunits of the heterotrimeric G protein activated downstream from the cAMP chemoattractant receptor cAR1 [24, 25].

2 Materials

Prepare all solutions with ultrapure water, and sterilize all solutions and culture wares that come in contact with cells.

2.1 Generating Rluc and GFP Fusion Constructs

1. *Dictyostelium* expression vectors, such as the integrative pEXP4 or extrachromosomal pDM304, as well as shuttle vectors to be used in combination with pDM expression vectors [26], such as pDM344, are available through the Dicty Stock Center [27].
2. cDNA of the chosen BRET^2 donor and acceptor pair [Rluc (PerkinElmer), Rluc2, or Rluc8 (from S. S. Gambhir, Stanford University, CA); and GFP^2 (PerkinElmer) or GFP10 (from Michel Bouvier, Université de Montréal, Québec, Canada)].
3. cDNA or gene encoding the proteins of interest and of those that will be used as controls (*see* **Note 1**).

2.2 Dictyostelium Cell Culture and Transformation

1. HL5 medium including glucose (ForMedium™): 14 g/l peptone, 7 g/l yeast extract, 13.5 g/l glucose, 0.5 g/l KH_2PO_4, and 0.5 g/l Na_2HPO_4. pH should be 6.4–6.7. If necessary adjust pH with HCl. Autoclave for 20 min to sterilize, cool, and supplement with antibiotics/antimycotic if desired. Store media at 4 °C, protect from light, and warm up to 22 °C before use.
2. 12 mM Na/K phosphate buffer: 2.5 mM Na_2HPO_4, 9.5 mM KH_2PO_4, pH 6.1. To prepare a 1 l 10× solution, dissolve 3.4 g Na_2HPO_4 and 13 g KH_2PO_4 in dH_2O. Dilute 1:10 with dH_2O to make 1× buffer. Autoclave for 20 min to sterilize. Store at 22 °C.
3. Electroporation buffer: 10 mM Na/K phosphate, 50 mM sucrose, pH 6.1. To make 250 ml, use 20.8 ml 10× 12 mM Na/K phosphate buffer and dissolve 4.28 g sucrose in dH_2O. Autoclave for 20 min to sterilize. Store at 4 °C.
4. Appropriate antibiotics for selection of transformed cells (e.g., G418, Hygromycin).

5. 0.4 cm electroporation cuvettes.
6. Electroporator that enables transformation of microorganisms (e.g., MicroPulser™ Electroporator from Bio-Rad).
7. 100 and 150 mm cell culture dishes.

2.3 Generation of cAMP-Responsive, Chemotactically Competent Cells

1. 30 mM cAMP stock solution: Dissolve cAMP in 12 mM Na/K phosphate buffer pH 6.1. Store aliquots at −20 °C. Dilute in 12 mM Na/K phosphate buffer pH 6.1 to make all working solutions.
2. 125 ml Erlenmeyer flasks.
3. Dispensing pump: any high accuracy and programmable multichannel dispensing pump with variable speed (e.g., Ismatec® IPC 16-Channel Low-Speed Digital Dispensing Pump, from IDEX Health & Science SA).
4. Gyratory shaker with appropriate clamps.

2.4 Validating the BRET Constructs

1. SDS-PAGE and western blotting equipment and solutions.
2. Rluc and GFP antibodies (e.g., Mouse Anti-*Renilla* luciferase antibody clone 5B11.2, EMD Millipore; Living Colors® GFP monoclonal antibody, Clontech).
3. Luciferase substrate: e.g., coelenterazine h or coelenterazine 400a (Biotium). Dissolve in anhydrous ethanol to make a 1 M stock. Protect from light and store at −20 °C. Dilute with 12 mM Na/K phosphate buffer to make a 30 μM coelenterazine working solution.
4. 96-well white and black microplates.
5. Microplate reader(s) capable of measuring luminescence and fluorescence (*see* **Note 2**).
6. 12 mM Na/K phosphate agar plates: 12 mM Na/K phosphate, 2 mM $MgSO_4$, 0.5 mM $CaCl_2$, 15 g/l agar. Autoclave for 20 min to sterilize and pour into culture dishes. After the agar has solidified, seal to prevent drying and store at 4 °C.
7. Null strains of the proteins of interest (*see* **Note 3**).
8. Dissection microscope (e.g., Nikon SMZ-U).

2.5 BRET^2 Assay

1. 96-well white microplates (Greiner) (*see* **Note 4**).
2. BRET^2-specific luciferase substrate coelenterazine 400a (Biotium) or Prolum Purple (NanoLight™ Technologies). Dissolve coelenterazine 400a in anhydrous ethanol and Prolum Purple in NanoFuel solvent (NanoLight™ Technologies) to make a 1 M stock. Protect from light and store at −20 °C. Dilute with 12 mM Na/K phosphate buffer to make a 40 μM coelenterazine substrate working solution.

3. Microplate luminometer equipped with an injector and the ability of measuring light through two different filters, ideally simultaneously (e.g., LUMIstar or POLARstar Omega from BMG Labtech) (*see* **Note 5**).
4. Appropriate optical filters for detection of BRET^2 donor and acceptor emission signals (e.g., 370–450 and 500–530 nm).

3 Methods

The example given in this protocol is using the BRET^2 method, but the protocol is easily adaptable for other BRET methods.

3.1 Generating Luc and GFP Fusion Constructs

1. Choose an appropriate expression system (integrative versus extrachromosomal, two separate plasmids versus a single one). This choice will partly depend on the nature of the proteins studied, their known relationship, and the presence of competing, endogenous interacting partners (*see* **Note 6**).
2. The cDNA or gene encoding the proteins of interest is cloned in-frame with either the BRET^2 donor or acceptor in the chosen expression vector system, using conventional DNA recombination techniques. When joining the acceptor and donor sequences to those of the proteins of interest, the stop codon between them is removed and replaced with a flexible linker (*see* **Note 7**).
3. If possible, generate multiple different conformations of fusion constructs to test, both N- and C-term fusions. If adding the donor/acceptor to the protein of interest at the N- or C-terminus is found to affect the protein's function, it may be possible to insert them within the protein of interest's sequence, ideally in a loop facing outward, but this would require knowledge of the protein's structure.
4. As much as possible, especially if the protein–protein interaction under investigation is unknown, prepare both positive and negative controls to be tested in parallel with the proteins of interest (*see* **Note 1**).

3.2 *Dictyostelium* Cell Culture and Transformation

1. Grow *Dictyostelium* cells at 22 °C, either attached to a plastic substrate in culture dishes or in suspension in Erlenmeyer flasks shaken at ~140 rpm on a gyratory shaker. For transformation and all experiments, cells taken from dishes should always be sub-confluent and the cells grown in suspension should be at ~2×10^6 cells/ml.
2. Suspend the cells grown in culture dishes by gently pipetting media up and down on the cells, and determine the density of the cell suspension using a hemocytometer (need 8×10^6 cells per transformation).

3. Pellet the cells by centrifugation at 500 × *g* for 3 min at 4 °C, and wash the cells once with cold 12 mM Na/K phosphate buffer.
4. Pellet the cells again and discard buffer. Incubate the cell pellet on ice 5 min.
5. Resuspend the cells at 10^7 cells/ml in electroporation buffer by gently pipetting up and down.
6. Mix 800 μl of cell suspension (8×10^6 cells) with 20 μg plasmid DNA by gently pipetting up and down and incubate on ice 1 min (*see* **Note 8**).
7. Transfer to cold electroporation cuvette and pulse twice at 1 kV and 3 mF, with a time constant of ~0.8 ms (*see* **Note 9**).
8. Transfer cells to a 100 mm dish containing 10 ml HL5 medium and add appropriate selection antibiotic 24 h after electroporation (*see* **Note 10**).
9. Once cells grow to confluency in the 100 mm dishes, transfer them to 150 mm dishes or start a suspension culture and grow until desired number of cells is reached (*see* **Note 11**).

3.3 Validating the BRET Constructs

1. It is crucial to confirm that the BRET^2 donor and acceptor are successfully fused to the proteins of interests and generate functional fusion proteins that are appropriately expressed in cells. Verify the integrity of the fusion constructs by performing a Western Blot of total cell lysates using Rluc and GFP antibodies.
2. Verify BRET^2 acceptor functionality using a fluorimeter equipped with appropriate filters (e.g., excitation 450–470 nm and emission 500–530 nm). To do this, collect the transformed cells and the parental strain as control, wash twice with 12 mM Na/K phosphate buffer (pelleting the cells by centrifugation at 500 × *g* for 3 min between each wash), and resuspend at 12×10^6 cells/ml. Dispense 125 μl of cell suspension (1.5×10^6 cells) in duplicate in a 96-well black microplate and assess fluorescence.
3. Verify BRET^2 donor functionality using a coelenterazine substrate and luminometer equipped with the appropriate filter (e.g., 520–550 nm). Prepare cells as described for the measurement of acceptor fluorescence described above. Dispense 125 μl of cell suspension in duplicate in a 96-well white microplate and add 25 μl of a freshly prepared coelenterazine substrate working solution (5 μM final concentration). Mix well, protect from light, and incubate 10 min before assessing luminescence (*see* **Note 12**).
4. Verify that the fusion proteins are functional using appropriate phenotypic rescue and/or activity assays. In the case of many

proteins important in chemotaxis, lack of protein expression results in the inability of *Dictyostelium* cells to aggregate and develop into fruiting bodies when plated on non-nutrient agar. To assess the rescue of an aggregation null phenotype by the fusion constructs, express these in their corresponding null strain, wash the cells twice and resuspend at 20×10^6 cells/ml in 12 mM Na/K phosphate buffer. Apply 50 μl of cell suspension on a 12 mM Na/K phosphate agar plate and monitor the cells' development using a dissection microscope.

3.4 Generation of cAMP-Responsive, Chemotactically Competent Cells

1. Suspend the cells grown in culture dishes by gently pipetting media up and down on the cells, and determine the density of the cell suspension using a hemocytometer.
2. Transfer 50×10^6 cells to a 50 ml centrifuge tube, pellet by centrifugation at $500 \times g$ for 3 min, and wash the cells twice with 40 ml of 12 mM Na/K phosphate buffer.
3. Resuspend the cells in 10 ml of 12 mM Na/K phosphate buffer to obtain a cell suspension of 5×10^6 cells/ml, transfer to a 125 ml Erlenmeyer flask, and place the flask on a gyratory shaker.
4. Using a programmable dispensing pump, induce the development of cells by stimulating them with cAMP to a final concentration of 30 nM (e.g., if using a working solution of 15 μM of cAMP, set the pump to deliver 20 μl) every 6 min for 5.5 h. Keep the cells at 22 °C and in suspension by shaking at ~140 rpm.
5. Transfer the developed cells to 50 ml centrifuge tubes, pellet by centrifugation at $500 \times g$ for 3 min at 4 °C, and wash the cells twice with cold 12 mM Na/K phosphate buffer.
6. Resuspend the cells at 12×10^6 cells/ml in 12 mM Na/K phosphate buffer (4.2 ml) and keep the cells in suspension by shaking at ~140 rpm on a gyratory shaker, either at 22 or 4 °C. Keeping the cells at 4 °C will inhibit cAMP secretion and, thereby, autocrine stimulation. If cells are kept at 4 °C, incubate samples at 22 °C for 5 min prior to stimulation with the chemoattractant. Alternatively, cells can be treated with 2 mM caffeine for 20 min, which inhibits cAMP synthesis [28, 29] (*see* **Note 13**).

3.5 BRET^2 Assay

1. Before assaying BRET^2, determine the relative expression levels of the donor and acceptor fusion constructs by measuring the luminescence and fluorescence, respectively, as described under Subheading 3.3 (*see* **Note 12**).
2. Dispense 125 μl of the 12×10^6 cells/ml cell suspension (1.5×10^6 cells) in duplicate in a 96-well white microplate, and add 25 μl of a freshly prepared 40 μM BRET^2 coelenterazine

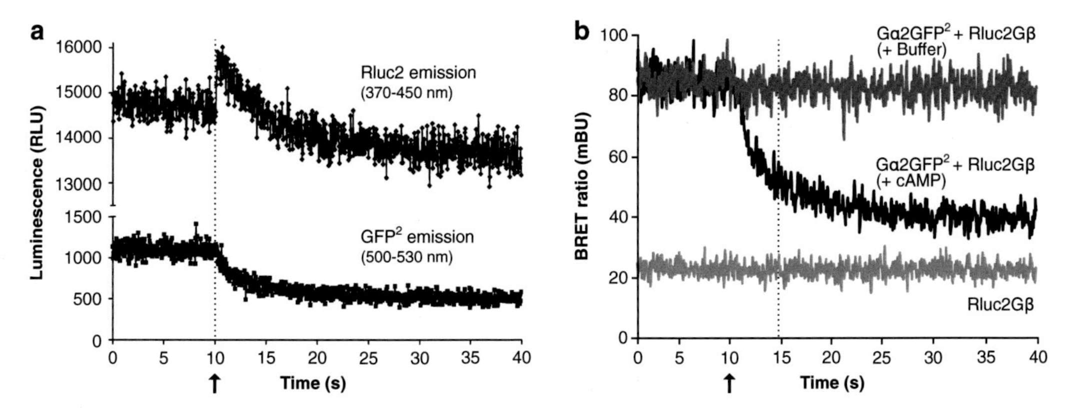

Fig. 2 Kinetics of cAMP-induced dissociation of Gα2 and Gβ monitored by BRET². (**a**) Example of BRET² measurements collected in real-time before, during, and after cAMP stimulation of cells expressing Gα2GFP² and Rluc2Gβ. Measurements were performed with a POLARstar Omega (BMG Labtech) every 20 ms for 40 s. cAMP was injected at 10 s (*block arrow*). *RLU* relative light units. (**b**) The graph displays the BRET ratios calculated for each measurement of the collected data shown in (**a**) for the cAMP-stimulated Gα2GFP²/Rluc2Gβ-expressing cells. The calculated BRET ratios obtained with the same cells but where only 12 mM Na/K phosphate buffer was injected, as well as the background BRET obtained with cells expressing Rluc2Gβ alone was also plotted. Fitting of the Gα2GFP²/Rluc2Gβ dissociation kinetics curve with the GraphPad Prism software revealed a $t_{1/2}$ of 2.5 s. *mBU* milli-BRET units

substrate (400a or Prolume Purple) working solution (5 μM final concentration).

3. Mix well, place the microplate in a luminometer, and immediately proceed to BRET² signal detection by measuring light through two appropriate filters (e.g., 370–450 and 500–530 nm). Make sure the temperature of the luminometer stays at ~22 °C (*see* **Note 14**).
4. For chemoattractant (cAMP) stimulation, load the injector with a 4× cAMP stock solution and program it to deliver 50 μl (final volume of samples is 200 μl) (*see* **Note 15**).
5. Set up the luminometer to take BRET² measurements before and after injection of the chemoattractant, and monitor the BRET² signal in real-time (e.g., Fig. 2). Repeat with different concentrations of cAMP to determine dose–response effects (e.g., Fig. 3) (*see* **Note 16**).
6. Using the same protocol as that for cAMP stimulation, treat samples with 12 mM Na/K phosphate buffer only, which provides a background signal control.

3.6 BRET² Signal Analysis

1. For each measurement, divide the acceptor emission signal (e.g., GFP²; fluorescence) by that of the donor (e.g., Rluc2; luminescence) to calculate the BRET² ratio (*see* **Note 17**).
2. Subtract the BRET² ratio obtained with cells expressing only the donor construct (background control) from that of cells

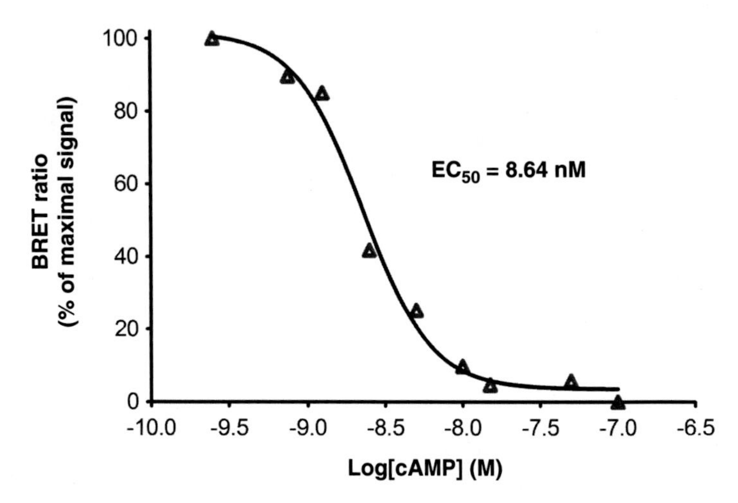

Fig. 3 cAMP concentration-dependent dissociation of Gα2GFP² and Rluc2Gβ. For different concentrations of cAMP, the induced dissociation of Gα2GFP² and Rluc2Gβ was monitored in real time as described in Fig. 2. For each measurement, the net BRET was calculated and the maximal responses were plotted against the logarithmic concentration of cAMP and expressed as relative % of the highest measured BRET value, which corresponds to the associated subunits. Fitting of the cAMP-induced Gα2GFP²/Rluc2Gβ dissociation dose–response curve with GraphPad Prism revealed an EC_{50} of 8.6 nM, extremely similar to what was previously reported [4]

expressing both the donor and acceptor fusion constructs to calculate the "net BRET²" [15].

3. For chemoattractant-induced BRET² signals, subtract the BRET² ratio obtained for buffer-treated samples from that of cAMP-stimulated samples.
4. Plot the BRET² signals against time to determine real-time kinetics profiles, which are used to derive apparent association/dissociation rate constants.
5. Plot the BRET² signals against the logarithmic concentration of chemoattractant to generate dose–response curves and identify the half maximal effective concentration (EC50), a measure of the chemoattractant's potency to produce the observed effect.

4 Notes

1. When studying previously uncharacterized binding partners, it is important to include at least one negative control, where BRET between the proteins under study and another protein known not to interact with them is assessed in parallel. An ideal negative control construct is one whose localization and behavior is similar to that of one of the proteins under study, but that

is known to function independently and not to bind the same partners. In addition, in the eventuality that the proteins under study fail to produce a BRET signal, positive controls can help determine if the fusion constructs are functional and if the obtained negative result is meaningful.

2. Whereas detection of luminescence and BRET only requires a luminometer, assessment of the acceptor's functionality by fluorescence requires a fluorimeter. These can be two separate plate-readers or two-in-one (e.g., POLARstar Omega, BMG Labtech).
3. If possible, create the single and double null strains of the genes encoding the proteins of interest. This is a great advantage of using *Dictyostelium* in BRET-based, or FRET-based, protein–protein interaction studies since these can then be performed in the absence of the endogenous proteins that compete for interaction with the fusion constructs. In addition, these strains are a great tool to assess the functionality of the fusion proteins through phenotypic rescue.
4. Any white microplate will do, however, different ones can produce varying background noise. Therefore, we recommend testing different microplates with the chosen luminometer to determine which one(s) are optimal. For the BMG Labtech luminometers, we have found that the Greiner plates are best.
5. Several luminometers are capable of BRET detection, including the VICTOR Light (PerkinElmer), Mithras LB 940 (Berthold Technologies), and FLUOstar, LUMIstar or POLARstar Omega (BMG Labtech). We have found the Mithras LB 940 to be slightly more sensitive than the other ones, however, the LUMIstar and POLARstar Omega revealed better suited for the study of the rapid chemoattractant signal transduction events in *Dictyostelium* because these luminometers are equipped with a dual emission detection system allowing simultaneous measurements of the donor and acceptor signals. This can make a significant difference when monitoring extremely rapid association/dissociation events, where sequential readings of the donor and acceptor emission signals could occur in different parts of the curve and, consequently, generate apparent BRET ratios that are either under- or over-estimated.
6. Maximal energy transfer theoretically occurs when the donor is saturated with the acceptor. Thus, optimal BRET is often reached when higher levels of acceptor are expressed compared to the donor. However, not every protein can be functionally over-expressed, especially when part of multi-protein complexes. If similar expression levels are preferred, both constructs should be expressed from the same vector (e.g., engineered using the pDM shuttle and expression vectors), insuring that all cells expressing a donor also express the acceptor.

7. To fuse two proteins together, flexible linkers of five to six residues such as glycines and serines (e.g., GGSGG) are often used. Then, optimization of the linker composition and length may be required if the fusion proteins are not functional, or generate no or little BRET signal.
8. If the donor and acceptor constructs are being expressed from different vectors, it is important to test the transformation of various amounts and ratios of donor and acceptor DNA in order to determine their optimal expression levels. As a starting point, we suggest testing 10 and 20 μg of total DNA to transform 8×10^6 cells, and donor to acceptor ratios of 1:2, 1:5, and 1:10. In addition, the donor construct should always be transformed alone, in parallel to the transformations of the BRET pair. Cells expressing only the donor will later be used as background control. Here, we also recommend testing the transformation of different amounts of donor DNA alone to ensure obtaining proper control cells in which the expression of the donor corresponds to its expression in cells expressing both the donor and acceptor constructs.
9. Some electroporators come with programmed protocols for *Dictyostelium* electroporation (e.g., MicroPulser™ from Bio-Rad: in the "Fungi" settings, choose "Dicty").
10. For selecting transformed cells with G418 or hygromycin, we suggest using 20 μg/ml of G418 and 50 μg/ml of hygromycin. However, lower concentrations of selection antibiotic can be used to obtain lower expression levels. This can be particularly useful as a tool to test various expression levels and donor to acceptor ratios.
11. Depending on the expression system used, as well as the nature of the proteins expressed, the time required for selecting transformed cells can vary. Typically, cells transformed with extrachromosomal vectors (e.g., pDM vectors) take less time (~2–5 days) than cells transformed with integrative vectors (e.g., pEXP4; ~1–2 weeks).
12. To determine the relative expression of the donor through the measurement of luminescence, any coelenterazine substrate can be used. An incubation time of 10 min after addition of the coelenterazine substrate is recommended because the produced luminescence is initially very high and decays rapidly, but reaches a steady state after 10 min. This incubation is not necessary when measuring BRET because the calculated BRET ratio is independent of signal intensity.
13. *Dictyostelium* cells developed for 5.5 h secrete cAMP in an autocrine manner, on a 6 min time period, which serves to relay the chemoattractant signal to other cells in order to promote aggregation [30]. Such autocrine chemoattractant

stimulation raises the basal activity levels of chemoattractant signal transduction pathways. If reduction of basal chemotactic pathway activity is desired, keep cells at 4 °C to prevent cAMP secretion and incubate samples at room temperature for 5 min before stimulating with the chemoattractant and assaying BRET. Alternatively, treating cells with 2 mM caffeine for 20 min can also be used to inhibit cAMP secretion. However, caffeine treatment could potentially interfere with the signaling pathway under study and the effect of such treatment on the function and interaction of the proteins under study should be tested and characterized prior to use.

14. Optimal BRET measurement parameters need to be empirically determined for every protein–protein interaction under study. The measurement time can range from 20 ms to minutes. Whereas smaller measurement times allow obtaining highly resolved kinetics, longer measurement times will provide higher, and potentially more accurate, measurement values. The time between measurements is then limited by the set measurement time in addition to the time necessary to switch the filters, if applicable. If necessary and applicable, the signals can be increased by adjusting the gain. Brief shaking of the plate is recommended before starting a measurement and between wells.
15. We have found that injecting a fairly large volume of cAMP (e.g., 50 μl) at a high speed (e.g., 300 μl/s) provides the most reproducible results when performing measurements at very small time intervals (e.g., ≤200 ms). Presumably, this is due to better mixing of the cAMP with the cell sample in such conditions.
16. After injection, the measured signals normally increase slightly, proportionally to the volume of injected liquid, which is most obvious for the donor emission signal (*see* Fig. 2a). This effect is likely due to the fact that an increase in sample volume reduces the distance between the site of signal emission and the detector(s) above. However, because BRET is expressed as a ratio of acceptor over donor emission signals, the change in detected signal intensities due to increased sample volume does not affect the calculated BRET (*see* Fig. 2b).
17. Because the donor emits so much more light than the acceptor, BRET ratios are always fractions, typically anywhere between ~0.01 and 0.50. Consequently, although ligand-induced changes in BRET ratios can appear minor, they are often significant (e.g., BRET ratio of 0.08 between Gα2GFP^2 and Rluc2Gβ in resting conditions, decreased to 0.04 after chemoattractant stimulation; Fig. 2b). To facilitate interpretation, BRET ratios are sometimes expressed as milli-BRET units (mBU), which represent the obtained BRET ratios multiplied by 1000.

Acknowledgements

We are thankful to Michel Bouvier for providing BRET[2] donor and acceptor cDNAs, to Chris Jenetopoulos for providing the *Dictyostelium* Gα2 and Gβ cDNAs, and to Billy Breton for constructive discussions and critical reading of the manuscript.

References

1. King JS, Insall RH (2009) Chemotaxis: finding the way forward with *Dictyostelium*. Trends Cell Biol 19:523–530
2. Loomis WF (2014) Cell signaling during development of *Dictyostelium*. Dev Biol 391:1–16
3. Artemenko Y, Lampert TJ, Devreotes PN (2014) Moving towards a paradigm: common mechanisms of chemotactic signaling in *Dictyostelium* and mammalian leukocytes. Cell Mol Life Sci 71:3711–3747
4. Janetopoulos C, Jin T, Devreotes P (2001) Receptor-mediated activation of heterotrimeric G-proteins in living cells. Science 291: 2408–2411
5. Gardiner EM, Pestonjamasp KN, Bohl BP et al (2002) Spatial and temporal analysis of Rac activation during live neutrophil chemotaxis. Curr Biol 12:2029–2034
6. Wong K, Pertz O, Hahn K, Bourne H (2006) Neutrophil polarization: spatiotemporal dynamics of RhoA activity support a self-organizing mechanism. Proc Natl Acad Sci U S A 103:3639–3644
7. Han JW, Leeper L, Rivero F, Chung CY (2006) Role of RacC for the regulation of WASP and phosphatidylinositol 3-kinase during chemotaxis of *Dictyostelium*. J Biol Chem 281: 35224–35234
8. Bagorda A, Das S, Rericha EC et al (2009) Real-time measurements of cAMP production in live *Dictyostelium* cells. J Cell Sci 122: 3907–3914
9. Xu X, Brzostowski JA, Jin T (2009) Monitoring dynamic GPCR signaling events using fluorescence microscopy, FRET imaging, and single-molecule imaging. Methods Mol Biol 571: 371–383
10. Xu X, Brzostowski JA, Jin T (2006) Using quantitative fluorescence microscopy and FRET imaging to measure spatiotemporal signaling events in single living cells. Methods Mol Biol 346:281–296
11. Herman B, Krishnan RV, Centonze VE (2004) Microscopic analysis of fluorescence resonance energy transfer (FRET). Methods Mol Biol 261:351–370
12. Hamdan FF, Percherancier Y, Breton B, Bouvier M (2006) Monitoring protein-protein interactions in living cells by bioluminescence resonance energy transfer (BRET). Curr Protoc Neurosci Chapter 5:Unit 5
13. Loening AM, Fenn TD, Wu AM, Gambhir SS (2006) Consensus guided mutagenesis of *Renilla* luciferase yields enhanced stability and light output. Protein Eng Des Sel 19: 391–400
14. Xu Y, Piston DW, Johnson CH (1999) A bioluminescence resonance energy transfer (BRET) system: application to interacting circadian clock proteins. Proc Natl Acad Sci U S A 96:151–156
15. Angers S, Salahpour A, Joly E et al (2000) Detection of β2-adrenergic receptor dimerization in living cells using bioluminescence resonance energy transfer (BRET). Proc Natl Acad Sci U S A 97:3684–3689
16. Bertrand L, Parent S, Caron M et al (2002) The BRET2/arrestin assay in stable recombinant cells: a platform to screen for compounds that interact with G protein-coupled receptors (GPCRs). J Recept Signal Transduct Res 22:533–541
17. Zhang L, Xu F, Chen Z et al (2013) Bioluminescence assisted switching and fluorescence imaging (BASFI). J Phys Chem Lett 4:3897–3902
18. De A, Ray P, Loening AM, Gambhir SS (2009) BRET3: a red-shifted bioluminescence resonance energy transfer (BRET)-based integrated platform for imaging protein-protein interactions from single live cells and living animals. FASEB J 23:2702–2709
19. Achour L, Kamal M, Jockers R, Marullo S (2011) Using quantitative BRET to assess G protein-coupled receptor homo- and heterodimerization. Methods Mol Biol 756:183–200
20. Kocan M, Pfleger KD (2011) Study of GPCR-protein interactions by BRET. Methods Mol Biol 746:357–371
21. Mizuno N, Suzuki T, Kishimoto Y, Hirasawa N (2013) Biochemical assay of G protein-coupled receptor oligomerization: adenosine

A1 and thromboxane A2 receptors form the novel functional hetero-oligomer. Methods Cell Biol 117:213–227

22. Borroto-Escuela DO, Flajolet M, Agnati LF et al (2013) Bioluminescence resonance energy transfer methods to study G protein-coupled receptor-receptor tyrosine kinase heteroreceptor complexes. Methods Cell Biol 117:141–164
23. Ciruela F (2008) Fluorescence-based methods in the study of protein-protein interactions in living cells. Curr Opin Biotechnol 19: 338–343
24. Kumagai A, Hadwiger JA, Pupillo M, Firtel RA (1991) Molecular genetic analysis of two Gα protein subunits in *Dictyostelium*. J Biol Chem 266:1220–1228
25. Wu L, Valkema R, Van Haastert PJ, Devreotes PN (1995) The G protein β subunit is essential for multiple responses to chemoattractants in *Dictyostelium*. J Cell Biol 129:1667–1675
26. Veltman DM, Akar G, Bosgraaf L, Van Haastert PJ (2009) A new set of small, extrachromosomal expression vectors for *Dictyostelium discoideum*. Plasmid 61:110–118
27. Fey P, Dodson RJ, Basu S, Chisholm RL (2013) One stop shop for everything *Dictyostelium*: dictyBase and the Dicty Stock Center in 2012. Methods Mol Biol 983: 59–92
28. Theibert A, Devreotes PN (1983) Cyclic 3′, 5′-AMP relay in *Dictyostelium discoideum*: adaptation is independent of activation of adenylate cyclase. J Cell Biol 97:173–177
29. Brenner M, Thoms SD (1984) Caffeine blocks activation of cyclic AMP synthesis in *Dictyostelium discoideum*. Dev Biol 101: 136–146
30. Mahadeo DC, Parent CA (2006) Signal relay during the life cycle of *Dictyostelium*. Curr Top Dev Biol 73:115–140

Chapter 6

Wave Patterns in Cell Membrane and Actin Cortex Uncoupled from Chemotactic Signals

Günther Gerisch and Mary Ecke

Abstract

When cells of *Dictyostelium discoideum* orientate in a gradient of chemoattractant, they are polarized into a protruding front pointing toward the source of attractant, and into a retracting tail. Under the control of chemotactic signal inputs, Ras is activated and PIP_3 is synthesized at the front, while the PIP_3-degrading phosphatase PTEN decorates the tail region. As a result of signal transduction, actin filaments assemble at the front into dendritic structures associated with the Arp2/3 complex, in contrast to the tail region where a loose actin meshwork is associated with myosin-II and cortexillin, an antiparallel actin-bundling protein. In axenically growing strains of *D. discoideum,* wave patterns built by the same components evolve in the absence of any external signal input. Since these autonomously generated patterns are constrained to the plane of the substrate-attached cell surface, they are optimally suited to the optical analysis of state transitions between front-like and tail-like states of the membrane and the actin cortex. Here, we describe imaging techniques using fluorescent proteins to probe for the state of the membrane, the reorganization of the actin network, and the dynamics of wave patterns.

Key words Actin waves, Arp2/3 complex, Cell polarity, Myosin, Phosphoinositides, PI3-kinase, PTEN, Ras, Self-organization

1 Introduction

Chemotaxis in *Dictyostelium discoideum* is based on signal transduction that translates an external gradient of chemoattractant into a graded intracellular signal, which causes the cell to protrude pseudopods preferentially into the direction of the gradient (for review, see [1, 2]). On the molecular level, this response is reflected in the spatial differentiation of the plasma membrane and the underlying actin cortex of the cell. At the cytoplasmic face of the membrane, Ras is activated and phosphatidyl-inositol(3,4,5)trisphosphate (PIP_3) is synthesized at the front, whereas the PIP_3-degrading PI3-phosphatase PTEN is bound to the tail region of the cell. At the actin cytoskeleton, the Arp2/3 complex is localized to the front, indicating nucleation of dendritic actin filament assemblies, whereas filamentous myosin II and the antiparallel

Tian Jin and Dale Hereld (eds.), *Chemotaxis: Methods and Protocols*, Methods in Molecular Biology, vol. 1407,
DOI 10.1007/978-1-4939-3480-5_6, © Springer Science+Business Media New York 2016

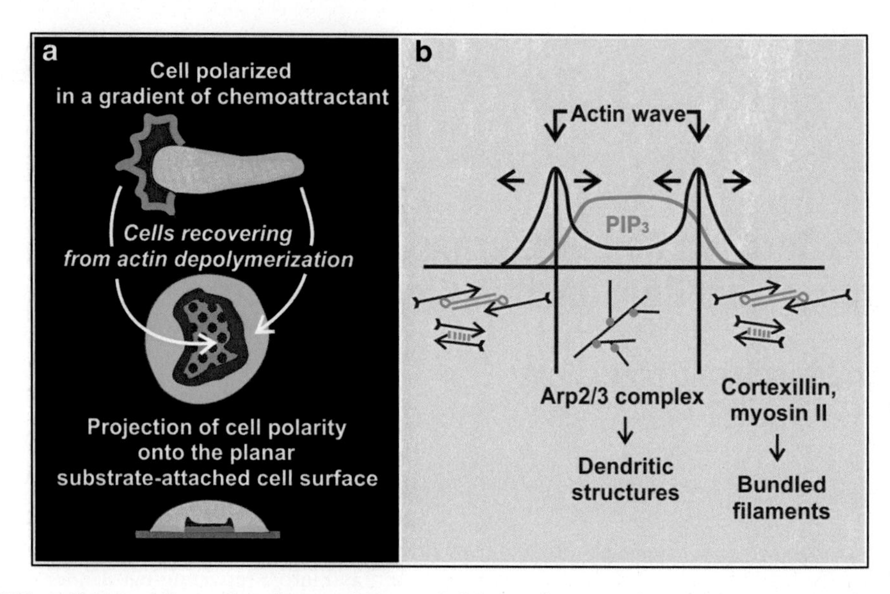

Fig. 1 Differentiation of the cell surface and the underlying actin cortex in polarized chemotaxing cells, as compared to wave patterns on the substrate-attached cell membrane in *Dictyostelium discoideum*. (**a**) A cell moving in a gradient of chemoattractant (*top*) is polarized into a front and a tail. The membrane of the front region is rich in PIP_3 (*blue*) and the actin network beneath the membrane consists of a dense fabric of filaments associated with the Arp2/3 complex (*red*). In contrast, the actin cortex of the tail region consists of a loose meshwork of filaments associated with filamentous myosin-II. In wave-forming cells of the AX2 strain of *D. discoideum* (*middle*) a similar differentiation is observed at the substrate-attached cell surface. In the wave pattern, actin waves encircle an inner territory corresponding to the front of a motile cell and separate this area from an external one that corresponds to the tail of the cell. A side view (*bottom*) illustrates that the wave pattern can be visualized in one plane of focus by confocal or TIRF microscopy. Modified from [3]. (**b**) Different structures of actin assembly on the two sides of a propagating wave. The PIP_3-rich inner territory encircled by a wave is dominated by the Arp2/3 complex responsible for the nucleation of branched actin-filament structures. The external area is rich in proteins responsible for the antiparallel bundling of actin filaments: cortexillin and the conventional myosin-II. Data from [4]

actin-bundling protein cortexillin is enriched within the retracting tail region.

Independent of chemotactic stimulation, a differentiation of the substrate-attached cell surface into regions that correspond to the front-to-tail polarity of chemotaxing cells is observed in axenically growing strains of *Dictyostelium discoideum* (Fig. 1a). This pattern consists of propagating actin waves that separate two regions of the plasma membrane from each other, a PIP_3-rich inner territory encircled by the actin wave, and a PTEN-decorated external area [3]. These differences are correlated with differences in actin network organization [4], which in turn are reflected in the proteins that preferentially bind to the actin cortex in one or the other region (Fig. 1b; Table 1).

The autonomous formation of wave patterns is well suited to analyze the state transitions in the plasma membrane and the actin network that underlie the conversion of a "front state" into a

Table 1
Membrane-binding (A) and cytoskeleton-associated (B) constituents of wave patterns

	Constituent	Localization	Fluorescent construct	Reference
A	PI(3,4,5)P_3 (and PI(3,4) P_2)	Inner territory	GFP-PHcrac or PHcrac-mRFP	[23]
	PI(4,5)P_2	Higher in external area than in inner territory	PLCδ1-GFP	[3]
	Activated Ras	Inner territory	RBD of human Raf1	[3]
	PTEN	External area	PTEN-GFP	[5]
B	Filamentous actin	Actin wave and enriched in inner territory	LimEΔ-GFP or mRFP-LimEΔ	[7]
	Arp2/3 complex	Actin wave and inner territory	GFP-Arp3	[6]
	Myosin-IB	Front of actin wave	GFP-myosin-IB heavy chain	[6]
	Coronin	Back and roof of actin wave	GFP-coronin	[6]
	Myosin-II	External area	GFP-myosin-II heavy chain	[4]
	Cortexillin I, full-length or actin-bundling fragment 352–435	External area	GFP-cortexillin I or GFP-CI (352–435)	[4]

"tail state" and vice versa. The mode of transitions from the tail to the front state differs from the mode of front-to-tail transitions. As shown in Fig. 2, tail-to-front transitions are based on the local dissociation of PTEN from the membrane. In some of the "PTEN holes" thus produced, PIP_3 synthesis is amplified [5]. This mode of transition introduces a stochastic element to pattern dynamics. Subsequently, the PIP_3-rich area expands with an almost constant rate of progression of its border [6]. In contrast, front-to-tail transitions occur in a programmed way by the continuous decay of PIP_3 at the back of a propagating wave. Eventually, the PIP_3-depleted membrane area is being decorated with PTEN, most often beginning at the very border of the substrate-attached cell surface.

Here we focus on imaging of the components relevant for wave patterns, and on the experimental analysis of state transitions and their spatiotemporal dynamics. The wave patterns are observed in cells of axenically growing strains of *D. discoideum* without any pretreatment, in particular during the first 3 h of starvation [7, 8]. However, they are most profusely formed in cells recovering from actin depolymerization [9]. Therefore, we describe the pretreatment of cells with latrunculin A, a blocker of actin polymerization. The wave formation can be reversibly inhibited by LY294002, an inactivator of PI3-kinases that synthesize PIP_3 [10, 11]. Since the fully developed patterns exceed the size of normal cells, we also cover the generation of giant cells by electric-pulse-induced fusion as a tool to study pattern evolution under less restricted spatial conditions [12].

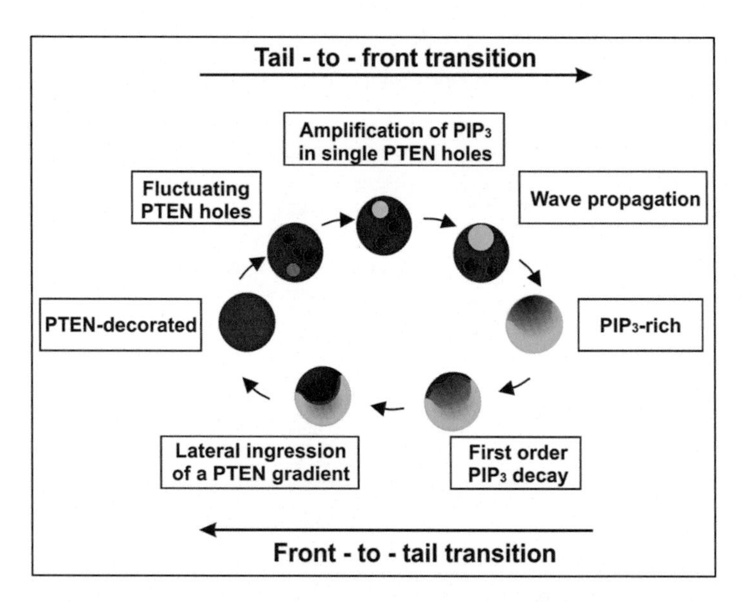

Fig. 2 Diagram of transitions between a PTEN-decorated and a PIP_3-rich state of the membrane. The PTEN-decorated state (*red*) corresponds to the "tail-like" state of the membrane, and the PIP_3-rich state (*green*) to its front-like state. The transition from a fully PTEN-decorated to a PIP_3-rich state begins with the local depletion of PTEN. "PTEN holes" (*black*) are sites where a cascade of PIP_3 synthesis can be initiated. Beyond a threshold of amplification, a PIP_3 wave propagates from the site of initiation over the area of the substrate-attached cell membrane (upper part of the cycle). This wave is followed by a zone of PIP_3 depletion. The depleted area can be occupied by PTEN that typically ingresses laterally in the form of a gradient, beginning at the non-attached PTEN-decorated portion of the membrane (lower part of the cycle). Data from [5]

1.1 Visualizing Fluorescent Proteins in Wave Patterns of Dictyostelium Cells

Dictyostelium cells can be transfected to permanently express fluorescent proteins. Control by the actin 15 promoter enables proteins to be expressed in the growth phase as well as the aggregation-competent stage of starved cells. The expression of multiple proteins with fluorescent tags that are distinguishable by their excitation and emission spectra makes it possible to relate in one cell different spatial and temporal patterns to each other by dual-wavelength imaging. All constituents of wave patterns are either associated with the substrate-attached plasma membrane or integrated into the membrane-anchored actin cortex (Table 1). The fluorescent markers used are, in the unbound state, cytosolic proteins. They are responsible for a cytoplasmic background against which the label at the membrane or in the cell cortex has to be detected.

Since *Dictyostelium* cells contract in response to light [13], care should be taken to minimize this response by the lowest light exposure possible (*see* **Note 1**). These requirements are met by total internal reflection fluorescence (TIRF) and by confocal laser-scanning microscopy. In TIRF microscopy, a shallow layer close to a reflecting surface is illuminated by an evanescent field. Appropriate equipment is available from various suppliers. The equipment used in our own studies has been specified in detail [9, 14]. In the following sections, advantages and drawbacks of each of the imaging techniques are outlined.

1.2 Choice of Imaging Techniques

Confocal microscopy using spinning-disc or line-scan equipment is the technique of choice for recording structures apart from the substrate surface and for evaluating protein distributions in z-direction (*see* **Note 2**). Current spinning-disc or line-scan confocal microscopes with an inverted stage are suitable for the imaging of wave patterns. The microscope should have oil immersion optics. Spinning-disc microscopes, other than line-scan microscopes, make it possible to illuminate a microscopic field almost synchronously in the confocal mode. However, taking into account that a single actin filament has a diameter of 8 nm and even with high-aperture objectives the point-spread function in *z*-direction has a width of ≥0.7 μm, it is obvious that delicate actin structures are less precisely portrayed than by TIRF.

A TIRF microscope is preferable for fluorescence imaging close to the substrate-attached plasma membrane or in the substrate-to-cell interspace, since cytosolic fluorescence due to unbound marker proteins is most efficiently suppressed when only a layer of the cell close to the substrate is illuminated by the short-range evanescent field used in TIRF microscopy. The evanescent field declines exponentially with increasing distance from the reflecting surface. In practice, fluorescence will be recorded from a layer extending about 100 nm from the plasma membrane of an attached cell into its interior: this means with minimal background contributed by the bulk of the cytoplasm. TIRF microscopy visualizes the actin network structures with a brilliance that is unmatched by any other technique [15]. Most of the cortical actin network of *Dictyostelium* cells is located within the layer illuminated by the evanescent field. However, actin waves may extend 1 μm or more into the cytoplasmic space. Therefore, these waves will not be represented with their full height in TIRF images. For chemotaxing cells one has to keep in mind that the very front of a cell is not always attached to the substrate and therefore may not be seen in TIRF images.

Disadvantages of TIRF microscopy reside in the ambiguous relation of fluorescence intensity to the amount of fluorescent protein present. Since the fluorescence excitation in an evanescent field depends on the distance from the reflecting surface, only the intensities of proteins at the same distance can be directly compared, for instance the intensities at regions of the plasma membrane that adhere to the substrate. Since the penetration of the evanescent field into the cell depends on the wavelength of excitation, structures remote from the membrane may be seen at 568 nm but not 488 nm excitation. This ambiguity is avoided if two fluorophores are excited at the same wavelength and discriminated by their emission spectra. mRFP is weakly but often sufficiently excited at 488 nm for this wavelength to be used for both mRFP and GFP. For dual-wavelength TIRF imaging it is therefore recommended to combine GFP with mRFP rather than Cherry.

1.3 Imaging Specific Constituents of the Wave Pattern

At the membrane, PIP_3 is strongly enriched within the inner territory encircled by an actin wave, whereas its precursor, PIP_2, is about twofold more abundant in the external area than in the inner one. Ras, primarily the RasG isoform that is recognized by the Ras-binding domain (RBD) of human Raf 1 [16, 17], is activated within the inner territory similar to, but not identical with, the region occupied by PIP_3 [3]. Ras activation is considered to be a co-activator of PI3-kinases and thus upstream of PIP_3 in chemotactic signal transduction [18].

1.3.1 $PI(3,4,5)P_3$

This phosphoinositide is characteristic of the front of chemotaxing cells [19, 20]. In wave patterns, PIP_3 is a selective marker of the inner territory [21–23]. In large cells produced by electric-pulse-induced fusion, PIP_3 forms traveling bands of 5 μm width that reflect the dynamics of its coupled synthesis and degradation [12]. PIP_3 is synthesized in *Dictyostelium* cells by at least five PI3-kinases [24] and degraded to $PI(4,5)P_2$ by at least two PI3-phosphatases [25]. PIP_3 is recognized by the PH domains of a variety of proteins. Crucial for the choice of a PH domain as a PIP_3 marker is its sufficient affinity in balance with a dissociation rate high enough for accommodating to the dynamics of PIP_3 synthesis and degradation. The PH domain of the cytoplasmic regulator of adenylate cyclase (CRAC) is an appropriate one [19]. PH-CRAC can be tagged at its N-terminus with GFP or mRFP. In our hands, superfolding (sf) GFP [26, 27] proved to be an excellent tag [12].

1.3.2 $PI(4,5)P_2$

This substrate of PI3-kinases and product of PI3-phosphatases is a major phosphoinositide in the plasma membrane. In contrast to PIP_3, PIP_2 is enriched in the external area but is less stringently distributed in the wave patterns than PIP_3 [3].

1.3.3 Phosphatase and Tensin Homolog (PTEN)

This PI3-phosphatase shuttles between an inactive state in the cytoplasm and an activated state bound to the plasma membrane [28]. In chemotaxing cells, PTEN binding is restricted to the membrane of the tail region [29, 30]. Binding to the membrane requires dephosphorylation of C-terminal serine and threonine residues [31] and is mediated by two binding sites, one of them specific for PIP_2 [32]. The binding to its product establishes a positive feedback circuit of PTEN activation.

1.3.4 Activated Ras

Using Ras-binding domains (RBDs) of human Raf1, activation in the inner territory encircled by an actin wave is recognized [3]. Raf1-RBD labels in *Dictyostelium* specifically the activated, GTP-bound state of RAS G [16, 17].

1.3.5 Actin

The actin 8 gene encodes the predominant actin species in *Dictyostelium* cells. With GFP attached to its N-terminus through an appropriate linker, this actin polymerizes together with untagged actin. If the ratio of the GFP-tagged actin to native actin is below 10 %, no major reduction in the rate of polymerization is observed in vitro, and cells expressing the fluorescent actin are not obviously defective except for a slight impairment of cytokinesis [33]. However, the possibility remains that the tag may alter the interaction of actin with regulatory proteins.

Since about half of the actin exists in an unpolymerized state in *Dictyostelium* as in other cells, the direct fusion of actin to GFP or mRFP is unavoidably deteriorated by a high cytoplasmic background. Therefore, the direct labeling of actin is only advisable if the purpose of the experiment does demand it. Measuring the turnover rate of filamentous actin structures by FRAP is one of these purposes [6].

An optimal label for filamentous actin structures should bind selectively and with high affinity to filamentous actin. On the other hand, the dissociation rate should be high enough for the label to accommodate to the dynamics of actin structures in vivo. For the actin waves in *Dictyostelium* cells, we found fluorescent constructs of the LimE protein most appropriate [7, 34], although it should be taken into account that LimE constructs may not label all filamentous actin structures equally well (*see* **Note 3**). The LimE protein of *Dictyostelium* is an actin-binding protein comprised of three domains: a Lim domain, a glycine-rich region, and a coiled-coil tail. The two first domains are essential for actin binding, whereas deletion of the tail even improves the quality of labeling. The construct lacking the C-terminal coiled-coil domain is designated as LimEΔ. It has a molecular mass of 15 kDa and can be tagged at its N- or C-terminus with GFP. For dual-wavelength imaging in concert with GFP-tagged proteins, mRFP-LimEΔ or Cherry-LimEΔ can be used (*see* **Note 4**).

1.3.6 Actin-Associated Proteins Shared by Actin Waves and the Front of Chemotaxing Cells

The three proteins discussed in this section label different sites at a leading edge and respond differently to the protrusion and retraction of leading edges during the reorientation of cells in gradients of chemoattractant. All three proteins associate also with phagocytic cups and sites of clathrin-associated endocytosis [14].

1.3.7 Myosin-IB at the Front of Actin Waves

The single-headed myosin-IB can be tagged at its N-terminus with GFP. GFP-myosin-IB localizes transiently to the very border of leading edges which are protruded in response to chemoattractant [35], and delineates the front of actin waves [6, 36].

1.3.8 Subunits of the Arp2/3 Complex

The Arp2/3 complex is concentrated in regions of the cell cortex that are densely packed with filamentous actin. In wave patterns, the Arp2/3 complex is enriched in the actin waves and throughout the inner territory encircled by these waves [7]. Among the seven members of this complex two subunits, Arp3 and ARPC1, have been successfully tagged with fluorescent proteins at their N-terminus without abolishing localization to actin structures in *Dictyostelium* cells [14]. These subunits differ from each other in their association with the Arp2/3 complex: whereas Arp3 is tightly bound to the complex, ARPC1 is the only subunit that readily exchanges between the bound and free state [37]. Therefore, we recommend GFP-Arp3 as a label for the Arp2/3 complex.

1.3.9 Coronin, an Indicator of Actin Depolymerization

Coronins are WD40-repeat proteins that form seven-bladed propellers [38]. *Dictyostelium* coronin is strongly enriched at leading edges [39]. Detailed analyses of its localization in phagocytosis [40, 41], cell spreading [14], and chemotaxis [39] indicated that this coronin is recruited to actin structures that are in the process of disassembly. The coronin localizes to the roof and the back of an actin wave [6]. The coronin can be tagged at its N- or C-terminus without impairing its localization.

1.4 Practical Considerations for Imaging Wave Patterns

1.4.1 Expression Levels

Since actin structures are assembled from G-actin present in the cytoplasm, and most of the proteins associated with these structures are also recruited from the cytosolic pool, a cytoplasmic background of these proteins will always be present. The rule is to keep the expression of any fluorescent protein not higher than necessary for the imaging of its structure-bound fraction. Strong expression would not only cause the risk of overexpression artifacts but would also increase the cytoplasmic background. For instance, incorporation of the GFP-tagged Arp3 subunit into the functional Arp2/3 complex will be limited by the endogenous concentration of the other six members of the complex. Since the expression level usually varies from cell to cell even in a cloned cell population, it is easy to choose the appropriate cells for imaging.

Small proteins may pass the nuclear pores. For instance, mRFP-LimEΔ slightly accumulates in the nucleus relative to the cytoplasm. It should also be taken into account that a tagged protein may localize appropriately, but lacks function and even interferes with the activity of its untagged endogenous counterpart. An example is a GFP-actin construct that is incorporated into actin filaments but blocks the movement of myosin along these filaments [42].

The integrating vector constructs we are using to express fluorescent proteins under control of the actin 15 promoter integrate randomly into the genome and the level of expression depends on the site of integration. If green and red fluorescent proteins are expressed simultaneously by independent integration of the vectors, their levels fluctuate independently. Therefore, the experimentalist has an opportunity to choose cells within a population that exhibit the optimal levels of expression.

1.4.2 Background Fluorescence

Intrinsic fluorescence of the cytoplasm at the wavelengths used for green or red fluorescent proteins is low enough for being neglected in the imaging of actin structures, provided that the fluorescent protein of interest is sufficiently expressed. Endosomes filled with nutrient medium are the major source of intrinsic fluorescence. Their fluorescence is of no concern in TIRF images, since the endosomes are located to the interior of the cells out of the range of the evanescent field. In confocal images the contents of the endosomes will be recognized in cells cultivated axenically in HL5 or similar media containing yeast extract. If this is a problem, low-fluorescent media [43] may be used, though the cells grow then more slowly or saturate at lower density. Alternatively, the cells may be cultivated on bacteria such as *E. coli* B/2.

1.4.3 Marking the Cell-to-Substrate Interspace

Only if the cells are in close contact with the glass surface, all fluorescent components at the membrane or in the actin cortex will be represented in one plane of focus. To check whether there is extracellular space left between glass and cell surface, this space can be labeled with fluorescent dextran of molecular mass 3000 or higher that does not penetrate the plasma membrane [9, 44]. Alexa Fluor 680-dextran can be combined with GFP- and mRFP-tagged proteins for imaging three fluorescent channels together. Cells capable of growing axenically gradually take up the dextran by macropinocytosis until after an hour the contents of vesicles in the entire endosomal pathway are labeled.

1.5 Production of Giant Cells by Electric-Pulse-Induced Cell Fusion

The evolution of wave patterns in normal *Dictyostelium* cells is limited by the size of the cells of not more than 20 μm length. To study the shape of waves, their velocity of propagation, and interference upon collision, a large surface accessible to imaging is required. Cells with a substrate-attached surface of sufficient size can be produced by inducing cohering cells to fuse with each other [45], using electric pulses in an electroporator commercially available for the transfer of DNA [12].

Mutant strains that require attachment to a solid substrate for cytokinesis can be cultivated in shaken suspension in order to produce giant cells. This technique is applicable to mutants deficient in myosin-II or PTEN. Myosin-II-null cells have been shown to form actin waves similar to those observed in wild-type cells fused by electric pulses [12].

2 Materials

2.1 Cell Strains Expressing Fluorescent Proteins

Cells of the axenically growing strain AX2-214 of *Dictyostelium discoideum*, transformed to express fluorescent proteins, are used. GFP is the enhanced S65T version, sf-GFP is superfolding GFP [26, 27], and mRFP is mRFPMars optimized for expression in *D. discoideum* [46]. Most of the transformed cells or expression vectors needed for the experiments are available from the Dicty Stock Center at http://dictybase.org/StockCenter/StockCenter.html [47].

2.2 Cell Culture Components

1. Nutrient medium (1 L): 7.15 g Yeast Extract (Oxoid, Lenexa, KS, USA), 14.3 g Bacteriological Peptone L37 (Oxoid) or Neutralized Peptone (Oxoid), 18.0 g maltose, 0.486 g KH_2PO_4, 0.616 g $Na_2HPO_4 \cdot 2H_2O$, adjusted to pH 6.7, and autoclaved at 120 °C for 20 min (*see* **Note 5**).
2. PB: 17 mM phosphate buffer, pH 6.0 (1 L): 2.0 g KH_2PO_4, 0.356 g $Na_2HPO_4 \cdot 2H_2O$, autoclaved at 120 °C. This buffer is generally used for starving and imaging of *D. discoideum* cells.
3. Blasticidine S hydrochloride (Invitrogen, Life Technologies, Grand Island, NY, USA) is dissolved in PB at 10 mg/mL, aliquoted, and stored at −20 °C.
4. Geneticin (Sigma-Aldrich, St. Louis, MO, USA) is dissolved in PB at 20 mg/mL, aliquoted, and stored at −20 °C.
5. Hygromycin B (Calbiochem, EMD Bioscience, Temecula, CA, USA) is dissolved in PB at 33 mg/mL, aliquoted, and stored at −20 °C.
6. Petri dishes, 10 cm diameter (e.g., Corning, Union City, CA, USA).

2.3 Fluorescence Microscopy Components

1. Inverted fluorescence microscope, equipped for confocal or TIRF microscopy.
2. For TIRF imaging, an oil immersion objective of at least 1.45 NA is required (*see* **Note 6**).
3. For confocal or TIRF microscopy, commercial glass-bottom dishes (MatTek Corp., Cat.No. P35G-1.5-20-C Ashland, MA, USA or WPI, FluoroDish, FD35-100, Sarasota, FL, USA) may be used as imaging chambers. For custom-made chambers, *see* **Note 7**.

2.4 Components for Latrunculin A Treatment of Normal Cells

1. Latrunculin A (Invitrogen, Life Technologies, Grand Island, NY, USA), dissolved in anhydrous dimethyl sulfoxide (DMSO) to 1 mM (100 μg in 237 μL). Aliquots of 5 μL are kept at −20 °C, thawed, and diluted to 5 μM in PB before the experiment (*see* **Note 8**).
2. Anhydrous DMSO (Sigma-Aldrich).
3. PB, *see* Subheading 2.2, **item 2**.

2.5 Components for Inhibiting Waves with LY294002

1. LY294002 (Sigma-Aldrich), dissolved in anhydrous DMSO to 20 mM. This stock solution is thawed and diluted to 50–70 μM in PB before the experiment.
2. Anhydrous dimethyl sulfoxide (DMSO) (Sigma-Aldrich).

2.6 Component for Marking the Cell-to-Substrate Interspace

Fluorescent dextran of a molecular mass 3000 or higher like Alexa Fluor Dextran 680 (Molecular Probes, Invitrogen, Life Technologies, Grand Island, NY, USA, Cat. No. D34680) is used. Prepare a 5 mg/mL stock solution in phosphate buffer and freeze aliquots at −20 °C.

2.7 Components for Electric-Pulse-Induced Cell Fusion

1. Gene-Pulser (BioRad, No.1657077, Hercules, CA, USA) as used for the electroporation of cells.
2. 4 mm gap Electroporation Cuvettes (Molecular BioProducts, Thermo Fisher Scientific Inc., Waltham, MA, USA).
3. Filter tips for 1 and 0.2 mL pipettes. The tips should be cut off to reduce shear and thus avoid dissociation of cell aggregates.
4. 50 mL centrifuge tubes (BD-Falcon, BD Biosciences, Bedford, MA, USA).
5. Roller shaker (New Star Environmental, Roswell, GA, USA), or custom-made shaker that gently agitates the cells to form suspended clumps.
6. PB with 2 mM $CaCl_2$ and 2 mM $MgCl_2$. A stock solution containing 1 M $CaCl_2$ and 1 M $MgCl_2$ is prepared in PB, filter sterilized, and stored in the refrigerator. Shortly before use dilute 1:500 with PB. 2 mL is needed per glass bottom dish.

2.8 Data Analyzing Programs

1. Fiji (ImageJ) (http://fiji.sc/Fiji or http://imagej.nih.gov/ij/).
2. Excel (Microsoft Office).

3 Methods

3.1 Cultivation of Cells

Axenically growing laboratory strains like AX2 or AX3 efficiently take up liquid by macropinocytosis. The cells can be cultivated in shaken suspension or attached to the bottom of Petri dishes. In nutrient medium they multiply with a generation time of 8–9 h at 21–23 °C. There is a strict upper temperature limit of 25 °C below which the cells can be used for experiments.

1. Day 1: Spores or cells of the *D. discoideum* strain AX2 expressing fluorescent proteins, or mutants derived from this strain, are incubated with 10 mL nutrient medium in a Petri dish of 10 cm diameter. An inoculate of about 5×10^4 cells per mL will be appropriate.
2. Day 2: To apply selection pressure on the expression of fluorescent proteins, antibiotics (*see* Subheading 2.2, **items 3–5**) are added depending on the resistance cassettes in the transformation vectors used.
3. Day 3 or 4: The cells are harvested before monolayers become confluent. They are gently suspended in the 10 mL medium

within a Petri dish. 1 mL of cell suspension is pipetted into an imaging chamber (*see* Subheading 2.3, **item 3**). Subsequently the cells are allowed to settle for 20 min.

4. The cells are then washed twice with 1 mL PB by exchanging the liquid gently from the side.
5. After 3 h of starvation in PB the cells are imaged.

3.2 Treatment of Cells with Latrunculin A

1. Cells are prepared as in Subheading 3.1, **items 1–4**.
2. Dilute latrunculin A to 5 μM by adding 995 μL of PB to a thawed 5 μL aliquot of the 1 mM stock solution.
3. Bring cells to the microscope and make all adjustments, e.g., for fluorophores, time series settings.
4. Then add 5 μM latrunculin A by replacing the buffer with latrunculin A (*see* **Note 8**).
5. Image cells within the latrunculin A solution.
6. Remove latrunculin A after 15 min and add PB.
7. Within 30 min after the removal of latrunculin A, waves should reappear. Image the waves by focusing to the substrate-attached area of the cells.

3.3 Inhibition of Wave Formation by LY294002

1. Cells are prepared as in Subheading 3.2, **steps 1**–7.
2. Dilute the 20 mM stock solution of LY294002 to 50–70 μM with PB (*see* **Note 9**).
3. Add LY294002 by replacing the buffer on top of the cells with the diluted LY294002 solution.
4. Check by imaging the disappearance of waves.
5. For the recovery of wave formation, the LY294002 can be removed after 30 min and replaced with PB.

3.4 Marking the Cell-to-Substrate Interspace with Fluorescent Dextran

1. Cells are prepared as in Subheading 3.1, **steps 1–4**.
2. A stock solution of 5 mg/mL of Alexa Fluor dextran, MW about 10,000 in PB, is stored frozen. Before use, an aliquot is thawed and diluted to 0.5 mg/mL.
3. Excess buffer is removed from the cells and the diluted dextran is added carefully.
4. Image the space between the substrate surface and the bottom of an attached cell by TIRF or confocal microscopy of the substrate-attached area of the cells.

3.5 Electric-Pulse-Induced Cell Fusion for the Production of Giant Cells

Aggregates are produced in a roller shaker (*see* Subheading 2.7, **item 5**). Rolling Falcon tubes at 30 rpm with an angle of about 15° relative to the axis of rotation will be appropriate. Since the cells have to be in contact with each other for successful fusion, disruption of the aggregates should be prevented. Therefore all pipetting steps have to be done carefully and with cut pipette tips. The aggregated cells are fused by the application of electric pulses. This protocol applies to cells of *D. discoideum* strain AX2-214. For myosin-II-null or PTEN-null cells it is sufficient to cultivate them in shaken suspension to produce large cells by the inhibition of cytokinesis.

1. Day 1: Spores, or cells of the non-sporulating myosin-II-null strains, are transferred to 10 mL of nutrient medium in a 10 cm diameter Petri dish and incubated at 21–23 °C. A concentration of about 5×10^4 spores or cells per mL is appropriate to start with. For an experiment with fused cells, at least eight Petri dishes are needed.
2. Day 2: To apply selection pressure on the expression of fluorescent proteins, appropriate antibiotics are added (see Subheading 2.2, **items 3–5**).
3. Day 3 or 4: The cells are harvested before monolayers become confluent. Usually one gets 1×10^7 cells per dish at this stage. They are gently suspended in the 10 mL of medium of a Petri dish and are centrifuged in a plastic tube for 4 min at 4 °C in a swing-out rotor at $200 \times g$. Wash the cells twice. Each time the pelleted cells are resuspended in 10 mL ice-cold PB and centrifuged as before.
4. After the second wash the pelleted cells are adjusted to a density of 1.5×10^7 cells per mL in PB, transferred into a 50 mL Falcon tube, and gently shaken on a roller-shaker (*see* Subheading 2.7, **item 5**).
5. After 3 h of starvation the cells are electro-fused. Transfer 800 μL of the cell suspension to a 4 mm gap electroporation cuvette using a cut 1 mL tip.
6. Place cuvette into an electroporation apparatus, the voltage and capacity of which are set to 1 kV and 1 or 3 μF (*see* **Note 10**). Apply three pulses at intervals of 1 s.
7. Transfer 25–50 μL suspension of the fused cells to a glass-bottom dish or a custom-made imaging chamber using a cut 200 μL tip.
8. After settling of the fused cells for 5 min, gently add 2 mL of PB supplemented with 2 mM $CaCl_2$/2 mM $MgCl_2$.
9. Allow cells to spread on the glass and to recover for 30 min prior to imaging.

3.6 Data Analysis

3.6.1 Line Scans

We routinely record wave propagation with a frame-to-frame interval of 1 s. The spatiotemporal relationship of two components that participate in pattern formation may provide information relevant for the modeling of molecular interactions. One informative approach is the analysis of data in dual-fluorescence-labeled cells by line scans that are orientated in the direction of wave propagation. Draw a line using the *line tool* with a width between 10 and 30 pixels and apply the plug-in *make-profile-movie* in ImageJ [48].

An immediate view of the dynamics of individual components in space and time during the initiation and propagation of a wave can be obtained by the plotting of a kymograph (examples in [5]). This can be done in ImageJ with the *Image-stack-reslice* command.

3.6.2 Point Scans

To quantify temporal changes at a fixed point within the two-dimensional wave pattern, the region of interest (ROI) should be made as small as compatible with a sufficiently high signal-to-noise ratio. If the ROI is large, measures taken at different positions in the area will be averaged. With the size of the ROI the risk increases that two events, which are actually separated in time, seem to overlap. A convenient ROI is a nearly circular area of 12 pixels, generated by the use of the *point tool* of ImageJ specified to *oval* with a *height and width of 4*. Copy the measured numbers from the *result window* in ImageJ in a Microsoft Excel work sheet and plot intensities over time.

3.6.3 Phase Plots

When different components involved in wave propagation are measured at a defined point of the cell membrane, the interrelationship of their changes can be displayed in a phase plot, which may serve as a basis for modeling the molecular network that generates the wave pattern. Fluorescence intensities of the two proteins of interest are acquired by a point scan using ImageJ. The data are copied from the *result window* in ImageJ and pasted in a Microsoft Excel work sheet and plotted against each other. An example is given in Fig. 5b of [5].

4 Notes

1. Cells of *D. discoideum* tend to contract when blue or green light is turned on, apparently by signal transduction from photoreceptors to the actin system [13]. This light sensitivity limits fluorescence excitation more severely than photobleaching of the fluorophores. Therefore, light-induced changes in actin structure and dynamics should be seriously taken into account by avoiding exposure to the laser light prior to imaging and using the first image as a reference for potential changes in the subsequent ones. The formation of wave patterns is especially light sensitive. It may help to increase the frame-to-frame intervals from 1 to 2 s or more. In myosin-II-null mutants the contraction is reduced, but not necessarily other alterations in the actin system.

2. For deconvolution of *z*-stacks, a distance of 200 nm between the optical sections would be optimal. However, for the 3-dimensional reconstruction of a *Dictyostelium* cell, about 60 sections would be needed (for a cell of 10 μm height, including five sections each below and above the cell). Double that number will be required for dual-wavelength imaging. Since chemotaxing cells are moving during the acquisition of an image stack, their shape will appear distorted in *z*-direction. Therefore, one has to compromise about the number of optical sections. Much information on the cortical actin system is already obtained by alternating between two planes, one through the cortex close to the substrate-attached cell surface, and the other about 2–4 μm above through the leading edge and the middle of the cell body.
3. Another construct that binds selectively to filamentous actin is the calponin-like actin-binding domain of ABP120, the *Dictyostelium* ortholog of filamin [49]. With GFP added to the N-terminus of this domain, the cytoplasmic background has been somewhat higher in our hands than with LimEΔ-GFP. The domain from ABP120 has the benefit of being known to work also in mammalian cells. Lemieux et al. [50] recommend GFP-LifeAct [51] as an actin label in *Dictyostelium*, which we have found to label also the actin waves.
4. Any protein that binds exclusively or preferentially to polymerized actin should, by mass action law, shift the equilibrium from monomeric to polymeric actin. Therefore, F-actin markers should be expressed at the lowest level compatible with image acquisition. To avoid artifacts owing to overexpression, either full-length or truncated LimE constructs may be expressed in LimE-null cells [15].
5. Batches of yeast extract and peptone are of varying quality. Each combination of batches should be tested. If found appropriate for growth and development, a sufficiently large quantity of the same two batches should be ordered. Short autoclaving followed by rapid cooling is recommended.
6. *Dictyostelium* cells are obligatory aerobes that must be continuously supplied with oxygen; otherwise the cells bleb, round up, and die within a couple of minutes. Therefore, an inverted microscope is recommended for imaging through a glass cover slip on which the cells are migrating in a fluid layer of about 3 mm height exposed to air, or are covered by an agarose layer. The specimen should be exposed to air or closed on top with a teflon or other membrane permeable for oxygen.
7. As an alternative to commercial glass-bottom dishes, custom-made chambers with specially cleaned glass surfaces can be prepared as follows: (1) Put 40 × 24 mm glass cover slips one

by one in 1 N HCl and shake overnight. For TIRF microscopy, the cover slips should have a controlled thickness of 0.17 ± 0.01 mm (Assistant, Cat. No. 1014/4024, Sondheim, Germany). (2) Wash the cover slips four times with distilled water. (3) Rinse them with absolute ethanol. (4) Take them out one by one, rinse them again by dipping in ethanol, and air-dry. (5) Keep them in a clean Petri dish until use. (6) Fix a plexiglass ring, outer diameter 24 mm, wall thickness 1 mm, and height 4 mm onto the cleaned cover slip with silicone grease.

8. The concentration of 5 μM latrunculin A efficiently blocks actin polymerization, such that the evolution of wave patterns from a uniform state can be followed by fluorescence microscopy. However, cryo-electron microscopy has shown that some actin filaments are still persisting under these conditions [52]. For more rigorous elimination of filamentous actin, 30 μM latrunculin A or a mixture of 10 μM latrunculin A and 10 μM cytochalasin A may be used, although not every cell will recover after this treatment. It should be taken into account that the efficacy of latrunculin A batches varies, even when they are purchased from the same supplier.
9. These concentrations of LY294002 are a compromise at which about 5 % of the PI3-kinase activities are left. At higher concentrations other kinases may be inhibited such that the effect is no longer PI3-kinase specific [10, 11].
10. These values may be altered by trial and error. There is only a small window of pulse sizes that cause fusion without killing the cells.

Acknowledgements

We thank Annette Müller-Taubenberger, LMU München, for PTEN-sf-GFP and are grateful for funds of the Max Planck Society.

References

1. Wang Y, Chen CL, Iijima M (2011) Signaling mechanisms for chemotaxis. Dev Growth Differ 53:495–502
2. Jin T (2013) Gradient sensing during chemotaxis. Curr Opin Cell Biol 25:532–537
3. Gerisch G, Ecke M, Wischnewski D, Schroth-Diez B (2011) Different modes of state transitions determine pattern in the phosphatidylinositide-actin system. BMC Cell Biol 12:42–57
4. Schroth-Diez B, Gerwig S, Ecke M et al (2009) Propagating waves separate two states of actin organization in living cells. HFSP J 3: 412–427
5. Gerisch G, Schroth-Diez B, Müller-Taubenberger A, Ecke M (2012) PIP_3 waves and PTEN dynamics in the emergence of cell polarity. Biophys J 103:1170–1178
6. Bretschneider T, Anderson K, Ecke M et al (2009) The three-dimensional dynamics of actin waves, a model of cytoskeletal self-organization. Biophys J 96:2888–2900
7. Bretschneider T, Diez S, Anderson K et al (2004) Dynamic actin patterns and Arp2/3

assembly at the substrate-attached surface of motile cells. Curr Biol 14:1–10

8. Taniguchi D, Ishihara S, Oonuki T et al (2013) Phase geometries of two-dimensional excitable waves govern self-organized morphodynamics of amoeboid cells. Proc Natl Acad Sci U S A 110:5016–5021
9. Gerisch G, Bretschneider T, Müller-Taubenberger A et al (2004) Mobile actin clusters and traveling waves in cells recovering from actin depolymerization. Biophys J 87: 3493–3503
10. Dormann D, Weijer G, Dowler S, Weijer CJ (2004) In vivo analysis of 3-phosphoinositide dynamics during *Dictyostelium* phagocytosis and chemotaxis. J Cell Sci 117:6497–6509
11. Loovers HM, Postma M, Keizer-Gunnink I et al (2006) Distinct roles of PI(3,4,5)P3 during chemoattractant signaling in *Dictyostelium*: a quantitative in vivo analysis by inhibition of PI3-kinase. Mol Biol Cell 17:1503–1513
12. Gerhardt M, Ecke M, Walz M et al (2014) Actin and PIP_3 waves in giant cells reveal the inherent length scale of an excited state. J Cell Sci 127:4507–4517
13. Häder DP, Claviez M, Merkl R, Gerisch G (1983) Responses of *Dictyostelium discoideum* amoebae to local stimulation by light. Cell Biol Int Rep 7:611–616
14. Heinrich D, Youssef S, Schroth-Diez B et al (2008) Actin-cytoskeleton dynamics in non-monotonic cell spreading. Cell Adh Migr 2:58–68
15. Diez S, Gerisch G, Anderson K et al (2005) Subsecond reorganization of the actin network in cell motility and chemotaxis. Proc Natl Acad Sci U S A 102:7601–7606
16. Kae H, Lim CJ, Spiegelman GB, Weeks G (2004) Chemoattractant-induced Ras activation during *Dictyostelium* aggregation. EMBO Rep 5:602–606
17. Sasaki AT, Chun C, Takeda K, Firtel RA (2004) Localized Ras signaling at the leading edge regulates PI3K, cell polarity, and directional cell movement. J Cell Biol 167:505–518
18. Charest PG, Firtel RA (2007) Big roles for small GTPases in the control of directed cell movement. Biochem J 401:377–390
19. Parent CA, Devreotes PN (1999) A cell's sense of direction. Science 284:765–770
20. Parent CA (2004) Making all the right moves: chemotaxis in neutrophils and *Dictyostelium*. Curr Opin Cell Biol 16:4–13
21. Arai Y, Shibata T, Matsuoka S et al (2010) Self-organization of the phosphatidylinositol lipids signaling system for random cell migration. Proc Natl Acad Sci U S A 107:12399–12404
22. Shibata T, Nishikawa M, Matsuoka S, Ueda M (2012) Modeling the self-organized phosphatidylinositol lipid signaling system in chemotactic cells using quantitative image analysis. J Cell Sci 125:5138–5150
23. Gerisch G, Ecke M, Schroth-Diez B et al (2009) Self-organizing actin waves as planar phagocytic cup structures. Cell Adh Migr 3:373–382
24. Hoeller O, Kay RR (2007) Chemotaxis in the absence of PIP_3 gradients. Curr Biol 17:813–817
25. Lusche DF, Wessels D, Richardson NA et al (2014) PTEN redundancy: overexpressing lpten, a homolog of *Dictyostelium discoideum* ptenA, the ortholog of human PTEN, rescues all behavioral defects of the mutant ptenA-. PLoS One 9:e108495
26. Pédelacq JD, Cabantous S, Tran T et al (2006) Engineering and characterization of a superfolder green fluorescent protein. Nat Biotechnol 24:79–88
27. Müller-Taubenberger A, Ishikawa-Ankerhold HC (2013) Fluorescent reporters and methods to analyze fluorescent signals. Methods Mol Biol 983:93–112
28. Vazquez F, Matsuoka S, Sellers WR et al (2006) Tumor suppressor PTEN acts through dynamic interaction with the plasma membrane. Proc Natl Acad Sci U S A 103:3633–3638
29. Iijima M, Devreotes P (2002) Tumor suppressor PTEN mediates sensing of chemoattractant gradients. Cell 109:599–610
30. Funamoto S, Meili R, Lee S et al (2002) Spatial and temporal regulation of 3-phosphoinositides by PI 3-kinase and PTEN mediates chemotaxis. Cell 109:611–623
31. Vazquez F, Devreotes P (2006) Regulation of PTEN function as a PIP_3 gatekeeper through membrane interaction. Cell Cycle 5: 1523–1527
32. Iijima M, Huang YE, Luo HR et al (2004) Novel mechanism of PTEN regulation by its phosphatidylinositol 4,5-bisphosphate binding motif is critical for chemotaxis. J Biol Chem 279:16606–16613
33. Westphal M, Jungbluth A, Heidecker M et al (1997) Microfilament dynamics during cell movement and chemotaxis monitored using a GFP-actin fusion protein. Curr Biol 7: 176–183
34. Schneider N, Weber I, Faix J et al (2003) A Lim protein involved in the progression of cytokinesis and regulation of the mitotic spindle. Cell Motil Cytoskeleton 56:130–139
35. Brzeska H, Guag J, Preston GM et al (2012) Molecular basis of dynamic relocalization of

Dictyostelium myosin IB. J Biol Chem 287: 14923–14936
36. Brzeska H, Pridham K, Chery G et al (2014) The association of myosin IB with actin waves in *Dictyostelium* requires both the plasma membrane-binding site and actin-binding region in the myosin tail. PLoS One 9:e94306
37. Mullins RD, Pollard TD (1999) Structure and function of the Arp2/3 complex. Curr Opin Struct Biol 9:244–249
38. Appleton BA, Wu P, Wiesmann C (2006) The crystal structure of murine coronin-1: a regulator of actin cytoskeletal dynamics in lymphocytes. Structure 14:87–96
39. Bretschneider T, Jonkman J, Köhler J et al (2002) Dynamic organization of the actin system in the motile cells of *Dictyostelium*. J Muscle Res Cell Motil 23:639–649
40. Maniak M, Rauchenberger R, Albrecht R et al (1995) Coronin involved in phagocytosis: dynamics of particle-induced relocalization visualized by a green fluorescent protein Tag. Cell 83:915–924
41. Clarke M, Maddera L (2006) Phagocyte meets prey: uptake, internalization, and killing of bacteria by *Dictyostelium* amoebae. Eur J Cell Biol 85:1001–1010
42. Aizawa H, Sameshima M, Yahara I (1997) A green fluorescent protein-actin fusion protein dominantly inhibits cytokinesis, cell spreading, and locomotion in *Dictyostelium*. Cell Struct Funct 22:335–345
43. Liu T, Mirschberger C, Chooback L et al (2002) Altered expression of the 100 kDa subunit of the *Dictyostelium* vacuolar proton pump impairs enzyme assembly, endocytic function and cytosolic pH regulation. J Cell Sci 115: 1907–1918
44. Gingell D, Todd I, Bailey J (1985) Topography of cell-glass apposition revealed by total internal reflection fluorescence of volume markers. J Cell Biol 100:1334–1338
45. Gerisch G, Ecke M, Neujahr R et al (2013) Membrane and actin reorganization in electropulse-induced cell fusion. J Cell Sci 126:2069–2078
46. Fischer M, Haase I, Simmeth E et al (2004) A brilliant monomeric red fluorescent protein to visualize cytoskeleton dynamics in *Dictyostelium*. FEBS Lett 577:227–232
47. Fey P, Dodson RJ, Basu S, Chisholm RL (2013) One stop shop for everything *Dictyostelium*: dictyBase and the Dicty Stock Center in 2012. Methods Mol Biol 983:59–92
48. Schindelin J, Arganda-Carreras I, Frise E et al (2012) Fiji: an open-source platform for biological-image analysis. Nat Methods 9:676–682
49. Pang KM, Lee E, Knecht DA (1998) Use of a fusion protein between GFP and an actin-binding domain to visualize transient filamentous-actin structures. Curr Biol 8:405–408
50. Lemieux MG, Janzen D, Hwang R et al (2014) Visualization of the actin cytoskeleton: different F-actin-binding probes tell different stories. Cytoskeleton (Hoboken) 71:157–169
51. Riedl J, Crevenna AH, Kessenbrock K et al (2008) Lifeact: a versatile marker to visualize F-actin. Nat Methods 5:605–607
52. Heinrich D, Ecke M, Jasnin M et al (2014) Reversible membrane pearling in live cells upon destruction of the actin cortex. Biophys J 106: 1079–1091

Chapter 7

Chemotactic Blebbing in *Dictyostelium* Cells

Evgeny Zatulovskiy and Robert R. Kay

Abstract

Many researchers use the social amoeba *Dictyostelium discoideum* as a model organism to study various aspects of the eukaryotic cell chemotaxis. Traditionally, *Dictyostelium* chemotaxis is considered to be driven mainly by branched F-actin polymerization. However, recently it has become evident that *Dictyostelium*, as well as many other eukaryotic cells, can also employ intracellular hydrostatic pressure to generate force for migration. This process results in the projection of hemispherical plasma membrane protrusions, called blebs, that can be controlled by chemotactic signaling.

Here we describe two methods to study chemotactic blebbing in *Dictyostelium* cells and to analyze the intensity of the blebbing response in various strains and under different conditions. The first of these methods—the cyclic-AMP shock assay—allows one to quantify the global blebbing response of cells to a uniform chemoattractant stimulation. The second one—the under-agarose migration assay—induces directional blebbing in cells moving in a gradient of chemoattractant. In this assay, the cells can be switched from a predominantly F-actin-driven mode of motility to a bleb-driven chemotaxis, allowing one to compare the efficiency of both modes and explore the molecular machinery controlling chemotactic blebbing.

Key words Blebbing, Chemotaxis, *Dictyostelium*, Agarose overlay, 3D migration, Bleb-driven motility, Cyclic-AMP shock

1 Introduction

Researchers in the field of eukaryotic cell chemotaxis have developed multiple experimental techniques to study the efficiency and molecular mechanisms of cellular gradient sensing, polarization, and motility. Most of these approaches are based on the use of devices where a steady-state gradient of chemoattractant is created by its diffusion through the liquid bridge connecting two reservoirs with a buffer containing different concentrations of a chemoattractant (Zigmond chamber [1, 2], Dunn chamber [3, 4], Insall chamber [5]). The cells are normally placed on the surface of a cover slip representing the bottom of this bridge and are therefore covered with a buffer. Although this system has proven itself as a powerful tool to study certain aspects of chemotaxis, in nature cells rarely happen to move through liquid environments and on

Tian Jin and Dale Hereld (eds.), *Chemotaxis: Methods and Protocols*, Methods in Molecular Biology, vol. 1407,
DOI 10.1007/978-1-4939-3480-5_7, © Springer Science+Business Media New York 2016

2D surfaces. More often, they migrate through resistive 3D structures—such as tissues, extracellular matrix, or soil. It has become evident in the last decade that the principles of cell migration, and especially the nature of the forces driving cell locomotion, differ dramatically between these two modes of migration [6–8]. Cells under a liquid buffer normally generate motile force by branched actin polymerization at the cell front, while in resistive environments intracellular hydrostatic pressure, which pushes the membrane forward in the form of blebs, becomes the main motor of cell motility. The set of techniques for studying three-dimensional chemotaxis has been limited so far mainly to transwell assays where the efficiency of 3D chemotaxis is evaluated by the number of cells that traverse the porous membrane separating two wells containing a buffer with different chemoattractant concentrations (Boyden chamber assay [9, 10], Transwell® invasion assays [11]). However, in these assays one cannot precisely observe cellular events accompanying the migration. In the other category of experiments, researchers observe the cells moving in vivo within natural three-dimensional tissues, but in this case they cannot control the chemoattractant gradients faced by the cells, which can also readily move out of the plane of focus.

Here we describe a method allowing the precise observation of *Dictyostelium* cells moving through a resistive environment and employing mainly the bleb-driven mode of migration. In this method the cells are attracted under a stiff agarose overlay by a cyclic-AMP gradient, which is formed across the agarose gel by diffusion from a cyclic-AMP-filled trough. Initially, an under-agarose migration assay was described by the Knecht group as a method to flatten the vegetative *Dictyostelium* cells for a better quality of microscopy [12, 13]. Thus, when applied for the developed *Dictyostelium* cells, this method not only switches them to the blebbing mode of migration but also allows the observation of the accompanying events with a high spatiotemporal resolution [14].

Since F-actin polymerization has been long considered the basic mechanism driving the chemotactic response, many researchers measured F-actin polymerization dynamics after a uniform saturating chemoattractant stimulation of cells [15, 16]. But given that blebbing is the other and similarly important component of the chemotactic response, there is a need to measure the blebbing responses as well. Here we describe a method for evaluating the levels of cellular blebbing induced by chemoattractant stimulation in *Dictyostelium*. By evaluating the number and dynamics of blebs in this "cyclic-AMP shock assay" one can compare the blebbing responses in different conditions and between different strains of *Dictyostelium*, thus examining the environmental and molecular factors that control blebbing [14, 17].

2 Materials

1. Axenic medium: HL5 plus glucose medium (Formedium), 200 μg/ml dihydrostreptomycin. Sterile.
2. KK2 buffer: 16.5 mM KH_2PO_4, 3.9 mM K_2HPO_4, 2 mM $MgSO_4$, 0.1 mM $CaCl_2$, pH 6.1. If sterile, add magnesium and calcium after autoclaving to avoid precipitation.
3. Cyclic-AMP solutions in KK2: 9 μM for cell pulsing, 15 μM for cyclic-AMP shock assay, and 4 μM for under-agarose chemotaxis. A 100 mM cyclic-AMP stock solution can be prepared from free acid by neutralization with KOH.
4. Low-melting-point ultrapure agarose.
5. Two- and eight-well chambered slides (Nunc Lab-Tek, Thermo Scientific), well area—4.0 and 0.7 cm^2, respectively.
6. Inverted microscope with a 40× or 63× objective lens.

3 Methods

3.1 Preparation of Developed Dictyostelium Cells

1. Grow *Dictyostelium* cells in a shaking suspension at 22 °C in axenic medium to a density of $1–5 \times 10^6$ cells/ml (*see* **Note 1**). For this, we recommend using conical flasks and shaking the suspension at 180 rpm. Alternatively, the cells can be grown in association with bacteria on nutrient agar plates (*see* **Note 2**).
2. Transfer the cell suspension into a 50 ml Falcon tube and harvest the cells by centrifugation in a bench-top centrifuge at 1200 rpm ($300 \times g$), 2 min, room temperature.
3. Discard the supernatant and wash the cells three times by resuspending them in KK2 and centrifuging at 1200 rpm ($300 \times g$), 2 min. Every time discard the supernatant and resuspend the cells in fresh KK2.
4. After the last wash, discard the supernatant, resuspend the cells in KK2, and count the cell density using a cell counter or a hemocytometer.
5. Pellet the cells by centrifugation, discard the supernatant, and resuspend the cells in KK2 at 2×10^7 cells/ml.
6. Transfer 10 ml of the cell suspension to a 50 ml conical flask (flask volume of at least three times the volume of the cell suspension is required for proper aeration).
7. Place the flask with the cell suspension on a shaker at 180 rpm and incubate for 1 h at 22 °C (the cell starvation step).
8. After 1-h starvation, start stimulating the cells with pulses of cyclic-AMP (using an automated pump) to drive developmental gene expression: every 6 min inject 100 μl of a 9 μM cyclic-AMP

solution into the flask to produce a final concentration of 60–90 nM after each pulse (*see* **Note 3**). To get a good blebbing response, pulse the cells with cyclic-AMP for 4.5–5.5 h (*see* **Note 4**). After this time, small clumps of cells can be observed sticking to the walls of the flasks, giving a morphological check for adequate development.

9. Pellet the developed cells by centrifugation (to get rid of cyclic-AMP present in the suspension and other secreted metabolites), and resuspend them in fresh KK2.

3.2 Uniform Chemoattractant Stimulation ("Cyclic-AMP Shock" Assay)

1. Dilute the suspension of developed *Dictyostelium* cells to a final density of about 1×10^6 cells/ml (in KK2 buffer).
2. Plate the cells in an eight-well chambered slide by placing 210 μl of the diluted cell suspension per well. This gives a cell density of ~2×10^5 cells per well. Let the cells settle for about 15 min, and move the slides under the microscope.
3. Start time-lapse imaging of the cells in a particular well of the slide (*see* **Note 5**), and then gently add dropwise 15 μl of a 15 μM cyclic-AMP solution to the well to provide a final concentration of 1 μM (if the cells respond poorly to this stimulation, try increasing the final concentration of cyclic-AMP to 2–4 μM). Image the cells for the following 60–100 s. The time of cyclic-AMP addition is apparent as a disturbance in the image, and normally, intensive blebbing follows 10–50 s later (Fig. 1).
4. To characterize and compare the blebbing response in different strains (or under different conditions), analyze the time-lapse movies and extract such quantitative characteristics of the response as the percentage of cells that demonstrate blebbing, average number of blebs per cell, or duration of blebbing.

3.3 Under-Agarose Chemotaxis Assay

1. Preheat a two-well chambered slide to 50–60 °C by placing it on a heated metal thermoblock. This prevents the slide from cracking when the melted agarose is poured in.
2. Carefully heat a 0.5–2 % suspension of low-melting-point ultrapure agarose in KK2 (*see* **Note 6**). During heating, mix it regularly until the agarose melts completely, and the solution becomes transparent and homogeneous. Optionally: add a fluorescent dye into the melted agarose (e.g., 0.5 mg/ml rhodamine B isothiocyanate-dextran) to negatively stain the cells. Use dyes that do not easily diffuse through or bind to the plasma membrane, or get endocytosed by the cell.
3. Let the agarose cool down to ~50–60 °C and then pour a thin layer of melted agarose into each preheated two-well chambered slide. Use 800 μl of agarose per well to get a ~2 mm layer (*see* **Note 7**). Hereinafter keep the slides in a moist chamber (a closed box with a wet tissue in it) to prevent the agarose drying.

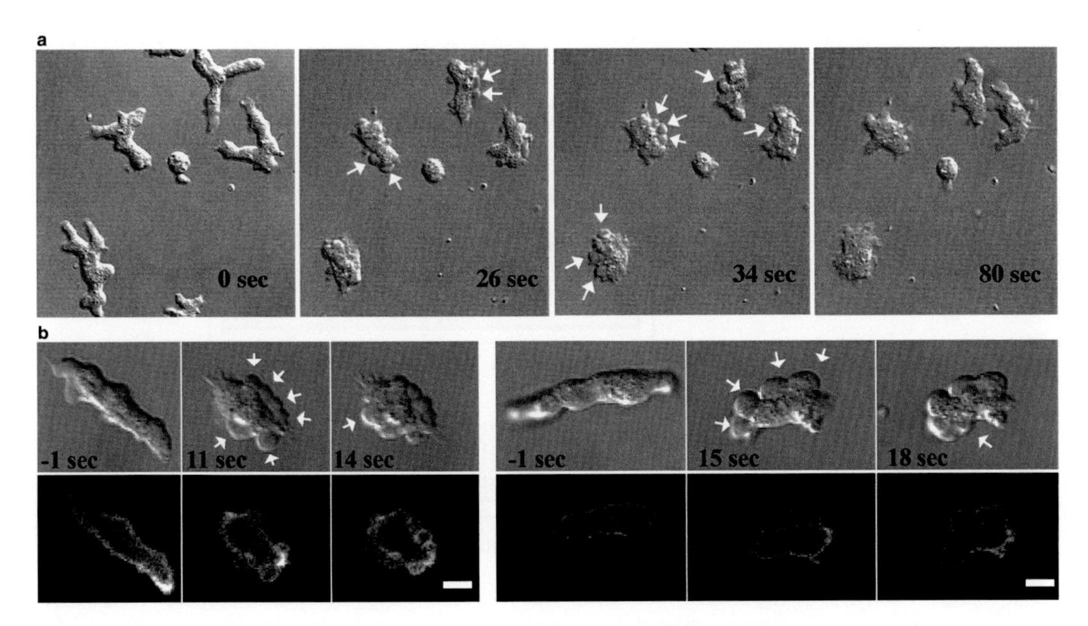

Fig. 1 Blebbing response of *Dictyostelium* cells to a uniform cyclic-AMP stimulation (cyclic-AMP shock). (**a**) DIC images of a group of cells before and after the cyclic-AMP stimulation (cyclic-AMP is added immediately after the "0 sec" time point). (**b**) Two sample cells shown at higher magnification to illustrate the appearance of blebs: *upper panels*—DIC images, *lower panels*—confocal images of the ABD-GFP fluorescent reporter for F-actin [19]. The images were collected using Zeiss 710 laser scanning confocal microscope with a 63× oil emersion objective. For the experiment, Ax2 strain *Dictyostelium discoideum* cells were developed for 6 h and stimulated with 1 μM cyclic-AMP. Scale bars—5 μm

4. When the agarose stiffens, gently cut rectangular troughs in it using a sharp scalpel (*see* **Note 8** for important details). We normally cut three parallel troughs (~1.5×8 mm^2 in size), about 5 mm apart, fill the middle one with chemoattractant, and put two different strains of *Dictyostelium* that we would like to compare in the side ones (Fig. 2). But if only one type of cell is to be analyzed two troughs will suffice.
5. Fill one of the troughs (the central one if you have three troughs, or any one if you have two) with 4 μM cyclic-AMP. Fill the remaining one(s) with the developed *Dictyostelium* cells suspended in KK2 (~10^5 cells per trough).
6. Wait for 30–40 min before starting imaging the cells. During this time the cyclic-AMP gradient will form in the agarose layer, the cells will settle on the slide, and start migrating under the agarose along this gradient.
7. Start time-lapse imaging of the cells migrating under the agarose. For acquisition, use an inverted microscope with a sufficient temporal and spatial resolution to allow bleb detection (*see* **Note 5**). Make sure that the agarose gel does not dry during imaging (otherwise, this will not only interfere with the imaging because of the changes in optical properties, but also,

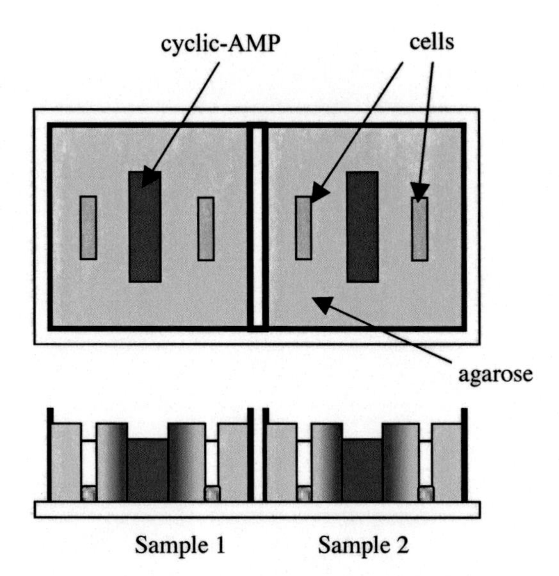

Fig. 2 Scheme of a two-well microscopy chambered slide prepared for an under-agarose chemotaxis experiment: *upper image*—top view, *lower image*—vertical cross section. Agarose is poured into the chambers, and then, after solidification, parallel troughs are cut in the agarose and filled either with 4 μM cyclic-AMP (central troughs) or cell suspension (side troughs)

when drying, the gel may squash the cells underneath and thereby affect their appearance and behavior). If necessary, add a few drops of KK2 into the troughs and onto the surface of the agarose gel to keep it moist. Figure 3 shows an example of a *Dictyostelium* cell migrating under the agarose in the blebbing mode.

8. Analyze the movies of the cells chemotaxing under agarose to measure their overall blebbing activity, localization of the blebs on the cell surface, as well as speed and directionality of migration. When analyzing the high spatiotemporal resolution movies, it is important to use software optimized for accurate cell segmentation and tracking of small and short-lived cellular protrusions (for example, we use electrostatic contour migration-based program QuimP11 [18]).

4 Notes

1. We recommend using at least 2×10^8 cells per experiment; otherwise the cell suspension gets too dilute during the cyclic-AMP pulsing step. Such a number of cells can be obtained from approximately 50 ml of cell suspension (grown to ~ 4×10^6 cells/ml), or from two clearing bacterial plates (*see* **Note 2**).
2. Working stocks of *Dictyostelium* should be renewed from spores or frozen stocks every month to prevent genetic drift. If necessary

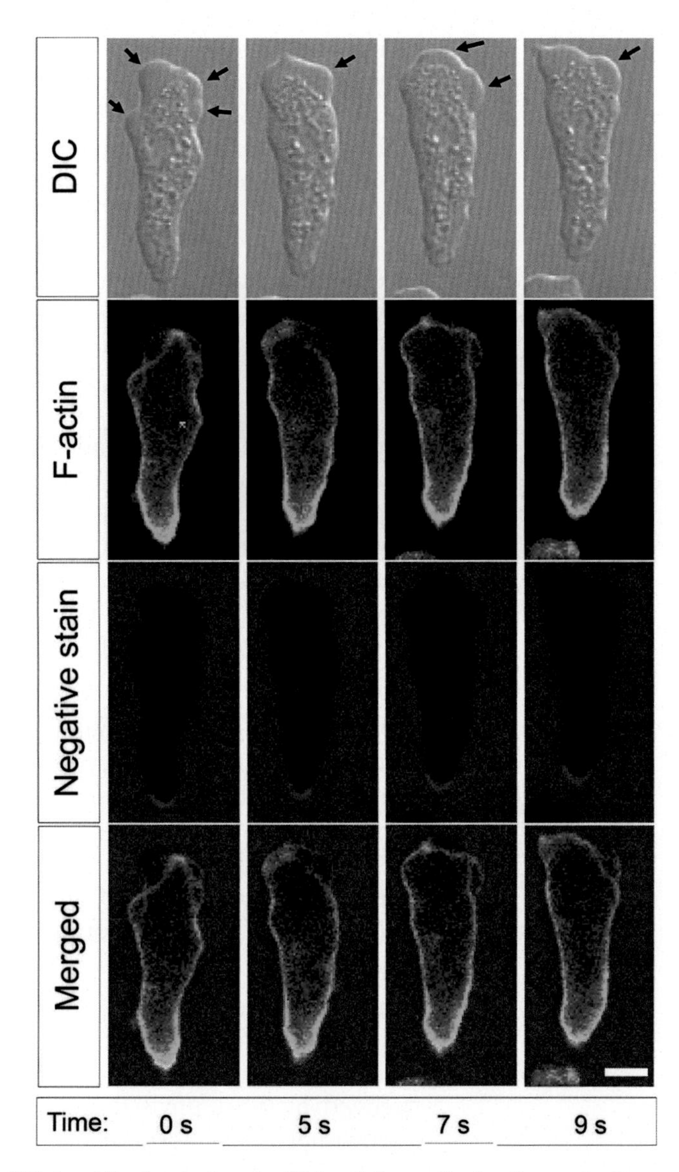

Fig. 3 Blebs at the front edge of a *Dictyostelium* cell migrating under the agarose. ABD-GFP is used as a reporter for F-actin [19], and 0.5 mg/ml rhodamine B isothiocyanate-dextran is added into the agarose as a fluorescent dye for negative staining. The images were collected using Zeiss 780 laser scanning confocal microscope with a 63× oil emersion objective. For the experiment, Ax2 strain *Dictyostelium discoideum* cells were developed for 6 h and chemoattracted under 0.7 % agarose overlay. Scale bar—5 μm

(e.g., when using a strain of *Dictyostelium* that does not grow well in axenic medium), *Dictyostelium* cells can be grown on nutrient agar plates in association with an overnight culture of *Klebsiella aerogenes* bacteria—on so-called clearing plates. To set up a clearing plate, mix *Dictyostelium* cells with a bacterial suspension and plate them out together by spreading on SM agar plates (1.5 % agar with 1 % peptone, 0.1 % yeast extract, 0.22 % KH_2PO_4, 0.1 % Na_2HPO_4, 1 % glucose, 8 mM $MgSO_4$). The bacteria initially outgrow the *Dictyostelium* amoebae and form a lawn but then they exhaust their nutrient and are consumed by the amoebae. When this happens the plate "clears" (goes translucent). Experiments are normally done with cells from plates that have started to look translucent, or have "half-cleared." The initial *Dictyostelium* inoculum is adjusted so that this happens 42–48 h after plating (for example, for the *Dictyostelium* Ax2 strain, 2×10^5 cells are mixed with 200 μl of the overnight bacterial culture and evenly spread on an SM agar plate). The plates should be of a dryness such that all surface liquid dries after no more than about 1 h; excess liquid causes poor growth. Harvest the cells from clearing plates by scraping and resuspend in KK2 buffer. Repeatedly wash them with KK2 by low-speed centrifugation as above.

3. If the cells are known or suspected to be defective in cyclic-AMP relay, use 400 nM cyclic-AMP pulses (final concentration after each pulse), instead of 70–90 nM.
4. During development the cells become increasingly blebby [14]; therefore, slightly longer cyclic-AMP pulsing times are recommended for blebbing experiments, compared to traditional chemotaxis studies. However, with longer pulsing, the cells tend to become stickier and aggregate into large clumps, which makes the observation of individual cells more challenging. We therefore do not recommend pulsing the cells for more than 5.5 h.
5. For time-lapse imaging of blebbing we recommend using an inverted microscope with a 40× or 63× objective lens and a frame rate of 0.5–2 frames per second, as the blebs are characterized by a ~1 μm size and a rapid expansion (the whole growth phase takes less than a second) [14].
6. As the agarose concentration increases from 0.5 to 2 %, the *Dictyostelium* cells switch gradually from the hybrid mode of migration (blebbing + F-actin-driven pseudopodia) to a predominantly bleb-driven motility [14].
7. We normally use a ~2 mm layer of agarose because thicker layers absorb more laser light in the microscopy experiments and also may prevent oxygen diffusion to the cells that move underneath. At the same time, thinner agarose layers are harder to work with, especially, when cutting the wells (in particular, for lower agarose concentrations).
8. Proper cutting of the troughs in the agarose gel is a critical step in this protocol. One must be careful not to scratch the bottom of the slide when making the cuts, as this may abolish cell migration. It is also important not to detach the gel from

the slide's bottom when removing the pieces of agarose from the troughs. Otherwise, the cells will easily migrate through the gap between the agarose and the bottom in the areas of detachment without meeting a sufficient resistance, and therefore without switching to the blebbing mode of migration. We normally use a curved scalpel blade to cut the long sides of the rectangular troughs with a rocking motion, and an elongated triangular one to cut the short sides and remove the pieces of agarose. The scalpel blade must be sharp to make the walls of the troughs clear-cut and avoid agarose detachment during cutting.

Acknowledgements

We thank Richard Tyson and Till Bretschneider for discussions and the Medical Research Council (MRC file reference number U105115237) and a Herchel Smith Fellowship for support.

References

1. Zigmond SH (1977) Ability of polymorphonuclear leukocytes to orient in gradients of chemotactic factors. J Cell Biol 75:606–616
2. Zigmond SH (1988) Orientation chamber in chemotaxis. Methods Enzymol 162:65–72
3. Zicha D, Dunn GA, Brown AF (1991) A new direct-viewing chemotaxis chamber. J Cell Sci 99:769–775
4. Zicha D, Dunn G, Jones G (1997) Analyzing chemotaxis using the Dunn direct-viewing chamber. Methods Mol Biol 75:449–457
5. Muinonen-Martin AJ, Veltman DM, Kalna G, Insall RH (2010) An improved chamber for direct visualisation of chemotaxis. PLoS One 5:e15309
6. Petrie RJ, Gavara N, Chadwick RS, Yamada KM (2012) Nonpolarized signaling reveals two distinct modes of 3D cell migration. J Cell Biol 197:439–455
7. Vu LT, Jain G, Veres BD, Rajagopalan P (2015) Cell migration on planar and three-dimensional matrices: a hydrogel-based perspective. Tissue Eng Part B Rev 21:67–74
8. Charras G, Sahai E (2014) Physical influences of the extracellular environment on cell migration. Nat Rev Mol Cell Biol 15:813–824
9. Boyden S (1962) The chemotactic effect of mixtures of antibody and antigen on polymorphonuclear leucocytes. J Exp Med 115:453–466
10. Falasca M, Raimondi C, Maffucci T (2011) Boyden chamber. Methods Mol Biol 769:87–95
11. Marshall J (2011) Transwell(®) invasion assays. Methods Mol Biol 769:97–110
12. Laevsky G, Knecht DA (2001) Under-agarose folate chemotaxis of *Dictyostelium discoideum* amoebae in permissive and mechanically inhibited conditions. Biotechniques 31:1140–1149
13. Woznica D, Knecht DA (2006) Under-agarose chemotaxis of *Dictyostelium discoideum*. Methods Mol Biol 346:311–325
14. Zatulovskiy E, Tyson R, Bretschneider T, Kay RR (2014) Bleb-driven chemotaxis of *Dictyostelium* cells. J Cell Biol 204:1027–1044
15. Hall AL, Schlein A, Condeelis J (1988) Relationship of pseudopod extension to chemotactic hormone-induced actin polymerization in amoeboid cells. J Cell Biochem 37:285–299
16. Langridge PD, Kay RR (2007) Mutants in the *Dictyostelium* Arp2/3 complex and chemoattractant-induced actin polymerization. Exp Cell Res 313:2563–2574
17. Langridge PD, Kay RR (2006) Blebbing of *Dictyostelium* cells in response to chemoattractant. Exp Cell Res 312:2009–2017
18. Tyson RA, Zatulovskiy E, Kay RR, Bretschneider T (2014) How blebs and pseudopods cooperate during chemotaxis. Proc Natl Acad Sci U S A 111:11703–11708
19. Pang KM, Lee E, Knecht DA (1998) Use of a fusion protein between GFP and an actin-binding domain to visualize transient filamentous-actin structures. Curr Biol 8:405–408

Chapter 8

Dissecting Spatial and Temporal Sensing in *Dictyostelium* Chemotaxis Using a Wave Gradient Generator

Akihiko Nakajima and Satoshi Sawai

Abstract

External cues that dictate the direction of cell migration are likely dynamic during many biological processes such as embryonic development and wound healing. Until recently, how cells integrate spatial and temporal information to determine the direction of migration has remained elusive. In *Dictyostelium discoideum*, the chemoattractant cAMP that directs cell aggregation propagates as periodic waves. In light of the fact that any temporally evolving complex signals, in principle, can be expressed as a sum of sinusoidal functions with various frequencies, the *Dictyostelium* system serves as a minimal example, where the dynamic signal is in the simplest form of near sinusoidal wave with one dominant frequency. Here, we describe a method to emulate the traveling waves in a fluidics device. The text provides step-by-step instructions on the device setup and describes ways to analyze the acquired data. These include quantification of membrane translocation of fluorescently labeled proteins in individual *Dictyostelium* cells and estimation of exogenous cAMP profiles. The described approach has already helped decipher spatial and temporal aspects of chemotactic sensing in *Dictyostelium*. More specifically, it allowed one to discriminate the temporal and the spatial sensing aspects of directional sensing. With some modifications, one should be able to implement similar analysis in other cell types.

Key words Chemotaxis, *Dictyostelium*, cAMP waves, MFCS, Gradient sensing, Temporal-sensing

1 Introduction

Chemotaxis of crawling cells such as *Dictyostelium* and neutrophils is thought to be dictated largely by extension of plasma membrane towards the direction facing higher concentrations of chemoattractant molecules. In the so-called spatial sensing scheme, cells sense direction by comparing the attractant concentrations across the cell body [1]. Additionally, there appears to be temporal aspects in the cellular sensing. Most notably, many of the responses that characterize the leading edge such as small GTPase Ras /PI3Kinase activity and the formation of F-actin are transiently elevated uniformly at the plasma membrane upon application of spatially uniform persistent stimuli. The existence of such transient response and spatial

Tian Jin and Dale Hereld (eds.), *Chemotaxis: Methods and Protocols*, Methods in Molecular Biology, vol. 1407,
DOI 10.1007/978-1-4939-3480-5_8, © Springer Science+Business Media New York 2016

sensing of stationary gradient can be consistently understood by the adaptive sensing scheme mediated by fast activation that is spatially localized coupled with slow inhibitory reaction that takes place uniformly in space and thus acts globally. However, how the transient of the leading edge response is utilized in chemotaxis has received minimal attention, partly due to technical limitations.

Until recently, a majority of chemotaxis assays were performed using chemoattractant gradients that are either not well defined in time or semi-stationary at best. Varying the gradients in time using a pressurized point source requires manual maneuvering of the micropipette [2–4]. Zigmond chamber and the likes combined with change in the source concentration [5, 6], enzymatic degradation [7], or displacement of the attractant reservoirs [8] can also create dynamic gradients. However, these approaches are all subjected to large trial-to-trial variations and require trained hands.

The recent development in micro- and milli-fluidics is now enabling researchers to generate dynamically changing chemoattractant gradients and to study how their spatial and temporal components are being read and interpreted by the cells to determine the direction of movement. Continuously applied flow of attractant in combination with Percoll density gradients [9] supports mechanically stable concentration gradients that can be monotonically increased or decreased in time; however gradient reversal is difficult by design. Two-layered pyramidal mixer [10] allows rapid reversal of gradient orientation; however it may not be suitable for experiments where transient stimulus during the switch should be avoided. On the other hand, the so-called flow-focusing approach where laminar flows from three independent inlets are combined and focused generates gradients that could be displaced continuously in time [11]. The system has allowed detailed analysis of directional sensing response under dynamically increasing/decreasing gradients as well as bell-shaped wave stimuli. The technique has uncovered how temporal sensing is integrated together with spatial sensing to realize directional movement against the propagating waves of chemoattractant cAMP in *Dictyostelium* [12]. In this chapter, we describe a step-by-step instruction on the preparation of the experimental setup, data acquisition, and analysis. Due to the simplicity of the experimental design, similar analysis should be applicable to other cell types including immune cells.

2 Materials

2.1 General Media

1. *Culture medium*: PS medium (modified HL5 medium). Solution A: 10 g Special peptone (LP0072, OXOID), 7 g yeast extract (LP0021, OXOID), 0.12 g $Na_2HPO_4{\cdot}7H_2O$ (191441, MP Biomedicals), 1.4 g KH_2PO_4 (169-04245, Wako). Dissolve in milliQ water to 900 mL final volume and autoclave.

Solution B: 10 mg cyanocobalamin (vitamin B12) (224-00344, Wako), 20 mg folic acid (060-01802, Wako) in 10 mL of milliQ water. Store solution B at −30 °C. Solution C: 15 g Difco™ Dextrose (215530, BD) dissolved in milliQ water to 90 mL final volume, 10 mL of antibiotic-antimycotic (100×) (15240-062, Gibco), 40 μL of solution B. Filter sterilize. Mix solutions A and C to obtain 1 L medium.

2. *Phosphate buffer* (PB): 20 mM Na/K-phosphate buffer. Prepare 10× (0.2 M pH 6.5) solution by mixing approximately 1:2 of 0.2 M Na_2HPO_4 solution and 0.2 M KH_2PO_4 solution to obtain 10× pH 6.5 buffer. Check pH using a pH meter. Autoclave and store at room temperature. Use at 1× concentration.
3. *Fluorescein solution*: Prepare 1 mM fluorescein (213-00092, Wako) in PB. Store at 4 °C.
4. *Chemoattractant*: Dissolve 500 mg cAMP (sodium salt) (A6885-500MG, Sigma) in 13.54 mL water to obtain 100 mM stock solution. Store at −30 °C. Working stock solution is 100 μM in PB.

2.2 Equipment

1. Fluidics chamber (μ-slide 3-in-1 ibiTreat ib80316, Ibidi).
2. Air compressor (2522C-05, Welch).
3. Pneumatic pressure regulator (MFCS™-FLEX, Fluigent).
4. Flow controller (Flowell, Fluigent; 0-69 mBar or 0-345 mBar pressure range recommended): Four channels of which three are equipped with flow meters.
5. Laptop PC with MAESFLO (Fluigent) software: This is used together with MFCS to monitor and control pressures applied to the four input reservoirs. Rate of flow from up to three reservoirs can be monitored in real time.
6. A terminal emulator and a UNIX (-like) shell environment in the same computer controlling the MFCS or in another computer (e.g., Terminal.app on a Macintosh OSX or Cygwin on a PC).

2.3 Tubing and Connectors

1. Polyetheretherketone (PEEK) tubing (OD 1/32″ ID 0.25 mm; NPK-007, ARAM), 20 cm length × 3 pieces: Cut out three pieces 20 cm in length using a tubing cutter (JC-2, GL Sciences).
2. Tubing sleeve (OD 1/32″ to OD 1/16″; F-247X, Upchurch Scientific).
3. 1/16″ Fitting nut (Flowell accessory).
4. 1/32″ Fitting nut (M-645X, Upchurch Scientific; Flowell accessory).
5. PEEK tubing (OD 0.5 mm ID 0.3 mm; ICT-30P, Institute of Microchemical Technology), 40 cm length × 3, 60 cm length × 1: Cut using a tubing cutter (ICT-CT).

6. PEEK tubing (OD 1/16″ ID 0.01″; #1531, Upchurch Scientific) 35 mm length × 2 pieces: Cut using a tubing cutter (JC-2).
7. Polyethylene (PE) tubing (OD 2.33 mm No.7, HIBIKI), 15 mm length × 2 pieces: Cut using a tubing cutter (JC-2).
8. Tubing sleeve (OD 0.5 mm to OD 1/32″; ICT-32S, Institute of Microchemical Technology).
9. Luer fitting, male luer for ID 1.5 mm tubings (VRS106, ISIS).
10. Union 1/32″ OD to 1/16″ OD (P-881, Upchurch Scientific).
11. Tubing sleeve (OD 0.5 mm to OD 1/16″; ICT-16S, Institute of Microchemical Technology).
12. Union (1/16″ OD to 1/16″ OD; P-702, Upchurch Scientific).
13. Luer adaptor (P-605, Upchurch Scientific).

3 Methods

3.1 Cell Culture

1. Grow *Dictyostelium discoideum* AX4 cells expressing RFP-Raf1RBD [12] in PS medium containing an appropriate selection drug (10 μg/mL G418) at 22 °C. Typically, 10 mL cell culture is shaken in a 50 mL conical tube placed horizontally with a slight angle on a rotary shaker. It is important to propagate the cells below 3×10^6 cells/mL.
2. Collect the growing cells (2.5×10^7 cells total) by centrifugation at $700 \times g$ and resuspend in PB. Wash by repeating this twice and resuspend the washed cells in 5 mL PB at a density of 5×10^6 cells/mL. Incubate by shaking at 22 °C.
3. After 1 h, while shaking, pulse the cells with 50 nM cAMP (final concentration) every 6 min for 4–4.5 h. Typically, a peristaltic pump (Masterflex L/S, Cole-Parmer) is used to dispense 50 μL of 5 μM cAMP solution in 5 mL cell suspension.

3.2 Flowell Setup

1. Put together three sets of tubing assemblies each in the following order [from left to right in Fig. 1a (left panel)]:
   (a) Reservoir tube (Flowell accessory)
   (b) 1/16″ Fitting nut (Flowell accessory)
   (c) Tubing sleeve (OD 1/32″ to 1/16″)
   (d) PEEK tubing (OD 1/32″ ID 0.25 mm. 20 cm length)
   (e) 1/32″ Fitting nut (Flowell accessory)
2. Screw in the reservoir tubes to the reservoir ports 1, 2, and 3 (Fig. 1a, right panel).
3. Connect the 1/16″ fitting nut of the assembly to the reservoir port, and the 1/32″ fitting nut to the flowmeter port [(A, B, C; indicated by green arrows on the device; Fig. 1a (right panel)].

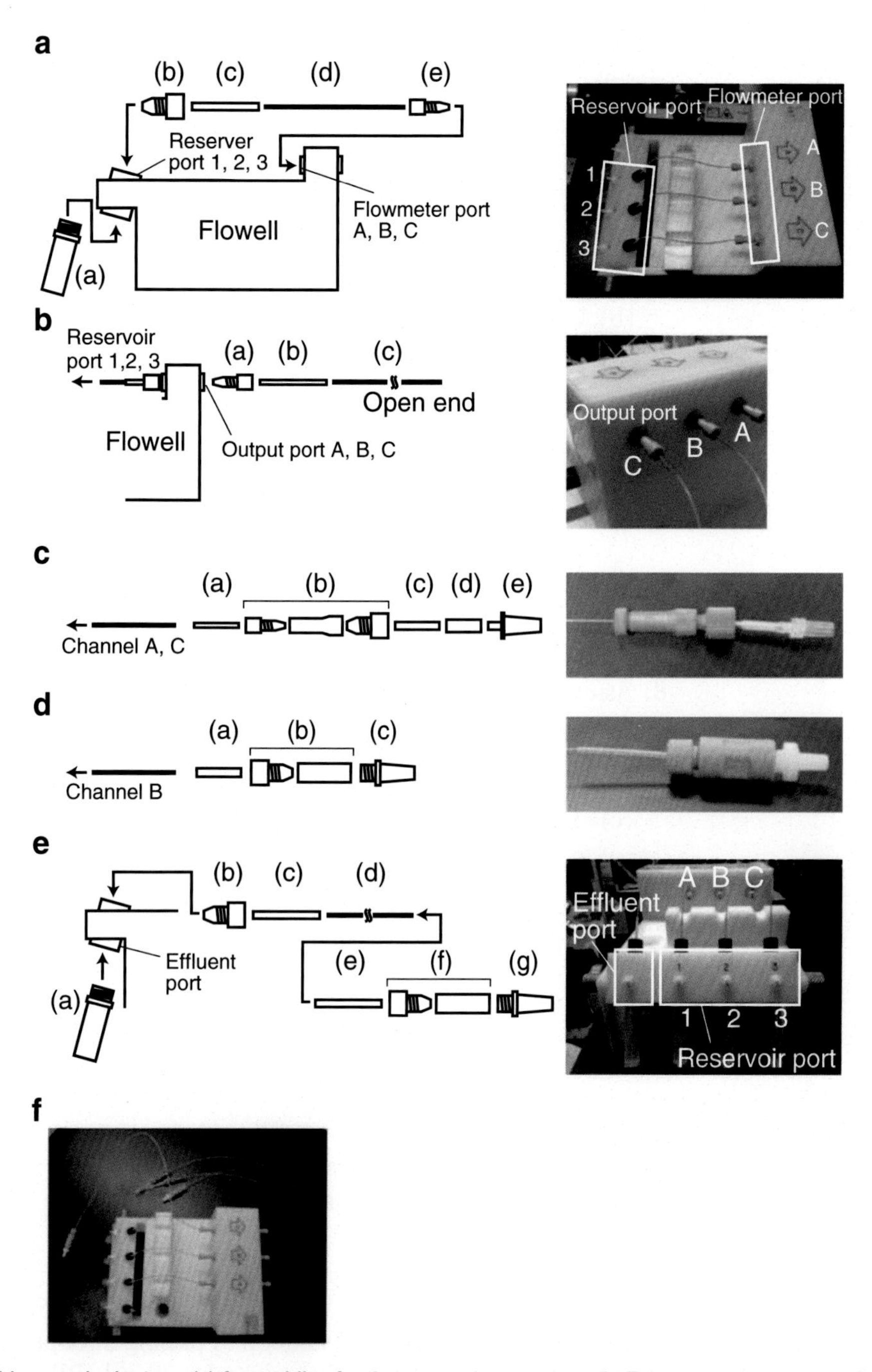

Fig. 1 Tubings and adaptors. (**a**) Assemblies for the reservoir ports 1–3. (**b**) Tubings and connectors for Flowell output ports. (**c**–**e**) Connectors for the chamber-side inlets A′ and C′ (**c**). Center inlet B′ (**d**), and the output (**e**). (**f**) The overview of the Flowell device and the tubings

4. Assemble three sets of the tubing parts in the following order (Fig. 1b left panel):
   (a) 1/32″ Fitting nut

  (b) Tubing sleeve (OD 0.5 mm to OD 1/32″)
  (c) OD 0.5 mm ID 0.3 mm PEEK tubing (40 cm length)
5. Connect the fitting nut to the Flowell output ports (A, B, and C; Fig. 1b, right panel).
6. To each open end of the PEEK tubing on channels A and C, attach connectors and tubing in the following order (Fig. 1c):
  (a) Tubing sleeve (OD 0.5 mm to OD 1/32″)
  (b) Union (OD 1/32″ to OD 1/16″)
  (c) PEEK tubing (OD 1/16″ ID 0.01″ , 35 mm length)
  (d) PE tubing (OD 2.33 mm, 15 mm length)
  (e) Luer fitting (male for ID 1.5 mm tubing)
7. To the open end of the PEEK tubing on channel B, attach connectors and tubing in the following order (Fig. 1d):
  (a) Tubing sleeve (OD 0.5 mm to OD 1/16″)
  (b) Union (1/16″ OD to 1/16″ OD)
  (c) Luer adaptor (P-605)
8. Assemble in the following order (Fig. 1e, d (right panel)):
  (a) Reservoir (attached to the Flowell effluent port)
  (b) 1/16″ Fitting nut (attached to the Flowell effluent port)
  (c) Tubing sleeve (OD 0.5 mm to OD 1/16″)
  (d) PEEK tubing (OD 0.5 mm 60 cm length)
  (e) Tubing sleeve (OD 0.5 mm to OD 1/16″)
  (f) Union (1/16″ OD to 1/16″ OD)
  (g) Luer adaptor (P-605)
9. Check that all tubings are assembled and connected correctly to Flowell (Fig. 1f).

3.3 MFCS Setup

1. Connect both the MFCS-FLEX and the Flowell to a laptop computer with USB cables. Turn on the laptop PC, the MFCS, and the air compressor. Run the MAESFLO software (ver. 2.1.1, Fluigent).
2. While MFCS is preheating (~10 min), prepare 10 mL of 1 μM cAMP 3 μM fluorescein solution in PB. Remove dusts and small particles in the cAMP solution by passing through a 0.22 μm pore size syringe filter. Similarly, take 20 mL PB and pass it through a syringe filter.
3. Fill reservoir tubes #1 and #3 (connected to channels A and C, respectively) with 1.5 mL of filtered PB. Fill reservoir tube #2 (connected to channel B) with 1.5 mL of filtered cAMP/fluorescein solution. Fill the leftmost reservoir for effluent (unmarked; not connected to the flowmeter) with 0.5 mL of buffer.

4. Place the chamber (μ-slide 3-in-1, ibidi) on the microscope stage holder. Make sure that the chamber is held tightly in place.
5. Rinse the chamber by pipetting up and down 200 μL PB two to three times from the inlet.

3.4 Live-Cell Imaging

1. Collect the starved cells by centrifugation at 700 ×*g*. Resuspend the cells in PB at 2×10^5 cells/mL.
2. Introduce 100 μL of the suspended cells into the chamber by manual pipetting using a yellow pipette tip. Let the cells settle and attach to the surface for 30 min.
3. Connect between the Flowell channels A, B, and C and the chamber inlets (A′, B′, and C′) with the tubing assembly (from Subheading 3.2, **steps 6** and 7) (Fig. 2a). Similarly, connect the chamber outlet and the tubing assembly (from Subheading 3.2, **step 8**) for the effluent.

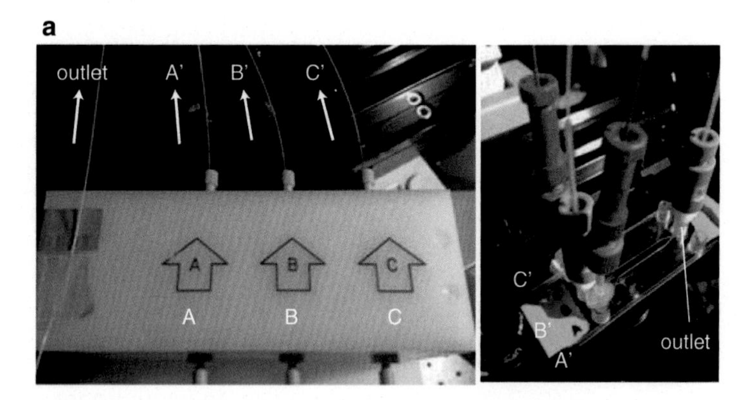

Fig. 2 Connections between MFCS™-FLEX pressure regulator and the fluidics chamber. (**a**) Tubings from Flowell (*left panel*) are connected to the inlets and the outlet of the chamber (*right panel*). (**b**) The pressure regulator and the Flowell ports (*red squares*) are connected using soft tubings (*red squares*)

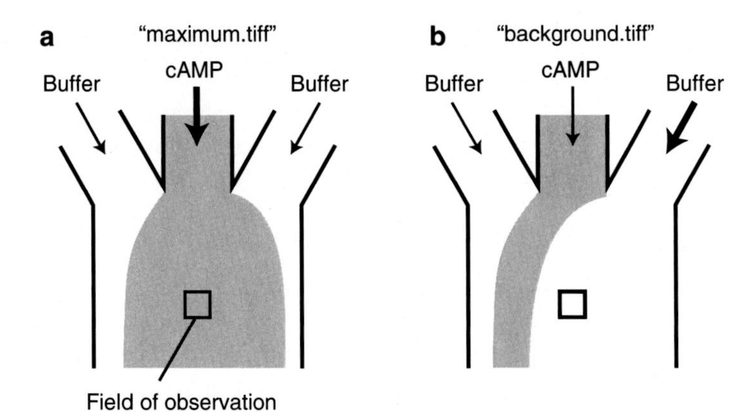

Fig. 3 Flow profiles used to obtain calibration images. (**a**, **b**) "maximum.tiff" (**a**) and "background.tiff" (**b**)

4. Connect the pressure regulator (MFCS™-FLEX; Fluigent) and the Flowell reservoir ports with soft tubings (included in the MFCS system) (Fig. 2b).
5. By moving a slider icon in the MAESFLO software, initiate the flow by applying pressure to the buffer reservoirs (channels A and C) and the cAMP/fluorescein reservoir (channel B). Start from 10 mbar and increase slowly to about 30 mbar.
6. For each channel, adjust the pressure to obtain the target flow rate of 10 μL/min.
7. Allow approximately 5 min for the flow rate to stabilize.
8. While waiting on **step 7**, move the microscope stage so that the field of view is located 2–4 mm downstream from the junction of the three inlets and equidistant from the chamber side walls.
9. Acquire calibration images as follows:
   (a) Adjust the reservoir pressure so that the flow rate for left (channel A; buffer), center (B; cAMP/fluorescein), and right (C; buffer) channels are 4 μL/min, 26 μL/min, and 3 μL/min, respectively (*see* **Note 1**) (Fig. 3a). Record the adjusted pressure in your notebook. The pressure should typically be between 15 and 50 mbar. Wait for 1 min for the laminar flow to stabilize. Acquire fluorescence images at 3-s intervals for 90 s (*see* **Note 2**). The images will serve as the reference for the "maximum" intensity. Save the images as a single-stacked TIFF file "maximum.tiff."
   (b) Adjust the pressure of each channel to obtain the flow rate of 4 μL/min, 3 μL/min, and 26 μL/min for left, center, and right channels, respectively (Fig. 3b). Record the pressure. Wait for 1 min for the flow to become stable. Capture fluorescence image at 3-s intervals for 90 s duration. The images will be used later to correct for nonuniformity

of the background. These images are saved as stacked TIFF file "background.tiff."

10. Adjust the pressure of each channel to obtain the flow rate of 26 μL/min, 3 μL/min, and 4 μL/min for left, center, and right channels, respectively. Record the pressure in a notebook.
11. Using a text editor (e.g., TextEdit, Emacs; *see* **Note 3**), open a script file ("travelingwave.sh"), and assign the values for parameters "initial pressure" and "end pressure" according to the recorded pressure in **steps 10** and **9b**, respectively. Save the file.
12. Using a command-line interface (e.g., Terminal.app in Mac, Cygwin in PC), run the scriptfile by executing

    ./travelingwave.sh > wavestimulus.txt

    This generates another script file "wavestimulus.txt" that can be read in the MAESFLO software.
13. If "wavestimulus.txt" is generated on a separate PC, copy the file to the PC connected to MFCS. A USB link cable (KB-USB-LINK3M, Sanwa) is convenient for data transfer. A USB flash drive can also be used.
14. Select the "Script Editor" tab in MAESFLO software, and select "OPEN" to load the script file (Fig. 4a).
15. Wait for 5 min before starting image acquisition. Run the script file by selecting "PLAY Script" button at the right bottom of the MAESFLO window (Fig. 4, panels b–d).
16. Acquire fluorescence images for RFP-Raf1RBD and fluorescein sequentially at 12-s intervals for 15 min (Fig. 5). In this example, the respective images are saved as stacked TIFF files "RBD.tiff" and "fluorescein.tiff."

3.5 Data Analysis to Estimate the Spatiotemporal Profiles of Exogenous cAMP Concentration

1. Apply flat-field correction as follows (*see* **Note 4**). Using ImageJ, open stack image files "maximum.tiff" and "background.tiff." From the menu bar, select "Image" > "Stacks" > "Z Project…." Select "Average Intensity" for "projection type" and press "OK." Save the averaged images in preferred file names. These data will be referred to as $I_{max}(x, y)$ and $I_{back}(x, y)$, respectively (Fig. 6a).
2. Using ImageJ, open the fluorescein image file "fluorescein.tiff" ($I_{fl}(x, y, t)$). From the ImageJ menu bar, select "Process" > "Image Calculator… ." In the "Image Calculator" window that appears, select image "fluorescein.tiff" $I_{fl}(x, y, t)$ and $I_{back}(x, y)$ from the drop-down menus "Image1" and "Image2," respectively. Choose "Subtract" for "Operation" and click the "OK" button to obtain subtracted data $I_{f2}(x, y, t)$.
3. Similarly, obtain $I_{max2}(x, y, t)$ by subtracting image $I_{back}(x, y)$ from image $I_{max}(x, y)$.

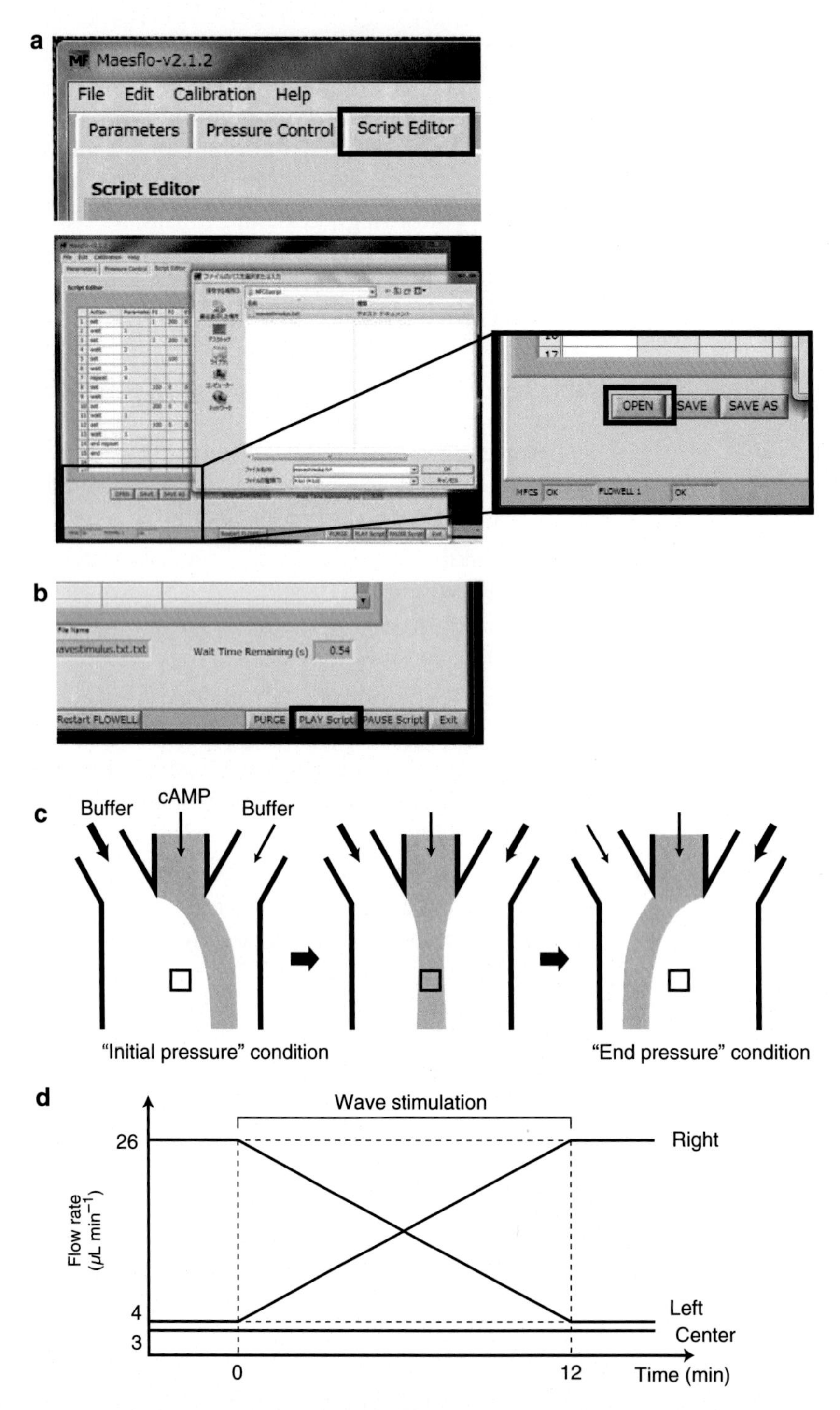

Fig. 4 Automated pressure control using the MAESFLO software. (**a**) A script file for wave stimulation is imported using the "Script Editor" tab. (**b**) The script is then executed by pressing "PLAY script" button. (**c**) Flow profiles during wave stimulation. (**d**) The script automates linear ramping of the rate of flow from the left channel (inlet A′; 26 to 4 μL/min) and the right channel (inlet C′; 4–26 μL/min) while keeping the flow rate from the center channel fixed (inlet B′; 3 μL/min). As a result, the position of the bell-shaped gradient is continuously displaced at a constant speed from right to left

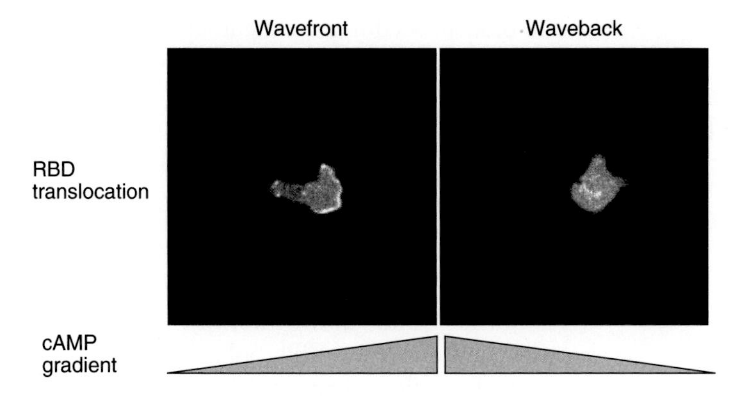

Fig. 5 Membrane translocation of fluorescently labeled Raf1-RBD in a *Dictyostelium* cell under wave stimulation

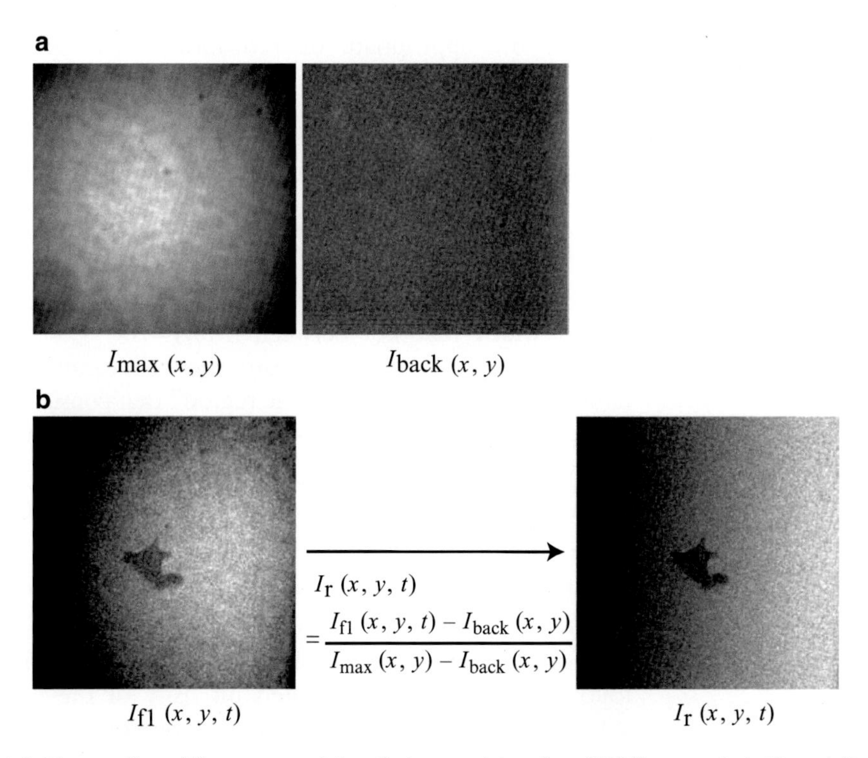

Fig. 6 Flat-field correction of fluorescence intensity images taken for cAMP/fluorescein buffer solution. (**a**) The averaged image taken for the maximum cAMP concentration (in this example 1 μM) ($I_{max}(x, y)$; *left panel*) and the background ($I_{back}(x, y)$; *right panel*). (**b**) Image intensity from the fluorescein channel ($I_{f1}(x, y, t)$) is normalized by $I_{max}(x, y)$ and $I_{back}(x, y)$

4. Select "Process" > "Image Calculator… ." In the ImageJ pop-up window, select $I_{f2}(x, y, t)$ for "Image1," and $I_{max2}(x, y)$ for "Image2.". Choose "Divide" for "Operation," and make sure that "32-bit (float) result" check box is checked. Click "OK" to execute. The generated image, which we shall refer to as $I_r(x, y, t)$ (Fig. 6b), provides a background-corrected estimate of the concentration profile of exogenous cAMP (in μM) (*see* **Note 5**).

3.6 Data Analysis for Quantification of Directional Sensing Response during Wave Chemotaxis

1. The basic procedures for the image processing are described in the following steps. Details of the implementation would depend on the platform of choice (e.g., ImageJ, MATLAB, or Metamorph). The calculations should be relatively straightforward for anyone with some experience in image processing.
2. Apply Gaussian filter (standard deviation $\sigma = 2$ pixels) on "RBD.tiff" with a kernel size of 3×3.
3. Binarize the image by Otsu's method.
4. Fill in the "holes" in the interior of the cell region by flood-fill operation (e.g., Flood-fill tool in ImageJ). The obtained region is hereafter referred to as "cell mask" (Fig. 7a).
5. Define the direction θ_i from the center ($0 < \theta_i < 2\pi$) to pixels that constitute the outline $R(\theta_i)$ (μm) of the cell (Fig. 7b). Choose the appropriate discretization steps for the angle $\theta_i = \frac{\pi}{N} i$ for $1 \leq i \leq 2N$. In the example shown N=90 is chosen. For the sake of analysis, it is convenient to define the coordinate so that the cAMP wave propagates from the direction of $\theta = \frac{\pi}{2}$ and propagates away from the cell in the direction of $\theta = \frac{3\pi}{2}$.
6. Define a region of interest (ROI) "membrane region" ΔR_{mem} μm in width that lies between $R(\theta_i) = \Delta R_{\mathrm{margin}} = \Delta R_{\mathrm{mem}}$ and $R(\theta_i) = \Delta R_{\mathrm{margin}}$ in distance from the cell center (Fig. 7b). Similarly define a ROI "cytosolic region" that consists of all pixels within the distance $R(\theta_i) = \Delta R_{\mathrm{margin}} = \Delta R_{\mathrm{mem}}$ from the center. In the example shown, $\Delta R_{\mathrm{margin}} = 0.3 \mu\mathrm{m}$ and $\Delta R_{\mathrm{mem}} = 1.0 \mu\mathrm{m}$.
7. Extract the maximum fluorescence intensity in the ROI "membrane region" as a function of the angle θ_i; $I_{\mathrm{mem}}(\theta_i, t)$.
8. Calculate the mean intensity of the ROI "cytosolic region": $I_{\mathrm{cyt}}(\theta_i, t)$.
9. Quantify membrane translocation of RFP-Raf1RBD as follows. Compute the averaged intensity of the membrane-cytosolic ratio weighted by the wave direction,
$$J^{+}_{\mathrm{mem}}(t) = \left(\frac{1}{N} \sum_{\{i | 0 < \theta_i < \pi\}} \frac{I_{\mathrm{mem}}(\theta_i, t)}{I_{\mathrm{cyt}}(t)} |\sin\theta_i| \right) - \frac{2}{\pi},$$
for the positive half of a cell periphery ($0 < \theta_i < \pi$), and
$$J^{-}_{\mathrm{mem}}(t) = \left(\frac{1}{N} \sum_{\{i | \pi < \theta_i < 2\pi\}} \frac{I_{\mathrm{mem}}(\theta_i, t)}{I_{cyt}(t)} |\sin\theta_i| \right) - \frac{2}{\pi},$$
for the remaining half ($\pi < \theta_i < 2\pi$) (Fig. 7c).
10. Plot fluorescence intensity of fluorescein and RBD translocation vs. time (Fig. 7d).

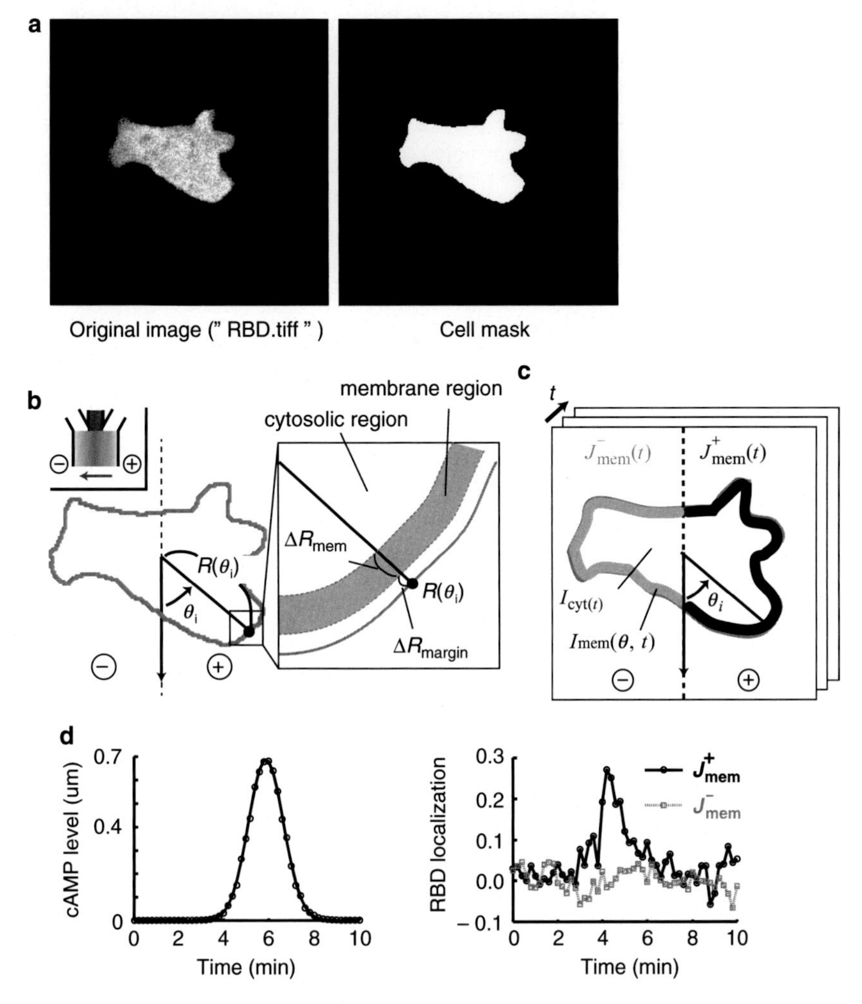

Fig. 7 Quantification of membrane-bound RFP-Raf1RBD. (**a**) Captured fluorescent images ("original image"; *left panel*) are binarized to define the cell region ("cell mask"; *right panel*). (**b**) The "membrane region" is specified and extracted from the "cell mask." (**c**) A schematic illustration of RFP-Raf1RBD quantification. (**d**) Example time plots of estimated exogenous cAMP concentrations (*left panel*) and membrane translocation of RFP-Raf1RBD (*right panel*)

4 Notes

1. When obtaining the image for the maximum cAMP/fluorescein concentration, the flow rate of the center channel (=26 μL/min) is set much higher than left (=4 μL/min) and right (=3 μL/min) channels. The small difference in the flow rate between left and right side channels is not essential for this purpose. Rather, choice of this particular set of flow rates

(3, 4, and 26 μL/min) minimizes the effort to find the appropriate pressure in the subsequent wave stimulation (the initial and end flow rate for the wave stimulus is 26 μL/min, 3 μL/min, and 4 μL/min, and 4 μL/min, 3 μL/min, and 26 μL/min, respectively).

2. Note that, under this condition, fluorescein concentration is expected to be spatially uniform throughout the field of captured image so that the intensity profile only reflects systematic bias from nonuniformity of illumination and to a lesser extent data capture.

3. Copy the lines below and save as an executable shell script file named "travelingwave.sh." Alternatively, this file and its updates are available at http://sawailab.c.u-tokyo.ac.jp/protocols.

```
#!/bin/sh
###############################################
### Generating MFCS script file
### written by Akihiko NAKAJIMA, 20141230
###############################################
Period=720        # Stimulus changing time (sec)
Dt=2          # Update interval (sec)
Nloop=`expr "  Period" / "  Dt"` # Loop number
Nrepeat=1         # Repeat number
Tw_init=1         # Waiting time before stimulation(sec)
Tw_end=600         # Waiting time after stimulation(sec)
#initial pressure (mbar)
Pr1_init=21.0
Pr2_init=19.0
Pr3_init=39.0
Pr4_init=0.0
#end pressure (mbar)
Pr1_end=40.0
Pr2_end=19.0
Pr3_end=22.0
Pr4_end=0.0
#final pressure
Pr1_final=0.0
Pr2_final=0.0
Pr3_final=0.0
Pr4_final=0.0
#pressure increase/decrease per step
D_Pr1_0=`echo "  Pr1_end"-"  Pr1_init" |bc -l`
D_Pr1=`echo "  D_Pr1_0"/"  Nloop" |bc -l`
D_Pr3_0=`echo "  Pr3_end"-"  Pr3_init" |bc -l`
D_Pr3=`echo "  D_Pr3_0"/"  Nloop" |bc -l`
#!/bin/sh
###############################################
### Generating MFCS script file
```

```
### written by Akihiko NAKAJIMA, 20141230
##############################################
Period=720          # Stimulus changing time (sec)
Dt=2           # Update interval (sec)
Nloop=`expr "   Period" / "   Dt"` # Loop number
Nrepeat=1          # Repeat number
Tw_init=1          # Waiting time before stimulation(sec)
Tw_end=600          # Waiting time after stimulation(sec)
#initial pressure (mbar)
Pr1_init=21.0
Pr2_init=19.0
Pr3_init=39.0
Pr4_init=0.0
#end pressure (mbar)
Pr1_end=40.0
Pr2_end=19.0
Pr3_end=22.0
Pr4_end=0.0
#final pressure
Pr1_final=0.0
Pr2_final=0.0
Pr3_final=0.0
Pr4_final=0.0
#pressure increase/decrease per step
D_Pr1_0=`echo "   Pr1_end"-"   Pr1_init" |bc -l`
D_Pr1=`echo "   D_Pr1_0"/"   Nloop" |bc -l`
D_Pr3_0=`echo "   Pr3_end"-"   Pr3_init" |bc -l`
D_Pr3=`echo "   D_Pr3_0"/"   Nloop" |bc -l`
```

4. Obtain normalized background-corrected image of fluorescence intensity profiles by calculating $I_r(x,y,t)=\frac{I_{f1}(x,y,t)-I_{back}(x,y)}{I_{max}(x,y)-I_{back}(x,y)}$ where $I_{fl}(x, y, t)$ is the fluorescence intensity profiles of the fluorescein ("fluorescein.tiff").
5. The concentration profile of exogenous cAMP is estimated by calculating $C_{cAMP}(x,y,t)=(C_{max}-C_{min})I_r(x,y,t)+C_{min}$. C_{max} and C_{min} are the maximum and minimum concentrations of cAMP at the source, respectively. In this example, $C_{max}=1\mu M$ and $C_{min}=0\mu M$.

Acknowledgements

This work was supported by grants from the Japan Society for the Promotion of Science (JSPS) Grant-in-Aid for Scientific Research on Innovative Areas (23111506, 25111704) (to S.S.), JSPS Grant-in-Aid for Young Scientists (A) (22680024, 25710022) (to S.S.),

JSPS Grant-in-Aid for Young Scientists (Start-up) and (B) (23870006, 25840069) (to A.N.), Japan Science and Technology Agency (JST) Precursory Research for Embryonic Science and Technology (PRESTO) (to S.S.), Platform for Dynamic Approaches to Living System from the Ministry of Education, Culture, Sports, Science and Technology, Japan, and in part by the Human Frontier Science Programme (RGY 70/2008) and JSPS Grant-in-Aid for Scientific Research on Innovative Areas (25103008) (to S.S.).

References

1. Lauffenburger D, Farrell B, Tranquillo R et al (1987) Gradient perception by neutrophil leucocytes, continued. J Cell Sci 88:415–416
2. Futrelle RP (1982) *Dictyostelium* chemotactic response to spatial and temporal gradients. Theories of the limits of chemotactic sensitivity and of pseudochemotaxis. J Cell Biochem 18: 197–212
3. Xu X, Meier-Schellersheim M, Yan J, Jin T (2007) Locally controlled inhibitory mechanisms are involved in eukaryotic GPCR-mediated chemosensing. J Cell Biol 178:141–153
4. Zhang S, Charest PG, Firtel RA (2008) Spatiotemporal regulation of Ras activity provides directional sensing. Curr Biol 18: 1587–1593
5. Fisher PR, Merkl R, Gerisch G (1989) Quantitative analysis of cell motility and chemotaxis in *Dictyostelium discoideum* by using an image processing system and a novel chemotaxis chamber providing stationary chemical gradients. J Cell Biol 108:973–984
6. Vicker MG (1994) The regulation of chemotaxis and chemokinesis in *Dictyostelium* amoebae by temporal signals and spatial gradients of cyclic AMP. J Cell Sci 107:659–667
7. van Haastert PJ (1983) Sensory adaptation of *Dictyostelium discoideum* cells to chemotactic signals. J Cell Biol 96:1559–1565
8. Tani T, Naitoh Y (1999) Chemotactic responses of *Dictyostelium discoideum* amoebae to a cyclic AMP concentration gradient: evidence to support a spatial mechanism for sensing cyclic AMP. J Exp Biol 202:1–12
9. Ebrahimzadeh PR, Högfors C, Braide M (2000) Neutrophil chemotaxis in moving gradients of fMLP. J Leukoc Biol 67:651–661
10. Irimia D, Liu SY, Tharp WG et al (2006) Microfluidic system for measuring neutrophil migratory responses to fast switches of chemical gradients. Lab Chip 6:191–198
11. Meier B, Zielinski A, Weber C et al (2011) Chemotactic cell trapping in controlled alternating gradient fields. Proc Natl Acad Sci U S A 108:11417–11422
12. Nakajima A, Ishihara S, Imoto D, Sawai S (2014) Rectified directional sensing in long-range cell migration. Nat Commun 5:5367

Chapter 9

Employing *Dictyostelium* as an Advantageous 3Rs Model for Pharmacogenetic Research

Grant P. Otto, Marco Cocorocchio, Laura Munoz, Richard A. Tyson, Till Bretschneider, and Robin S.B. Williams

Abstract

Increasing concern regarding the use of animals in research has triggered a growing need for non-animal research models in a range of fields. The development of 3Rs (replacement, refinement, and reduction) approaches in research, to reduce the reliance on the use of animal tissue and whole-animal experiments, has recently included the use of *Dictyostelium*. In addition to not feeling pain and thus being relatively free of ethical constraints, *Dictyostelium* provides a range of distinct methodological advantages for researchers that has led to a number of breakthroughs. These methodologies include using cell behavior (cell movement and shape) as a rapid indicator of sensitivity to poorly characterized medicines, natural products, and other chemicals to help understand the molecular mechanism of action of compounds. Here, we outline a general approach to employing *Dictyostelium* as a 3Rs research model, using cell behavior as a readout to better understand how compounds, such as the active ingredient in chilli peppers, capsaicin, function at a cellular level. This chapter helps scientists unfamiliar with *Dictyostelium* to rapidly employ it as an advantageous model system for research, to reduce the use of animals in research, and to make paradigm shift advances in our understanding of biological chemistry.

Key words *Dictyostelium*, Mechanism of action, Pharmacogenetics, Random cell movement

1 Introduction

Dictyostelium discoideum is a soil amoeba that consumes bacteria, but when starved it initiates multicellular development, a morphogenetic process that involves chemotaxis, intercellular signaling, and cell type differentiation; these are also the attributes of its life cycle that are most studied. However, *Dictyostelium* has becoming increasingly useful as a simple model organism to dissect the molecular mechanism of action of poorly characterized chemicals with therapeutic benefits, including new drugs and natural products. It is easy and cheap to grow and maintain, its small 34 Mb genome has been sequenced, and it is simple to manipulate genetically to

Tian Jin and Dale Hereld (eds.), *Chemotaxis: Methods and Protocols*, Methods in Molecular Biology, vol. 1407, DOI 10.1007/978-1-4939-3480-5_9,

either ablate or overexpress a gene of interest, all attributes that make it ideal as a pharmacogenetic model [1].

Drug discovery and development is a costly, lengthy process of identifying chemicals that are not only specific and effective, but are also well tolerated by patients. Many drugs have nausea as a side effect, which can lead to vomiting (emesis), and this adversely impacts patient compliance. Thus, investigating the emetic liability of new drugs is important, but it is expensive and has animal welfare implications because it typically involves vomiting or food avoidance experiments in sentient animals like dogs and ferrets. We have used *Dictyostelium* to study the emetic liability of compounds previously reported to be emetic or aversive in in vivo animal studies [2], and identified a G-protein-coupled receptor and its likely mammalian counterpart as the molecular target of the bitter compound phenylthiourea [3]. We have also used *Dictyostelium* to study the molecular mechanism of the action of the dietary flavonoid naringenin, a component of citrus fruits that has antiproliferative and chemopreventive actions *in vitro* and in animal models of carcinogenesis. Our studies in *Dictyostelium* identified a Ca^{2+}-permeable nonselective cation channel, TRPP2 or polycystin-2, as the target of naringenin [4]. This channel is implicated in the development of autosomal dominant polycystic kidney disease (ADPKD) [5, 6], and accounts for about 15 % of the mutations in ADPKD sufferers. The channel is the target of naringenin in mammalian kidney cells too, and represents a potential novel treatment for the 85 % of ADPKD patients where TRPP2 is present and can be activated by naringenin.

These studies used a simple, rapid assay based on random cell movement of chemotactically-competent amoebae as a phenotypic readout, but growth and development have also been used to dissect the cellular uptake mechanism for the bipolar disorder drug valproic acid [7]. Here we describe the random cell movement assay in *Dictyostelium* cells treated with the pungent alkaloid capsaicin, which is responsible for the spicy taste of chilli peppers of the *Capsicum* genus, and is of biomedical interest because of its potential therapeutic effects for the treatment of a variety of human conditions, including pain relief, weight reduction, anti-carcinogenic properties, and cardiovascular, gastric, urological, and dermatological conditions (reviewed in [8]). The gustatory effects of capsaicin in mammals are through its action as an agonist of a cation channel, transient receptor potential vanilloid subfamily member 1 (TRPV1), which is present in nociceptive neurons in the oral cavity, skin, eyes, and gastrointestinal mucosa, but is also present in non-neuronal cells including muscle, keratinocytes, and endothelium (reviewed in [9]). Prolonged exposure to capsaicin causes TRPV1 receptor desensitization [10]; yet extended effects on cellular signaling suggest that capsaicin can act independently of TRPV1. Evidence of these effects includes its cocarcinogenic effect in TRPV1 knockout mice [10], that it alters sucrose preference in TRPV1 knockout mice [11], and that capsaicin inhibition of collagen-induced aggregation of platelets is unaffected by a TRPV1 antagonist [12]. Additionally, capsaicin has a significant acute effect on *Dictyostelium* cell behavior, including loss of velocity, cell shape, and angular movement [2], even though clear homologues of human, mouse, or worm TRPV1 proteins have not been identified in the *Dictyostelium* genome. This result suggests that capsaicin is exerting these behavioral effects in *Dictyostelium* by a TRPV1-independent mechanism, making this organism a useful model to identify the molecular identity of this alternate pathway.

2 Materials

2.1 Equipment

A small-volume peristaltic pump (e.g., Watson Marlow 505Di, Cornwall, UK) is required to deliver cAMP to the cells to make them chemotaxis-competent.

2.2 Cell Culturing and Preparation

1. *Dictyostelium* cells (*see* **Note 1**) are grown axenically in HL5 medium (Formedium) in sterilized glass conical flasks with aeration on a temperature-controlled platform shaker (21 °C, 120 rpm). Medium is supplemented with streptomycin and penicillin (100× stock of 5000 U/ml, Gibco) and either blasticidin S (10 μg/ml; Apollo Scientific) or G418 (10 μg/ml; Invitrogen).
2. KK2 buffer: 20 mM Potassium phosphate buffer, pH 6.1.
3. cAMP: A 200 mM stock solution is prepared in water; adjust to pH 6.3 with NaOH. Store in aliquots at −20 °C.

2.3 Preparation of Capsaicin

1. A 50 mM stock solution of capsaicin is prepared in the solvent dimethyl sulfoxide (DMSO). Store in aliquots at −20 °C.

2.4 Microscopy

1. Lab-tek 8-well chambered coverglasses (Thermo Scientific Nunc).
2. Olympus IX71 inverted microscope with 40× objective.
3. QImaging RetigaExi Fast 1394 digital camera.

2.5 Time-Lapse Video Analysis

The following four software programs are used to capture images, outline individual cells, quantify cell behavior parameters, and perform statistical analysis.

1. Images are captured using ImagePro software (Media Cybernetics, Buckinghamshire, UK).
2. The QuimP11 plug-ins for the Fiji distribution of ImageJ are used to outline individual cells and to quantify cell behavior parameters (available for download at http://go.warwick.ac.uk/quimp).

3. Two custom scripts in MATLAB (MathWorks, USA) are used to first read the output generated by Fiji cell segmentation and then combine the individual cell values for each parameter (displacement, motility, circularity, and protrusion formation) (described in [13]). The final MATLAB output is an Excel file containing all the information of cell segmentation for each video analyzed (individual values, mean and standard deviation).
4. Graphpad Prism v5.02 (GraphPad Software Inc, San Diego, USA) is used to perform statistical analysis (data normalization, one-way ANOVA with Bonferroni posttest) and to plot the data for the generation of graphs.

3 Methods

3.1 Preparation of Cells and Compounds

1. *Dictyostelium* strains are grown for at least 2 days on a shaking platform at 22 °C.
2. Cells are harvested during exponential growth (between 1 and 5×10^6 cells/ml) by centrifugation at $500 \times g$ for 3 min, and washed twice with KK2 buffer.
3. Cells (2×10^7) are resuspended in 6 ml KK2 buffer in a 50 ml conical flask, and pulsed with 20 nM cAMP every 6 min for 5 h at 22 °C (giving a final cAMP concentration of 1 μM).
4. These cells are chemotactically-competent, and can be used to monitor acute changes in cell behavior.

3.2 Imaging Cell Behavior

1. Starved, chemotactically-competent cells are diluted tenfold with KK2, and 250 μl of this dilution is added to individual wells of 8-well Lab-tek chambers. Allow 15 min for the cells to adhere before imaging.
2. Cell behavior is visualized using a 40× objective on an inverted bright-field microscope, and images are captured every 15 s with an attached CCD camera.
3. As an internal control for each experiment, images are captured every 15 s for 5 min before the addition of the compound to be analyzed (a 2× stock is prepared in 250 μl KK2 and added gently to the well), or vehicle alone for control experiments (*see* **Note 2**).
4. The camera shutter is closed for compound/vehicle addition, to provide one black image to indicate compound addition in the resultant movie.
5. Images are then captured for a further 10 min every 15 s to record cell behavior.

3.3 Analysis of Random Cell Movement

1. Cell outlines are established by hand on the first image of each series, and cells tracked through the series using Fiji. Two custom scripts in MATLAB are used to first read the output generated by Fiji and then combine the individual cell values for each of the four cell behavior parameters: cell shape (circularity), motility, protrusion number, and displacement.
2. The MATLAB output is transferred to Graphpad Prism for statistical analysis and graph production. The data for each compound concentration may be normalized to the highest value which is defined as 100 %. Data can be presented as an average cell response (in velocity, displacement, cell shape, or protrusion formation) during the time period assessed (e.g., 15 min) (*see* Fig. 1a, b), with a dose-dependent effect on all cell behavior parameters. Alternatively, a change in cell behavior can be visualized by plotting cell response over a period after addition of compound (Fig. 1c, d). In regard to capsaicin, cells treated with 2–3 μM changed shape and produced fewer protrusions, with the result that both displacement and motility were reduced.

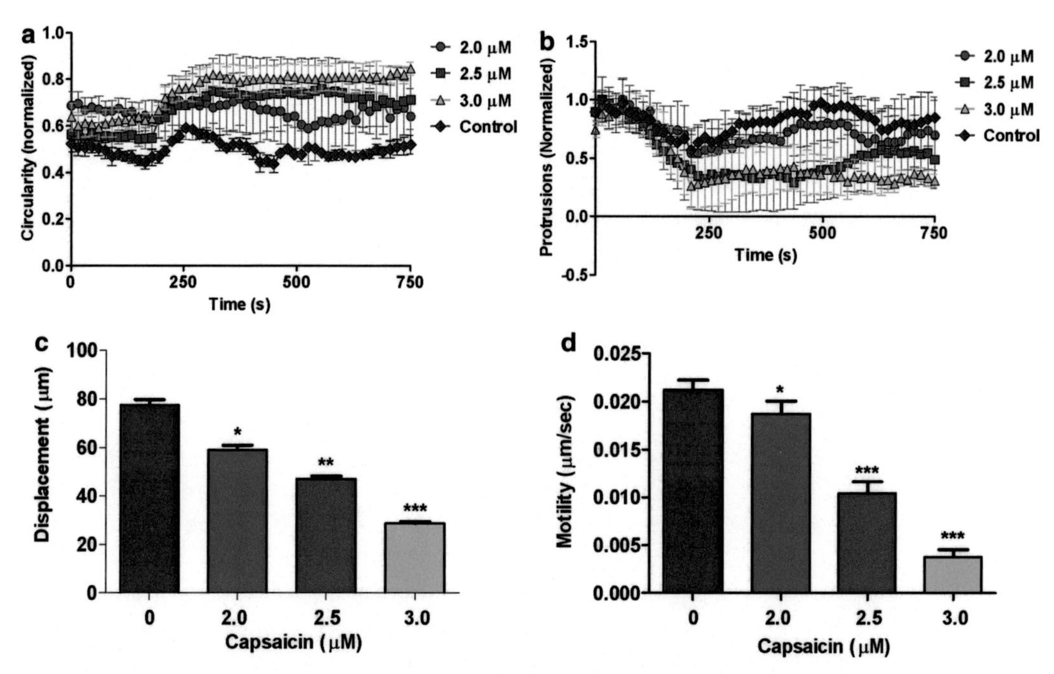

Fig. 1 The effect of capsaicin on random cell movement parameters in parental *Dictyostelium* cells. Adherent Ax2 cells in Lab-Tek chambers were imaged every 15 s for 5 min before the addition of capsaicin (2.0 μM, 2.5 μM, or 3.0 μM), followed by a further 10-min imaging in the presence of the compound. The effect of capsaicin on four parameters of cell behavior was analyzed. There was a dose-dependent effect on the degree of rounding up or circularity of the cells (**a**), the number of protrusions formed (**b**), displacement (**c**), and motility (**d**). Note that displacement and motility are the averages for the final 288 s of imaging, and statistical analysis was between control and each capsaicin concentration. $^{*}p<0.05$, $^{**}p<0.01$, $^{***}p<0.001$. Measurements were derived from an average of 30 cells analyzed from triplicate experiments. Error bars represent ±S.E.M.

3. A two-tailed paired student t-test is performed to compare between the 5 min preceding addition of compound and the final 5 min of image capture for all four parameters.
4. A one-way analysis of variance (ANOVA) with subsequent Bonferroni test is conducted to determine whether there is any statistical difference in cell behavior between different concentrations of compound or between wild-type and mutant cells at the same compound concentration.
5. Several factors should be taken into consideration regarding the selection of concentrations of a given compound to be analyzed (*see* **Note 3**).

3.4 Identifying the Molecular Identity of Chemical Detection Pathways

1. To identify the molecular basis of the action of a chemical, a library of mutants is screened for resistance to the effects of the compound on growth, development, or cell behavior, as described elsewhere [3, 4, 7] (*see* **Note 4**).
2. Mutants identified in one type of screen (e.g., growth) can then be analyzed for resistance to a compound in other assays (e.g., development or behavior), thus providing additional information on the molecular pathway affected by each compound. Alternatively, a range of published mutants are widely available if a candidate approach is taken (*see* **Note 1**).
3. A recent library screen to characterize the molecular mechanism of the bitter tastant phenylthiourea [3] identified two RacGEF mutants that were partially resistant to the action of capsaicin on cell behavior: *gxcP*– (dictybase ID: DDB_G0285859) and *gxcKK*– (dictybase ID: DDB_G0293340) (*see* Fig. 2).

4 Notes

1. Strains: Many *Dictyostelium* strains are available from the searchable *Dictyostelium* strain database run by Dictybase (www.dictybase.org; *see* ref. [14]) and can be obtained for a nominal fee and the cost of shipping.
2. Vehicle for drugs/compounds: Commonly used vehicles include DMSO, DMF, ethanol, methanol, and acetic acid. We have conducted experiments to determine the highest concentrations that can be used that do not significantly disrupt cell behavior in the random cell movement assay (e.g., up to 1 % DMSO, 3 % EtOH).
3. The choice of concentrations to be analyzed is based on those used in published studies on whole animals or tissue culture cells, and two- to fivefold lower and higher concentrations can also be tested. If there is no published data for a chemical, then

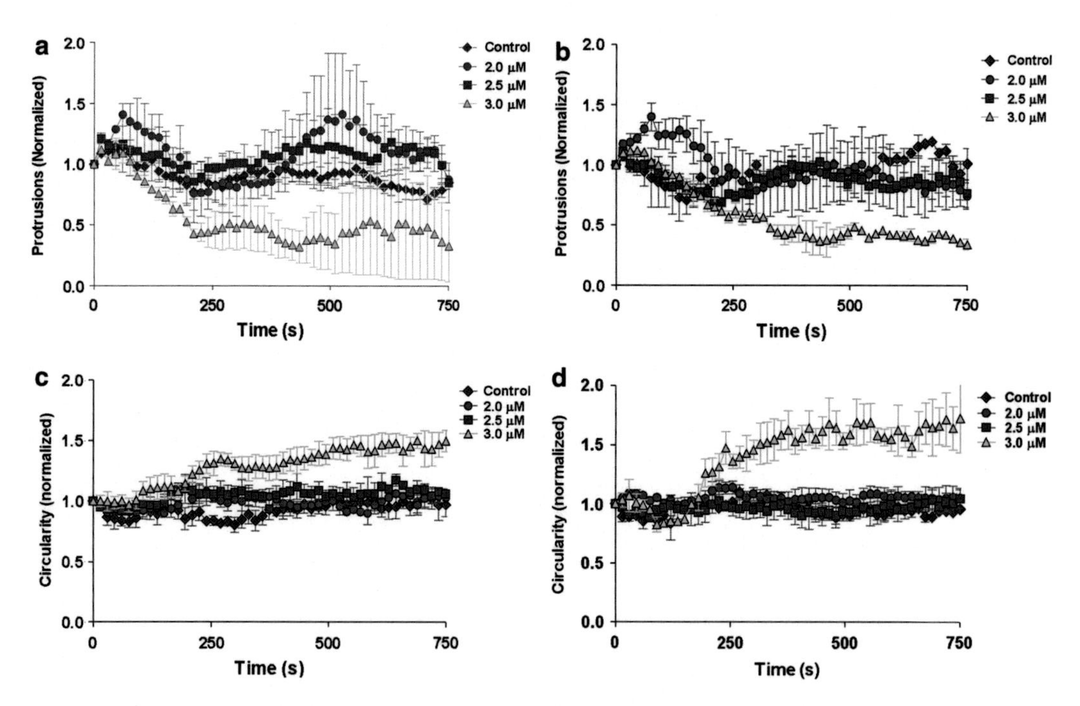

Fig. 2 Behavior of *Dictyostelium gxcP–* and *gxcKK–* cells in random cell movement assays. Both *gxcP–* and *gxcKK–* display partial resistance to 2 and 2.5 μM capsaicin, both in the number of protrusions formed (**a**, **b**, respectively) and the circularity of cells (**c**, **d**, respectively). Measurements were derived from 50 cells analyzed from a minimum of triplicate experiments. Error bars represent ±S.E.M.

a concentration range for related compounds is used, as long as it is within the solubility range of the chemical and does not have acute effects on cell viability (e.g., cell lysis). This can be simply tested by visual inspection under a microscope.

4. Where possible, it is advantageous to use appropriate controls during these experiments. For example, when screening a library of mutants for resistance to a compound, the wild-type parental strain containing the antibiotic resistance cassette but without gene ablation should be used. When studying individual mutants, at a minimum the immediate parental strain should be employed because variability in compound sensitivity is found between laboratory wild-type strains.

Acknowledgements M.C. is funded by GlaxoSmithKline. G.P.O. is funded by the Dr. Hadwen Trust (DHT) and did not participate in experiments involving animals, or cells or tissues from animals or from human embryos. The Dr. Hadwen Trust is the UK leading medical research charity that funds and promotes exclusively human-relevant research that encourages the progress of medicine with the replacement of the use of animals in research. Authors G.P.O. and M.C. contributed equally.

References

1. Williams RS (2005) Pharmacogenetics in model systems: defining a common mechanism of action for mood stabilisers. Prog Neuropsychopharmacol Biol Psychiatry 29: 1029–1037
2. Robery S, Mukanowa J, Percie du Sert N et al (2011) Investigating the effect of emetic compounds on chemotaxis in *Dictyostelium* identifies a non-sentient model for bitter and hot tastant research. PLoS One 6:e24439
3. Robery S, Tyson R, Dinh C et al (2013) A novel human receptor involved in bitter tastant detection identified using *Dictyostelium discoideum*. J Cell Sci 126:5465–5476
4. Waheed A, Ludtmann MH, Pakes N et al (2014) Naringenin inhibits the growth of *Dictyostelium* and MDCK-derived cysts in a TRPP2 (polycystin-2)-dependent manner. Br J Pharmacol 171:2659–2670
5. González-Perrett S, Kim K, Ibarra C et al (2001) Polycystin-2, the protein mutated in autosomal dominant polycystic kidney disease (ADPKD), is a Ca^{2+}-permeable nonselective cation channel. Proc Natl Acad Sci U S A 98:1182–1187
6. Vassilev PM, Guo L, Chen XZ et al (2001) Polycystin-2 is a novel cation channel implicated in defective intracellular Ca^{2+} homeostasis in polycystic kidney disease. Biochem Biophys Res Commun 282:341–350
7. Terbach N, Shah R, Kelemen R et al (2011) Identifying an uptake mechanism for the antiepileptic and bipolar disorder treatment valproic acid using the simple biomedical model *Dictyostelium*. J Cell Sci 124:2267–2276
8. Sharma SK, Vij AS, Sharma M (2013) Mechanisms and clinical uses of capsaicin. Eur J Pharmacol 720:55–62
9. Fernandes ES, Fernandes MA, Keeble JE (2012) The functions of TRPA1 and TRPV1: moving away from sensory nerves. Br J Pharmacol 166:510–521
10. Hwang MK, Bode AM, Byun S et al (2010) Cocarcinogenic effect of capsaicin involves activation of EGFR signaling but not TRPV1. Cancer Res 70:6859–6869
11. Costa RM, Liu L, Nicolelis MA, Simon SA (2005) Gustatory effects of capsaicin that are independent of TRPV1 receptors. Chem Senses 30(Suppl 1):i198–i200
12. Mittelstadt SW, Nelson RA, Daanen JF et al (2012) Capsaicin-induced inhibition of platelet aggregation is not mediated by transient receptor potential vanilloid type 1. Blood Coagul Fibrinolysis 23:94–97
13. Tyson RA, Zatulovskiy E, Kay RR, Bretschneider T (2014) How blebs and pseudopods cooperate during chemotaxis. Proc Natl Acad Sci U S A 111:11703–11708
14. Fey P, Dodson RJ, Basu S, Chisholm RL (2013) One stop shop for everything *Dictyostelium*: dictyBase and the Dicty Stock Center in 2012. Methods Mol Biol 983: 59–92

Chapter 10

Identification of Associated Proteins by Immunoprecipitation and Mass Spectrometry Analysis

Xiumei Cao and Jianshe Yan

Abstract

Protein-protein interactions play central roles in intercellular and intracellular signal transduction. Impairment of protein-protein interactions causes many diseases such as cancer, cardiomyopathies, diabetes, microbial infections, and genetic and neurodegenerative disorders. Immunoprecipitation is a technique in which a target protein of interest bound by an antibody is used to pull down the protein complex out of cell lysates, which can be identified by mass spectrometry. Here, we describe the protocol to immunoprecipitate and identify the components of the protein complexes of ElmoE in *Dictyostelium discoideum* cells.

Key words Immunoprecipitation, Mass spectrometry, Protein-protein interaction, *Dictyostelium discoideum*

1 Introduction

Protein-protein interactions are the heart of cell signaling, which is the essential part of communication that governs the cells not only to properly respond to each other but also to their environments. It has been estimated that over 80 % of human genome-coding proteins function in complexes but do not operate alone [1]. Dysregulation of protein-protein interactions is responsible for many, if not all, diseases. For instance, p53R273H is one of the mutants of p53 with single base change at amino acid position of 273. This mutant, but not wild-type p53, is found to specifically bind to nardilysin, and thereby promotes an invasive response to heparin-binding-epidermal growth factor-like growth factor in H1299 cells [2]. von Hippel–Lindau (VHL) disease is characterized by generation of blood vessel tumors in brain, kidney, eye, and spinal cord. VHL protein is involved in targeting hypoxia-inducible factor (HIF) for ubiquitination and degradation. HIF is a transcription factor that prompts the expression of angiogenesis-related factors. Mutation of the VHL protein, such as pVHL(Y98H), abolishes its binding to HIF and consequently causes the

Tian Jin and Dale Hereld (eds.), *Chemotaxis: Methods and Protocols*, Methods in Molecular Biology, vol. 1407,
DOI 10.1007/978-1-4939-3480-5_10, © Springer Science+Business Media New York 2016

accumulation of HIF, which in turn results in the development of blood vessel [3]. In addition, protein-protein interactions are involved in the pathogenesis of cardiomyopathies, diabetes, microbial infections, and genetic and neurodegenerative disorders (as reviewed in [4]).

During the last decades, many techniques and methods have been developed to identify protein-protein interactions. One of the most established techniques is the yeast two-hybrid (Y2H) system. In this system, two tested proteins are individually fused to the independent DNA-binding and transcriptional activation domains of the Gal4p transcription factor. If these two tested proteins bind to one another, they form a functional transcriptional activator which then induces transcription of a downstream reporter gene as readout [5]. Y2H system is an easy assay that can be carried out in any lab without sophisticated equipment required. However, like any other system, Y2H system has some drawbacks that prevent it from being more widely used in protein-protein interaction screening. Y2H system screening generates a high chance (as high as 50 %) of false-positive identifications [6]. In addition, due to the characteristics of this method, Y2H system can only be used to study binary interactions but not those involving more than two proteins. Therefore, once protein-protein interacting partners are identified by a certain method such as Y2H system, it is always necessary to verify the interaction by other approaches. Recently, thanks to the amazing advances in mass spectrometry (MS), immunoprecipitation-coupled mass spectrometry analysis has been developed and widely used for identification of the components of the protein complexes. In most of the cases, a target protein of interest bound by an antibody is used as a "bait" to pull down the protein complex out of cell lysates. The antibody is directly covalently immobilized, or indirectly through protein A or protein G, to sepharose or magnetic beads. Immunoprecipitation is performed by binding the lysate to the beads, followed by a series of washes. Finally, the bait and its associated proteins are eluted, and then be identified directly by MS.

In this chapter, we describe a protocol to identify the components of the protein complexes in *Dictyostelium discoideum* cells. *D. discoideum* is a well-established model system for studying eukaryotic chemotaxis [7, 8]. This organism utilizes the GPCR cAMP receptor 1 coupled with the heterotrimeric G-protein Gα2Gβγ to detect the chemoattractant cAMP and control cell migration [9, 10]. The method described in this chapter enabled us to reveal a new signaling pathway in which a novel Gβγ effector, ElmoE, interacts with Gβγ and Dock-like proteins to activate the small GTPase Rac, and also associates with Arp2/3 complex and F-actin. Thus, ElmoE serves as a link between chemoattractant GPCRs, G-proteins, and the actin cytoskeleton during chemotaxis [11]. The protocol has been successfully modified to study other signaling pathways in *Dictyostelium discoideum* cells and mammalian systems [12, 13].

2 Materials

2.1 Solutions

1. HL5 growth medium: 5 g Proteose peptone, 5 g thiotone E peptone, 10 g glucose, 5 g yeast extract, 0.35 g $Na_2HPO_4 \cdot 7H_2O$, 0.35 g KH_2PO_4; bring to a volume of 1 l with distilled water, adjust pH with HCl to pH 6.4–6.7, and autoclave for 20 min to sterilize and store at room temperature.
2. Ampicillin/tetracycline/chloramphenicol (ATC) antibiotics stock solution (50×): 5 g Ampicillin, 1.25 g tetracycline, and 1.25 g chloramphenicol; bring to a volume of 1 l with distilled water, sterile-filter, and freeze at −20 °C.
3. Blasticidin S (Invitrogen) (1000×): 10 mg/ml with distilled water, sterile-filtered, freeze at −20 °C.
4. Geneticin (Invitrogen) (1000×): 20 mg/ml with distilled water, sterile-filtered, freeze at −20 °C.
5. Development buffer (DB): 7.4 mM $NaH_2PO_4 \cdot H_2O$; 4 mM $Na_2HPO_4 \cdot 7H_2O$; 2 mM $MgCl_2$; 0.2 mM $CaCl_2$; pH 6.5. Bring to a volume of 1 l with distilled water.
6. cAMP (Sigma): 10 mM Stock solution in distilled water, store in aliquots at −20 °C.
7. Caffeine (Sigma), 100 mM stock solution in distilled water, store in aliquots at −20 °C. The final concentration in experiments is 2 μM.
8. Lysis buffer: 150 mM NaCl, 1 % Triton X-100, 50 mM Tris–HCl, pH 8.0.
9. Wash buffer 1: 150 mM NaCl, 1 % Igepal CA-630 (formerly NP-40), 0.5 % sodium deoxycholate, 0.1 % SDS, 50 mM Tris–HCl, pH 8.0.
10. Wash buffer 2: 20 mM Tris–HCl, pH 7.5.
11. Elution buffer: 50 mM Tris–HCl (pH 6.8), 50 mM DTT, 1 % SDS, 1 mM EDTA, 0.005 % bromophenol blue, 10 % glycerol.
12. Protease inhibitor (Roche, Chicago, IL): Add one tablet of complete protease inhibitor cocktail to 50 ml of buffer.
13. NuPage 4–12 % Bis-Tris Gel (Invitrogen).
14. Cell lines: *elmoE* null cells expressing ElmoE-YFP; *elmoE* null cells expressing YFP.

2.2 Instruments

1. Sterile Erlenmeyer flasks.
2. Platform shaker.
3. Centrifuge.
4. Heated block.
5. Miniplus 3 peristaltic pump (Gilson, Middletown, WI).
6. ChronTrol XT programmable timer (ChronTrol Corp, San Diego, CA).

7. μMACS Separator (Miltenyi Biotec, Auburn, CA).
8. MultiStand (Miltenyi Biotec, Auburn, CA).
9. μ Column (Miltenyi Biotec, Auburn, CA).
10. Quadrupole ion trap mass spectrometer (Thermo, Palo Alto, CA).
11. The non-redundant protein database: NCBI's GenBank.

3 Methods

3.1 Cell Culturing and Development

1. *D. discoideum* cells expressing ElmoE-YFP are grown to $1–3 \times 10^6$ cells/ml in HL5 growth medium with ATC antibiotics at 22 °C (*see* **Notes 1** and **2**). Cells expressing YFP alone are served as a control. For cells containing a blasticidin- or geneticin-resistant cassette, the drug of blasticidin or geneticin is added at the final concentration of 10 μg/ml or 20 μg/ml, respectively.
2. For development to a chemotactically competent stage, 2×10^8 cells are harvested by centrifugation ($400 \times g$) at 4 °C for 5 min, washed in 50 ml DB twice, and then resuspended in 10 ml DB for a final concentration of 2×10^7 cells/ml (*see* **Note 3**).
3. For synchronous development in shaking suspension, cells are placed in a 125 ml sterile flask and rotated at 100 rpm on a platform shaker at 22 °C for the initiation of starvation (*see* **Note 4**).
4. After 1 h of starvation (*see* **Note 5**), 75 nM pulses of cAMP are added into the cell suspension using a Miniplus 3 peristaltic pump that delivers cAMP to the cell suspension through tubing, which is controlled by a ChronTrol XT programmable timer. The pump is programed to deliver 100 μl cAMP every 6 min for 4–5 h (*see* **Note 6** and Fig. 1).

3.2 cAMP Stimulation and Immunoprecipitation

1. After development with cAMP pulses, cells are washed, resuspended at 4×10^7 cells/ml in DB, and shaken with 2 mM caffeine at 200 rpm for 20 min to inhibit endogenous cAMP signaling to bring intracellular responses to basal levels [14].
2. Cells are stimulated or unstimulated with a saturating dose of 50 μM cAMP. 5 s after the stimulation, cells are lysed by adding an equal volume of 2× cold lysis buffer. Lysates are kept on ice for 30 min with occasional mixing (*see* **Note 7**).
3. Immunoprecipitation is performed by using the μMACS GFP isolation kit (Miltenyi Biotec, Auburn, CA) according to the manufacturer's protocol with modifications.
4. Cleared lysates are obtained by centrifugation at $10{,}000 \times g$ for 10 min at 4 °C (*see* **Note 8**).
5. The supernatant is transferred to a precooled 1.5 ml Eppendorf tube and incubated with 50 μl μMACS anti-GFP MicroBeads

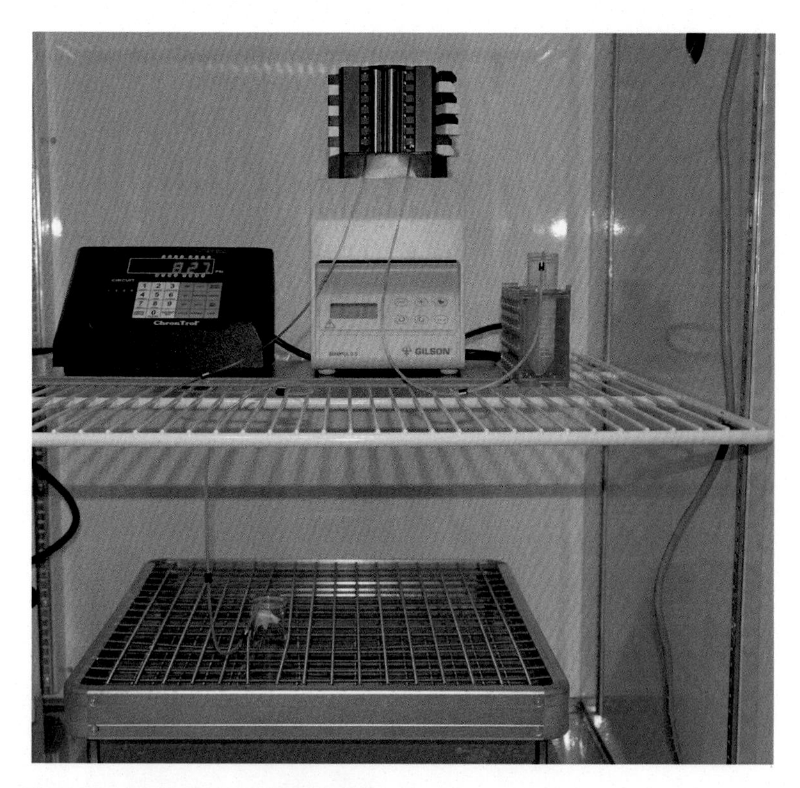

Fig. 1 The system to deliver the cAMP pulses for development of cells in suspension. For synchronous development in shaking suspension, *Dictyostelium* cells are placed in a 125 ml sterile flask and rotated at 100 rpm on a platform shaker at 22 °C. 75 nM pulses of cAMP are added into the cell suspension using a Miniplus 3 peristaltic pump that delivers cAMP to the cell suspension through tubing, which is controlled by a ChronTrol XT programmable timer. The pump is programed to deliver 100 μl cAMP every 6 min for 4–5 h

to label the YFP-tagged protein. Mix well and incubate on ice for 30 min (*see* **Note 9**).

6. Set the μ Column onto the μMACS Separator. Wash the μ Column by applying 200 μl lysis buffer on the column (*see* **Note 10**).
7. After the labeling incubation, the cell lysate is applied onto a μ Column. Let the lysate run through (*see* **Note 11**).
8. Wash the columns with 200 μl Wash Buffer 1; repeat three times.
9. Wash the columns with 200 μl Wash Buffer 2 (*see* **Note 12**).
10. Apply 20 μl of preheated 95 °C elution buffer to the columns and keep at room temperature for 5 min.
11. Apply 50 μl of preheated 95 °C elution buffer to the columns and collect the eluate as the immunoprecipitate.

3.3 Mass Spectrometry Analysis

1. Run the eluates on a 4–12 % Bis-Tris gel until the dye front reaches the bottom of the gel.
2. The gel is fixed with 50 % methanol and 10 % acetic acid for 1 h and then stained overnight with 0.1 % Coomassie Brilliant Blue with 50 % methanol and 10 % acetic acid (*see* **Note 13**).
3. The gel is then destained with distilled water and then with 50 % methanol and 10 % acetic acid until the background is clear.
4. Gel slices are excised all the way down the track of the gel, typically generating around 30 gel bands (*see* **Note 14** and Fig. 2).

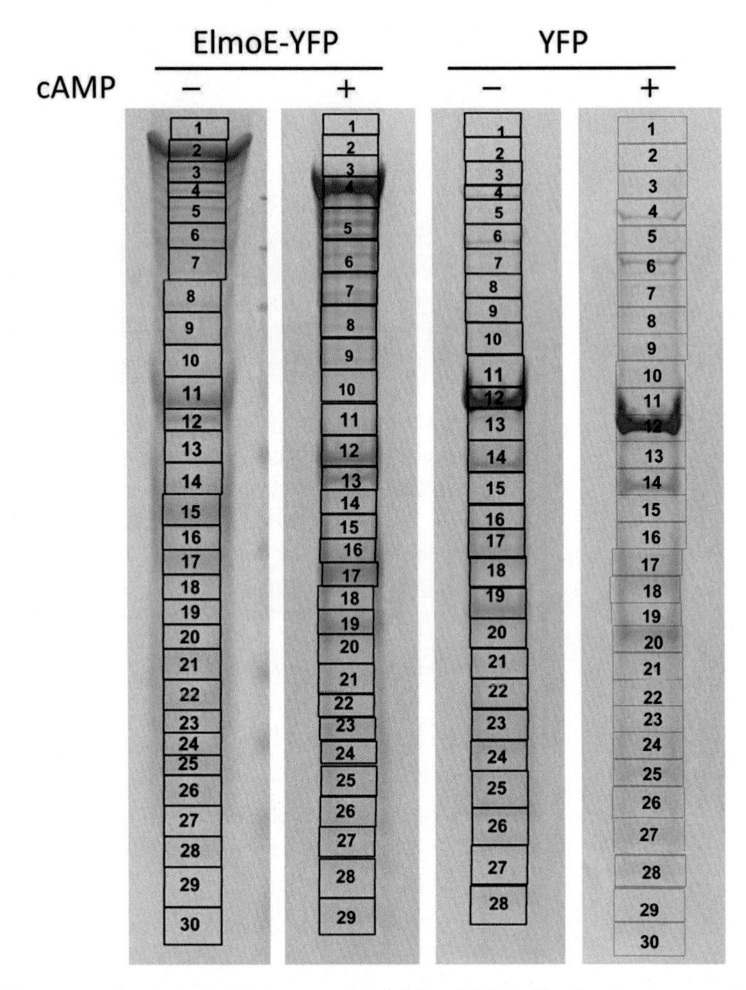

Fig. 2 Identification of proteins that associate with ElmoE using immunoprecipitation and NanoLC-MS/MS peptide sequencing. *Dictyostelium* cells expressing ElmoE-YFP or YFP alone (as a control) were stimulated with or without 50 μM cAMP, suspended in lysis buffer, and then incubated with μMACS anti-GFP MicroBeads. After incubation, the beads were washed. The eluted proteins were subjected to SDS-PAGE. Shown is a representative protein gel stained with Coomassie blue. The indicated *boxes* show where the bands were excised for the subsequent mass spectrometric analysis

5. Each gel slice is subsequently digested in-gel with sequencing grade-modified trypsin (*see* **Note 15**).
6. The resulted peptide mixture is analyzed by an LC-MS/MS system, in which a high-pressure liquid chromatography (HPLC) with a 75 μm inner diameter reverse-phase C18 column is online coupled with an ion trap mass spectrometer (Thermo, Palo Alto, CA).
7. The mass spectrometric data acquired are used to search the most recent non-redundant protein database from NCBI's GenBank. A specific search is done for the *D. discoideum* database (www.dictybase.org).

4 Notes

1. It is important to remain the growth of *Dictyostelium* cells in an exponential growth phase (cell density is lower than 4×10^6 cells/ml) rather than growing them to reach stationary phase. Cells at high densities start development that changes the physiology of the culture, does not favor synchronous development, and thus results in inconsistent results.
2. Other antibiotics, such as ampicillin and streptomycin solution (GIBCO, Grandland, NY), work fine as well for the growth of *Dictyostelium* cells. The antibiotics described here are routinely used in our lab.
3. Use a 50 ml flask for pulsing if the cell volumes are less than 5 ml, and a 250 ml flask for pulsing if the cell volumes are more than 10 ml.
4. Keep in mind that development of *Dictyostelium* cells is induced by starvation, so any bacteria or medium leftover in the DB may cause abnormal development. It is therefore important to maintain sterility throughout this process.
5. In this process the cells will accumulate the secreted phosphodiesterase enzyme so that the added cAMP during pulsing will be degraded and will not be accumulated.
6. Check the cells under microscopy to make sure that they are aggregation competent. Variations of time may happen due to different cell lines and slight difference of the experimental conditions.
7. In case of co-immunoprecipitations with weak protein-protein interactions, a low-salt lysis buffer (1 % NP-40, 50 mM Tris–HCl at pH 8.0) may be used instead. On the contrary, if strong unspecific binding is expected, high-salt lysis buffer (500 mM NaCl, 1 % NP-40, 50 mM Tris–HCl at pH 8.0) is suggested. Proteinase inhibitors should be added to the lysis and/or wash

buffers if a background smear is observed following SDS-PAGE analysis of the eluates at the end of immunoprecipitation.

8. High-speed centrifugation is necessary to remove cell debris which often causes occluding the column. In case blocking still occurs after centrifugation, higher speed and/or longer centrifugation is suggested.
9. Optionally, the lysate can be stored at −80 °C. Thawing of the lysate prior to use should be on ice and followed by immediate use in the subsequent assay.
10. The column occasionally runs slowly maybe due to air bubble formation. To solve this problem, use room-temperature buffers or degas the buffers before use.
11. Do not run dry the columns.
12. High concentrations of residual salt and detergent should be removed from the immune complex prior to elution at this step; otherwise they may interfere with the subsequent SDS-PAGE analysis.
13. In addition to Coomassie Blue, colloidal Coomassie can also be used for gel staining. However, traditional silver staining is not recommended because of the presence of glutaraldehyde, which is not compatible to the subsequent mass spectrometric analysis.
14. For proteins with intense staining, gel bands can be easily cut out. However, for those proteins that only appear as very faint bands on the gel, it is hard to cut out the bands of interest for protein identification. We therefore suggest this whole-gel analysis to avoid missing those important proteins at low concentration.
15. Be very careful to avoid contamination. Do not touch the gel with bare hands. Keratin of human skin is one of the major resources of contamination which may lead to failure of protein identification.

Acknowledgement

This work was supported by grants of National Basic Research Program of China (2014CB541804), Shanghai Pujiang Program (14PJ1407700), and Health and Family Planning Commission Foundation of Shanghai (201440300) to J.Y. and by National Nature Science Foundation of China (81201977) to X.C.

References

1. Berggård T, Linse S, James P (2007) Methods for the detection and analysis of protein-protein interactions. Proteomics 7:2833–2842
2. Coffill CR, Muller PA, Oh HK et al (2012) Mutant p53 interactome identifies nardilysin as a p53R273H-specific binding partner that promotes invasion. EMBO Rep 13:638–644
3. Ohh M, Park CW, Ivan M et al (2000) Ubiquitination of hypoxia-inducible factor requires direct binding to the beta-domain of the von Hippel-Lindau protein. Nat Cell Biol 2:423–427
4. Kuzmanov U, Emili A (2013) Protein-protein interaction networks: probing disease mechanisms using model systems. Genome Med 5:37
5. Fields S, Song O (1989) A novel genetic system to detect protein-protein interactions. Nature 340:245–246
6. Deane CM, Salwiński Ł, Xenarios I, Eisenberg D (2002) Protein interactions: two methods for assessment of the reliability of high throughput observations. Mol Cell Proteomics 1:349–356
7. Jin T, Xu X, Hereld D (2008) Chemotaxis, chemokine receptors and human disease. Cytokine 44:1–8
8. Van Haastert PJ, Devreotes PN (2004) Chemotaxis: signalling the way forward. Nat Rev Mol Cell Biol 5:626–634
9. Kimmel AR, Parent CA (2003) The signal to move: *D. discoideum* go orienteering. Science 300:1525–1527
10. Parent CA, Devreotes PN (1999) A cell's sense of direction. Science 284:765–770
11. Yan J, Mihaylov V, Xu X et al (2012) A Gβγ effector, ElmoE, transduces GPCR signaling to the actin network during chemotaxis. Dev Cell 22:92–103
12. Cao X, Yan J, Shu S et al (2014) Arrestins function in cAR1 GPCR-mediated signaling and cAR1 internalization in the development of *Dictyostelium discoideum*. Mol Biol Cell 25:3210–3221
13. Li H, Yang L, Fu H et al (2013) Association between Gαi2 and ELMO1/Dock180 connects chemokine signalling with Rac activation and metastasis. Nat Commun 4:1706
14. Brenner M, Thoms SD (1984) Caffeine blocks activation of cyclic AMP synthesis in *Dictyostelium discoideum*. Dev Biol 101: 136–146

Chapter 11

Biochemical Responses to Chemically Distinct Chemoattractants During the Growth and Development of *Dictyostelium*

Netra Pal Meena and Alan R. Kimmel

Abstract

Dictyostelium discoideum has proven an excellent model for the study of eukaryotic chemotaxis. During growth in its native environment, *Dictyostelium* phagocytose bacteria and fungi for primary nutrient capture. Growing *Dictyostelium* can detect these nutrient sources through chemotaxis toward the metabolic by-product folate. Although *Dictyostelium* grow as individual cells, nutrient depletion induces a multicellular development program and a separate chemotactic response pathway. During development, *Dictyostelium* synthesize and secrete cAMP, which serves as a chemoattractant to mobilize and coordinate cells for multicellular formation and development. Separate classes of GPCRs and Gα proteins mediate chemotactic signaling to the chemically distinct ligands. We discuss common and separate component responses of *Dictyostelium* to folate and cAMP during growth and development, and the advantages and disadvantages for each. As examples, we present biochemical assays to characterize the chemoattractant-induced kinase activations of mTORC2 and the ERKs.

Key words cAMP, Folate, GPCR, ERK, mTORC2, Deamination, Methotrexate

1 Introduction

Dictyostelium discoideum cells chemotax to chemically distinct ligands at different stages of the life cycle [1]. During growth as single cells, *Dictyostelium* respond to the metabolic by-product folate as a nutrient sensor for bacteria and yeast [1]. Folates are widely distributed in nature; its name derives from *folium*, indicative of high-level occurrence in dark green leafy vegetables. Folate sensitivity is enabled by a specific family of G protein-coupled, cell surface receptors (GPCRs) and Gα proteins [2]. Upon nutrient starvation, multicellular development is initiated, and cells express and display unique sets of GPCRs and Gα proteins [3–5]. These receptors (CARs) recognize secreted extracellular cAMP and mediate directed movement to coalesce (aggregate) cells at centers of cAMP signaling for multicellular formation [3–5].

Tian Jin and Dale Hereld (eds.), *Chemotaxis: Methods and Protocols*, Methods in Molecular Biology, vol. 1407,
DOI 10.1007/978-1-4939-3480-5_11, © Springer Science+Business Media New York 2016

During rapid growth, *Dictyostelium* respond to folate, but not cAMP [1]; conversely, during aggregation, *Dictyostelium* are chemotactic to cAMP, but not to folate ligands [1]. It should be emphasized that maximal chemotactic response to cAMP requires developmental competence [1]. Thus, developmentally defective cells, irrespective of cAMP sensitivity, may chemotax poorly [1, 3–5]. However, there is a growth period, as cell densities rise and nutrients begin to deplete, when cells are partially responsive to cAMP while retaining folate sensitivity [6]. In addition, many pathways downstream from the folate and cAMP receptors and their specific Gα protein couplings are common [6]. Thus, studies of both folate and cAMP response at different developmental periods provide separate advantage and supporting data for chemotactic understanding.

We had previously detailed methods to characterize cAMP binding to the multiple CARs under both physiological and non-physiological conditions [7]. Data allow estimation of relative Kd, receptor numbers per cell, and relative affinity for various analogs of cAMP.

Here we extend analyses to folate. Folate (folic acid) is a more complex molecule compared to cAMP and is biologically represented as a family of pteroylglutamates, comprised of a pteroyl moiety linked to glutamate. For *Dictyostelium*, the pteroyl domain contains the active chemoattractant entity. The folate form often used is [(2-amino-4-hydroxypteridin-6-ylmethylamino)benzoyl]-l(+)-glutamic acid, with the 2-amino group essential for chemotaxis. Similar to how *Dictyostelium* secrete a phosphodiesterase to inactivate extracellular cAMP in order to maintain a strong positive chemoattractant gradient during development, growing *Dictyostelium* secrete a deaminase that coverts the 2-amino folate form to an inactive 2-hydroxyl variant [8]. The deaminase can be partially inhibited by 8-azaguanine. We discuss methods to analyze folate deamination, folate binding [9], and application of folate and cAMP analogs.

Responses to folate and cAMP can be rapid (<15″) or slightly delayed (~30″) [6, 7]. Kinase assays for mTORC2 and the ERKs are presented that, respectively, represent examples for each.

2 Materials

2.1 Solutions

1. Growth medium: D3-T Medium (KD Medical #D3-T001).
2. DB buffer: 10 mM Na_2HPO_4, 10 mM NaH_2PO_4, 2 mM $MgSO_4$, 1 mM $CaCl_2$, adjust to pH 6.5.
3. PB buffer: 5 mM Na_2HPO_4, 5 mM NaH_2PO_4, adjust to pH 6.5.
4. cAMP: 1 mM in DB, dilute as needed.

5. [^{3}H]-Folate: Diammonium salt [3′,5′,7,9-^{3}H], 1 mCi/mL (2.2×10^9 dpm/mL), in EtOH:H_2O (4:6); 10–30 Ci/mmol, 30–100 μM (Moravek Biochemicals and Radiochemicals # MT783).
6. 20 % SDS in H_2O.
7. 50 mM Folate, neutralize as needed with 1 N NaOH. Folate is light sensitive; for most consistent results, prepare fresh and keep in a lighttight vial.
8. 2 mM Caffeine; for most consistent results, prepare fresh.
9. 2-Mercaptoethanol (βME).
10. Thin-layer chromatography solvent: 5 % K_2CO_3.
11. Silicon Oil AR 20 (Sigma-Aldrich #10836).
12. Silicon Oil AR 200 (Sigma-Aldrich #85419).
13. Protein gel loading buffer: 4×-Laemmli sample buffer (Bio-Rad #161-0747).
14. Protein gel running buffer: 10×-Tris/glycine/SDS buffer (Bio-Rad #161-0732).
15. 20 % Sucrose in H_2O.
16. Scintillation counting fluid: Cytoscint-Esliquid Scintillation Cocktail (MP Biomedicals #0188245301).
17. SuperSignal West Dura Extended Duration Substrate (Life Technologies #34075).
18. Blocker BLOTTO (nonfat dry milk) in TBS (Thermo Scientific #37530)
19. 20× TBS (Tris-buffered saline), pH 7.4 (Crystalgen #300-800-4000).
20. Tween 20 (Sigma-Aldrich #P2287-500 mL).

2.2 Supplies

1. TLC Cellulose Matrix (Sigma-Aldrich #Z122866-25EA).
2. UV Transilluminator (VWR).
3. Precast protein gel electrophoresis: 4–15 % Criterion TGX Gel, 26 well (Bio-Rad #567-1085).
4. Actin-HRP antibody (Santa Cruz Biotechnology #SC-1616).
5. pThr202/pTyr204-ERK antibody (Cell Signaling Technology #9101); this antibody will recognize the phospho-forms of *Dictyostelium* ERK1 and ERK2.
6. Donkey anti-rabbit IgG-HRP (Santa Cruz Biotechnology #SC-2313).
7. Phospho-p70 S6 Kinase (Thr389) (1A5) Mouse mAb (Cell Signaling Technology #9206); this antibody will recognize pT470-PKBR1 and pT435-AKT of *Dictyostelium*, which are phosphorylated by mTORC2.

8. Goat anti-mouse IgG-HRP (Santa Cruz Biotechnology #SC-2005).
9. Nitrocellulose membrane: Trans-Blot Turbo Midi Nitrocellulose Transfer Packs, 0.2 μm pore size (Bio-Rad #170-4159).
10. Protein gel electrophoresis chamber: Criterion Cell, Vertical midi-format electrophoresis cell (Bio-Rad #165-6001).
11. The Trans-Blot turbo transfer system for semidry blotting protein transfers (Bio-Rad #170-4155).
12. 8-Azaguanine (Santa Cruz Biotechnology #SC-207194).
13. Microtube cutter, to 9.4 mm diameter (Denville Scientific #C19125).
14. Scintillation counting vials: Kimble 20 mL Glass Screw-Thread Scintillation Vials (Fischer Scientific #03-340-129).
15. Beckman LS 6500 Liquid Scintillation Counter (Beckman Coulter Inc.).
16. Peristaltic Pump (Gilson).
17. Timer (ChronTrol).

3 Methods

3.1 Preparation of Growing Dictyostelium

1. Grow cells in shaking culture (at ~200 rpm) to ~1–3 × 10^6 cells/mL at 22 °C.
2. Centrifuge cells at 22 °C for 5 min at 500 × *g*.
3. Wash cells once with PB at 22 °C.
4. Resuspend cells in PB at 2–3 × 10^7 cells/mL.
5. Incubate cells at ~150 rpm at 22 °C for 30 min.
6. Wash cells once in PB at 22 °C and resuspend in PB at ~3–5 × 10^7 cells/mL, as needed.

3.2 Preparation of Aggregation-Competent Dictyostelium in Shaking Culture (See Note 1)

1. Grow cells to ~2–3 × 10^6 cells/mL at 22 °C.
2. Centrifuge cells at 22 °C for 5 min at 500 × *g*.
3. Wash cells twice with DB at 22 °C.
4. Resuspend cells in DB at 2 × 10^7 cells/mL.
5. Incubate cells at 100–125 rpm at 22 °C.
6. To develop cells in shaking culture, add cAMP to an immediate (within 10 s) final concentration of 75 nM every 6 min for 6 h. This can be automated by connecting a peristaltic pump to a programmable on/off timer; for a 6 mL cell suspension culture, add 100 μL of a 4.5 μM cAMP stock solution. After 6 h,

the cell volume will have doubled. The 50 % reduction in effective pulsed cAMP concentration is not problematic.

3.3 Detection and Preparation of Deaminated Folate (See Note 2)

1. Grow and harvest cells, as per Subheading 3.1.
2. Resuspend cells in PB at 5×10^7 cells/mL.
3. Unlabeled folate is added at 5 mM to 10 mL shaking cells.
4. 100 μL cell samples are taken at various times (e.g., 0, 5, 10, and 20 h), centrifuged at 4 °C for 30 s at 14,000 ×*g*. The entire sample is collected at ~20 h.
5. Collect the supernatants and incubate at 90 °C for 5 min. The supernatants can be reserved for re-assay.
6. Slowly spot 10 μL reaction mixtures for each time point on a TLC plate. Individual spots should be separated by ~1–1.5 cm, and spots should be 1.5 cm from the bottom edge, to prevent merging with solvent. Sample spots should air-dry completely. Mark the origin.
7. A small amount of solvent (5 % K_2CO_3) is placed at the bottom of a TLC separation chamber.
8. Carefully place the TLC plate in the chamber so that the sample spots do not contact the solvent; close the chamber top.
9. The plate should be removed from the chamber before the solvent front reaches top of the TLC plate, ~2 h. Mark the solvent front and dry the plate.
10. Folate (~10 μg) can be visualized by UV shadowing with a transilluminator. Mark each spot. There will be two distinct and separated spots. The spot closer to the origin at 0 time is unmodified folate, R_f~0.25. The spot closer to the solvent front is deaminated folate, R_f~0.42. With time, there will be conversion of folate to deaminated folate. Full conversion of 5 mM folate to deaminated folate can require ~20 h.
11. Deaminated folate can be a convenient and interesting control. After 20 h, the entire folate sample should have been converted to deaminated folate. Treatment at 90 °C for 5 min (*see* **step 5**) to inactivate cellular activities does not affect folate or deaminated folate. Deaminated folate (~5 mM; ~2 μg/μL) can be stored at 4 °C. Deaminated folate can be used as a separate migration marker on TLC plates and for stimulation assays.

3.4 Assaying Folate Deaminase Activity

1. Grow and harvest cells, as per Subheading 3.1.
2. Resuspend cells in PB at 5×10^7 cells/mL.
3. Unlabeled folate is added to shaking cells at 50 μM, with [^{3}H]-folate at 1–10 μL/mL (~2–20 × 10^6 dpm/mL).
4. 100 μL of cell samples are collected at various times (e.g., 0, 1, 3, 5, 10, and 30 min) and centrifuged at 4 °C for 30 s at 14,000 ×*g*.

5. Collect the supernatants. 5 μL is added immediately to a 5 μL mix of 10 μg folate + 10 μg deaminated folate (**step 11** of Subheading 3.3); the specific activity dilution effectively terminates further deamination of [^{3}H]-folate. The remaining supernatants are incubated at 90 °C for 5 min and can be reserved for re-assay.
6. Slowly spot the 10 μL reaction mixtures for each time point on a TLC plate. Individual spots should be separated by ~1–1.5 cm, and spots should be 1.5 cm from the bottom edge, to prevent merging with solvent. Sample spots should air-dry completely. Mark the origin.
7. A small amount of solvent (5 % K_2CO_3) is placed at the bottom of a TLC separation chamber.
8. Carefully place the TLC plate in the chamber so that the sample spots do not contact the solvent; close the chamber top.
9. The plate should be removed from the chamber before the solvent front reaches top of the TLC plate, ~2 h. Mark the solvent front and dry the plate.
10. Folate (~20 μg) can be visualized by UV shadowing with a transilluminator. Mark each spot. There will be two distinct and separated spots. The spot closer to the origin at 0 time is unmodified folate, R_f~0.25. The spot closer to the solvent front is deaminated folate, R_f~0.42.
11. Excise the marked folate and deaminated spots and determine relative radioactivity in each by scintillation counting. This will give a conversion rate. The rate of deamination can be ~10 nmol/mL/min; at 50 μM, folate would be deaminated to <10 nM within ~10 min.

3.5 Assaying Inhibitors of Folate Deaminase (See Note 3)

1. Follow **steps 1–3** of Subheading 3.4.
2. Add varying concentrations of deaminase inhibitor. The most common is 8-azaguanine at concentrations of 0, 0.1, 0.3, 0.5, 1.0, and 1.5 mM.
3. Follow **steps 4–11** of Subheading 3.4.
4. Deaminase activity is inhibited to ~70 % with 0.5 mM 8-azaguanine, and to ~90 % with 1.5 mM 8-azaguanine (*see* **Note 3**).

3.6 Binding of Cell Surface Receptors for Folate in Phosphate Buffer (See Note 4)

1. Grow and harvest cells, as per Subheading 3.1.
2. Resuspend the cells in PB at 5×10^7 cells/mL, at 4 °C.
3. Add 50 μL of 900 nM [^{3}H]-folate (~3.3×10^6 dpm) in 10 mM PB (*see* **Note 4**), containing 4.5 mM 8-azaguanine (to inhibit the deamination of folate) to a 100 μL cell aliquot, at 4 °C; the final folate concentration will be 300 nM. To determine non-

specific [^{3}H]-folate binding, include controls that contain 50 μM unlabeled folate (*see* **Note 4**).

4. Mix and incubate at 4 °C for <1 min.
5. To remove unbound [^{3}H]-folate, carefully transfer the reaction mixture to a 1.5 mL tube containing 20 μL of 20 % sucrose at the bottom, overlayered with 200 μL of Silicon Oil AR 20:200 (11:4), at 4 °C.
6. Centrifuge at 4 °C for 30 s at 14,000 × *g*. Immediately, place the tube tips in dry ice for 15 min. Further sample processing can be at convenience.
7. Collect the cell pellets by cutting the tube tips, using the tube cutter.
8. Dissolve only the pellets in 100 μL 1 % SDS, and determine the bound [^{3}H]-folate by scintillation counting.
9. Calculate the number of folate receptors per cell, from the dpm/cell number value compared to the input-specific activity of [^{3}H]-folate, and corrected for nonspecific background binding in the presence of >150-fold excess unlabeled folate (*see* **Note 4**).
10. Binding can also be followed, by varying the [^{3}H]-folate input, at **step 3**, from 30 nM to ~1 μM (*see* **Note 4**).

3.7 Folate Stimulation of Growing and Developed Cells

1. For folate stimulation of growing cells, follow steps per Subheading 3.1, and for folate stimulation of developed cells, follow steps per Subheading 3.2.
2. Resuspend the cells in PB (growth) or DB (development) at ~3×10^7 cells/mL.
3. Aliquot 1–2 mL of cells into a small plastic cup and shake at 120–150 rpm at 22 °C.
4. For a saturating stimulus, add folate to a final concentration of 50 μM; for each mL of cell suspension, add 10 μL of a 5 mM folate stock. By varying the folate concentration, it is possible to determine a dose–response curve. It is also sometimes advisable to have parallel reactions with 0.5 mM 8-azaguanine, to assess effects of folate deamination during the stimulus.

3.8 Stimulation with Folate Analogs and Derivatives

1. Several analogs of folate have proven to be advantageous for studying signal transduction. Folate is a complex compound comprised of a pteroyl moiety linked to glutamate. The essential folate activity resides in the pteroyl domain, but there are receptors in *Dictyostelium* for glutamate and the glutamate derivative γ-aminobutyric acid, GABA. In addition, *Dictyostelium* can show differential sensitivity to folate and folate analogs methotrexate, aminopterin, and 10-methyl

folate, at various developmental stages. Deaminated folate can be another control (**step 11** of Subheading 3.3).

2. Follow steps per Subheading 3.7, using various concentrations of a preferred analog.

3.9 cAMP Stimulation of Growing and Developed Cells

1. For cAMP stimulation of growing cells, follow steps per Subheading 3.1, and for cAMP stimulation of developed cells, follow steps per Subheading 3.2.
2. Resuspend the cells in PB (growth) or DB (development) at ~3×10^7 cells/mL.
3. Aliquot 1–2 mL of cells into a small plastic cup and shake at 120–150 rpm at 22 °C.
4. For a saturating stimulus, add cAMP to a final concentration of 10 μM; for each mL of cell suspension, add 10 μL of a 1 mM cAMP stock. By varying the cAMP concentration, it is possible to determine a dose–response curve. It is also sometimes advisable to have parallel reactions with 10 mM DTT, which inhibits the degradation of cAMP by the secreted phosphodiesterase (PDE).

3.10 Stimulation with cAMP Analogs

1. Several analogs of cAMP have proven to be advantageous for studying signal transduction. Relative to cAMP, 2′-H-cAMP, has a >100×-preferred affinity for the cAMP receptor than the regulatory subunit of PKA, whereas cell-permeable 8-Br-cAMP exhibits opposite binding characteristics: a >100×-fold affinity preference for the regulatory subunit of PKA relative to the cAMP receptor. (*Sp*)-cAMPS is resistant to degradation by extracellular PDE, and can be used to maintain non-fluctuating levels of a cAMP stimulus. Regardless, all the analogs have reduced CAR affinity relative to cAMP.
2. Follow steps per Subheading 3.9, using various concentrations of a preferred analog.

3.11 Assay for Chemoattractant Stimulation of ERK1 and ERK2 Phosphorylation

1. Follow steps in Subheading 3.7 or 3.8, for folate, and Subheading 3.9 or 3.10, for cAMP.
2. Remove 70 μL at 0, 10, 15, 30, 45, 60, and 180 s and add directly to 30 μL of 4×-Laemmli sample buffer, with 10 % βME.
3. Incubate at 95 °C for 10 min.
4. Load 15 μL samples onto 4–15 % Criterion TGX Gel and electrophorese, in duplicate gels.
5. Transfer the two gels to separate nitrocellulose membranes and pre-block in Blocker BLOTTO for 30 min at 22 °C.
6. Immunoblot one membrane with the primary α-pT202/pY204-ERK antibody (1:1000 dilution; 4 °C, overnight)

followed by the secondary antibody, goat anti-rabbit IgG-HRP (1:5000 dilution; room temperature, 1 h); immunoblot the second membrane with α-actin-HRP (1:10,000 dilution; room temperature, 1 h). Actin serves as a loading control and has a similar mobility on SDS gels to *Dictyostelium* pERK2.

7. The immunoblot signals are detected by incubation at 22 °C, in the dark, for 5 min, with the Supersignal West Dura enhanced chemiluminescent (ECL) substrate for horseradish peroxidase (HRP) activity.
8. Chemiluminescence is captured by exposure to X-ray film.
9. pERK1 and pERK2 migrate on SDS gels at ~65 kDa and ~42 kDa, respectively. Their activations are measured as an increase in levels of pERK1 and pERK2.

3.12 Assay for Chemoattractant Stimulation of mTORC2 Activity

1. Follow steps in Subheading 3.7 or 3.8, for folate, and Subheading 3.9 or 3.10, for cAMP.
2. Follow **steps 2** and **3** in Subheading 3.11.
3. Load 15 μL samples onto a 4–15 % Criterion TGX Gel and electrophorese.
4. Immunoblot the membrane with the primary Phospho-p70 S6 Kinase (Thr389) (1:1000 dilution; 4 °C, overnight) for pT470-PKBR1/pT435-AKT followed by the secondary antibody, goat anti-mouse IgG-HRP (1:5000 dilution; room temperature, 1 h).
5. The immunoblot signals are detected by incubation at 22 °C, in the dark, for 5 min, with the Supersignal West Dura ECL substrate for horseradish peroxidase (HRP) activity.
6. Chemiluminescence is captured by exposure to X-ray film.
7. pAKT1 and pPKBR1 migrate on SDS gels at ~50 kDa and ~75 kDa, respectively. mTORC2 activation is measured as an increase in levels of pAKT and pPKBR1.
8. Wash the membrane three times with TBS containing 0.1 % tween 20, for 5 min each, at 22 °C.
9. Immunoblot the washed membrane with Gα-actin-HRP (1:10,000 dilution; room temperature, 1 h).
10. Repeat **steps 5** and **6**.
11. Actin serves as a loading control and has a different mobility on SDS gels than *Dictyostelium* pAKT or pPKBR1.
12. Alternatively, duplicate gels can be used (*see* **steps 4–6** in Subheading 3.11) with separate immunoblots for either pAKT/pKBR1 or actin.

4 Notes

1. The differentiation of *Dictyostelium* in shaking culture with exogenous cAMP bypasses the requirement of cells to produce and secrete cAMP endogenously. This allows the study of many mutant cell types. Of course, all assays shown can also utilize cells that are allowed to undergo normal development.
2. The active form of folate has a 2-amino group. *Dictyostelium* secrete a deaminase that generates a 2-hydroxyl inactive variant of folate.
3. Inactivation of the deaminase has advantages for binding studies and for controlling absolute levels of added folate. Thus, it is important to understand the limits to minimize deaminase activity. Care should be noted when using 8-azaguanine for in vivo experiments, and various concentrations should be tested for secondary effects. 1.5 mM may inhibit general signaling aspects independently of the effects on deaminase.
4. To prepare the folate-binding cocktail in **step 3** of Subheading 3.6, add 1.5 μL of [^{3}H]-folate at a stock concentration of 30 μM, 2.2×10^9 dpm/mL, to 48.5 μL of PB; final concentration will be 900 nM folate, 3.3×10^6 dpm. Nonspecific background binding in the presence of >100-fold excess of unlabeled folate can approach 30–50 %. While background binding cannot be fully eliminated, it can sometimes be reduced further by inclusion of 100 μM latrunculin A, to inhibit pinocytosis, increase of unlabeled folate, and/or inclusion of excess deaminated folate (**step 11** of Subheading 3.3).

Acknowledgements

This work was supported by the Intramural Research Program of the National Institute of Diabetes and Digestive and Kidney Diseases, National Institutes of Health. We are indebted to the helpful comments and interest of Drs. Robert Kay and Carole Bewley.

References

1. Woznica D, Knecht DA (2006) Under-agarose chemotaxis of *Dictyostelium discoideum*. Methods Mol Biol 346:311–325
2. Hadwiger JA, Lee S, Firtel RA (1994) The Gα subunit Gα4 couples to pterin receptors and identifies a signaling pathway that is essential for multicellular development in *Dictyostelium*. Proc Natl Acad Sci U S A 91: 10566–10570
3. McMains VC, Liao XH, Kimmel AR (2008) Oscillatory signaling and network responses during the development of *Dictyostelium discoideum*. Ageing Res Rev 7:234–248
4. Artemenko Y, Lampert TJ, Devreotes PN (2014) Moving towards a paradigm: common mechanisms of chemotactic signaling in *Dictyostelium* and mammalian leukocytes. Cell Mol Life Sci 71:3711–3747

5. Veltman DM, Keizer-Gunnik I, Van Haastert PJ (2008) Four key signaling pathways mediating chemotaxis in *Dictyostelium discoideum*. J Cell Biol 180:747–753
6. Liao XH, Buggey J, Lee YK, Kimmel AR (2013) Chemoattractant stimulation of TORC2 is regulated by receptor/G protein-targeted inhibitory mechanisms that function upstream and independently of an essential GEF/Ras activation pathway in *Dictyostelium*. Mol Biol Cell 24:2146–2155
7. Liao XH, Kimmel AR (2009) Biochemical responses to chemoattractants in *Dictyostelium*: ligand-receptor interactions and downstream kinase activation. Methods Mol Biol 571: 271–281
8. Pan P, Wurster B (1978) Inactivation of the chemoattractant folic acid by cellular slime molds and identification of the reaction product. J Bacteriol 136:955–959
9. Segall JE, Bominaar AA, Wallraff E, De Wit RJ (1988) Analysis of a *Dictyostelium* chemotaxis mutant with altered chemoattractant binding. J Cell Sci 91:479–489

Chapter 12

Live Imaging of Border Cell Migration in *Drosophila*

Wei Dai and Denise J. Montell

Abstract

Border cells are a cluster of cells that migrate from the anterior tip of the *Drosophila* egg chamber to the border of the oocyte in stage 9. They serve as a useful model to study collective cell migration in a native tissue environment. Here we describe a protocol for preparing ex vivo egg chamber cultures from transgenic flies expressing fluorescent proteins in the border cells, and using confocal microscopy to take a multi-positional time-lapse movie. We include an image analysis method for tracking border cell cluster dynamics as well as tracking individual cell movements.

Key words Border cell, *Drosophila*, Egg chamber, Live imaging, Confocal, 3D tracking

1 Introduction

A general review of the anatomy and development of *Drosophila* ovaries can be found elsewhere [1]. Here we focus on stage 9 egg chambers (Fig. 1), which are composed of 1 oocyte and 15 nurse cells surrounded by several hundred epithelial follicle cells. At each end of each egg chamber, a special pair of cells called polar cells develops. The anterior polar cells recruit four to eight follicle cells to surround them and form a cluster called the border cells. In early stage 9 egg chambers (Fig. 1a), the border cell cluster rounds up, detaches from the anterior follicle cell layer, and migrates in between the nurse cells. As the cluster approaches the oocyte (Fig. 1b), it turns dorsally and stops close to the oocyte nucleus in late stage 9. Border cell migration serves as a useful model to study collective cell migration in a native environment [2]. The development of live imaging of border cell migration ex vivo [3] has opened the door to studying the dynamics of this process and has revealed new insights into the molecular and cellular mechanisms underlying this behavior (e.g., [4, 5]).

Here we describe a protocol for preparing stage 9 egg chambers with fluorescently labeled border cells for multi-positional time-lapse confocal imaging followed by 3D analysis to track the

Tian Jin and Dale Hereld (eds.), *Chemotaxis: Methods and Protocols*, Methods in Molecular Biology, vol. 1407,
DOI 10.1007/978-1-4939-3480-5_12,

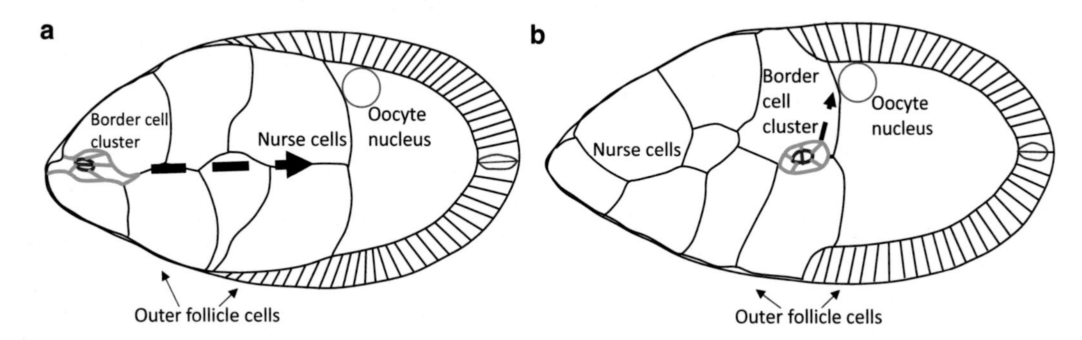

Fig. 1 Illustration of border cell cluster migration in stage 9 egg chambers. (**a**) Border cell cluster (*green*) surrounding anterior polar cells (*red*) detaching from the outer follicle cell layer during early stage 9. (**b**) Border cell cluster reaching oocyte border during late stage 9. *Arrows* indicate the direction of movement. Anterior is to the *left* and dorsal is *up*

whole cluster as well as individual cell movements. Oxygen, pH, and nutrients are critical factors for ensuring healthy ex vivo egg chamber development. Key factors to consider during imaging include avoiding phototoxicity and dynamic adjustment to keep the region of interest in focus. For image analysis, it is necessary to perform a drift correction to adjust for egg chamber motion and growth.

2 Materials

2.1 Fly Stocks That Express Fluorescent Reporters in Border Cells

1. Slbo-Gal4 can be used to drive UAS-fluorescent reporters in border cells [6] (Fig. 2a, c). An advantage of slbo-Gal4 is that it is expressed relatively specifically during oogenesis and is viable in combination with many transgenes, such as dominant-negative Rac, that are lethal in combination with most other Gal4 drivers. A disadvantage of slbo-Gal4 however is that its expression in border cells is relatively weak at the beginning of their migration and becomes stronger and stronger during the migration, so that exposure times need to be adjusted throughout the course of the experiment. Slbo-Gal4 also drives expression in a small group of posterior follicle cells and in centripetal follicle cells later in stage 10. Other Gal4 lines such as c306-Gal4 also drive high expression in border cells [7] (Fig. 2b, d). The advantage of c306-Gal4 is that the expression level does not change much during stage 9. A disadvantage of c306-Gal4 is that it drives expression earlier in development and is lethal in combination with many RNAi transgenes or transgenes driving dominant-negative proteins. This problem can be circumvented by combining c306-Gal4 with $Gal80^{ts}$ and growing the flies at 18 °C followed by a shift to 30 °C to inactivate $Gal80^{ts}$. We note

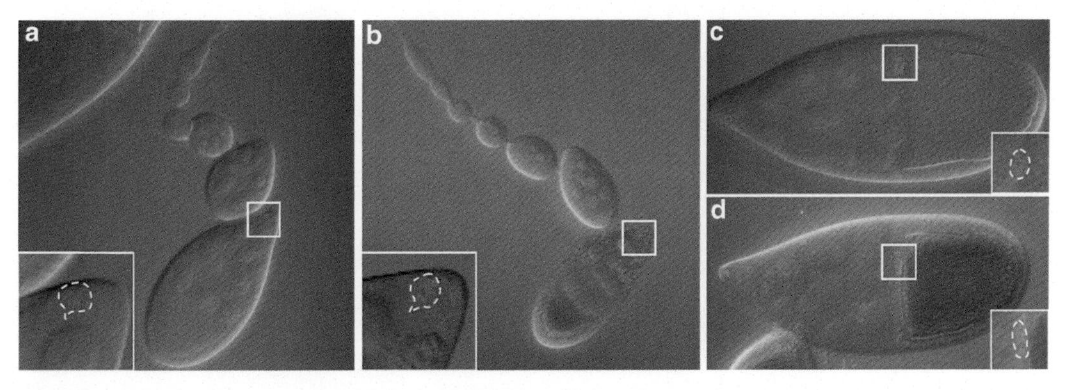

Fig. 2 Expression patterns of slbo-Gal4 and c306-Gal4. Maximum intensity projection of z stacks showing fluorescence superimposed on DIC images for slbo-Gal4 (**a** and **c**) or c306-Gal4 (**b** and **d**) driving UAS-dsRednls expression. The germarium through early stage 9 is shown in (**a**) and (**b**). Stage 10 egg chambers are shown in (**c**) and (**d**). *Inset* shows the magnified view of the *boxed area* with maximum intensity projection of the cropped z stacks for the border cell cluster. The outline of the border cell cluster is shown by the *dashed line*

that c306-Gal4 drives expression in a larger subset of follicle cells as compared to slbo-Gal4, which may be an advantage or a disadvantage, depending on the experiment.

2. After choosing the Gal4 driver, one needs to decide on the fluorescent protein to express. Nuclear fluorescent proteins such as dsRednls are useful to track individual cells, while F-actin-associated fluorescent proteins such as GFP::moesin and lifeact are useful to track the morphology of the whole cluster. Fluorescent reporters such as His2Av-mCherry for labeling all nuclei, or fluorescent dyes such as FM4-64 for labeling cell membrane, may also be used. We have generated transgenic flies expressing fluorescent proteins under the direct control of the slbo enhancer, which allows the use of Gal4 for other purposes. The lines we have found to be most useful are summarized in Table 1.
3. Fly food vials.
4. Dry yeast.

2.2 Imaging Medium

1. Schneider's *Drosophila* medium (Invitrogen, cat. no. 11720).
2. Fetal bovine serum (FBS; Invitrogen, cat. no. 16140).
3. Insulin (Sigma, cat. no. I1882).
4. HCl (1 M).
5. 0.22 μm filter (Millipore, cat. no. SCGP00525).
6. Dissection medium: Schneider's medium + 20 % FBS.
7. Live imaging medium: Dissection medium + 0.2 mg/ml insulin.

Table 1
Useful fluorescent labels for live imaging

Transgenic line	Purpose	Pattern[a]	Advantages	Disadvantages
Slbo-Gal4	Label border cells using the Gal4/UAS system	Stage 9 bc + pc + pfc + Stage 10 bc+++ pc++, pfc+++, cfc+++	Specific to adult stage	Expression level in border cells changes over time Expression in border cells can vary from one cell to the other Polar cell expression can be weak
C306-Gal4		bc+++ pc++ afc++ pfc+++	Expression level in border cells does not change over time	Start to express early in development and may be lethal when driving expression of RNAi or other transgenes
Upd-Gal4	Label polar cells using the Gal4/UAS system	pc +++	Specific for polar cells	Start to express early in development and may be lethal when driving expression of RNAi or other transgenes
Slbo-LifeactGFP	Label border cells without the Gal4/UAS system	Labels F-actin network under the Slbo promoter	Can be used in combination with Gal4/UAS system	
hsFLP; AyGal4, UAS-mCD8::GFP hsFLP; AyGal4, UAS-GFP::moesin	Mosaic labeling of the border cells in combination with RNAi or other transgenes	Labels cell membrane and cytoplasm in the FlipOut clone Labels F-actin network in the FlipOut clone	Can be used to drive RNAi or other transgenes in labeled cells	The number of border cells labeled is random

hsFLP, c306-Gal4, UAS-GFP; FRTXX, tub-Gal80	Mosaic labeling of the border cells in combination with mutant	Labels border cell in MARCM clone	Can be used to create labeled homozygous mutant clones	The number of border cells labeled is random
UAS-mCD8::GFP	Label cell membrane using the Gal4/UAS system	Membrane and cytoplasm		
UAS-GFP::moesin or UAS-mcherry::moesin UAS-LifeactGFP or UAS-lifeactRFP	Label F-actin network using the Gal4/UAS system	F-actin network in Gal4-expressing cells		
Ubi-DECad::GFP	Label all cell membranes	Labels all cell membranes, stronger in border cells	Do not have dye internalization issue compared with FM4-64	Signal weaker in nurse cell junctions compared with FM4-64
UAS-dsRed.nls or UAS-GFP.nls	Label nuclei using the Gal4/UAS system	Labels nuclei in the Gal4-expressing cells. Note that GFP.nls has some leaky cytoplasmic expression		
His2Av-mCherry	Label all cell nuclei	Labels all cell nuclei	Expression level is even in border cells and polar cells	Nurse cell nuclei are also labeled, which may be undesired in some cases

[a] *bc* border cells, *pc* polar cells, *pfc* posterior follicle cells, *cfc* centripetal follicle cells, *afc* anterior follicle cells

2.3 Dissection Tools

1. Anesthesia: CO_2 blow gun and fly pad.
2. Paintbrush.
3. 2-Well depression glass slides (Fisher Scientific, cat. no. 12-565B).
4. Dissecting microscope.
5. Dumont tweezers #5 Dumoxel (WPI, cat. no. 14098).
6. Transfer pipets (Thermo Scientific, cat. no. 231).

2.4 Mounting Tools

1. Micropipettes with disposable tips (2–20 μl, 20–200 μl, 100–1000 μl).
2. Lumox hydrophilic culture dishes (50 mm) (Sarstedt, cat. no. 94.6077.410).
3. Glass cover slips (Fisher, cat. no. 12-542-B, 12-545-C).
4. Kimwipe tissue.
5. Halocarbon oil 27 (Sigma, cat. no. H8773).

2.5 Imaging Tools

In order to take a 3D time-lapse movie, fast imaging speed is required for acquiring a z stack of the entire border cell cluster in less than 1 min. Border cell migration will be affected by phototoxicity; therefore a sensitive detector is needed so low light can be used. Because border cells are located 50–100 μm below the surface of the egg chamber and compression of the egg chamber will impede migration, a long-working-distance objective is needed. Also, water immersion is a more suitable medium than air or oil because it matches the refractive index of the mounting medium and tissue. The software should be able to handle live adjustment of the focus and location in a time series. We use the system below in the following Subheading 3 although we have been able to acquire live imaging on Nikon, Leica, and Zeiss 510 and Zeiss 710 laser scanning confocal systems, as well as a Zeiss Axioplan 2 widefield microscope. Confocal imaging gives substantially better results. We have not had success using spinning disk microscopy, which for reasons that are not entirely clear seems more phototoxic.

1. Zeiss 780 34Ch Spectral Confocal System with scanning stage.
2. 40× LD water objective (LD C-Apochromat 40×).
3. Zen software.
4. Multitime macro.

2.6 Data Analysis Tools

ImarisTrack or other segmentation software

3 Methods

3.1 Fatten Flies

1. Collect newly eclosed flies (8–12 females and a couple of males) of the desired genotype in a fresh vial with fly food and dry yeast for 2–3 days at room temperature (RT) or 25 °C (*see* **Note 1**).
2. The day before the experiment, transfer the flies to a fresh vial with fly food and dry yeast, put in 29 °C incubator for 16–18 h to increase the Gal4-driven gene expression level (*see* **Note 2**).

3.2 Prepare Live Imaging Medium

1. Aliquot 40 ml Schneider's medium in a 50 ml tube in a cell culture hood and store at 4 °C, aliquot 10 ml FBS per 15 ml tube in a cell culture hood, and store at −20 °C (*see* **Note 3**).
2. Make dissection medium: Warm Schneider's medium and FBS to room temperature, calibrate pH meter, and then measure pH of the mixed medium. The desired range is 6.85–6.95. Add HCl or NaOH if pH is too high or too low (*see* **Note 4**).
3. Filter sterilize and aliquot to 1.7 ml tubes in a cell culture hood to avoid contamination over multiple usages (*see* **Note 5**). Label with the date and store at 4 °C for up to 20 days.
4. Prepare insulin stock solution. Make acidified water (12 μl 1 M HCl to 1 ml of H_2O). Prepare 10 mg/ml insulin (5 mg powder to 500 μl acidified water). Aliquot 10 μl per 0.6 ml tube. Store at −80 °C.
5. Make live imaging medium: Before dissection, add 490 μl dissection medium to 10 μl of 10 mg/ml insulin. Final concentration of insulin is 0.2 mg/ml.

3.3 Dissect Ovary and Pull out Ovarioles

1. Clean up dissection bench using ethanol. Prepare wet tissue pad for disposal of fly body and unwanted ovary tissues. Fill the depression slide with dissection medium using the transfer pipette and place under a stereomicroscope.
2. Anesthetize flies using CO_2. Use forceps to pick up one female by its wing and hold it down in the middle of the depression well. Use one forcep to gently immobilize the body and the other forcep to pinch a small piece of abdominal cuticle and pull toward the posterior (Fig. 3a, b). The pair of ovaries should pop out (Fig. 3c, d). Discard the remaining carcass by wiping it onto the moist tissue pad. Carefully remove any undesired tissue and leave only the ovaries in the depression well. Dissect several flies until you get three to four ovary pairs (Fig. 3g, h) (*see* **Note 6**). Well-fattened ovaries are bigger than poorly fattened ones. A third of the ovaries should be composed of egg chambers stage 10 or younger in well-fattened ovaries (Fig. 3g). Well-fattened ovaries should not have more than one mature egg chamber per ovariole (Fig. 3h), nor should they contain egg chambers that are degenerating.

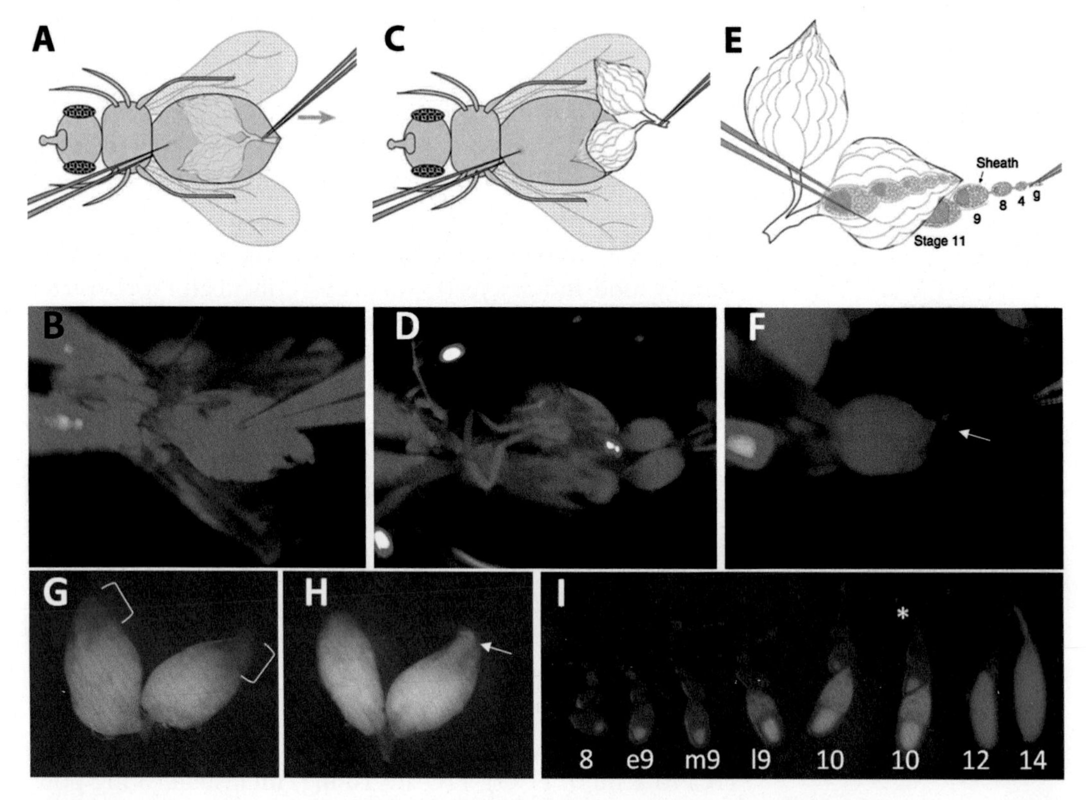

Fig. 3 Dissection of stage 9 egg chambers. (**a** and **b**) Hold the fly body and pinch the abdominal cuticle. (**c** and **d**) Pull until the ovaries come out. (**e** and **f**) Pull ovarioles out of muscle sheath. *Arrow* points to a stage 9 egg chamber emerging from the ovary. (**g**) In well-fattened ovaries, the anterior 1/3 of the volume of the ovaries should be composed of young egg chambers, which are not as opaque as mature eggs (*bracket*). (**h**) Ill-fattened ovaries are usually smaller, have less early-stage egg chambers, and have more mature egg chambers sometimes in the anterior tip of the ovary (*arrow*). (**i**) Different stage egg chambers. Stage 9 is characterized by the size of the egg chamber, outer follicle cell arrangement, oocyte proportion, and the amount of yolk deposits in the germ cells. *Asterisk* points to an ovariole that remains in muscle sheath

3. Use one pair of forceps to anchor the posterior part of the ovary where stage 10 and later egg chambers are located, and use the other forceps to pinch the anterior tip of the ovary where the germarium and early-stage egg chambers are located. Pull very slowly and the string of egg chambers will pop out of the muscle sheath (Fig. 3e, f). Repeat this procedure one ovariole at a time (*see* **Note 7**). Aim for early stage 9 egg chambers (Fig. 3i). Repeat this to get 20–30 ovarioles.
4. Separate ovarioles that are joined at the tip and egg chambers older than stage 9.
5. Move 20–30 desired ovarioles to one side of the slide and collect with a transfer pipette. Transfer to a 0.6 ml tube.
6. Let the egg chambers sink to the bottom, remove supernatant, and add 90 μl of live imaging medium (*see* **Note 8**).

3.3.1 Mount Egg Chambers

1. Break a 22 × 22 mm cover slip into half using the blunt end of the forceps, and remove glass debris.
2. Decide whether to mount on the inner side or bottom of the Lumox dish depending on the use of an upright or inverted microscope. Place two 5 μl drops of live imaging medium on the 50 mm Lumox dish separated by ~18 mm.
3. Put a cover slip fragment on each of the two small drops with the smooth edge facing the center. Tap and nudge the cover slip until there are no air bubbles in between the dish and the cover slip. Properly space the cover slip bridge so they are 18 mm apart. Use a Kimwipe to wick away excess medium (*see* **Note 9**).
4. Remove all but a few microliters of medium from the tube containing the egg chambers, then add 90 μl of fresh live imaging medium, pipet up and down gently a few times, then take up the egg chambers, and expel them onto the Lumox dish in a circular motion so they are not crowded or too close to the edge.
5. Use one forceps to pick up a 22 × 40 cover slip and another forceps end to guide one edge of the cover slip to line up with the bridge fragments. Slowly lower the cover slip right on top of the bridge. Gently tap the dish to let medium spread evenly between the cover slip and the membrane.
6. Remove excess medium from the two sides of the bridge. Apply a thin layer of halocarbon oil and remove any excess oil with tissue paper (Fig. 4) (*see* **Note 10**).

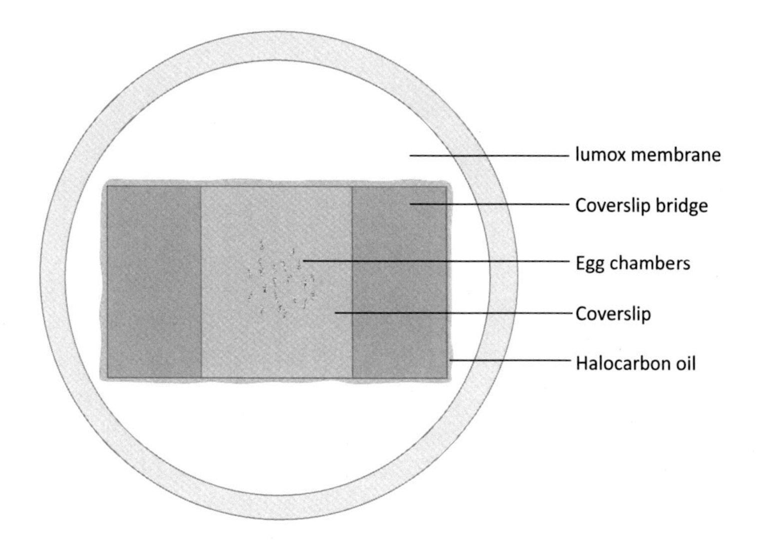

Fig. 4 Mounted egg chambers in Lumox dish

Table 2
Troubleshooting for live imaging

Problem	Possible reason	Solutions
Migration does not complete	Phenotype of mutant	Confirm phenotype in fixed samples
	Ovary was dissected from a fly that was old or ill fattened	Use young and well-fattened flies
	Dissection damage	Dissect quickly and do not touch stage 9 with the forceps
	Medium is old or of wrong pH	Make new medium
	Surrounded by late-stage egg chambers or too crowded	Remove late-stage egg chambers and mount sparsely
	Compressed by cover slip	Leave enough medium for mounting
	Phototoxicity	Reduce laser power or scan time
Signal gets weaker later	Laser power too strong	Reduce laser power and open up pinhole
	Scanning time too long	Increase scan speed
Egg chamber drifts a lot	Excess medium	Remove excess medium so stage 9 moves slowly when the dish is tilted
	Big air bubbles	Carefully put on the cover slip to avoid big bubbles
	Excess halocarbon oil	Apply a thin layer of halocarbon oil
	Dish not loaded horizontally in the sample holder	Fit the dish well into the sample holder
	Stage moves too fast	Choose positions that are not too far away
	Stage 9 is close to germarium or ovariole has remaining muscle sheath	Avoid such egg chambers
	Ovariole is twisted and egg chamber moves when adjusting focus	Avoid such egg chambers
Egg chamber degenerates later	Dissection damage	Dissect quickly and do not touch stage 9 with sharp tip of the forceps
	Medium is old or of wrong pH	Make new medium
	Surrounded by late-stage egg chambers or too crowded	Remove late-stage egg chambers and mount sparsely

3.4 Confocal Imaging

Potential image acquisition problems that might be encountered and their solutions are summarized in Table 2.

3.4.1 Locate Positions

1. Turn on the microscope following Zeiss procedures. Log on to the computer and load the Zen software.
2. Load the Lumox dish into the sample holder.
3. Select configuration: "Acquisition" tab: Open acquisition configuration "Live_Pos_1." Check "Positions" and pull up the positions list (*see* **Note 11**).
4. Locate positions using the 10× objective lens: "Locate" tab: Use brightfield to focus, and then use GFP to find the desired

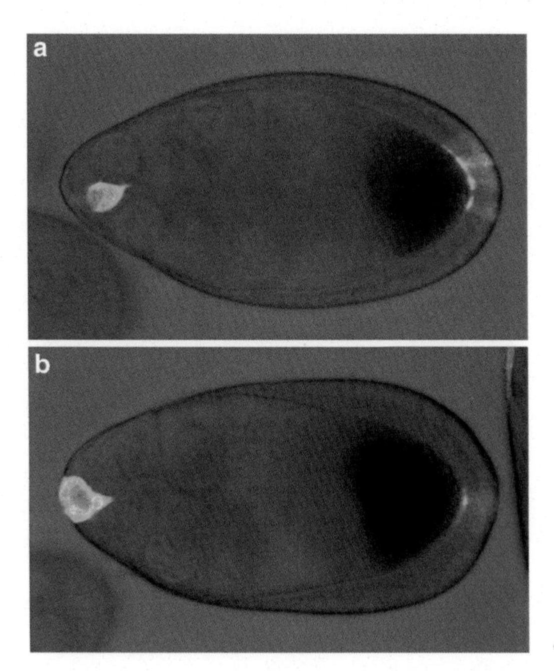

Fig. 5 DIC images with GFP and nuclear dsRed channels superimposed. (**a**) Early stage 9 egg chamber with normal morphology and border cell cluster detaching. (**b**) Stage 9 egg chamber with slightly delayed border cell migration. Note that the outer follicle cells and the proportion of the oocyte indicate that the egg chamber is at a later stage than (**a**), but the border cells are less detached. In this particular example, the border cells never detached indicating abnormal development

egg chambers that are intact, early stage 9, and in which border cells are just detaching (Fig. 5). Egg chambers should not be too close to a germarium or have residual muscle sheath attached. If the scanner head cannot be rotated, then a horizontally positioned egg chamber is preferable because it is faster to scan than a vertically positioned one. "Add" locations in the position list (*see* **Note 12**).

5. Fine adjust position: Change to the 20× objective. Lower the illuminator brightness at higher magnifications. "Update" positions for four to six best egg chambers that show a clear front protrusion, which is a characteristic of normal migrating border cell cluster during early stage 9. "Save" a copy of the positions to prevent accidental deletion.
6. Switch to the 40× water objective. Manually move the objective to the side to put on water or immersion fluid, rotate the objective back, and make sure that the immersion water touches the cover slip (*see* **Note 13**). Remove any location that moves when adjusting the focus. It is best to select egg chambers that are relatively close to each other to avoid large stage movements between positions, which can cause the egg chambers to move and no longer be at the marked position.

3.4.2 Set Acquisition Configuration

1. Modify acquisition settings: “Acquisition” tab: Move to the first position, and set scan area to “0.6” and frame size to “512×512.”
2. Center the sample: “Live” and “Stop,” and click “Stage” to center the egg chamber.
3. Define z stack: Check “Z-Stack” and select “First/Last.” “Live” and set first and last and then go to center, “Stop.” Select “Center” and set interval to “1.5” and slices to “28.” The z stack acquired will be larger than the border cell cluster to allow for movement of the cluster in the *z* direction over time.
4. Adjust laser power and pinhole: Go to range indicator and adjust the laser power so that the signal is bright but not saturated. If the laser power is already at the upper limit for not inducing significant phototoxicity, increase the pinhole if the signal is not bright enough (*see* **Note 14**).
5. Crop the sample: Click “Crop” and set scan area to “0.6” and frame size to “512×512.” Crop to select the egg chamber. “Update” position and “Save” configuration.
6. Repeat for all other samples you have and save as different configurations.

3.4.3 Set Mutlitime Macro

1. Select Multitime macro: A new window should pop up. Hit “refresh.”
2. “Saving” tab: Enter the base file name and select folder.
3. “Acquisition” tab: Choose “Scan Configuration” for each location, and check “Z Stack.”
4. “Timing” tab: Enter wait interval and number of repetition.
5. Check “Recipe Summary” and then go back to “Saving,” and hit “Start” and “Open Original Folder.”
6. Adjust focus and field of view during the time course: Check how much you need to adjust. Wait till the wait interval countdown, “Pause,” move to the location you want to adjust, adjust *z* position in the “Focus” window, and adjust *x y* position in the “Stage” window, “Replace XYZ.” Move to the first location, “Resume.”
7. “Finish” if you want to finish the time series early. “Stop” if you do not want to save the experiment.

3.5 Imaris Analysis

3.5.1 Correct Egg Chamber Drift

1. OPEN raw data in Imaris and SAVE AS an “ims” file.
2. Crop TIME to remove unwanted frames.
3. FREE ROTATE to position egg chamber anterior to the left (Fig. 6a–d) (*see* **Note 15**).
4. Start the SPOTS algorithm. Select Track Spots over time.

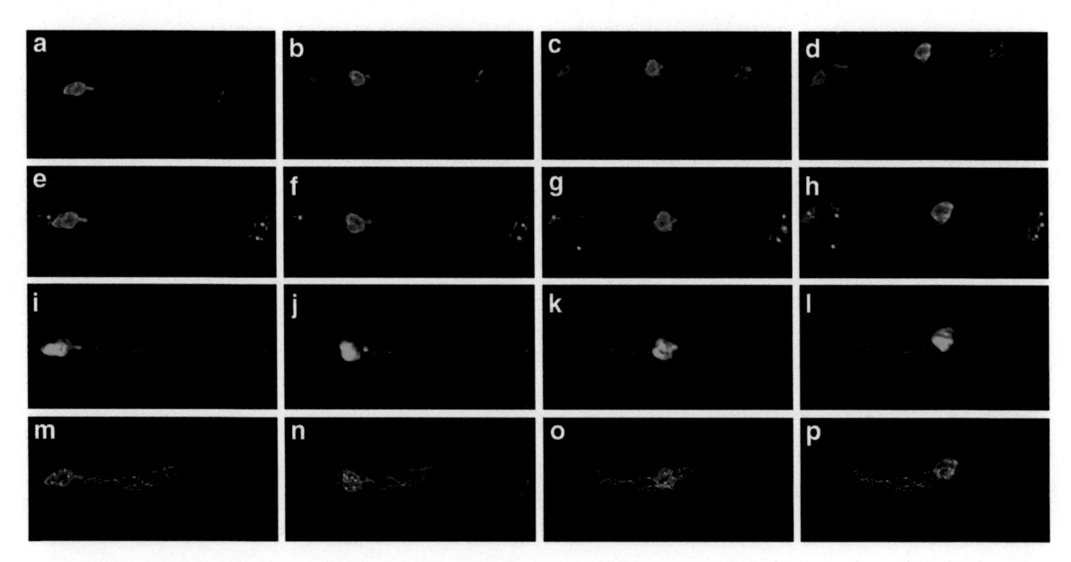

Fig. 6 Stage 9 slboGal4; UASmCD8GFP, UAS-dsRednls egg chambers at four time points in a time-lapse series. From *left* to *right* are images showing border cells detaching, in the first half of migration, in the second half of migration, and reaching oocyte border. (**a**–**d**) Before drift correction. (**e**–**h**) After drift correction. (**i**–**l**) Track of the border cell cluster. (**m**–**p**) Tracks of border cell nuclei

5. Select the region of interest to cover main body follicle cell nuclei with dsRed signal that you plan to use as immobile spots.
6. Choose the channel for follicle cell nuclei, estimate the diameter of the nuclei, and enter a number.
7. Adjust the quality filter so that the majority of the follicle cell nuclei you plan to track are detected automatically over the time course.
8. Use Brownian motion as the tracking algorithm.
9. The number of tracks should match the number of border cell nuclei. Finish the Spots algorithm.
10. Go to the Edit Tracks tab. Each track should represent one follicle cell nuclei. Select one track at a time and go frame by frame through the movie to ensure that the spots localize to the same nuclei over each frame. Sometimes when there is a sudden drift due to change of focus or position when imaging, the tracks will be incorrect. Manually disconnect and reconnect the right spots to get the correct track.
11. Correct drift and choose the "Translational and rotational drift" algorithm. Select "Include entire result" (Fig. 6e–h).

3.5.2 Track Border Cell Cluster and Nuclei

1. CROP 3D to select only the egg chamber.
2. Create SURFACES to track border cell cluster. (Fig. 6i–l).
3. Start the SPOTS algorithm. Select Track Spots over time.

4. Select the region of interest to cover the border cell migration path over the time course.
5. Choose the channel for the border cell nuclei, estimate the diameter of the nuclei, and enter a number.
6. Adjust the quality filter so that the majority of the border cell nuclei are detected automatically over the time course. Write down the time frames when there is a problem in nuclei detection. If the problem is that some nuclei are not detected in some time frames, try going back to **step 3** and enter a smaller number for the diameter and vice versa. Adjust estimate diameter and quality filter so that most of the border cell nuclei can be detected automatically.
7. Edit spots for the time frames that have problems in nuclei detection. Use ESC to switch between the viewing and editing mode. Use Shift + Left click to add or delete a spot in the editing mode.
8. Use Brownian motion as the tracking algorithm.
9. The number of tracks should match the number of border cell nuclei. Finish the Spots algorithm.
10. Go to the Edit Tracks tab. Each track should represent one border cell nuclei at each time frame. Manually added spots will show up as a nick in the track. Select one track at a time and go frame by frame through the movie to ensure that the spots localize to the same nuclei over each frame (Fig. 6m–p).

4 Notes

1. Always include a wild-type control when analyzing mutants.
2. If fattening needs to be done for longer than 18 h, flip once more halfway through incubation time.
3. Opened bottle of Schneider's medium tends to have salt precipitates over time. Aliquoting reduces the chance of contamination and salt composition change due to precipitation.
4. Freshly opened bottles of Schneider's medium usually have lower pH while the pH increases upon storage. pH of mixed medium can range from 6.6 to 7.1. After adding 1 N HCl or 5 N NaOH, mix thoroughly before measuring pH again.
5. To avoid contamination of the live imaging medium, these precautions should be taken: (1) Before using the cell culture hood, UV sterilize for 30 min. (2) Use sterile tips and tubes that have not been previously opened outside the hood. (3) When transferring materials to the hood, use ethanol to completely sanitize gloves and the outside of other materials. (4) Filter sterilize the medium in the cell culture hood.

6. Be careful with the sharp tip of the forceps. Do not press or drop on any hard surface as it will bend the tip and render the forceps useless.
7. If you pull too fast, the ovarioles will not come out of the muscle sheath, or will break off too soon. If you anchor the ovary too far at the posterior of the ovary, later stage egg chamber will also come out and will take time to separate later. When two to three ovarioles come out at once, there is usually more muscle sheath attached which should be avoided. Never touch the stage 9 egg chambers with the forceps because this will cause damage.
8. Keep stage 9 intact. Do not collect debris or broken egg chambers as they will affect the clarity of the buffer. Crowding and too many older stage chambers are undesirable as the stage 9 will not get sufficient nutrients and oxygen.
9. Do not touch the Lumox membrane with sharp edges of the cover slip or forceps. If the Lumox membrane of the dish breaks, do not use.
10. Avoid big bubbles in between the cover slip and membrane as they will affect the imaging. Too much medium will be bad for imaging while too little medium will be bad for the egg chambers. Stage 9 egg chambers should be able to move slowly when the dish is tilted. Halocarbon oil prevents evaporation. Make sure that the edges are sealed completely. Do not use excess oil as it will cause sample drift. Lumox dish has a gas-permeable base to help oxygen transfer. It can be reused multiple times as long as the membrane is still intact. After the experiment, lift the cover slip carefully, wash away the oil with Windex, and rinse thoroughly in distilled water. Air-dry the membrane and keep in a clean box.
11. Acquisition configuration: To increase speed: use simultaneous instead of sequential scanning of the 488 and 561 lasers, use bidirectional scanning, decrease frame size, use higher scan speed and less averaging, and crop the area of interest. To reduce phototoxicity: use sensitive detector, use lower laser power, open up pinhole, and increase gain. To improve resolution and image quality: increase frame size, reduce scan speed, and decrease gain. To monitor the morphology of the egg chamber and outer follicle cells: use T-PMT.
12. To reduce bleaching, adjust brightness of the illuminator, and close the shutter when not using the eye pieces.
13. Make sure not to mix different types of immersion solutions and use appropriate immersion solution for the objective.
14. To test for phototoxicity, compare samples imaged and not imaged. Also the fluorescent intensity of Slbo-Gal4-driven reporters should increase over time. If signal is reduced over time then photobleaching has occurred.

15. If you have a large file which takes a long time to render, choose a single time point and perform a 3D crop on this point to see if it works. If it does, then apply it to the whole image.

Acknowledgements

This work was supported by NIH grants R01-GM046425 and R01-GM073164 to DJM.

References

1. Spradling AC (1993) Developmental genetics of oogenesis. In: Martinez-Arias B (ed) The development of *Drosophila melanogaster*. Cold Spring Harbor Laboratory, Cold Spring Harbor, NY, pp 1–70
2. Montell DJ, Yoon WH, Starz-Gaiano M (2012) Group choreography: mechanisms orchestrating the collective movement of border cells. Nat Rev Mol Cell Biol 13:631–645
3. Prasad M, Jang AC, Starz-Gaiano M et al (2007) A protocol for culturing *Drosophila melanogaster* stage 9 egg chambers for live imaging. Nat Protoc 2:2467–2473
4. Wang X, He L, Wu YI et al (2010) Light-mediated activation reveals a key role for Rac in collective guidance of cell movement in vivo. Nat Cell Biol 12:591–597
5. Cai D, Chen SC, Prasad M et al (2014) Mechanical feedback through E-cadherin promotes direction sensing during collective cell migration. Cell 157:1146–1159
6. Rørth P, Szabo K, Bailey A et al (1998) Systematic gain-of-function genetics in *Drosophila*. Development 125:1049–1057
7. Manseau L, Baradaran A, Brower D et al (1997) GAL4 enhancer traps expressed in the embryo, larval brain, imaginal discs, and ovary of *Drosophila*. Dev Dyn 209:310–322

Chapter 13

shRNA-Induced Gene Knockdown In Vivo to Investigate Neutrophil Function

Abdul Basit, Wenwen Tang, and Dianqing Wu

Abstract

To silence genes in neutrophils efficiently, we exploited the RNA interference and developed an shRNA-based gene knockdown technique. This method involves transfection of mouse bone marrow-derived hematopoietic stem cells with retroviral vector carrying shRNA directed at a specific gene. Transfected stem cells are then transplanted into irradiated wild-type mice. After engraftment of stem cells, the transplanted mice have two sets of circulating neutrophils. One set has a gene of interest knocked down while the other set has full complement of expressed genes. This efficient technique provides a unique way to directly compare the response of neutrophils with a knocked-down gene to that of neutrophils with the full complement of expressed genes in the same environment.

Key words RNA interference, Short hairpin RNA, Neutrophils, Retroviral vectors, Transplantation, Gene knockdown

1 Introduction

Neutrophil recruitment is the primary cellular response in acute inflammation. However, the study of neutrophils has been limited due to their unique property of being short-lived and terminally differentiated and hence difficult to transfect. Gene knockout techniques have been successfully used to investigate neutrophil biology but knockout mice are expensive and time consuming to generate. For these reasons, we exploited endogenous RNA interference (RNAi) to develop a rapid and efficient system of gene knockdown. It has been shown that short hairpin RNA (shRNA) can be expressed in hard-to-transfect cells and in whole organisms [1]. We used shRNA to induce gene knockdown in hematopoietic stem cells and successfully transplanted these stem cells into irradiated mice. We showed that this technique allows the generation of mice with circulating neutrophils with a specific gene knocked down. We investigated in vivo responses in these mice and in vitro responses by using neutrophils harvested from these mice [2, 3].

Tian Jin and Dale Hereld (eds.), *Chemotaxis: Methods and Protocols*, Methods in Molecular Biology, vol. 1407, DOI 10.1007/978-1-4939-3480-5_13,

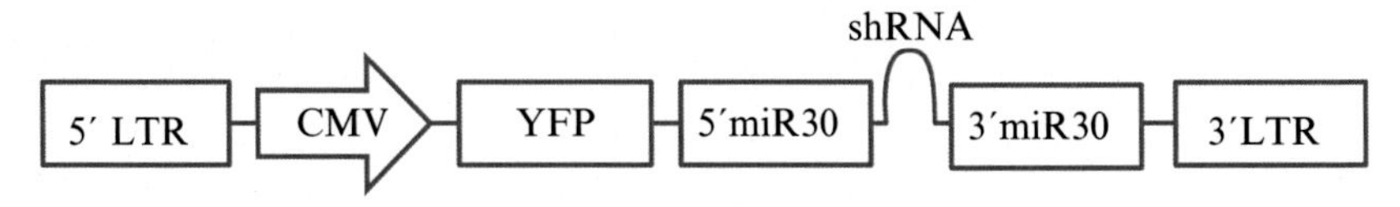

Fig. 1 Schematic of shRNA construct adapted from [2]

We discuss the principle of this efficient technique and describe the detailed protocol to perform these experiments.

Compared to a naked shRNA, the embedding of an shRNA within the sequence of a microRNA (miRNA) has been shown to improve the efficiency of gene knockdown [4]. We designed a cassette that included a sequence for venus fluorescent protein (a variant of YFP) and a sequence for shRNA directed at the β2 subunit of G proteins embedded within the sequence of human miR-30. This resulted in the cassette YFP-mir30-shGβ2. To transfect mouse hematopoietic stem cells, we chose a retroviral vector MIGR1, a mouse stem cell-based vector [5], and subcloned the YFP-miR30-shGβ2 cassette into it. This resulted in the vector LTR-YFP-miR30-shGβ2. To further improve the efficiency, we incorporated a cytomegalovirus (CMV) promoter with a globin intron between 5′ LTR and YFP-miR30-shGβ2. This resulted in the vector LTR-CMV-YFP-miR30-shGβ2. Similarly, a control vector construct was designed with shRNA directed at Luciferase and was called LTR-CMV-YFP-miR30-shLuc. Figure 1 shows a schematic of this construct.

We used the PHE (Phoenix Ecotropic) cell line for packaging [5, 6]. These cells produce viral gag-pol and envelope proteins. We used PCL-ECO as packaging vector. PCL-ECO and LTR-YFP-miR30-shGβ2 or LTR-YFP-miR30-shLuc were co-transfected into PHE cells and resulting virus was tested for transfection efficiency in 3T3 cells. Virus was then transfected into hematopoietic stem cells harvested from wild-type (C57BL/6) mice bone marrow. The mice had already been treated with 5-fluorouracil (5-FU) to enrich the stem cells. The transfected stem cells were then transplanted into wild-type (C57BL/6) mice that had been irradiated. Four weeks after transplantation, the blood collected from tail veins was tested for the circulating neutrophils by flow cytometry. There were two sets of circulating neutrophils: YFP positive and YFP negative. As the shRNA were co-induced with YFP, the YFP-positive cells are with knocked-down gene while YFP-negative cells have a full complement of expressed genes and are an excellent internal control.

These mice were used for in vivo studies and neutrophils harvested from bone marrow were used for in vitro studies, and we showed that shRNA-induced gene knockdown is an efficient method to investigate neutrophil biology in vivo [2].

The protocol to perform these experiments is spread over 8 days. After 8 days mice are monitored for 8 weeks and then used.

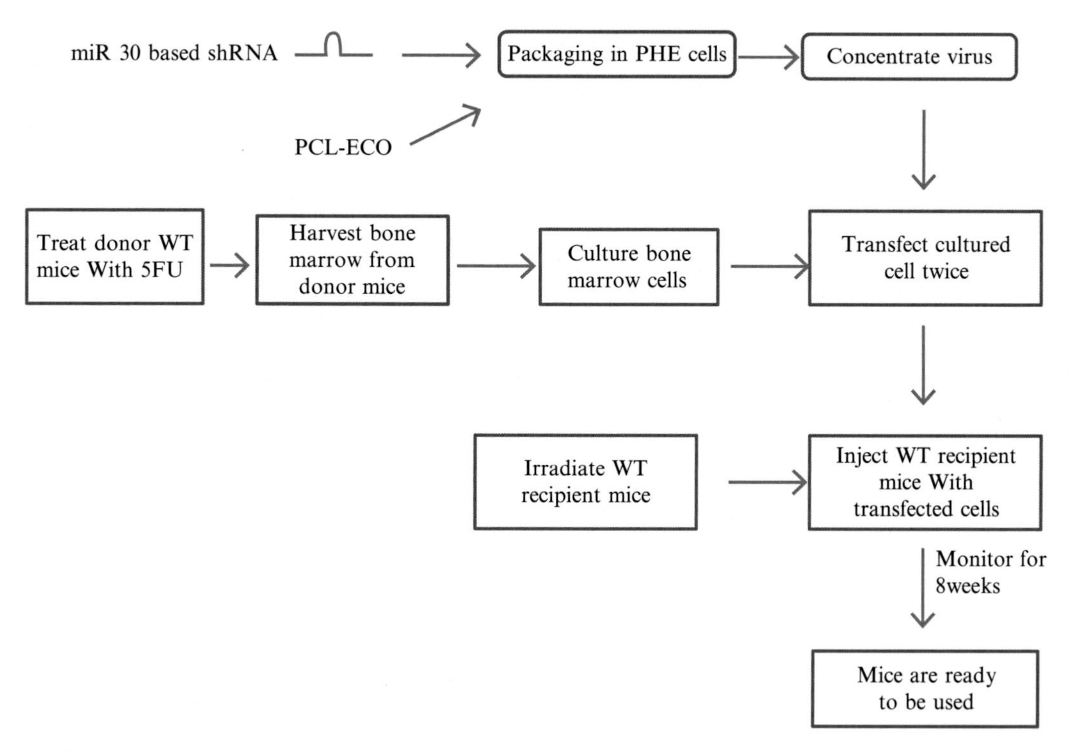

Fig. 2 Outline of the experimental protocol adapted from [2]

At times, the protocol requires parallel procedures being performed in cells and animals. Figure 2 shows an outline of the experiment protocol.

2 Materials

2.1 Plasmids and Transfection Reagents

1. MIGR1 plasmid with shRNA cassette (Addgene) (*see* **Note 1**).
2. PCL-ECO plasmid (Addgene).
3. Lipofectamine® LTX and Plus™ Reagent (Life Technologies).

2.2 Cells

1. PHE cells (Phoenix-ECO from ATCC).
2. 3T3 cells (ATCC).

2.3 Media, Solutions, and Cytokines

1. Iscove's modified Dulbecco's medium (IMDM).
2. Recombinant murine stem cell factor (mSCF; Invitrogen).
3. Recombinant murine interleukin-3 (IL-3; Peprotech).
4. Recombinant human interleukin-6 (IL-6; Peprotech).
5. Dulbecco's modified Eagle's medium (DMEM; Gibco).
6. Fetal bovine serum (Cat # 10082-139 Gibco).
7. Dulbecco's phosphate-buffered saline (Gibco).

8. Penicillin (Sigma).
9. Streptomycin (Sigma).
10. HEPES (Gibco).
11. 0.25 % Trypsin (Gibco).
12. Polybrene (Sigma).
13. ACK RBC lysis buffer: 0.15 M ammonium chloride, 10 mM potassium bicarbonate, 0.1 mM EDTA.
14. Normal saline USP 10 ml vials (Hospira).

2.4 Chemicals

1. 5-FU (Sigma).
2. 70 % Ethyl alcohol.

2.5 Equipment

1. T-75 flasks.
2. 12-Well plates.
3. 24-Well plates.
4. 50 ml tubes.
5. 15 ml tubes.
6. 1.5 ml Eppendorf tubes.
7. 0.2 μm Filter.
8. Virus concentrating filter (Merck Millipore cat #910024).
9. CO_2 chamber to euthanize mouse.
10. Surgical instruments: Scissors, forceps.
11. Centrifuge capable of high-speed centrifugation of up to $500 \times g$ for both tubes and plates.
12. Flow cytometer.
13. Incubator to culture cells at 37 °C under 5 % CO_2 in humidified air.
14. Access to irradiator to radiate mice.

2.6 Mice

1. C57BL/6 wild-type mice.

3 Methods

3.1 Preparative Steps (Day 1)

1. Inject mice with 5-FU, dose: 150 mg/kg, route intraperitoneal (*see* **Note 2**).
2. Seed a T-75 flask with 3×10^6 PHE cells in a medium of 90 % DMEM, 10 % FBS, 100 units/ml penicillin, 0.1 mg/ml streptomycin, and 25 mM HEPES and incubate at 37 °C under 5 % CO_2 in humidified air (*see* **Note 3**).

3.2 Transfect PHE Cells (Day 2)

1. Set up two 15 ml tubes as follows:
   (a) In "tube 1" combine shRNA plasmid (8 μg), PCL-ECO (10 μg), Plus reagent (35 μl), and serum-free medium (1.75 ml).
   (b) In "tube 2" combine Lipofectamine (35 μl) and serum-free medium (1.75 ml).
2. Incubate both tubes for 15 min at room temperature.
3. Mix the contents of tube 1 with tube 2 and wait for another 15 min.
4. Add 4 ml serum-free medium, bringing the total volume to 8 ml.
5. Aspirate the medium from the flask of PHE cells and wash them once with serum-free medium without dislodging the cells. Then transfer the above 8 ml transfection complex into the flask.
6. After 3 h, replace the transfection complex with 30 ml of complete medium (90 % DMEM, 10 % FBS, 100 units/ml penicillin, 0.1 mg/ml streptomycin, 25 mM HEPES).
7. Culture in incubator at 37 °C under 5 % CO_2 in humidified air.

3.3 Prepare 3T3 Cells for Viral Titer Measurement (Day 3)

1. Check PHE cell transfection efficiency by looking at the flask under microscope for fluorescent cells.
2. Grow 3T3 cells in a 24-well plate for measurement of viral titer. You will need five wells. In general, 1.5×10^6 cells are needed to seed all the wells of a 24-well plate.

3.4 Concentrate Virus (Day 4)

1. Collect supernatant of PHE cells into a 50 ml tube (approx. recovery: 30 ml).
2. Centrifuge at 500 × *g* for 3 min to remove the cell debris.
3. Load 14 ml of supernatant into concentrating filter.
4. Centrifuge at 500 × *g* for 10 min at room temperature.
5. Repeat **steps 3** and **4** until 30 ml supernatant has been concentrated to less than 500 μl.
6. Transfer the supernatant into new 1.5 ml Eppendorf and store at 4 °C. This is the virus stock solution.

3.5 Infect 3T3 Cells for Measurement of Viral Titers

1. Make Polybrene medium by mixing 4 μl of polybrene with 4 ml of full medium (90 % DMEM, 10 % FBS, 100 units/ml penicillin, 0.1 mg/ml streptomycin, 25 mM HEPES).
2. Use this Polybrene medium to dilute virus stock solution to make the following dilutions: 1:1000, 1:10,000, 1:30,000, 1:90,000, and one negative control.
3. Aspirate medium from wells of 24-well plate of 3T3 cells from Subheading 3.3 and load 500 μl of each dilution onto 3T3 cells.

4. Centrifuge 24-well plate at 500×*g* for 90 min at room temperature.
5. After centrifugation, incubate plate in the incubator.

3.6 Bone Marrow Extraction (See Note 4)

1. Set up 10 ml syringe, 27G needle, 22G needle, scissors and forceps, ethanol, 12-well pate, 50 ml tube, 15 ml tubes, DMEM free, FBS, PBS, penicillin, streptomycin, and cytokines and a 70 μm filter (*see* **Note 5**).
2. Euthanize mice with CO_2 exposure. Clean mouse surface with ethanol and bring it into the hood.
3. Set up medium: 4 % FBS in PBS with 100 units/ml penicillin and 0.1 mg/ml streptomycin and fill wells of a 12-well plate with the medium.
4. Remove long bones (tibia and femur). After removing the bone, rinse immediately with the medium. Repeat rinse twice (*see* **Note 6**).
5. Flush marrow with the medium and collect bone marrow in a 50 ml tube.
6. Filter bone marrow through a 70 μm filter into a new 50 ml tube.
7. Centrifuge at 500×*g* for 5 min at room temperature, collect pellet, wash with 5 ml medium, and transfer into a new 15 ml tube.
8. Centrifuge again and remove supernatant.
9. Lyse RBCs by resuspending the pellet with 3 ml ACK buffer. Allow lysis for 3 min at room temperature (*see* **Note 7**).
10. After 3 min, add new medium to make up to 15 ml and centrifuge at 500×*g* for 3 min at room temperature.
11. Remove supernatant and resuspend the pellet in 10 ml medium.
12. Count the number of cells (*see* **Note 8**).
13. Centrifuge the cell suspension at 500×*g* for 5 min. While the cell suspension is in the centrifuge, prepare the pre-stimulation medium in Subheading 3.7, **step 1**, below.
14. Aspirate the supernatant. The pellet of the cells will be dissolved in the pre-stimulation medium from Subheading 3.7, **step 1**.

3.7 Culture Bone Marrow Cells in 6-Well Plate

1. Prepare pre-stimulation medium by supplementing 20 % FBS in IMDM with the following cytokines at the indicated final concentrations: mSCF (50 ng/ml); IL3 (10 ng/ml); and IL6 (10 ng/ml).
2. Resuspend the bone marrow cells in the pre-stimulation medium at a concentration of 7.5 million cells/ml.

3. Add 2 ml of cell suspension to each well of 6-well plate and culture in the incubator.

3.8 Check Viral Titers (Day 5)

1. Remove supernatant from the wells of 3T3 cells (from Subheading 3.5).
2. Wash once with DPBS.
3. Add 100 μl of trypsin into each well.
4. Incubate for 30 s at 37 °C.
5. Add 400 μl of medium to each well and resuspend cells.
6. Aspirate cells into 1.5 ml Eppendorf tubes.
7. Determine the percentage of YFP-positive cells by flow cytometry.
8. Good transfection efficiency is represented by 20 % positive cells in the dilution 1:90,000 and 60 % positive cells in the dilution 1:30,000 (*see* **Note 9**).

3.9 Infect Bone Marrow Cells: First Infection (Day 6)

1. Add 40 μl of the viral stock solution into each well of 6-well plate with cultured bone marrow cells (*see* **Note 10**).
2. Add 2 μl of Polybrene to each well.
3. Centrifuge the plate at 500 × *g* for 90 min at room temperature.
4. Incubate plate overnight.
5. Replace the remaining viral stock solution to 4 °C.

3.10 Infect Bone Marrow Cells: Second Infection (Day 7)

1. Add 40 μl from the remaining viral stock solution to each well of the 6-well plate with cultured bone marrow cells.
2. Centrifuge the plate at 500 × *g* for 90 min at room temperature.
3. This time no need to use Polybrene.
4. Incubate plate overnight in the incubator at 37 °C under 5 % CO_2 in humidified air.

3.11 Inject Mice with Transfected Bone Marrow Cells (Day 8)

1. Collect all the bone marrow cells. To maximize cell recovery, wash the wells with PBS.
2. Count total number of cells and determine percentage of YFP-positive cells by flow cytometry.
3. Centrifuge cells at 500 × *g* for 5 min.
4. Remove supernatant and resuspend in normal saline (*see* **Note 11**).
5. Repeat **steps 3** and **4**.
6. Resuspend the final pellet in normal saline to leave a cell concentration of ten million/ml.

7. Anesthetize recipient mice.
8. Irradiate recipient mice (*see* **Note 12**).
9. Inject 120 μl of cell suspension into one mouse via retro orbital vein (*see* **Note 13**).
10. When mice recover from anesthesia, return them to cages.
11. Mice should only receive autoclaved food and water supplemented with Sulfatrim according to local animal care policies.
12. The mice are ready to be used for experiments 8 weeks post-transplantation.

4 Notes

1. shRNA can be obtained from commercial vendors. Alternatively, one can design shRNA. We use the program at the following website to design shRNA for our gene of interest:
   http://cancan.cshl.edu/RNAi_central/RNAi.cgi?type=shRNA.
   We then modify the sequence to make its ends cohesive and compatible for ligation into vector. We then design the complementary strand of the shRNA. These oligonucleotides are synthesized chemically by commercial service, annealed, and then ligated into the vector that had already been digested with appropriate restriction enzymes.
2. Warm 5-FU at 37 °C for at least 10 min.
3. We expect PHE cells to be 70 % confluent at the time of transfection.
4. Cells harvested from mouse bone marrow will be injected into irradiated mice. These mice are highly susceptible to infection. It is critical that aseptic technique is maintained during harvesting and preparation of cells for infection.
5. To avoid cell contamination, it is preferable to use autoclaved surgical instruments for harvesting bone marrow.
6. After removing long bones, remove the mouse body from the hood immediately.
7. Filter RBC lysis solution through a 0.2 μm syringe filter before use.
8. Counting cells at this stage helps determine the volume of prestimulation medium to make. We count cells by flow cytometry. An aliquot of 200 μl from the 10 ml cell suspension is used for counting, and the remaining cell suspension is centrifuged. If a flow cytometer is not available for counting, counting on a hemocytometer under microscope should work.

9. If the transfection efficiency is less than 20 % positive cells in the dilution 1:90,000 or less than 60 % in the dilution 1:30,000, successful outcome is unlikely.
10. Just add viral stock solution and Polybrene. There is no need to mix. Try not to interfere with cells.
11. From this stage onwards, washes are done in sterile normal saline.
12. We use radiation dose of 9.5 grays. The dose may vary depending upon the type of equipment available.
13. We do not inject into tail vein, but if expertise is available it should work fine.

Acknowledgements

We acknowledge the contributions of Yong Zhang, Ph.D., Research Scientist, Serono Inc., Rockland, MA, in the initial setting up of the system during his research fellowship in the Wu Lab.

References

1. Paddison PJ, Caudy AA, Sachidanandam R, Hannon GJ (2004) Short hairpin activated gene silencing in mammalian cells. Methods Mol Biol 265:85–100
2. Zhang Y, Tang W, Jones MC et al (2010) Different roles of G protein subunits β1 and β2 in neutrophil function revealed by gene expression silencing in primary mouse neutrophils. J Biol Chem 285:24805–24814
3. Tang W, Zhang Y, Xu W et al (2011) A PLCβ/PI3Kγ-GSK3 signaling pathway regulates cofilin phosphatase slingshot2 and neutrophil polarization and chemotaxis. Dev Cell 21:1038–1050
4. Silva JM, Li MZ, Chang K et al (2005) Second-generation shRNA libraries covering the mouse and human genomes. Nat Genet 37:1281–1288
5. Pear WS, Miller JP, Xu L et al (1998) Efficient and rapid induction of a chronic myelogenous leukemia-like myeloproliferative disease in mice receiving P210 bcr/abl-transduced bone marrow. Blood 92:3780–3792
6. Swift S, Lorens J, Achacoso P, Nolan GP (2001) Rapid production of retroviruses for efficient gene delivery to mammalian cells using 293T cell-based systems. Curr Protoc Immunol Chapter 10:Unit 10.17C

Chapter 14

Studying Neutrophil Migration In Vivo Using Adoptive Cell Transfer

Yoshishige Miyabe, Nancy D. Kim, Chie Miyabe, and Andrew D. Luster

Abstract

Adoptive cell transfer experiments can be used to study the roles of cell trafficking molecules on the migratory behavior of specific immune cell populations in vivo. Chemoattractants and their G protein-coupled seven-transmembrane-spanning receptors regulate migration of cells in vivo, and dysregulated expression of chemoattractants and their receptors is implicated in autoimmune and inflammatory diseases. Inflammatory arthritides, such as rheumatoid arthritis (RA), are characterized by the recruitment of inflammatory cells into joints. The K/BxN serum transfer mouse model of inflammatory arthritis shares many similar features with RA. In this autoantibody-induced model of arthritis, neutrophils are the critical immune cells necessary for the development of joint inflammation and damage. We have used adoptive neutrophil transfer to define the contributions of chemoattractant receptors, cytokines, and activation receptors expressed on neutrophils that critically regulate their entry into the inflamed joint. In this review, we describe the procedure of neutrophil adoptive transfer to study the influence of neutrophil-specific receptors or mediators upon the their recruitment into the joint using the K/BxN model of inflammatory arthritis as a model of how adoptive cell transfer studies can be used to study immune cell migration in vivo.

Key words Neutrophil, Adoptive transfer, Arthritis

1 Introduction

Neutrophils are critical innate immune cells in host defense and contribute importantly to many inflammatory diseases. The ability of the neutrophil to exert its functions relies heavily on their ability to traffic to sites of infection and inflammation [1]. In general, neutrophil migration into inflamed tissue occurs in postcapillary venules and can be described by the adhesion cascade, which begins with the capture of free-flowing leukocytes to the vessel wall, followed by (1) *rolling* on the vessel wall in the direction of flow, (2) *arrest* on the endothelium, (3) release from adhesion and *crawling* in all directions on the vessel to locate a receptive location for, (4) *transmigration* (TEM), followed then by (5) *swarming* to a specific location within the tissue [1]. These discrete steps are controlled by a combination of molecular signals, including the

Tian Jin and Dale Hereld (eds.), *Chemotaxis: Methods and Protocols*, Methods in Molecular Biology, vol. 1407,
DOI 10.1007/978-1-4939-3480-5_14, © Springer Science+Business Media New York 2016

selectins (rolling), integrins (arrest, crawling), and chemoattractants (arrest, possibly crawling, and TEM).

Chemoattractants regulate the trafficking of immune cells, including neutrophils, between tissues as well as their positioning and interactions within tissues. Even though chemoattractants consist of a wide range of diverse molecules with differing regulation and biophysical properties, including secreted proteins (e.g., chemokines), proteolytic fragments of serum proteins (e.g., complement component C5a), and bioactive lipids (e.g., leukotrienes), they all induce directed migration by activating seven-transmembrane-spanning G protein-coupled receptors (GPCRs) [2]. Differential expression of chemoattractant receptors on cells results in selective recruitment of specific cell types under particular conditions, providing appropriate and efficient immune responses tailored to the foreign insult [2]. Beyond their pivotal role in the coordinated migration of innate immune cells to inflamed tissue, chemoattractants play important roles in the development of lymphoid tissues, in the maturation of immune cells, and in the generation and delivery of adaptive immune responses [2]. Altered expression of chemoattractants and their receptors is implicated in a broad range of human diseases, including cancer and autoimmune and inflammatory diseases [2, 3].

Inflammatory arthritis, including rheumatoid arthritis (RA), is an autoimmune disease characterized by neutrophil recruitment into the diseased joint [4]. To analyze the mechanism of inflammatory cell recruitment into the joint, and the role of chemoattractants and their receptors in inflammatory arthritis in vivo, several arthritic model animals, including the K/BxN serum transfer model and type II collagen-induced arthritis model, have been studied [5, 6]. K/BxN mice are T cell receptor transgenic mice that spontaneously develop arthritis due to the development of autoantibodies in the ubiquitous protein glucose-6-phosphate isomerase [7]. Transfer of serum from K/BxN mice into recipient wild-type mice results in the deposition of autoantibody-antigen complexes in joints, which induce neutrophil recruitment and subsequent arthritis [7]. We and others have previously found that at least four chemoattractant receptors—CCR1, CXCR2, C5aR, and BLT1—contribute to the recruitment of neutrophils into the joint in K/BxN serum transfer model [6, 8, 9]. The neutrophil is a key effector cell in this model, and this model can be used to study the role of individual chemoattractants in neutrophil recruitment into the inflamed joint in vivo.

Adoptive cell transfer experiments are helpful for determining the role of certain molecules, including cytokines, chemokines, and their receptors, on individual immune cell populations, such as neutrophils, T cells, and B cells, in various animal models of diseases. For example, our previous research demonstrated that leukotriene B_4 (LTB_4) receptor 1 (BLT1)-deficient mice did not develop arthritis, and that adoptive transfer of wild-type (WT)

BLT1-expressing neutrophils into BLT1-deficient mice restored arthritis [6]. Further, mice deficient in 5-lipoxygenase and LTA_4 hydrolase, enzymes required to synthesize LTB_4, also did not develop arthritis, and adoptive transfer of WT neutrophils into these enzyme-deficient mice also resorted arthritis [9]. These adoptive transfer studies established that the neutrophil is the critical LTB_4-producing and -responding cell type required for neutrophil entry into the joint and for the development of arthritis. Thus, adoptive cell transfer experiments can be used to explore the role of a particular cell type as well as individual molecules on that cell type in the activation and migration of that cell type in vivo in models of inflammation and disease.

Here, we describe the method of neutrophil isolation from mouse bone marrow (BM) and adoptive BM neutrophil transfer in a model of inflammatory arthritis to study neutrophil recruitment into the joint. We describe how to quantitatively and qualitatively describe neutrophil migratory behavior into a defined peripheral tissue site in vivo.

2 Materials

(Selected supplies and instruments are depicted in Fig. 1.)

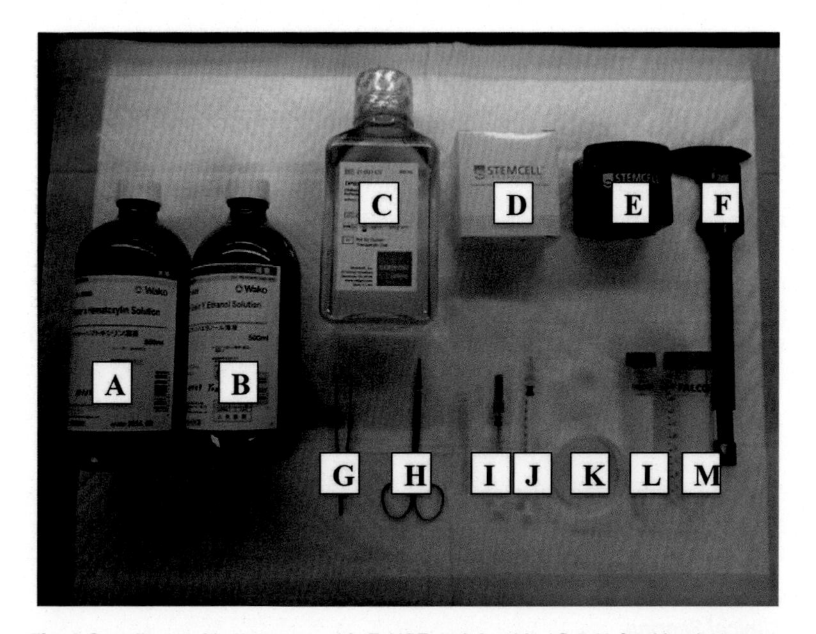

Fig. 1 Supplies and instruments. (*A*, *B*) H&E staining kit. (*C*) 2 % fetal bovine serum (FBS) + PBS without Ca^{2+} and Mg^{2+} + 1 mM EDTA. (*D*) EasySep mouse neutrophil isolation kit, containing normal rat serum, mouse neutrophil enrichment cocktail, biotin selection cocktail, and D magnetic particles. (*E*) Magnet. (*F*) Caliper. (*G*) Forceps. (*H*) Scissor. (*I*) 28-G × ½ insulin syringe. (*J*) 23-G needle with syringe. (*K*) 6 cm dish. (*L*) 15 ml Falcon tube. (*M*) 50 ml Falcon tube

2.1 K/BxN Serum Harvest

1. Eight- to ten-week-old K/BxN mice (*see* **Note 1**).
2. 23 G needles for cardiac puncture.
3. 1.5 ml Eppendorf tubes.

2.2 Neutrophil Isolation from Bone Marrow (BM)

1. Bone marrow neutrophil donors: 6–8-week-old C57/BL6 male or female mice or desired genetically modified mice, such as CCR1-, CXCR2-, C5aR-, or BLT1-deficient mice [6, 8, 9].
2. 1-cc plastic syringes with 23-G needle for intraperitoneal (i.p.) injection of serum from K/BxN mice.
3. 70 μm cell strainer.
4. 1-cc plastic syringes with 27-G needle.
5. EasySep mouse neutrophil enrichment kit (Stemcell technologies).
6. EasySep Magnet (Stemcell technologies).
7. Sterile Cell Isolation Medium: 2 % Fetal bovine serum (FBS) + PBS without Ca^{2+} and Mg^{2+} + 1 mM EDTA (*see* **Note 2**).
8. 15 and 50 ml Falcon tubes.
9. Falcon 5 ml polystyrene round-bottom tubes.
10. Timer.
11. Forceps.
12. Scissors.
13. 6 cm Tissue culture plates.
14. Razor blades.

2.3 Adoptive Transfer of Neutrophils and K/BxN Serum Transfer

1. Bone marrow neutrophil recipient mice: 6–8-week-old C57/BL6 male or female mice and/or age/sex-matched genetically modified mice (*see* **Note 3**).
2. Ketamine (80 mg/kg) and xylazine (12 mg/kg) for anesthesia.
3. 1-cc plastic syringe with 23-G needle for i.p. injection of ketamine and xylazine.
4. 28-G × ½ insulin syringes for intravenous (i.v.) injection of neutrophils.
5. 1 % FBS/PBS (sterile).
6. 1.5 ml Eppendorf tubes.

2.4 Arthritis Clinical Scoring and Measurement of Paw Thickness

1. Calipers.

2.5 Arthritis Histological Scoring

2.5.1 Decalcification

1. 10 % Formalin.
2. 70, 80, 90, and 100 % ethanol.
3. Mixed reagent (xylene:100 % ethanol = 1:3) for removing fat from tissue samples.
4. K-CX decalcification reagent (FALMA).

2.5.2 Deparaffinization

1. Xylene.
2. 100 % Ethanol.
3. 70 % Ethanol.
4. Distilled deionized water (DDW).

2.5.3 H&E Staining

1. Hematoxylin and eosin (H&E) staining kit (Wako Pure Chemical Industries).
2. Mount-Quick (Daido Sangyo) or other solvent-based mounting medium.

2.5.4 Dehydration

1. Xylene.
2. 100 % Ethanol.
3. 70 % Ethanol.

2.6 Immunohistochemistry for Quantification of Neutrophils in the Joints

1. Citrate buffer (Sigma-Aldrich).
2. PBS.
3. Skim milk.
4. Rat purified anti-mouse Ly-6G antibody (clone 1A8, Biolegend).
5. N-Histofine Simple Stain Mouse Max PO (Rat) (Nichirei Biosciences).
6. N-Histofine DAB 3S kit (Nichirei Biosciences).
7. H&E staining (optional).
8. Antibody Diluent (Dako Cytomation).
9. Coplin jars.
10. Slide holders.
11. Humidified slide chamber.

2.7 Flow Cytometry Analysis of Synovial Fluid Neutrophils

2.7.1 Synovial Fluid

1. 10 or 20 μl pipetman and tips.
2. Sterile 1 % FCS/PBS.
3. Hemocytometer.

2.7.2 Synovial Tissue

1. Surgical knife.
2. Collagenase type IV (Sigma-Aldrich).
3. 2 % FCS/PBS.

2.7.3 Staining with Ly-6G

1. Rat APC-conjugated anti-mouse Ly-6G antibody (clone 1A8, Biolegend).
2. Purified anti-mouse CD16/CD32 antibody (clone 2.4G2, BD Pharmingen) as Fc block.
3. 2 % FCS/PBS.
4. 2 % Paraformaldehyde.
5. Falcon 5 ml polystyrene round-bottom tubes.

2.8 MPO Assay

1. Mechanical homogenizer.
2. MPO assay kit (Hycult Biotech) containing anti-MPO antibody-coated 96-well plates, dilution buffer, streptavidin-peroxidase, biotinylated tracer antibody, TMB substrate, and stop solution.
3. PBS.
4. Plate reader.

2.9 Cell Tracking by Microscope

1. BM neutrophils.
2. Cell tracker Orange CMTMR (Life Technologies).
3. 1 % FCS/PBS (*see* **Note 4**).
4. FCS.
5. Water bath.

3 Methods

3.1 Isolation of Serum from K/BxN Mice

1. Bleed K/BxN mice by cardiac puncture immediately after CO_2 euthanasia and transfer the blood to 1.5 ml Eppendorf tubes (*see* **Note 5**).
2. Leave blood at room temperature (RT) for 1 h.
3. Centrifuge at 16,000 × *g* for 1 min at RT.
4. Transfer the supernatant (serum + blood) into new 1.5 ml Eppendorf tubes and discard the clotted blood.
5. Centrifuge the supernatant at 850 × *g* for 10 min at RT.
6. Remove the clarified supernatant (serum) and put it into new 1.5 ml Eppendorf tubes.
7. Store serum in 1 ml aliquots at −20 °C.

3.2 Neutrophil Isolation from BM

1. Dissect four bones (femurs and tibias) from bone marrow neutrophil donor mice, put in 50 ml tubes containing sterile PBS, and place on ice (*see* **Note 6**).
2. In a sterile hood, cut off the ends of each bone with a clean razor blade. Grasp the bone with forceps and insert a 27-G needle filled with 1 ml cell isolation medium into the marrow

space. Flush the marrow out into a 6 cm plate containing 5 ml of cell isolation medium (*see* **Note 7**).

3. Pass BM through a 27-G needle to break up any clumps.
4. Pass all the BM cells through a 70 μm cell strainer into a 50 ml Falcon tube to remove large pieces of bone and tissue.
5. Add cell isolation medium to the 6 cm plate used for BM harvest to rinse out any remaining cells, and put into a 50 ml Falcon tube through a 70 μm cell strainer.
6. Centrifuge for 5 min, 4 °C, $300 \times g$.
7. The method of neutrophil isolation described below is according to the EasySep manufacturer's protocol (from **steps 8** to **17**).
8. BM cells are resuspended at a concentration of 1×10^8 cells/ml in cell isolation medium.
9. Add normal rat serum at 50 μl/ml of cell suspension for blocking, and EasySep Mouse Neutrophil Enrichment Cocktail at 50 μl/ml of cells. Mix well and incubate on ice for 15 min.
10. Wash cells with cell isolation medium and centrifuge at $300 \times g$ for 10 min.
11. Discard the supernatant, resuspend the cells at 1×10^8 cells/ml in cell isolation medium, and put into Falcon 5 ml polystyrene round-bottom tubes.
12. Add EasySep Biotin Selection Cocktail at 50 μl/ml of cells, mix well, and incubate on ice for 15 min.
13. Vortex EasySep D Magnetic Particles for 30 s, add these particles at 150 μl/ml of cells, mix well, and incubate on ice for 10 min.
14. Bring the cell suspension to total volume of 2.5 ml by adding cell isolation medium.
15. Place the tube into the magnet. Set aside for 3 min.
16. Pour the tube while still in the magnet into a new 10 ml Falcon tube. Neutrophils are in the new tube.
17. Neutrophils derived from BM are resuspended at 1.0×10^8/ml in 100 μl sterile 1 % FCS/PBS and stored on ice until use.

3.3 Bone Marrow Neutrophil Adoptive Transfer and K/BxN Serum Transfer

To elucidate the function of genetically modified neutrophils, we recommend that the following four experimental groups be included: (1) WT neutrophil transfer into WT mice as recipient; (2) genetically modified neutrophil transfer into WT mice; (3) WT neutrophil transfer into genetically modified mice; and (4) genetically modified neutrophil transfer into genetically modified mice. However, if a given molecule is important in more than one immune cell type, WT neutrophil transfer may not restore disease in the knockout mouse recipient [9]. In this case, adoptively transferring WT neutrophils into a mouse strain that is known to become susceptible to disease after WT neutrophil transfer (e.g., IL-1α/β-deficient mice [8]) may be used as a positive control to ensure that the neutrophil transfer protocol is working correctly in the investigators' hands.

Perform the following three steps on both days 1 and 3.

1. Prior to injection of mice, calculate the amount of K/BxN serum required for the experiment as follows: (# mice + 1) × 150 μl = volume needed for experiment. Defrost the required amount of K/BxN serum needed for the experiment. If multiple K/BxN serum aliquots are required, combine and mix them prior to injection to ensure that all mice in a given experiment will receive the same serum.
2. Anesthetize recipient mice by i.p. injection of ketamine (80 mg/kg) and xylazine (12 mg/kg).
3. Introduce 1.0×10^7 cells/mouse in a volume of 100 μl sterile 1 % FCS/PBS by i.v. injection into the ophthalmic vein using a 28 G insulin syringe, followed by 150 μl K/BxN serum injected i.p. using a 26 G syringe.

3.4 Arthritis Clinical Scoring and Measurement of Paw Thickness

Arthritis clinical scoring and paw thickness are analyzed every 1–3 days after bone marrow neutrophil adoptive transfer.

1. Determine the arthritis clinical score as described previously [5]. Disease severity for each limb is recorded on a scale of 0–4 as follows: 0 = normal; 1 = erythema and swelling of one digit; 2 = erythema and swelling of two digits or erythema and swelling of ankle joint; 3 = erythema and swelling of more than three digits or swelling of two digits and ankle joint; and 4 = erythema and severe swelling of the ankle (*see* **Note 8**), foot, and digits with deformity. The clinical arthritis score is defined as the sum of the scores for all four paws of each mouse (maximum possible score 16).
2. Measure the thickness of each paw using a pair of digital slide calipers (Fig. 2).

3.5 Arthritis Histological Scoring

3.5.1 Preparation of H&E-Stained Tissue Specimens

1. Harvest mouse ankle joints, including the tibiotalar joint to the midfoot.
2. Fix samples with 10 % formalin overnight.
3. Remove fat by placing samples into the mixed reagent for at least 30 min (*see* **Note 9**).
4. Put each sample into 100, 90, 80, and 70 % ethanol, and DDW for 5 min at each step.
5. Place each sample into K-CX decalcification reagent overnight (*see* **Note 10**).
6. Wash samples with water for at least 12 h.

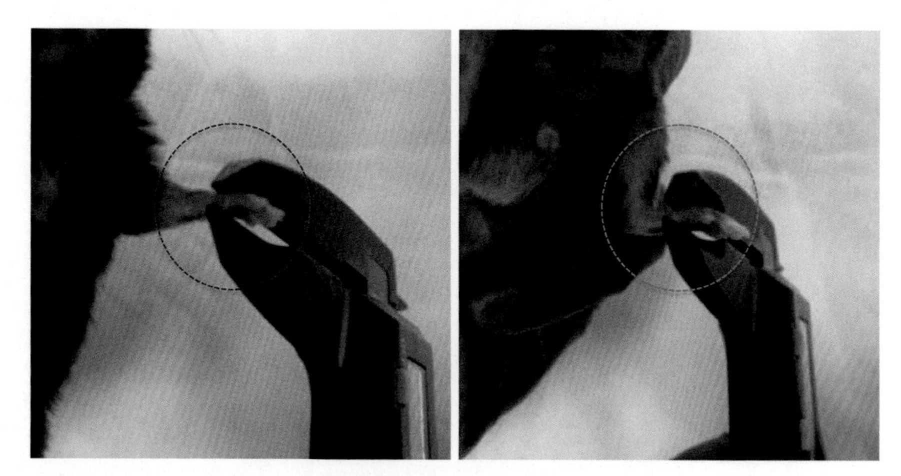

Fig. 2 Measurement of paw thickness. *Left panel*: The *red circle* shows the measurement of paw thickness of forefoot. *Right panel*: The *green circle* shows measurement of paw thickness from instep to plantar in the hind limb

7. Embed samples with paraffin, cut into 4 μm thick sections, and deposit them on glass slides.
8. Deparaffinize the specimens by placing them into xylene for at least 15 min, followed by an additional 5 min in new xylene.
9. Place slides into 100 % ethanol for 3 min. Perform this step twice.
10. Place samples into 70 % ethanol for 3 min. Perform this step twice.
11. Place samples into DDW for 3 min.
12. Place slides into hematoxylin for 5 min and then wash with water for 3 min.
13. Place slides into eosin for 5 min and then wash with water for 3 min.
14. To dehydrate the specimens, place the slides into 75 % ethanol for 5 min and repeat once.
15. Then place slides into 100 % ethanol for 5 min and repeat once.
16. Then place slides into xylene for 5 min and repeat once.
17. Finally, mount slides using Mount-Quick.

3.5.2 Histological Scoring

Examine the H&E-stained tissue section and assign histological score based on the following criteria: 0 = no inflammation, 1 = focal inflammatory infiltration, and 2 = severe and diffuse inflammatory infiltration (Fig. 3) (*see* **Note 11**).

3.6 Quantitation of Neutrophil Migration into the Joint

Immunohistochemistry and flow cytometry are used for evaluation of neutrophil infiltration into the inflamed lesion in vivo [6, 10]. In addition, enzymatic quantitation of the presence of neutrophil granule proteins in the tissue can also be used as a surrogate marker

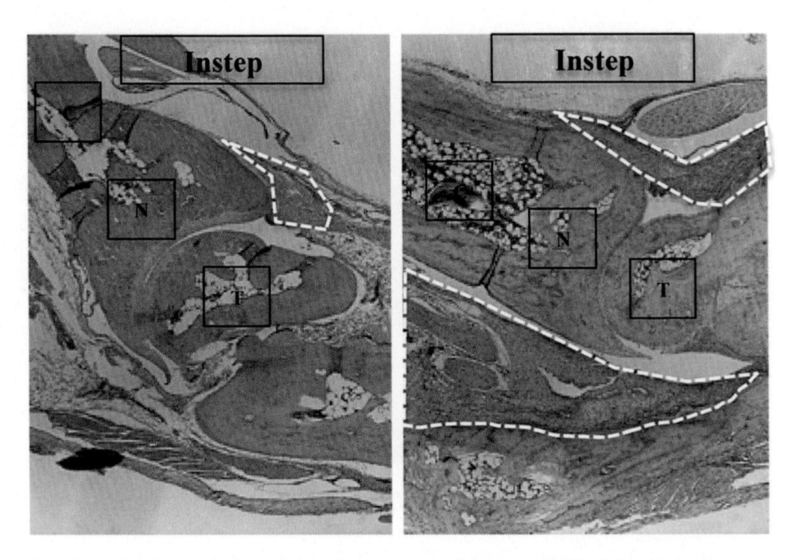

Fig. 3 Evaluation of the histological score of joints. H&E-stained sections are evaluated histologically and inflammation is scored using the following criteria: 0 = no inflammation; 1 = focal inflammation; 2 = severe and diffuse inflammatory infiltration. *Left panel* shows an example of focal inflammation where the inflammation is limited to one side of the ankle. *Right panel* shows an example of diffuse inflammatory infiltration throughout the ankle. *Broken line* indicates the inflamed site. *T* taulus, *N* naviculare, *C* cuneiform

to quantitate neutrophils in tissue [11]. Finally, labeled cell tracking dyes can be used to quantitate neutrophil migration into tissue [5, 10, 12].

3.6.1 Immunohistochemical Analysis of Neutrophils in Synovial Tissue (See Fig. 4)

1. Prepare tissue sections as described in Subheading 3.5.1, **steps 1–11**.
2. Put slides in Coplin jar filled with 10 mM citrate buffer for antigen retrieval and place in 65 °C hot water bath for 45 min (*see* **Note 12**).
3. Wash with PBS for 3 min. Each wash step is repeated three times (*see* **Note 13**).
4. Incubate slides in 0.3 % H_2O_2 for 30 min to block endogenous peroxidase activity.
5. Wash.
6. Block slides in 1 % skim milk for 45 min at RT (*see* **Note 14**).
7. Wash.
8. Staining: Cover tissue sections with 10 μg/ml purified anti-mouse Ly-6G antibody diluted with antibody diluent as first antibody and place slides in humidified chamber in 1 h at RT.
9. Wash.
10. Detect antibody binding using an N-Histofine Simple Stain kit according to the manufacturer's directions.

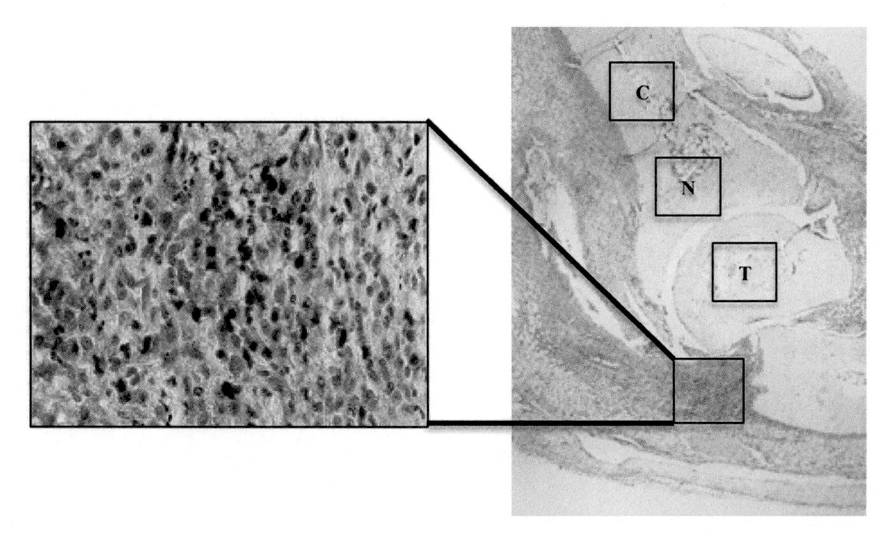

Fig. 4 Immunohistochemistry for Ly-6G-positive neutrophils into the joint. *Brown-colored cells* are Ly-6G-positive neutrophils in the joint. To quantitate neutrophils in the joint, the number of Ly-6G-positive neutrophils in three randomly selected fields per slide was counted. *Left* photomicrograph is a magnification of the *black square* in the *right* photomicrograph. *T* taulus, *N* naviculare, *C* cuneiform

11. Wash.
12. H&E staining (optional) (*see* Subheading 3.5.1).
13. Dehydration (*see* Subheading 3.5.1).
14. Evaluation: The number of Ly-6G-positive neutrophils in three randomly selected fields per slide is examined under light microscopy (*see* **Note 15**).

3.6.2 Analysis of Neutrophils in Synovial Fluid by Flow Cytometry (See Figs. 5 and 6)

1. Euthanize mice and harvest ankle joints, removing skin.
2. Using a surgical knife, make a small incision of about 3–5 mm on one side of the ankle (tibiotalar) joint.
3. Using a 10 or 20 μl pipetman and appropriate tip containing 5 μl sterile 1 % FCS/PBS, lavage through the ankle incision and transfer retrieved synovial fluid to an Eppendorf tube. Repeat this step until it appears that most cells have been retrieved from the joint space.
4. Synovial fluid cells are counted using a hemocytometer.
5. Resuspend 1×10^6 synovial fluid cells in a volume of 100 μl 1 % FCS/PBS, and transfer to a 5 ml polystyrene tube.
6. Incubate synovial fluid cells with 5 μg/ml anti-mouse CD16/CD32 antibody for 5 min at RT to block Fc receptors.
7. Stain with 2 μg/ml anti-mouse Ly-6G antibody for 10 min at RT.
8. Wash with 2 % FCS/PBS three times and resuspend in 2 % paraformaldehyde.
9. Store in the dark at 4 °C until analysis by flow cytometry.

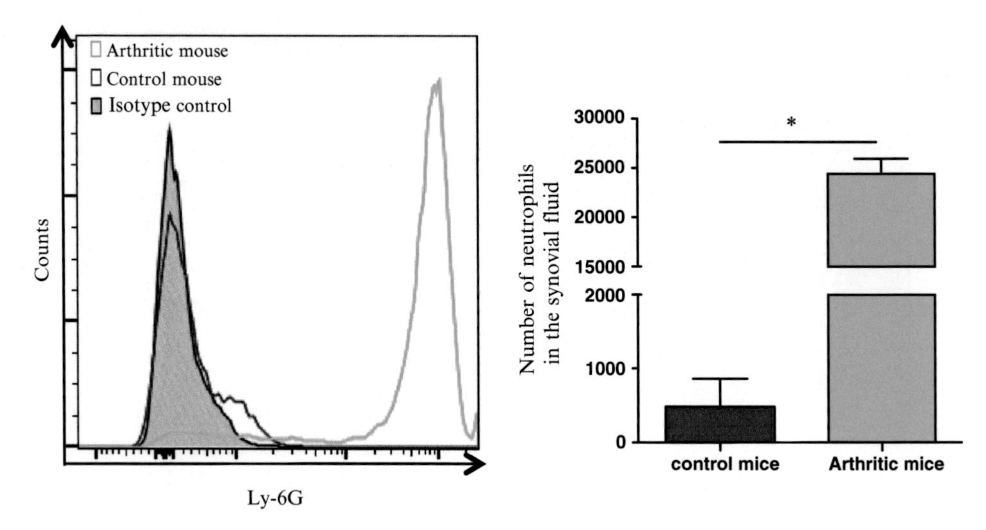

Fig. 5 Analysis of neutrophils in synovial fluid by flow cytometry. Synovial fluid neutrophils were obtained from a control mouse, and another mouse with established arthritis (days 7 after K/BxN serum injection). Cells were stained with Ly-6G antibody, and analyzed by flow cytometry. *Left panel*: *Green line* indicates synovial fluid neutrophils from a mouse with established arthritis, *red line* shows synovial fluid neutrophils from control mouse, and *black line* reveals staining of established arthritis using an isotype control. The arthritic synovial fluid contains a large population of Ly-6G-positive neutrophils while this population was absent in the synovial fluid of a control mouse. *Right panel*: Quantitation of the number of Ly-6G-positive cells in control and arthritic joints (control mice: $n=2$, arthritic mice: $n=2$). Data are the mean ± SEM; $*P<0.05$ vs. control mice

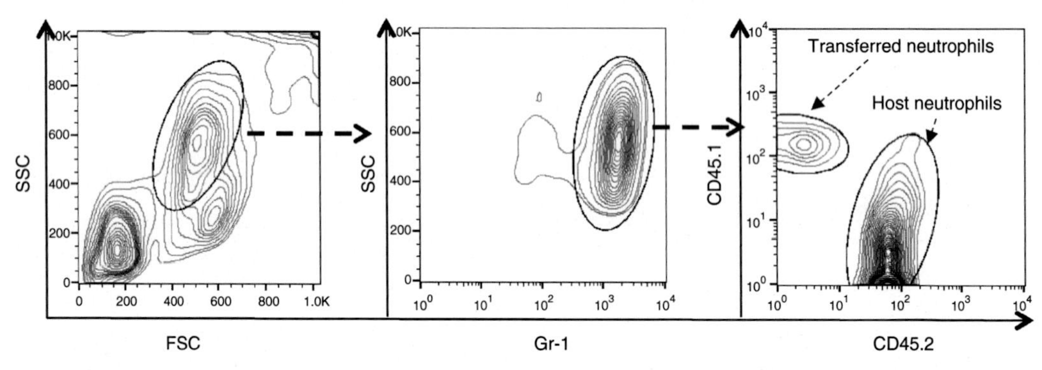

Fig. 6 Analysis of transferred and host-derived neutrophils in the same mouse using allotypic markers. CD45.1$^+$ wild-type neutrophils were transferred into CD45.2$^+$ BLT1-deficient mice on days 0 and 2. Arthritogenic serum was injected into these same mice on day 0 and synovial fluid neutrophils analyzed by flow cytometry on day 4. *Left panel*: Graph of FSC vs. SSC. *Middle panel*: Gr-1 (Ly-6G and Ly-6C) vs. SSC of the granulocyte gate. *Right panel*: CD45.1 vs. CD45.2 of the Gr-1$^+$ cell gate. CD45.1$^+$ cells reveal the presence of wild-type transferred neutrophils in the joint while CD45.2$^+$ cells reveals the presence of BLT1-deficient host-derived neutrophils in the joint (*see* Note 19)

3.6.3 Analysis of Neutrophils in Synovial Tissue by Flow Cytometry

1. Euthanize mice and harvest ankle joints, removing skin.
2. Dissect synovial tissue from mouse ankles by a sterile surgical knife.
3. Mince synovial tissue into small pieces using a sterile surgical knife.

4. Incubate synovial tissue with 1.5 mg/ml type IV-collagenase for 1 h at 37 °C (in hot water bath) (*see* **Note 16**).
5. Wash with 2 % FCS/PBS, and spin down twice.
6. Stain for Ly-6G-positive neutrophils and analyze by flow cytometry (*see* Subheading 3.6.2, **steps 4–9**).

3.6.4 Quantitation of the Neutrophil-Specific Granule Protein Myeloperoxidase in Synovial Tissue

Neutrophils can be quantitated in tissue by assaying for the presence of the neutrophil-specific granule protein myeloperoxidase (MPO) [11].

1. Dissect synovial tissue from mouse ankles (*see* Subheading 3.6.3, **steps 1** and **2**), and homogenize by a mechanical homogenizer. And then transfer to an Eppendorf tube (*see* **Note 17**).
2. Centrifuge synovial tissue homogenate at 15,000 × *g* for 20 min and remove supernatant, discarding cell pellet.
3. Use supernatant for MPO assays.
4. To specifically capture MPO, incubate supernatant in MPO ELISA dilution buffer on anti-mouse MPO antibody-coated 96-well plates for 1 h at RT.
5. Wash assay wells four times with washing buffer (PBS with 0.05 % Tween 20).
6. Incubate with 100 μl biotinylated tracer antibody for 1 h at RT.
7. Repeat **step 5**.
8. Incubate with 100 μl streptavidin peroxidase for 1 h at RT.
9. Repeat **step 5**.
10. Add 100 μl TMB solution for 30 min at RT.
11. Add 100 μl stop solution.
12. Analyze assay wells at 450 nm using a microplate reader.

3.6.5 Tracking Fluorescently Labeled Neutrophils into the Joint Using Microscopy and Flow Cytometry

Bone marrow neutrophils can be labeled ex vivo using cell tracker dyes [10] or can be harvested from transgenic mice genetically engineered to express a fluorescent protein in neutrophils [13]. LysM-GFP mice are commonly used to visualize neutrophils in vivo [13–15]. In these mice, the expression of the green fluorescent protein (GFP) has been engineered to be driven by the lysosomal M promoter, resulting in neutrophils and macrophages that express GFP and are thus green [13].

1. Isolate BM neutrophils (*see* Subheading 3.2) from wild-type mice for labeling with Cell Tracker or BM neutrophils from LysM-GFP mice.
2. Stain 1×10^7 BM neutrophils in a volume of 1 ml sterile 1 % FCS/PBS with Cell Tracker orange CMTMR Dye (2.5 μM) for 15 min at 37 °C (in hot water bath).

3. The staining reaction is stopped by putting cells on ice and then quickly spinning them over a cushion of 2 ml 100 % FCS followed by centrifugation at 1500 rpm for 5 min.
4. 1×10^7 BM neutrophils labeled ex vivo with Cell Tracker or harvested from LysM-GFP mice in 100 μl of sterile 1 % FCS/ PBS are intravenously transferred into arthritic mice (*see* Subheading 3.3, **step 3**).
5. Twenty-four hours after the transfer, the number of labeled cells in the joint may be counted under fluorescence microscopy after harvesting the ankle [5, 10, 12] or quantitated by flow cytometry (*see* **Note 18**).

4 Notes

1. KRN mice are bred with NOD mice (the Jackson Laboratory) to generate K/BxN F1 offspring for serum harvest [16].
2. Other medium, such as 1 % FCS/PBS, may also be used for this kit. However, the manufacturer recommends this medium.
3. The incidence of K/BxN serum transfer arthritis is almost 100 % in C57/BL6 mice. A wide range of mouse strains are susceptible to K/BxN serum-induced arthritis; however, the severity of arthritis may be different between mice strains [17].
4. High serum concentrations, such as 10 % FCS/PBS, reduce cell toxicity induced by cell tracker, but they also reduce staining intensity.
5. Both male and female K/BxN mice are acceptable donors.
6. Four bones from each mouse (two femurs and two tibias) should yield over 50 million BM cells. If more BM cells from one mouse are desired, two additional bones (humerus) may also be used.
7. Repeat this process with the same bone several times until the marrow appears to be cleared out. Repeat with all bones.
8. A severe swollen ankle is defined as a paw thickness of more than 4 mm.
9. This step is needed so that KC-X decalcification reagent will sufficiently penetrate through samples. Adequate fat removal will be indicated when the mixed reagent turns a yellow color.
10. The decalcification status can be checked by pricking a sample with a needle. The sample will be adequately decalcified if it feels soft.
11. It is recommended that two fields per slide, one around the instep area and the other around the Achilles tendon, be used to quantitate arthritis or neutrophil numbers. Severe and diffuse inflammatory infiltration is defined as inflammatory infiltration

in both the instep and Achilles tendon per slide. Focal inflammatory infiltration is defined as inflammatory infiltration in one field per slide.

12. Other methods of antigen retrieval, such as microwave and 1 mM EDTA, might also be useful.
13. PBS, PBS with tween 20, and TBS with tween 20 are appropriate as wash buffers.
14. 10 % Normal serum derived from the same animal species as the second antibody and 1 % BSA can also be used for blocking.
15. Counting all Ly-6G-positive neutrophils in the joint per slide may also be used for quantitation.
16. Collagenase is usually used for the isolation of single cells from tissue [18, 19]. However, collagenase can clip off the epitope of certain cell surface markers. This should be checked beforehand and if it does, several modifications can be tried to eliminate the cleavage of the specific epitope, such as changing the type of collagenase and reducing the concentration and/or reducing the incubation time of the collagenase digestion.
17. A sonicator may also be helpful for the homogenization.
18. References 8 and 10 show CD11b-positive splenocytes, which after transfer into arthritic mice then migrate into the inflamed joint, and ref. 9 demonstrates Ly-6G-positive splenocytes, which were transferred into vasculitic mice and then infiltrate into the inflamed vascular walls.
19. Fluorescently labeled neutrophils of different genotype, instead of using allotypic markers, can also be used for this experiment. Analysis by flow cytometry can be useful for both quantifying global inflammation in the joints and identifying the source of the neutrophil (transferred vs. endogenous).

Acknowledgements

ADL was supported by grants from the National Institutes of Health.

References

1. Mayadas TN, Cullere X, Lowell CA (2014) The multifaceted functions of neutrophils. Annu Rev Pathol 9:181–218
2. Griffith JW, Sokol CL, Luster AD (2014) Chemokines and chemokine receptors: positioning cells for host defense and immunity. Annu Rev Immunol 32:659–702
3. Chow MT, Luster AD (2014) Chemokines in cancer. Cancer Immunol Res 2:1125–1131
4. Wright HL, Moots RJ, Edwards SW (2014) The multifactorial role of neutrophils in rheumatoid arthritis. Nat Rev Rheumatol 10:593–601
5. Miyabe Y, Miyabe C, Iwai Y et al (2013) Necessity of lysophosphatidic acid receptor 1

for development of arthritis. Arthritis Rheum 65:2037–2047

6. Kim ND, Chou RC, Seung E et al (2006) A unique requirement for the leukotriene B4 receptor BLT1 for neutrophil recruitment in inflammatory arthritis. J Exp Med 203:829–835
7. Monach PA, Mathis D, Benoist C (2008) The K/BxN arthritis model. Curr Protoc Immunol Chapter 15:Unit 15.22
8. Chou RC, Kim ND, Sadik CD et al (2010) Lipid-cytokine-chemokine cascade drives neutrophil recruitment in a murine model of inflammatory arthritis. Immunity 33:266–278
9. Sadik CD, Kim ND, Iwakura Y, Luster AD (2012) Neutrophils orchestrate their own recruitment in murine arthritis through C5aR and FcγR signaling. Proc Natl Acad Sci U S A 109:E3177–E3185
10. Miyabe C, Miyabe Y, Miura NN et al (2013) Am80, a retinoic acid receptor agonist, ameliorates murine vasculitis through the suppression of neutrophil migration and activation. Arthritis Rheum 65:503–512
11. Pulli B, Ali M, Forghani R et al (2013) Measuring myeloperoxidase activity in biological samples. PLoS One 8:e67976
12. Nanki T, Urasaki Y, Imai T et al (2004) Inhibition of fractalkine ameliorates murine collagen-induced arthritis. J Immunol 173: 7010–7016
13. Faust N, Varas F, Kelly LM et al (2000) Insertion of enhanced green fluorescent protein into the lysozyme gene creates mice with green fluorescent granulocytes and macrophages. Blood 96:719–726
14. Devi S, Wang Y, Chew WK et al (2013) Neutrophil mobilization via plerixafor-mediated CXCR4 inhibition arises from lung demargination and blockade of neutrophil homing to the bone marrow. J Exp Med 210:2321–2336
15. Lämmermann T, Afonso PV, Angermann BR et al (2013) Neutrophil swarms require LTB4 and integrins at sites of cell death in vivo. Nature 498:371–375
16. Ji H, Ohmura K, Mahmood U et al (2002) Arthritis critically dependent on innate immune system players. Immunity 16:157–168
17. Christianson CA, Corr M, Yaksh TL, Svensson CI (2012) K/BxN serum transfer arthritis as a model of inflammatory joint pain. Methods Mol Biol 851:249–260
18. Deshane J, Zmijewski JW, Luther R et al (2011) Free radical-producing myeloid-derived regulatory cells: potent activators and suppressors of lung inflammation and airway hyperresponsiveness. Mucosal Immunol 4:503–518
19. Cortez-Retamozo V, Etzrodt M, Newton A et al (2012) Origins of tumor-associated macrophages and neutrophils. Proc Natl Acad Sci U S A 109:2491–2496

Chapter 15

Intravital Two-Photon Imaging of Lymphocytes Crossing High Endothelial Venules and Cortical Lymphatics in the Inguinal Lymph Node

Chung Park, Il-Young Hwang, and John H. Kehrl

Abstract

Lymphocyte recirculation through lymph nodes (LNs) requires their crossing of endothelial barriers present in blood vessels and lymphatics by means of chemoattractant-triggered cell migration. The chemoattractant-chemoattractant receptor axes that predominately govern the trafficking of lymphocytes into, and out of, LNs are CCL19/CCR7 and sphingosine 1-phosphate (S1P)/S1P receptor 1 (S1PR1), respectively. Blood-borne lymphocytes downregulate S1PR1 and use CCR7 signaling to adhere to high endothelial venules (HEVs) for transmigration. During their LN residency, recirculating lymphocytes reacquire S1PR1 and attenuate their sensitivity to chemokines. Eventually lymphocytes exit the LN by entering the cortical or medullary lymphatics, a process that depends upon S1PR1 signaling. Upon entering into the lymph, lymphocytes lose their polarity, downregulate their sensitivity to S1P due to the high concentration of S1P, and upregulate their sensitivity to chemokines. However, many of the details of lymphocyte transmigration across endothelial barriers remain poorly understood. Intravital two-photon imaging with advanced microscope technologies not only allows the real-time observation of immune cells in intact LN of a live mouse, but also provides a means to monitor the interactions between circulating lymphocytes and stromal barriers. Here, we describe procedures to visualize lymphocytes engaging and crossing HEVs, and approaching and crossing the cortical lymphatic endothelium to enter the efferent lymph in live mice.

Key words High endothelial venules (HEVs), Cortical lymphatics, Lymphocytes, Inguinal lymph node (LN), Intravital imaging, Two-photon microscopy

1 Introduction

Homeostasis of lymphocytes in LNs is maintained by two steps, the entry of circulating lymphocyte through HEVs into the lymph node parenchyma and the exit through lymphatic sinuses into the efferent lymph. In the first steps lymphocytes are required to cross blood vessel endothelial barriers and in the second step the lymphatic sinus endothelium. In both instances this requires chemoattractant-triggered cell migration. During lymph node

Tian Jin and Dale Hereld (eds.), *Chemotaxis: Methods and Protocols*, Methods in Molecular Biology, vol. 1407,
DOI 10.1007/978-1-4939-3480-5_15, © Springer Science+Business Media New York 2016

entry blood-borne lymphocytes are captured in the HEV. This leads to the lymphocyte rolling along the surface of endothelium. The initial engagement with the HEV endothelium depends upon lymphocyte CD62L expression and carbohydrate ligands (MECA-79) present on the endothelial cell surface [1, 2]. Next, the rolling lymphocyte must firmly adhere to the HEV. To do so the lymphocytes engage homeostatic chemokines present on the luminal surface of the HEVs with cognate chemokine receptors. Chemokine receptors are members of the G-protein-coupled receptor (GPCR) family and trigger firm adhesion by triggering the α-subunit of the heterotrimeric G protein G_i to exchange the nucleotide guanine diphosphate for guanine triphosphate. $G\alpha_i$ nucleotide exchange leads to effector protein activation, which results in the integrin LFA-1 adopting a high-affinity conformation. Inhibiting G-protein nucleotide exchange blocks integrin activation and firm adhesion.

Once lymphocytes adhere they crawl along the HEV endothelium searching for a transendothelial migration (TEM) site using adhesive filopodia that express high-affinity LFA-1. Upon locating a suitable site they transmigrate across the endothelium by engaging subluminal ICAM-1 [3]. After penetrating the endothelial basement membrane the cells must negotiate the perivenule space, which is created by overlapping pericytes. Finally, the lymphocytes reach the paracortical cords (PCCs), which originate in between and below LN follicles and extend into the medullary region [4–6]. Within the PCC corridors, the migratory paths of B and T lymphocytes diverge. T cells migrate along CCL19/21-expressing fibroreticular cells (FRCs) using their prominently expressed CCR7 to access the LN deep cortex [7]. However, since most T cells lack CXCR5 expression they avoid the LN follicle, where resident follicular dendritic cells express CXCL13, not CCL19/21. Initially B lymphocytes also migrate along CCL19/21-expressing FRCs using their expressed CCR7, albeit at lower levels than that found on T cells; however their prominent CXCR5 expression leads to their eventual localization in the LN follicle [8, 9]. Newly resident follicular B cells gradually enter the follicle centers, the site of high levels of CXCL13, while longer term residents move towards the follicle edges closer to nearby egress sites [10, 11].

To exit the LN lymphocytes must find the lymphatics and traverse the efferent lymphatic endothelium. The details of how lymphocytes find egress sites and cross the lymphatic endothelium are still poorly understood. The cortical lymphatics, which can be visualized on the basis of their expression of the hyaluronan receptor LYVE-1, are located adjacent to the LN follicles and T cell zone providing convenient egress sites for LN lymphocytes [11]. While chemokine receptor signaling has a prominent role in lymphocytes crossing HEVs, the high concentration of chemokines in the LN opposes lymphocyte egress [12]. Another $G\alpha_i$-coupled GPCR the S1PR1 has been implicated in facilitating lymphocyte egress into the lymph [13–15]. The LN parenchyma, while rich in homeostatic chemokines, has little S1P while the lymph and blood have high levels. A delicate balance between the synthesis, transport, and degradation of S1P achieves and maintains this gradient [16]. Since the observation that the administration of an S1P analogue FTY720 caused lymphopenia by preventing lymphocytes from exiting the LN into efferent lymphatics was reported [13], the mechanism by which FTY720 triggers lymphocyte retention has been intensively studied. A recent study has implicated S1P1 receptor cell-surface residency on lymphocytes as a crucial factor in lymphocyte egress kinetics following FTY720 treatment [17]. Less controversy surrounds the concept that the lymphocyte S1P1 receptor functions to facilitate normal lymphocyte egress; however, the precise mechanism by which it does so remains unresolved [11, 18].

The use of highly advanced two-photon laser scanning microscopy (TP-LSM) technology has allowed the imaging of fluorescently labeled cells in the LNs of live animals as well as the simultaneous delineation of microenvironment landmarks using fluorescently labeled antibodies [19–22]. The original studies, which examined lymphocyte LN homing, utilized intravital wide-field fluorescent microscopy to capture the rolling, tethering, adherence, and TEM of lymphocytes in HEVs outlined by fluorescent dextran. More recent studies have employed TP-LSM [23–25]. However, TP-LSM is limited by its slower frame acquisition rate, a problem for capturing images of lymphocytes rolling on the HEV endothelium. We found that by limiting the size of the imaging field we could boost our acquisition rate to three frames per second. While not fast enough to measure rolling velocities, it is sufficient to analyze the behavior of cells adherent to HEVs and to capture cells undergoing TEM. TP-LSM can also be used to image lymphocytes engaging and traversing the lymphatic endothelial, and to monitor lymphocytes in the cortical lymphatics and efferent lymph. A major concern in this type of imaging is that the method to expose LN for imaging must not disrupt LN innervation or the normal flow of blood and lymph. The TP-LSM imaging procedure detailed below is designed for imaging HEVs and lymphatics in the mouse inguinal LN and is a modification of a previously published method [26, 27].

2 Materials

2.1 Preparation of Lymphocytes from Mouse Spleen

1. 1× Phosphate-buffered saline (PBS), pH 7.4.
2. RPMI-1640 medium (Gibco).
3. Fetal calf serum (FCS), quantified (Gibco).
4. 40 and 100 μm Nylon Cell strainers (BD Falcon).

5. 25 × 5/8-gage needles (Kendall).
6. 12 ml syringe (Kendall).
7. ACK lysis buffer (BioWhittaker, Lonza).
8. Bovine serum albumin (BSA) fraction V.
9. Biotinylated mouse-Ab (MAb) (BD Pharmingen™).
10. Magnetic negative selection systems: Dynabeads and magnetic particle concentrator (Invitrogen).
11. Fluorescence-activated cell sorting (FACS) buffer: 1× PBS, pH 7.4, 1 % BSA.

2.2 Labeling Lymphocytes and Adoptive Transfer of Labeled Lymphocytes

1. CellTracker™ probes (Molecular Probes, Invitrogen) for long-term tracing of living cells: 5-Chloromethylfluorescein diacetate (CMFDA) and 5-(and-6)-(((4-chloromethyl)benzoyl) amino)tetramethylrhodamine (CMTMR), stock solution 5 mM in DMSO (stored at −20 °C).
2. 5 mM of eFluor450 in DMSO (eBioscience).
3. RPMI-1640 media (Gibco).
4. FCS, quantified (Gibco).
5. 1× PBS, pH 7.4.

2.3 Visualization of HEVs and Lymphatics in the LN

1. Evans Blue solution (0.5 μg/ml in PBS) (Sigma).
2. Purified rat anti-mouse PNAd carbohydrate epitope (BD Pharmingen™, clone: MECA-79) for HEVs.
3. LYVE-1 antibody (Clone# 223322, R&D System) for lymphatics.
4. Antibody Labeling Kits (Alexa Fluor® 488 or 647) (Cat# A20181, A20185, or A20189, Molecular probe) for fluorescent labeling.
5. 1× PBS, pH 7.4.

2.4 Multi-photon Imaging

1. Two-photon laser-scanning microscopy system: We use a Leica SP5 inverted five-channel confocal microscope, a Ti:Sappihire laser (Spectra Physics) with a 10 W pump tuned to 810 nm, a 25× water dipping objective with long working distance (for example, Leica 25× 0.95NA), an incubation cube chamber (Life Imaging Services, Basel, Switzerland) (*see* **Note 1** and Fig. 1a).
2. Heating mantles and blankets (Thermo Scientific): The heating blanket is set up to heat at 37 °C.
3. Anesthesia: 50× Avertin (tribromoethanol and 2-methyl-2-butanol, Sigma-Aldrich), stored at −20 °C. Diluted prior to use as 1×; see later.
4. Small animal clipper (Fisher Scientific).

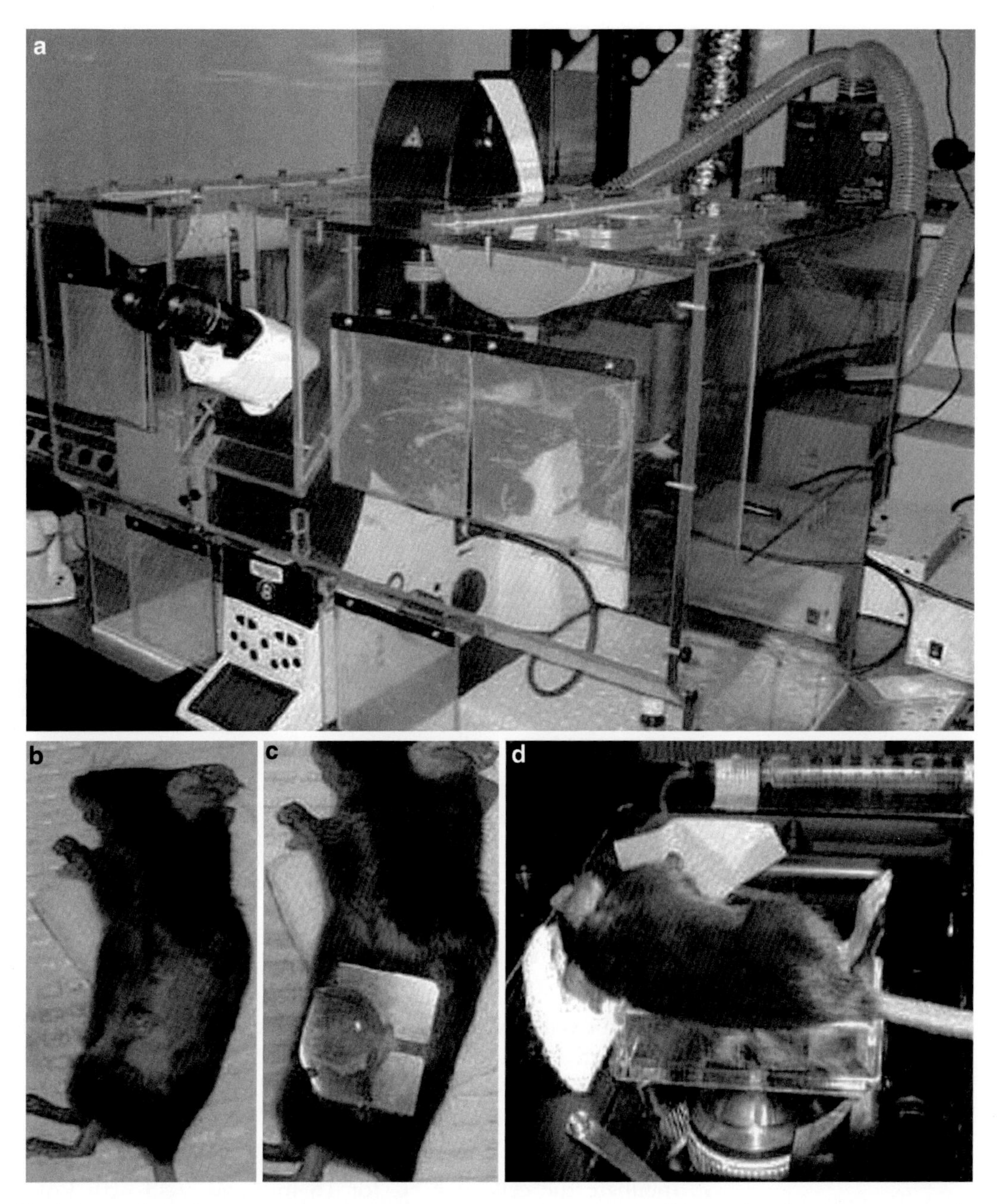

Fig. 1 A two-photon laser-scanning microscopy system and an example of the results of surgery to expose the inguinal LN. (**a**) Two-photon laser-scanning microscopy system including a Leica SP5 inverted five-channel confocal microscope. The cube and box, cube chamber, maintains temperature at 37.0 ± 0.5 °C. (**b**) Anesthetized mouse is shaved and has small incision on its flank. (**c**) Immobilized inguinal LN with cyanoacrylate glue on blade fitted to a chamber slide. To protect tissue from drying periodically add a few drops of PBS on exposed inguinal LN. (**d**) Complete setup for imaging is shown. Mouse is placed in a pre-warmed chamber slide. Blade can hold the exposed inguinal LN to minimize vibrations caused by breathing. Mouse is kept anesthetizing through intraperitoneal injection of Avertin using an infusion set

5. *Microsurgical instruments*: Roboz surgical scissor and forceps.
6. Cover-glass chamber slide (Nalgene, Nunc).
7. Classic Double Edge Blades 5-Pack (Wilkinson Sword).
8. 1× PBS, pH 7.4.
9. Cyanoacrylate glue.
10. Infusion set (Butterfly ST, 25× 3/8 32/2″ Tubing).

3 Methods

Here we describe details of procedures for real-time imaging of adoptively transferred lymphocytes visualized with fluorescent dyes in inguinal LN. We also describe how to visualize the LN microenvironments (HEVs and lymphatics). This method can also be used to observe the behavior of lymphocytes crossing endothelial barrier in HEVs and lymphatics.

3.1 Preparation of B and T Lymphocytes from Mouse Spleen (See Note 2 and [28])

1. Sacrifice mice and harvest the spleens.
2. The spleens are disrupted with forceps or needles in PBS and the cells dissociated by gentle teasing followed by filtering through a 100 and then a 40 μm nylon cell strainer to remove connective tissue cells.
3. Splenocytes are depleted of red blood cells with ACK buffer and subsequent washing with PBS.
4. Centrifuge the cell suspensions at 350 × *g* for 5 min at 4 °C, remove the supernatants, and resuspend the pellet in FACS buffer.
5. A count of the suspensions is made to determine cell number per milliliter (ml).
6. Pellet the cells and resuspend in FACS buffer. Stain the cells with biotinylated monoclonal antibodies (MAb) to non-B-cell markers such as CD4, CD8, GR-1, Mac-1, Terl19, CD11c, and DX5 in FACS buffer; to non-T-cell markers such as B220, GR-1, Mac-1, Terl19, CD11c, and DX5 in FACS buffer (*see* **Note 3**).
7. Incubate the cells at 4 °C for 15 min and wash with FACS buffer.
8. Resuspend the pelleted cells in FACS buffer. Wash the Dynabeads M-280 streptavidin beads with PBS twice and with FACS buffer twice. The bead-to-cell ratio is determined by the manufacturer's protocols [29].
9. Add the washed magnetic beads to the cell suspensions, incubate at 4 °C, and slowly rotate for 15 min. The suspensions are then attached to the magnet and allowed to separate.
10. The non-adherent suspension is collected and reapplied to the magnetic source, and again the non-adherent cells are col-

lected: This population is washed, counted, resuspended in RPMI/10 % FCS, and incubated at 37 °C for 30 min prior to any stain.

3.2 Labeling Lymphocytes and Their Adoptive Transfer

1. Resuspend 3–10 × 10^6 B lymphocytes/ml in RPMI/10 % FCS with 1–5 μM of CMFDA and CMTMR [30] or 5 μM of eFluor 450, respectively. Incubate the lymphocyte for 15 min at 37 °C and wash five times with PBS (*see* **Note 4**).
2. Resuspend labeled B lymphocytes in PBS and inject intravenously into lateral tail vein of recipient mice.
3. To visualize TEM in HEVs intravital imaging is best performed soon after cell injection. To visualize TEM in the cortical or medullary lymphatics intravital imaging is best performed 12–24 h after injecting B lymphocytes or 6–12 h after injecting T lymphocytes.

3.3 Visualization of Microvessels including HEVs and Lymphatics in the Inguinal LN

1. To visualize microvessels inject 50 μl of Evans Blue solution (0.5 μg/ml in PBS) (Sigma) into the tail or orbital vein.
2. To visualize HEVs using a direct marker inject 20 μl of fluorescently labeled anti-mouse PNAd (MECA-79) (0.25 mg/ml in PBS) (Sigma) into tail or orbital vein.
3. To visualize lymphatics using a direct marker inject 20 μl of fluorescently labeled anti-mouse LYVE-1 (0.25 mg/ml in PBS) (Sigma) into tail base or adjacent area of LN, subcutaneously.

3.4 Anesthesia

1. Make 50× Avertin (76.9 %) stock by dissolving 5 g tribromoethanol in 6.5 ml of 2-methyl-2-butanol.
2. Solution must be filtered with a 0.22 μm syringe filter (Millipore) before making aliquots of the solution in 1 ml stock vials. Can be maintained at −20 °C for 1 year.
3. To make 1× Avertin (1.51 %), mix 200 μl of 50× Avertin stock with the 10 ml pre-warmed PBS. This solution can be stored at 4 °C for 2 weeks.
4. Inject the mouse, intraperitoneal, with 1× Avertin at a dose of 300 mg/kg. Check the level of anesthesia by monitoring body and respiratory function [20].

3.5 Operation for Exposition of the Inguinal LN

For all of the following steps the mouse should be transferred onto the heated blanket. Inguinal LNs are prepared for intravital microscopy using a modification of a published method [26, 27].

1. After shaving the hair of the left or right flank with a small animal clipper, remove the skin and fatty tissue over the inguinal LN (*see* Fig. 1b).
2. The exposed inguinal LN of mouse is held with blade fitted to cover-glass chamber slide (*see* Fig. 1c).

3. Place the mouse in a pre-warmed cover-glass chamber slide and add 2 ml of pre-warmed PBS, 37 °C (*see* **Note 5** and Fig. 1d).
4. Place the chamber slide into the temperature control cube chamber on Leica SP5 microscope systems or equivalent (*see* **Notes 1** and **5**, and Fig. 1d). Distilled water is used as an immersion fluid.
5. To visualize both HEV and lymphatics in the same imaging field, image over the interfollicular channel, between LN follicles (where nearby blood vessels and cortical lymphatics are located) or at the interface between B and T cell zones (underneath the LN follicle).

3.6 Imaging Using Multi-photon Microscopy (See Fig. 1)

1. Tune the Mai Tai Ti:Sapphire laser to 790–830 nm.
2. For four-dimensional analysis of cell migration, set up the microscope configuration and filter set for the detection of the following dyes (Alexa Fluor® 488, 594, 647, CMFDA, CMTMR, eFluor 450, and second harmonics) (*see* **Note 6**).
3. Using mercury lamp illumination look through the eyepiece focusing onto the surface of the inguinal LN.
4. Start scanning using the Mai Tai Ti:Sapphire laser (*see* **Note 7**). Adjust the gain and offset of the detectors. Typically, labeled HEVs and lymphatics can be found in the interfollicular channels or at the bottom of the LN follicles. While scanning move the *x*-, *y*-, and *z*-axes to locate an area where a sufficient signal is detected with a specific marker for HEVs and/or lymphatics.
5. Once a suitable location is found, acquire *z* stacks of optical sections depending on the purpose of experiment. To examine lymphocyte adhesion and TEM on HEV set a volume that can be scanned every 1 or 2 s; for lymphocytes exiting through the lymphatic endothelium cortex to the lymphatic sinus set a volume that can be scanned every 5 or 10 s.
6. Check the level of anesthesia before starting a new image acquisition.

3.7 Analysis

1. Transform sequence of image stacks into volume-rendered four-dimensional movies using Imaris software (Bitplane) or similar software (*see* Fig. 2).
2. Use the spot analysis in Imaris for semiautomated tracking of cell motility in three dimensions using the parameters for autoregressive motion. The tracks of interested must be verified and corrected manually.
3. Calculation of cell motility parameters (track speed, speed variability, straightness, length, and displacement) can be performed using Imaris or other related software programs.

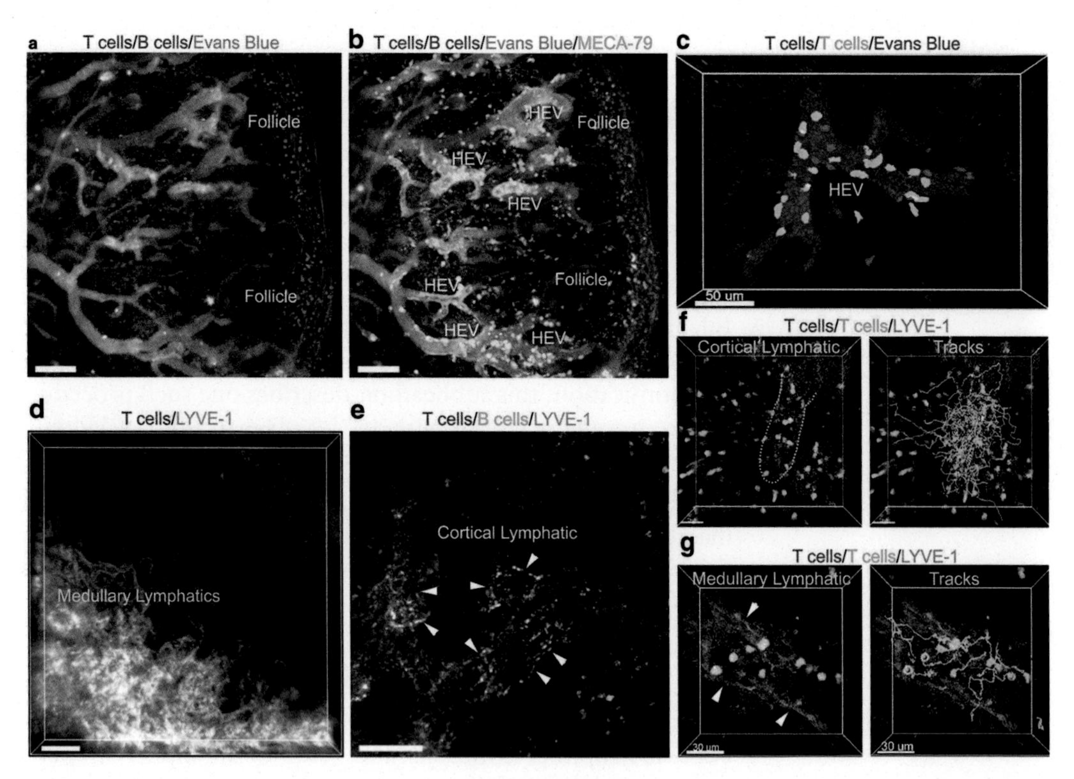

Fig. 2 Intravital images. (**a**, **b**) Shown are maximum projections of a *z* stack (100 μm, 3 μm slices). The blood vessels were visualized by the intravenous injection of Evans Blue (*gray*) and the HEVs were outlined by injection near the tail base of Alexa Fluor 488-conjugated MECA-79 antibody (*green*), shown in panel (**b**) only. Fluorescently labeled T cells (*red*) and B cells (*blue*) were adoptively transferred 1 day before the imaging. Scale bars are 100 μm. The HEVs and LN follicles are indicated. (**c**) TP-LSM image of an HEV showing adoptively transferred differentially labeled T lymphocytes. The still image is a snapshot from a 4D imaging stack, which was acquired from 3 min to 1 h after T cells were adoptively transferred. Blood vessels were outlined by injection of Evans Blue (*purple*) into the tail vein. Lymphocytes adherent to blood vessels demarked the location of HEVs in the LN. Scale bar is 50 μm. (**d**, **e**) Maximum projections of *z* stacks (60 μm) composed of images separated by 3 μm. The lymphatics were visualized by the subcutaneous tail base injection of Alexa Fluor 647-conjugated LYVE-1 antibody (*gray*). Fluorescently labeled T cells (*red*) and B cells (*green*) were adoptively transferred 1 day before imaging. Medullary lymphatics (panel **d**) and a cortical lymphatic (panel **e**) are indicated. Scale bars are 50 μm. (**f**, **g**) TP-LSM image and cell tracking near a cortical lymphatic and a medullary lymphatic. The *top left* panel shows an image from an area adjacent to a cortical lymphatic and the *bottom left* panel (**a**) medullary lymphatic. The *right* panels show superimposed cell tracks. Scale bars are 30 μm

4. The specific pattern analysis of TEM on HEV can be performed using the spot function in Imaris. On the basis of typical changes in cellular morphology that occur during TEM the duration of TEM of individual cells can be measured.
5. Statistical analysis and significance of statistics calculation can be performed with GraphPad Prism (GraphPad Software) or related software programs.
6. QuickTime movies can be generated using a maximum intensity projection of each *z* stack. Editing the movies can be performed with Adobe Premier Pro CS3 program (Adobe).

4 Notes

1. It is recommended the non-descanned detectors be used for collecting even very weak fluorescent signals. The air temperature in the cube chamber (The Cube and Box) (Life Imaging Services) was monitored and maintained at 37.0 ± 0.5 °C. Temperature control is very important for detecting proper cell movement [4].
2. B lymphocytes can be purified from bulk population by removing non-B-lymphocytes. There are multiple methods to achieve this purification; this subheading describes one such procedure (refer to methods published by Dynal, Inc., Miltenyi Biotec-MACS).
3. These biotinylated MAbs will bind CD4 and CD8 T cells, macrophages, erythrocytes, NK cells, granulocytes, and dendritic cells, or will bind B cells, macrophages, erythrocytes, NK cells, granulocytes, and dendritic cells.
4. When staining low cell concentrations FCS can protect cells from the toxic effect of the concentrated dye.
5. Inguinal LN is maintained at physiological temperature because lymphocyte motility and behavior are highly temperature dependent [19].
6. Wavelength separation was through a dichroic mirror at 560 nm and then separated again through a dichroic mirror at 495 nm followed by 525/50 emission filter for CMFDA or Alexa Fluor® 488 (Molecular probes); and the eFluor450 (eBioscience) or second harmonic signal was collected by 460/50 nm emission filter; a dichroic mirror at 650 nm followed by 610/60 nm emission filter for CMTMR or Alexa Fluor® 594; and the Evans blue or Alexa Fluor® 647 signal was collected by 680/50 nm emission filter.
7. Scrupulous care should be taken to minimize the laser power used for excitation as excessive power can cause phototoxicity resulting in cell movement arrest [31].

Acknowledgements

The authors would like to thank Dr. Anthony Fauci for his continued support. This research was supported by the intramural program of the National Institutes of Allergy and Infectious Diseases.

References

1. Butcher EC, Picker LJ (1996) Lymphocyte homing and homeostasis. Science 272:60–66
2. Rosen SD (2004) Ligands for L-selectin: homing, inflammation, and beyond. Annu Rev Immunol 22:129–156
3. Shulman Z, Shinder V, Klein E et al (2009) Lymphocyte crawling and transendothelial migration require chemokine triggering of high-affinity LFA-1 integrin. Immunity 30:384–396
4. Gretz JE, Anderson AO, Shaw S (1997) Cords, channels, corridors and conduits: critical architectural elements facilitating cell interactions in the lymph node cortex. Immunol Rev 156:11–24
5. Willard-Mack CL (2006) Normal structure, function, and histology of lymph nodes. Toxicol Pathol 34:409–424
6. Ma B, Jablonska J, Lindenmaier W, Dittmar KE (2007) Immunohistochemical study of the reticular and vascular network of mouse lymph node using vibratome sections. Acta Histochem 109:15–28
7. Bajénoff M, Egen JG, Koo LY et al (2006) Stromal cell networks regulate lymphocyte entry, migration, and territoriality in lymph nodes. Immunity 25:989–1001
8. Förster R, Mattis AE, Kremmer E et al (1996) A putative chemokine receptor, BLR1, directs B cell migration to defined lymphoid organs and specific anatomic compartments of the spleen. Cell 87:1037–1047
9. Reif K, Ekland EH, Ohl L et al (2002) Balanced responsiveness to chemoattractants from adjacent zones determines B-cell position. Nature 416:94–99
10. Park C, Hwang IY, Sinha RK et al (2012) Lymph node B lymphocyte trafficking is constrained by anatomy and highly dependent upon chemoattractant desensitization. Blood 119:978–989
11. Sinha RK, Park C, Hwang IY et al (2009) B lymphocytes exit lymph nodes through cortical lymphatic sinusoids by a mechanism independent of sphingosine-1-phosphate-mediated chemotaxis. Immunity 30:434–446
12. Pham TH, Okada T, Matloubian M et al (2008) S1P1 receptor signaling overrides retention mediated by Gα_i-coupled receptors to promote T cell egress. Immunity 28:122–133
13. Mandala S, Hajdu R, Bergstrom J et al (2002) Alteration of lymphocyte trafficking by sphingosine-1-phosphate receptor agonists. Science 296:346–349
14. Matloubian M, Lo CG, Cinamon G et al (2004) Lymphocyte egress from thymus and peripheral lymphoid organs is dependent on S1P receptor 1. Nature 427:355–360
15. Lo CG, Xu Y, Proia RL, Cyster JG (2005) Cyclical modulation of sphingosine-1-phosphate receptor 1 surface expression during lymphocyte recirculation and relationship to lymphoid organ transit. J Exp Med 201: 291–301
16. Schwab SR, Pereira JP, Matloubian M et al (2005) Lymphocyte sequestration through S1P lyase inhibition and disruption of S1P gradients. Science 309:1735–1739
17. Thangada S, Khanna KM, Blaho VA et al (2010) Cell-surface residence of sphingosine 1-phosphate receptor 1 on lymphocytes determines lymphocyte egress kinetics. J Exp Med 207:1475–1483
18. Grigorova IL, Schwab SR, Phan TG et al (2009) Cortical sinus probing, S1P1-dependent entry and flow-based capture of egressing T cells. Nat Immunol 10:58–65
19. So PT, Dong CY, Masters BR, Berland KM (2000) Two-photon excitation fluorescence microscopy. Annu Rev Biomed Eng 2:399–429
20. Miller MJ, Wei SH, Parker I, Cahalan MD (2002) Two-photon imaging of lymphocyte motility and antigen response in intact lymph node. Science 296:1869–1873
21. Miller MJ, Wei SH, Cahalan MD, Parker I (2003) Autonomous T cell trafficking examined in vivo with intravital two-photon microscopy. Proc Natl Acad Sci U S A 100:2604–2609
22. von Andrian UH, Mempel TR (2003) Homing and cellular traffic in lymph nodes. Nat Rev Immunol 3:867–878
23. von Andrian UH (1996) Intravital microscopy of the peripheral lymph node microcirculation in mice. Microcirculation 3:287–300
24. Park EJ, Peixoto A, Imai Y et al (2010) Distinct roles for LFA-1 affinity regulation during T-cell adhesion, diapedesis, and interstitial migration in lymph nodes. Blood 115:1572–1581
25. Boscacci RT, Pfeiffer F, Gollmer K et al (2010) Comprehensive analysis of lymph node stroma-expressed Ig superfamily members reveals redundant and nonredundant roles for ICAM-1, ICAM-2, and VCAM-1 in lymphocyte homing. Blood 116:915–925
26. Han SB, Moratz C, Huang NN et al (2005) Rgs1 and Gnai2 regulate the entrance of B lymphocytes into lymph nodes and B cell motility within lymph node follicles. Immunity 22:343–354
27. Park C, Hwang IY, Kehrl JH (2009) Intravital two-photon imaging of adoptively transferred B lymphocytes in inguinal lymph nodes. Methods Mol Biol 571:199–207

28. Moratz C, Kehrl JH (2004) In vitro and in vivo assays of B-lymphocyte migration. Methods Mol Biol 271:161–171
29. Dynabeads® Streptavidin Product Information and Protocol (2011) Invitrogen Life Technologies. https://tools.lifetechnologies.com/content/sfs/manuals/dynabeads_m280SAv_man.pdf. Accessed 5 Jun 2015
30. Mempel TR, Henrickson SE, Von Andrian UH (2004) T-cell priming by dendritic cells in lymph nodes occurs in three distinct phases. Nature 427:154–159
31. Shakhar G, Lindquist RL, Skokos D et al (2005) Stable T cell-dendritic cell interactions precede the development of both tolerance and immunity in vivo. Nat Immunol 6:707–714

Chapter 16

Flow Cytometry-Based Quantification of HIV-Induced T Cell Chemotactic Response

Yuntao Wu

Abstract

The human immunodeficiency virus (HIV) infects blood CD4 T cells through binding to CD4 and the chemokine co-receptor, CXCR4 or CCR5. This viral binding to CXCR4 or CCR5 also triggers the activation of a variety of signaling molecules such as LIMK/cofilin and WAVE2/Arp2/3 to promote actin dynamics, which are necessary for viral nuclear migration and the latent infection of blood resting CD4 T cells. In this chapter, we describe the methods for quantification of HIV-induced actin polymerization and cofilin phosphorylation in human T cells using flow cytometry.

Key words HIV-1, CD4 T cell, gp120, CXCR4, F-actin, Cofilin, Phalloidin

1 Introduction

HIV infects blood CD4 T cells and causes CD4 T cell depletion. In vitro, the virus infects blood CD4 T cells through the binding of the viral surface glycoprotein (gp120) to CD4 [1, 2] and the chemokine co-receptor, CXCR4 or CCR5 [3]. This interaction is essential to mediate the fusion of the viral membrane with the cell membrane, permitting viral entry [4]. The viral interaction with CXCR4/CCR5 also triggers chemotactic signaling that promotes actin dynamics to facilitate viral postentry nuclear migration [5–9]. The viral envelope glycoprotein gp120 is capable of activating a series of signaling molecules such as cofilin [5], Rac1/Pak/LIMK1 [7], and WAVE2 [9] that are commonly involved in T cell chemotactic responses. In the following protocols, we describe procedures on flow cytometry-based quantification of actin polymerization and cofilin activation that are initiated from viral attachment to the chemokine co-receptor. We also describe procedures for the assembly of HIV particles, the quantification of HIV infectivity, and the purification of resting blood CD4 T cells from the peripheral blood.

Tian Jin and Dale Hereld (eds.), *Chemotaxis: Methods and Protocols*, Methods in Molecular Biology, vol. 1407, DOI 10.1007/978-1-4939-3480-5_16, © Springer Science+Business Media New York 2016

2 Materials

2.1 Isolation of Human Resting CD4 T Cells from Peripheral Blood

1. RPMI 1640 medium supplemented with 10 % heat-inactivated fetal bovine serum (FBS), penicillin (50 U/ml), and streptomycin (50 μg/ml) (Invitrogen, Carlsbad, CA).
2. Lymphocyte Separation Medium (Mediatech, Inc., Manassas, VA).
3. PBS + 0.1 % BSA: 0.1 g of bovine serum albumin (BSA) (Sigma-Aldrich, St. Louis, MO) is dissolved in 100 ml of 1× PBS (no calcium, no magnesium; Invitrogen). The buffer is filtered through 0.45 μm filter.
4. Dynal T Cell Negative Isolation kit (Invitrogen).
5. Monoclonal antibodies against human CD4, CD11b, and CD19 (BD Biosciences, San Jose, CA).
6. Dynalbeads Pan Mouse IgG (Invitrogen, Carlsbad, CA).

2.2 HIV-1 Virus Preparation

1. DMEM supplemented with 10 % heat-inactivated FBS (Invitrogen).
2. Lipofectamine 2000 (Invitrogen).
3. HIV indicator cell line, Rev-CEM (GFP), from Virongy, Manassas, Virginia [10].
4. Alliance p24 antigen ELISA Kit (Perkin Elmer, Waltham, MA).

2.3 Stimulation of Cell Receptors with HIV-1, gp120, or Magnetic Beads Conjugated with Antibodies against CD4 and CXCR4

1. HIV-1 gp120 IIIB (Microbix Biosystems, Toronto, Canada).
2. Antibodies against human CD4 (clone RPA-T4) and CXCR4 (clone 12G5) (BD Biosciences).

2.4 Measurement of Actin Rearrangement Induced by HIV Stimulation

1. FITC-labeled phalloidin (Sigma-Aldrich). The reagent is dissolved in DMSO at a final concentration of 0.3 mM.
2. BD Cytoperm/Cytofix buffer and BD Perm/Wash buffer (BD Bioscience).
3. 1 % Paraformaldehyde in PBS (*see* **Note 1**).

2.5 Measurement of Cofilin Phosphorylation by Intracellular Staining and Flow Cytometry

1. PBS + 4 % BSA: 4 g of bovine serum albumin (BSA) (Sigma-Aldrich) is dissolved in 100 ml of 1× PBS.
2. Rabbit anti-human phospho-cofilin (ser3) antibody (Cell Signaling, Danvers, MA).
3. Alexa Fluor 647 goat anti-rabbit antibodies (Invitrogen).

3 Methods

To observe HIV-triggered actin activity and cofilin phosphorylation, resting CD4 T cells from peripheral blood are isolated through antibody-mediated negative depletion by removing non-CD4 and activated T cells. Purified cells can be stimulated with HIV particles to observe actin rearrangement and cofilin phosphorylation. Stimulated resting T cells need to be permeabilized and then stained with FITC-labeled phalloidin to detect actin polymers. FITC-phalloidin binds only to polymeric F-actin and not to monomeric G-actin. Stained cells can be analyzed with flow cytometry to measure F-actin intensity.

Cofilin phosphorylation (inactive form) or dephosphorylation (active form) at serine 3 can be observed following stimulation of resting CD4 T cells with HIV-1. This can be accomplished through intracellular staining with an anti-phospho-cofilin antibody followed by flow cytometry.

3.1 Isolation of Human Resting CD4 T Cells from Peripheral Blood

1. Pour peripheral blood collected from healthy donors into 50 ml Falcon tubes (up to 15 ml of blood per tube) and add an equal volume of PBS buffer to dilute the blood. Add approximately 15 ml of lymphocyte separation medium to the bottom of each tube. At room temperature, centrifuge for 20 min at $160 \times g$ with the brake off.
2. Remove approximately 20 ml of supernatant to deplete platelets, and then continue to centrifuge, this time at $350 \times g$, for an additional 20 min at room temperature with the brake off.
3. The white interface between the plasma and the lymphocyte separation medium contains the mononuclear cells. Recover and transfer the cells to a new 50 ml tube and wash them once with PBS + 0.1 % BSA at 4 °C by centrifugation at $400 \times g$ for 5 min.
4. Remove the supernatant and wash the cell pellet twice with cold PBS + 0.1 % BSA. Count the cell number before the last centrifugation. Resuspend the cell pellet in PBS + 0.1% BSA at a concentration of 10^8 cells/ml. Transfer the cells into a 5 ml Falcon tube, and add 0.2 ml of heat-inactivated FBS per ml of cells.
5. For negative depletion of non-T cells, using Dynal T cell Negative Isolation kit. Briefly, add antibody mix (20 μl per 1×10^7 cells) and incubate at 4 °C with gentle rocking for 15 min. Wash cells with PBS + 0.1 % BSA, and then pellet and resuspend the cells in PBS + 0.1 % BSA at a concentration of 1×10^7 cells per 900 μl. Add pre-washed depletion Dynal beads (100 μl beads per 1×10^7 cells) and incubate for 15 min at room temperature. Add PBS + 0.1 % BSA (1 ml per 1×10^7 cells) and then place on magnet for 2 min. Collect the supernatant and count the number of cells.

6. For further negative depletion of non-CD4 T cells, pellet cells and resuspend into PBS + 0.1 % BSA (100 μl per 10^7 cells), and then add 20 μl heat-inactivated FBS per 100 μl of cells. Add anti-human CD4 antibody (3 μl per 10^7 cells), anti-human CD11b antibody (3 μl per 10^7 cells), and anti-human CD19 antibody (2 μl per 10^7 cells), and incubate at 4 °C for 20 min with gentle rocking (*see* **Note 2**).
7. Wash cells with PBS + 0.1 % BSA as described above to remove unbound antibodies. Resuspend cells in PBS + 0.1 % BSA at a concentration of 1×10^7 cells per ml. Add pre-washed Dynal beads pan anti-mouse IgG (4 beads per cell) and incubate at 4 °C for 20 min with gentle rocking. Place the tube in magnet for 2 min and then transfer cell supernatant to a new tube. Count the number of cells and pellet them by centrifugation as described above.
8. Resuspend cells in RPMI 1640 + 10 % FBS at a concentration of 1×10^6 cell per ml. Culture cells in flasks at 37 °C, 5 % CO_2, overnight before treatment. An analysis of the purity of CD4 T cells is shown in Fig. 1.

3.2 HIV-1 Virus Preparation

1. Prepare virus stocks of HIV-1 by transfection of HeLa or HEK293T cells with cloned proviral DNA (*see* **Note 3**).
2. Culture cells in DMEM + 10 % FBS at 37 °C, 5 % CO_2, until they are 80–90 % confluent. One day before transfection, dislodge

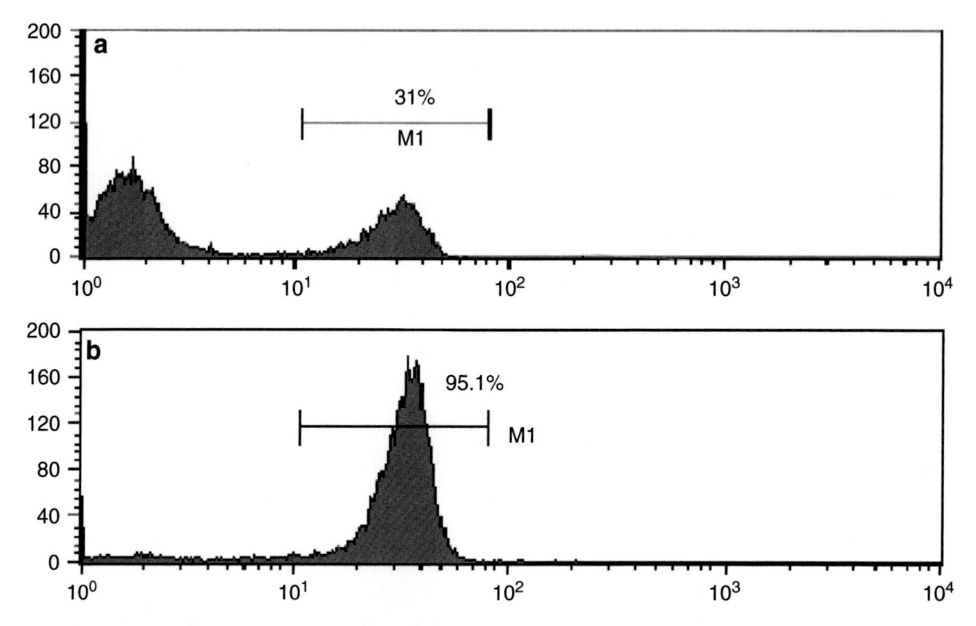

Fig. 1 Analysis of resting CD4 T cells purified from peripheral blood. Cells were purified from peripheral blood through antibody-mediated negative depletion. Shown are cells stained with FITC-labeled anti-human CD4 antibody before (**a**) and after (**b**) purification. In this example, unpurified blood mononuclear cells contain 31 % of CD4-positive cells, and following purification, the CD4-positive cells were enriched to 95.1 %

cells from T75 flasks and plate them into 10 cm petri dishes at 3×10^6 cells/dish.

3. Grow cells in petri dishes at 37 °C for 12–24 h until they are 80 % confluent. Remove medium from each petri dish, rinse with serum-free DMEM, and then add 5 ml serum-free DMEM.
4. For transfection of cells in each petri dish, mix 20 μg of plasmid DNA (such as pNL4-3) with 1.5 ml serum-free DMEM in a tube.
5. In another tube, mix 60 μl of Lipofectamine 2000 with 1.5 ml serum-free DMEM, and then incubate at room temperature for 5 min.
6. Mix DNA with Lipofectamine 2000. The total volume is 3 ml. Incubate at room temperature for 20 min.
7. Add 3 ml of the DNA/Lipofectamine 2000 mixture to each petri dish and incubate for 5–6 h. Remove the transfection supernatant and add DMEM + 10 % FSC (10 ml/dish) to continuously culture the transfected cells.
8. Harvest the supernatant at 48 h post-transfection by pelletting cells at 400 × *g* for 10 min. Save the supernatant and filter it through a 0.45 μM filter. Store the supernatant containing HIV-1 at −80 °C in 0.5–1.0 ml aliquots.
9. Virus titer ($TCID_{50}$) can be measured by infection of a Rev-dependent indicator cell line, Rev-CEM [10] (*see* **Note 4**). An example of the result using Rev-CEM is shown in Fig. 2.
10. Levels of viral p24 in the supernatant can also be measured using the Perkin Elmer Alliance p24 Antigen ELISA Kit.

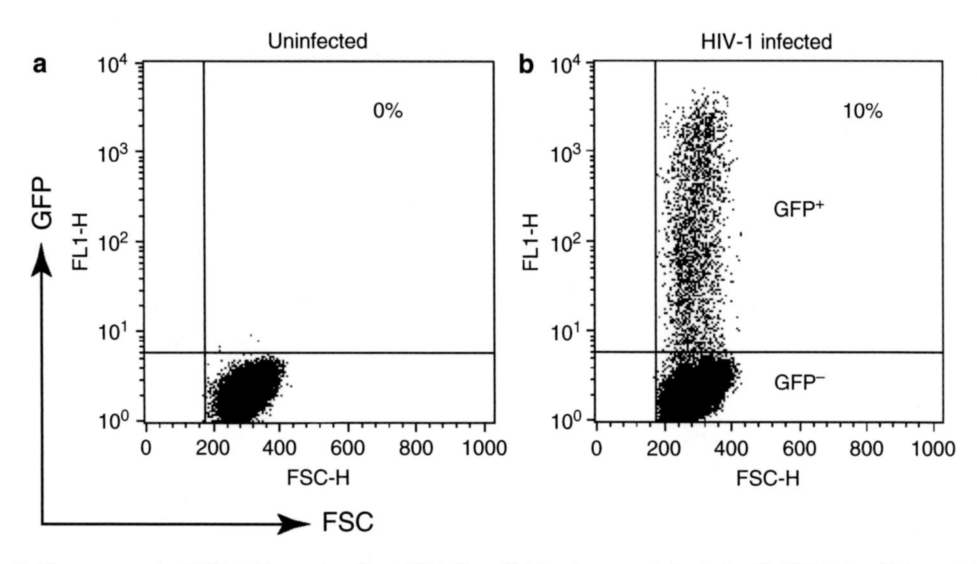

Fig. 2 Measurement of HIV-1 titer using Rev-CEM. Rev-CEM cells were infected with HIV-1 for 48 h, and then analyzed with flow cytometry to measure the number of GFP-positive cells. In this example, the HIV-1 preparation can productively infect 10 % Rev-CEM cells (**b**). Virus titer can be calculated based on the percentage of GFP-positive cells. Uninfected Rev-CEM cells did not generate any GFP-positive cells (**a**)

3.3 Stimulation of Resting T Cells with HIV-1, gp120, or Magnetic Beads Conjugated with Antibodies against CD4 and CXCR4

1. For the stimulation of resting CD4 T cells with HIV-1, normally $10^{3.5}$–$10^{4.5}$ $TCID_{50}$ units of HIV-1 are used to treat 10^6 cells ($2–4 \times 10^6$ per ml in medium) in 2 ml round-bottom tubes. Cells are incubated at 37 °C in an Eppendorf Thermomixer with gentle agitation (600–1200 rpm) to prevent cells to settle at the bottom. HIV-mediated chemokine receptor signaling can be transduced through both PTX-sensitive $G\alpha i$ and insensitive $G\alpha q$ [11, 12]. The treatment duration can last from 10 s to 2 h.
2. For the stimulation of resting CD4 T cells with purified gp120 protein, usually 5 pM to 50 nM gp120 is used to treat 10^6 cells at a concentration of $2–4 \times 10^6$ cells/ml. As described above, the treatment duration can last from 10 s to 2 h.
3. Resting CD4 T cells can also be stimulated with antibodies against the CD4 or CXCR4 receptors (*see* **Note 5**). For conjugation of antibodies to magnetic beads, 10 μg of antibodies are conjugated with 4×10^8 Dynal beads for 30 min at room temperature. Free antibodies are washed away with PBS + 0.5 % BSA, and the beads are resuspended in 1 ml of PBS + 0.5 % BSA. For stimulation of resting CD4 T cells, antibody-conjugated beads are washed twice, and then added to cell culture at a density of 2–4 beads per cell.

3.4 Measurement of Actin Rearrangement Induced by HIV Stimulation of Chemokine Co-receptors

1. One to two million cells are resuspended in medium at a density between 2 and 4×10^6 cells/ml. Cells are stimulated with HIV-1, gp120 or antibody-conjugated magnetic beads as described above.
2. F-actin staining using FITC-labeled phalloidin is carried out using $1–2 \times 10^6$ cells (*see* **Note 6**). Stimulated cells are pelleted in a microcentrifuge at $300 \times g$ for 2 min. The supernatant is removed and cells are fixed and permeabilized with 400 μl of BD Cytoperm/Cytofix buffer for 20 min at room temperature (*see* **Note 7**). Cells are washed with 2 ml cold BD Perm/Wash buffer twice, pelleted at $500 \times g$ for 5 min at 4 °C, followed by staining in the residual BD Perm/Wash buffer (approximately 100–150 μl) with 5 μl of 0.3 mM FITC-labeled phalloidin for 30 min on ice in the dark.
3. After washing twice with 2 ml cold BD Perm/Wash buffer, cells are pelleted as above, resuspended in 1 % paraformaldehyde, and analyzed on a flow cytometer. An example of the F-actin staining following treatment is shown in Fig. 3.

3.5 Measurement of Cofilin Activity by Flow Cytometry

1. Following stimulation as above, one to two million CD4 T cells are resuspended in formaldehyde for 10 min at room temperature.
2. Cells are pelleted at 600 x g in a microcentrifuge. Discard the supernatant and resuspend cells in cold methanol for 10 min at 4 °C.

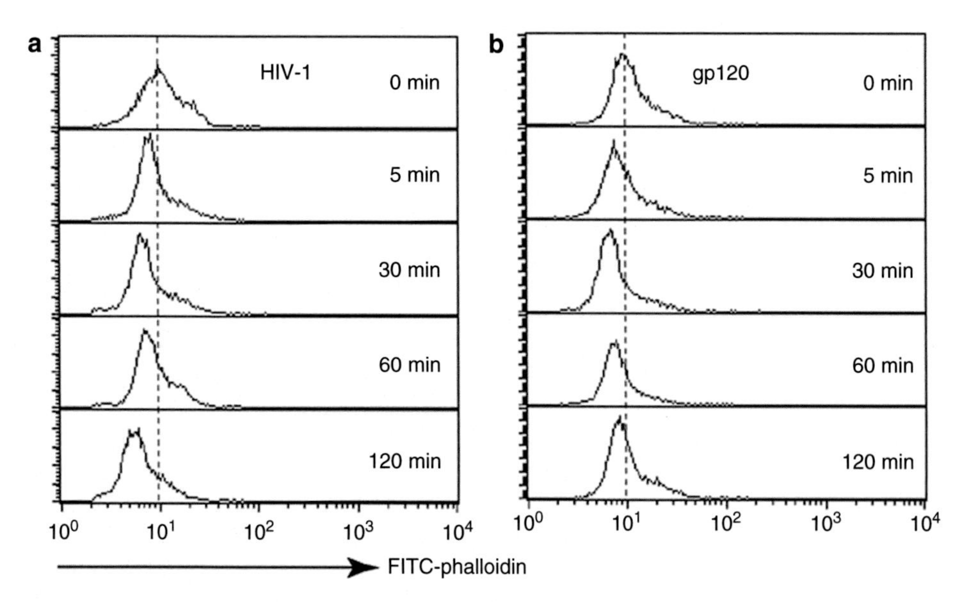

Fig. 3 HIV-1 envelope signaling triggers actin rearrangement. Resting CD4 T cells were treated with a laboratory-adapted viral strain, HIV-1_{NL4-3} (**a**), or with a primary viral isolate, HIV-$1_{93UG046}$ (**b**), or with gp120IIIB (100 pM) (**c**) for various time, fixed and permeabilized, and then stained with FITC-phalloidin for F-actin and analyzed by flow cytometry. Shown are histograms of F-actin staining

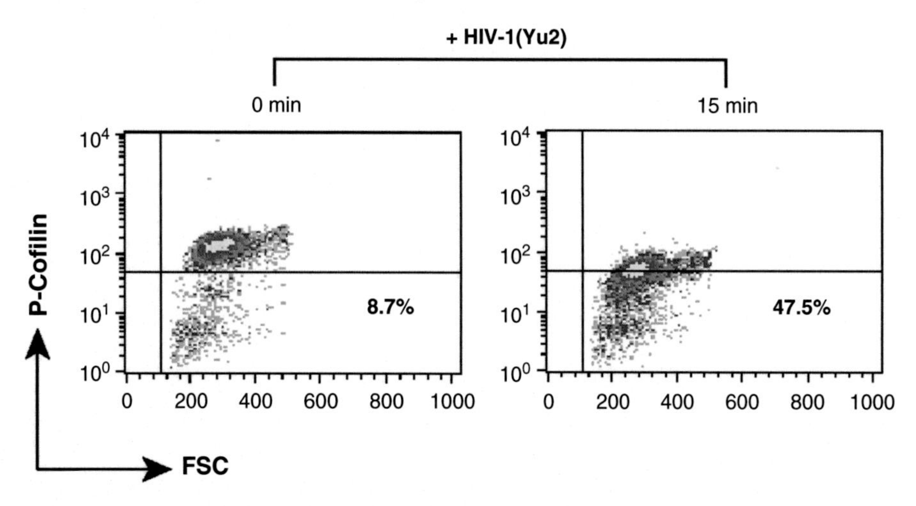

Fig. 4 Measuring cofilin phosphorylation in HIV-1 infection of resting CD4 T cells. Resting CD4 T cells were purified and cultured overnight without stimulation. Cells were treated with HIV-1 and stained for p-cofilin followed by flow cytometry

3. Cells are washed twice in PBS + 4 % BSA, and incubated with a rabbit anti-phospho-cofilin (ser3) antibody (1:50 dilution) for 60 min at room temperature.
4. Cells are washed three times with PBS + 4 % BSA, and incubated with Alexa Fluor 647-labeled goat anti-rabbit antibodies (1:2500 dilution) for 30 min at room temperature in the dark. Wash cells three times with PBS + 4 % BSA. Resuspend cells in wash buffer for flow cytometry. An example of the p-cofilin staining is shown in Fig. 4.

4 Notes

1. For preparing 1 % paraformaldehyde in PBS, all reagents and the pH meter should be placed in a chemical hood. Inside the hood, dissolve 1 g of paraformaldehyde in approximately 90 ml of distilled water by adjusting pH to 10 with 1 N NaOH, and then adjust pH to 7 with HCl. Add 10 ml 10× PBS and use distilled water to make the final volume to 100 ml.
2. For purification of resting CD4 T cells by negative depletion, two rounds of deletion will generate higher purity than a single round of depletion.
3. For transfection of cells, both HeLa and HEK293T cells can be used. However, between these two commonly used cell lines, HeLa cells generate lower virus titer and viral-free gp120 than HEK293T cells. For observing signaling at low viral dosages, we normally use HeLa cells to produce viruses. For observing strong actin polymerization at high viral dosages, HEK293T cells can be used and the virus can be harvested at day 3.
4. The Rev-CEM (GFP) cell line has no background GFP expression, and gives viral titers close to those obtained using PBMC [10]. For measuring viral titer, Rev-CEM can be used for $TCID_{50}$ assay or for flow cytometry. For $TCID_{50}$ assay, cells are resuspended into 1×10^6 cells/ml, and 100 μl of cells are used in each well of 24-well culture plates. Virus supernatant is serially diluted and 50 μl of the virus is added into each row (6 wells per row). The infection is carried out for 2 h, and then 1 ml fresh medium is added. The infected cells are cultured for 5–7 days and the number of GFP-positive wells is counted and calculated using the Reed-Muench method [10]. For using flow cytometry to measure viral titer, approximately $1–5 \times 10^5$ cells can be infected with 100 μl of virus, and following infection for 2 h, cells were washed once and resuspended into 1 ml fresh medium. To prevent secondary infection, protease inhibitors can be added and infected cells are cultured for 2–3 days and analyzed by flow cytometry to measure the percentage of infected cells.
5. Monoclonal antibodies against human CD4 (clone RPA-T4) or CXCR4 (clone 12G5) can be used to stimulate resting T cells. These antibodies are selected for their shared epitopes with gp120. The CD4 antibody, clone RPA-T4, binds to the D1 domain of the CD4 antigen and is capable of blocking gp120 binding to CD4 [2], whereas the anti-CXCR4 antibody, clone 12G5, interacts with the CXCR4 extracellular loops 1 and 2, which partially overlap domains for HIV-1 co-receptor function [13]. The 12G5 antibody has also been shown to block HIV-1-mediated cell fusion [14] and CD4-independent HIV-2 infection [15].

6. It is important to use a sufficient number of cells (at least one million) to obtain reliable and reproducible staining. Multiple washing and pelleting steps tend to decrease the number of cells and to deform them. A significant number of cells are not appropriate for flow cytometry analyses. These deformed cells and cell debris can be gated out based on the forward scatter and sideward scatter during the analysis.
7. After the first centrifugation at 300 × *g* for 2 min in a microcentrifuge, remove the supernatant with a P1000 pipette tip. Do it gently and do not touch the cell pellet. Leave some liquid at the bottom and lightly tapping the tube to resuspend cells, and then add approximately 400 μl Cytoperm/Cytofix buffer.

Acknowledgement

The preparation of this chapter was supported by the Public Health Service grant MH102144 to Y.W.

References

1. Klatzmann D, Champagne E, Chamaret S et al (1984) T-lymphocyte T4 molecule behaves as the receptor for human retrovirus LAV. Nature 312:767–768
2. Dalgleish AG, Beverley PC, Clapham PR et al (1984) The CD4 (T4) antigen is an essential component of the receptor for the AIDS retrovirus. Nature 312:763–767
3. Feng Y, Broder CC, Kennedy PE, Berger EA (1996) HIV-1 entry cofactor: functional cDNA cloning of a seven-transmembrane, G protein-coupled receptor. Science 272:872–877
4. Fields BN, Knipe DM, Howley PM, Griffin DE (2001) Fields virology, 4th edn. Lippincott Williams & Wilkins, Philadelphia
5. Yoder A, Yu D, Dong L et al (2008) HIV envelope-CXCR4 signaling activates cofilin to overcome cortical actin restriction in resting CD4 T cells. Cell 134:782–792
6. Wu Y, Yoder A (2009) Chemokine coreceptor signaling in HIV-1 infection and pathogenesis. PLoS Pathog 5:e1000520
7. Vorster PJ, Guo J, Yoder A et al (2011) LIM kinase 1 modulates cortical actin and CXCR4 cycling and is activated by HIV-1 to initiate viral infection. J Biol Chem 286: 12554–12564
8. Wang W, Guo J, Yu D et al (2012) A dichotomy in cortical actin and chemotactic actin activity between human memory and naive T cells contributes to their differential susceptibility to HIV-1 infection. J Biol Chem 287:35455–35469
9. Spear M, Guo J, Turner A et al (2014) HIV-1 triggers WAVE2 phosphorylation in primary CD4 T cells and macrophages, mediating Arp2/3-dependent nuclear migration. J Biol Chem 289:6949–6959
10. Wu Y, Beddall MH, Marsh JW (2007) Rev-dependent indicator T cell line. Curr HIV Res 5:394–402
11. Simon MI, Strathmann MP, Gautam N (1991) Diversity of G proteins in signal transduction. Science 252:802–808
12. Arai H, Charo IF (1996) Differential regulation of G-protein-mediated signaling by chemokine receptors. J Biol Chem 271: 21814–21819
13. Lu Z, Berson JF, Chen Y et al (1997) Evolution of HIV-1 coreceptor usage through interactions with distinct CCR5 and CXCR4 domains. Proc Natl Acad Sci U S A 94:6426–6431
14. Hesselgesser J, Liang M, Hoxie J et al (1998) Identification and characterization of the CXCR4 chemokine receptor in human T cell lines: ligand binding, biological activity, and HIV-1 infectivity. J Immunol 160:877–883
15. Endres MJ, Clapham PR, Marsh M et al (1996) CD4-independent infection by HIV-2 is mediated by fusin/CXCR4. Cell 87:745–756

Chapter 17

Visualizing Cancer Cell Chemotaxis and Invasion in 2D and 3D

Olivia Susanto, Andrew J. Muinonen-Martin, Max Nobis, and Robert H. Insall

Abstract

We describe three chemotaxis assays—Insall chambers, circular invasion assays, and 3D organotypic assays—that are particularly appropriate for measuring migration of cancer cells in response to gradients of soluble attractants. Each assay has defined advantages, and together they provide the best possible quantitative assessment of cancer chemotaxis.

Key words Chemotaxis, Melanoma, Metastasis, Invasion, Organotypic, Cell migration

1 Introduction

Chemotaxis is a central player in processes throughout medical science, in particular in embryogenesis and immune cell function [1]. More recently it has been recognized as a central player in the spread of many cancers [2]. Tumor cells must overcome two barriers before they can start to spread. First, they must acquire the ability to move, which often involves loss of adhesion to their neighbors and surroundings. However the ability to move is not alone sufficient—random migration is extremely inefficient, so effective spreading requires a directional stimulus. This is often provided by chemotaxis, in which cells migrate up gradients of diffusible chemicals towards a distant source.

Measuring chemotaxis is therefore a key tool in understanding the processes that underlie metastasis. However, although a wide range of different stimuli have been described, acting on tumors of almost every type, there have been surprisingly few advances that impact directly on medical practice. Two fundamental and related problems have held back discovery, each to do with the type of chemotaxis assays that are typically used.

Tian Jin and Dale Hereld (eds.), *Chemotaxis: Methods and Protocols*, Methods in Molecular Biology, vol. 1407, DOI 10.1007/978-1-4939-3480-5_17,

First, a number of widely used assays are subject to positive artifacts. In some cases they seem designed to give positive results whenever possible rather than dispassionately assess whether any single agent is a chemoattractant for a particular cell type. The widely used transwell assays, in particular, have been known for decades to be subject to misleading positive results unless carefully controlled [3]. Problems include chemokinesis (cells move faster, but without direction, in response to a stimulus), differential growth (cells in the high end of the stimulus gradient grow faster or cells in the low end die), and adhesion (cells move randomly, but stick in the presence of stimulus). Since commonly studied stimuli such as growth factors are known to induce random migration, growth, and adhesion, these artifacts are an obvious problem. Currently fashionable high-throughput assays [4] may be better, but their robustness has not been greatly studied.

The second problem—more subtle, but making interpreting the literature very difficult—is that the relative effectiveness of chemoattractants is almost never assessed. The literature tends to report efficient chemotaxis and slight responses equally, as chemotactic responses. But this misses a key point. Strong chemotaxis will efficiently move cells out of a tumor into surrounding tissues, lymph, or bloodstream. Weak responses are likely to be irrelevant in the context of cell:cell adhesion, basement membranes, and obstruction by surrounding tissues. To conclude anything physiological about the physiological importance of chemotaxis it is important to understand the relative strength of the response.

For these reasons, we assay chemotaxis wherever possible under conditions where the full details of cell movement are clearly and constantly visible. In 2D assays, this means using a continuous-view chemotaxis chamber. These were first detailed by Sally Zigmond [5], refined by Zicha & Dunn [6], and improved and updated by us [7]. For 3D chemotaxis—seen by some as a fundamentally different process—continuous view is less practical, so our focus is on "organotypic" assays [8] in which cells' shapes and distribution can be examined in detail after a fixed period. One compromise, which we have found helpful, is the semi-3D circular invasion assay [9], in which cells move in a single plane, so they can be viewed continuously, but must invade through matrix to do so.

2 Materials

All reagents to be stored at room temperature unless otherwise specified.

2.1 Chamber Assay Components

1. Insall chamber.
2. Cell culture media: both regular growth media and serum-free (SF) media. Store at 4 °C.

3. Chemoattractant.
4. Electrical tape.
5. VALAP: Weigh equal amounts of Vaseline, lanolin (Sigma), and paraffin. Place together in glass beaker, heat on heating block until all components have melted, and mix together.
6. Fine paintbrush for applying VALAP.
7. 22 × 22 mm glass cover slips (1.5 thickness).
8. 1 M Hydrochloric acid.
9. Bovine serum albumin. Store at 4 °C.
10. Fibronectin suitable for cell culture. Store at −20 °C.
11. 70 % Ethanol.
12. 0.45 μm Syringe filters.

2.2 Circular Invasion Assay Components

1. Matrigel basement membrane matrix (Corning). Store at −80 °C.
2. PBS.
3. Cell culture growth media. Store at 4 °C.
4. Ibidi 35 mm low μ-dish (Ibidi, 80209).
5. Ibidi culture inserts (Ibidi, 80136).

2.3 Organotypic Assay Components

1. Rat tails. Store at −80 °C.
2. Toothed and non-toothed forceps.
3. Large flat spatula.
4. Centrifuge capable of speeds of 30,000 × *g*.
5. 50–60 Beckman polypropylene centrifuge tubes, 50 ml capacity (Beckman Coulter, 357005).
6. Stainless steel grids (Sigma, Screens for CD-1, size: 40 mesh, S0770-5EA), cut into elevated tripods and autoclaved.
7. 12 × 35 mm Petri dishes.
8. 6 cm Petri dishes.
9. 24-Well dish.
10. T-75 flasks.
11. Dialysis tubing, 28.7 mm 14 kDa MWCO (Fisher Scientific, BID-010-040B).
12. Minimal essential medium (MEM) 10× (Invitrogen, 11430-030).
13. 70 % Ethanol.
14. 0.22 M NaOH.
15. 17.5 M (glacial) acetic acid (Sigma, 537020); diluted to 0.25 M, 0.5 M and 17.5 mM.
16. 10 % NaCl (w/v).

17. 4 % Paraformaldehyde (diluted from 16 % stock, Electron Microscopy Sciences, 15710).
18. FBS.
19. Primary human skin fibroblasts, or immortalized fibroblasts (such as telomerase-immortalized fetal fibroblasts, mouse embryonic fibroblasts, or cancer-associated fibroblasts), cultured in T75 flasks.

3 Methods

Carry out all procedures at room temperature unless otherwise specified.

We have recently altered the design of the chambers from that described in [7], which has bridge widths of 0.5 and 1.0 mm on the two sides and three viewing bridges on each side, to one with a consistent 0.6 mm width on both sides and four viewing bridges on each (Fig. 1a) [10]. We find that the 1.0 mm bridge is rarely used, but the 0.6 mm bridge gives a larger area for observing cells and a more stable gradient. The older chambers are designated mk. II, while the 0.6 mm bridges are found in mk. IV chambers.

3.1 Insall Chamber Assay

1. Acid wash glass cover slips by soaking them in a glass beaker with 1 M HCl for 15 min, and then wash them under running tap water for a further 15 min. Store acid-washed cover slips in 70 % ethanol.
2. Dilute fibronectin in PBS to a final concentration of 20 μg/ml. Store at 4 °C.
3. Prepare BSA. Make up a 0.5 % BSA solution in PBS, heat at 85 °C for 13 min, and then filter sterilize through a 0.45 μm syringe filter once cool. Store at 4 °C for up to 1 week.

The following steps are carried out in a tissue culture hood under sterile conditions:

4. Place a single acid-washed cover slip in each well of a 6-well plate, and air-dry until any residual ethanol has evaporated. To each well, add 2 ml of 20 μg/ml fibronectin. Incubate for 1 h.
5. Aspirate the fibronectin. Add 2 ml PBS to each well, and then aspirate, to wash the cover slips. Wash twice more.
6. Add 2 ml of sterile 0.5 % BSA solution to each well. Incubate for 1 h, and then wash cover slips three times with PBS. Coated cover slips can be stored in PBS at 4 °C for up to a week.
7. Trypsinize and count cells, and make up a cell suspension at 5×10^4 cells/ml in cell culture growth media. Remove PBS from coated cover slips, add 2 ml of cell suspension to each well, and shake plate gently to evenly distribute cells within the well.

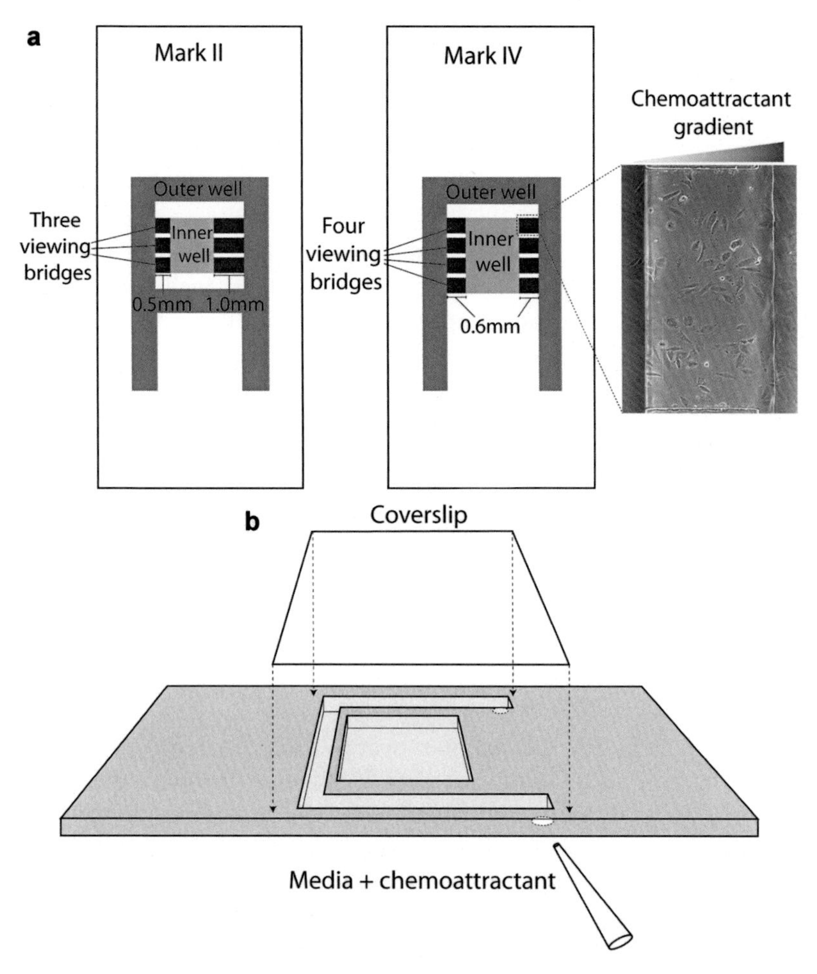

Fig. 1 (a) Schematic diagram showing alterations between the earlier Mark II design and latest Mark IV Insall chamber. *Inset* shows an example image of the viewing bridge, and the direction of the increasing chemotactic gradient. **(b)** Side view diagram of the Insall chamber setup. The inner well is filled with SF media, then the glass cover slip is placed cell side down so that it covers both the outer and inner wells. Media containing chemoattractant is added to the outer well through drilled holes on the underside

Incubate cells at 37 °C to adhere for 4+ hours (depending on cell type and how quickly they adhere).

8. Serum starve cells by removing cell culture media and replacing with SF media. Leave cells to incubate at 37 °C overnight.
9. Add 5–10 ml of SF media into a 10 cm dish. Place in 37 °C/5 % CO_2 incubator to warm and equilibrate overnight.

The following day:

1. Place beaker of VALAP on a heating block to liquefy.
2. Rinse each chamber thoroughly under running tap water to wash out any debris. Sterilize chambers by spraying thoroughly

with 70 % ethanol, then prop upright in a tissue culture hood, and leave to dry.

3. Take pre-incubated SF media and set aside 1 ml in an Eppendorf tube. Take another 1 ml of SF media and add the appropriate amount of chemoattractant for a suitable final concentration, to make media + chemoattractant.
4. Once the chamber has dried, add 55 μl SF media to the center well, or enough that the center well is full (preventing the creation of air bubbles) but not overflowing.
5. Using forceps, pick up a cover slip (while keeping note of which side the cells are adhered to) and gently touch the edge of the cover slip to a tissue to absorb excess media. Place the cover slip cell side down on the chamber, so that it fully covers both the center well and outer well (Fig. 1b).
6. While gently holding the cover slip in place, carefully dry the edges of the cover slip with a lint-free tissue to remove excess liquid. Dip a paintbrush in the liquid VALAP and paint along each side of the cover slip so that it is completely sealed while being careful not to drip any VALAP onto the center of the cover slip.
7. Flip the chamber over to expose the drilled holes on the underside of the outer wells. Take up 100 μl of media + chemoattractant with a pipette and then carefully angle the pipette tip to fit into one of the drilled holes. Add enough media + chemoattractant all the way through the entire U-shaped outer well, so that a small droplet of media comes out the other end but there are no bubbles within the well. Remove the pipette tip from the drilled hole before depressing the pipette.
8. Dry off any excess media around the drilled holes, and then cover both drilled holes with a small square of electrical tape.
9. Incubate the chamber at 37 °C for at least 15 min, cover slip side down.
10. Place the chamber in a slide holder (*see* **Note 1**) on a microscope equipped with a time-lapse camera and set each position to view the width of the bridge between the center well and the outer well. Keep note of which positions have the outer well (containing chemoattractant) on the left side of the image, or the right.

3.2 Circular Invasion Assay

The following steps are carried out in a tissue culture hood under sterile conditions:

1. Using a pair of forceps, place an Ibidi culture insert plug firmly glue side down into the center of an Ibidi dish (within the central circular depression), until it is well attached.
2. Trypsinize and count cells, adjusting the cell suspension to 1×10^6 cells/ml in cell culture media.

3. Add 1 ml (1×10^6 cells) of the cell suspension to the Ibidi dish around the culture insert and incubate at 37 °C overnight. The number of cells required for a confluent layer may vary according to cell type.

The following day:

1. Chill PBS and thaw Matrigel on ice. Warm cell culture media at 37 °C.
2. Once Matrigel has thawed, add 1 volume of chilled PBS to an equal volume of Matrigel and mix. Make up roughly 250 μl of Matrigel/PBS per dish, and keep on ice.
3. Carefully pipette off the media from the Ibidi dish without disturbing the adherent cells. The less liquid left, the more effectively the Matrigel will set to the bottom of the dish.
4. Using forceps, gently remove the plug from the dish while keeping the center of the plug area as dry as possible.
5. Add 200 μl of the Matrigel/PBS mix to the center of the Ibidi dish so that it covers both the dry plug area and the cells, taking care that it does not spill outside the central circular depression (Fig. 2).
6. Incubate the dish at 37 °C for 1 h to allow the Matrigel to set.

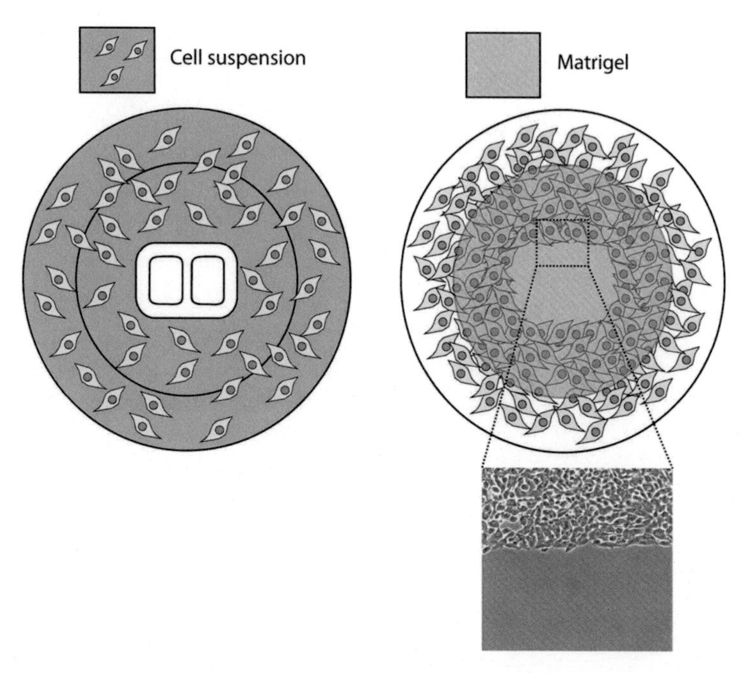

Fig. 2 Setup of a circular invasion assay. The cell suspension is added to the dish around the central plug (*left*). Once cells have adhered, the plug is removed and Matrigel added only to the inner circular depression (*right*). The edges of the plug area are imaged to visualize invasion of cells into the Matrigel (*inset*)

7. Once the Matrigel has set, carefully take the dish to the microscope. Add 1 ml of pre-warmed media to the dish, taking care not to disturb the Matrigel (*see* **Note 2**). Allow the dish to equilibrate on the microscope for 1 h before imaging along the boundary between the central plug area and the cells.

3.3 Organotypic Assay

3.3.1 Collagen Extraction Technique

1. Chill 1.5 l of 0.5 M acetic acid on ice.
2. Briefly bath the rat tails in 70 % ethanol before blotting dry.
3. Extract the tendons from approximately 15 rat tails. Slice each tail lengthwise with a scalpel, and then remove the skin by peeling it from the core of each tail. Detach the tendon from the core of the proximal end of the tail, and then use toothed forceps to pull off the tendon in the direction of the distal end of the tail. Avoiding the sheath, place the tendons in a petri dish with 70 % ethanol to keep hydrated.
4. To extract the collagen from the tendons, soak approximately 1 g tendons in a beaker containing 1.5 l chilled 0.5 M acetic acid for 48–72 h, stirring at 4 °C.
5. Transfer the extract to polypropylene centrifuge tubes and centrifuge the extract at 7500 × *g* for 30 min at 4 °C to remove the debris and undissolved tendons, discarding the pellet.
6. Re-precipitate the collagen by combining the supernatant with an equal volume of 10 % NaCl. Stir for 30–60 min at 4 °C.
7. Centrifuge again at 10,000 × *g* for 30 min and discard the supernatant. Redissolve the precipitate in 0.25 M acetic acid (at an approximate 1:1 ratio), stirring at 4 °C for 24 h.
8. Prepare the dialysis tubing by cutting two lengths of approximately 40 cm, and then soak in distilled water and microwave for 1 min before filling with the collagen supernatant.
9. Dialyze the supernatant against 17.5 mM acetic acid (1 ml glacial acetic acid/liter cold water), changing the dialysis solution four to five times (twice daily).
10. Centrifuge the dialyzed collagen at 30,000 × *g* for 1.5 h, and then transfer the supernatant to a sterile flask to remove any residual debris. The supernatant will be of a gel-like consistency, so it will need to be scraped out using a flat spatula.
11. The collagen solution is adjusted with 0.5 mM chilled acetic acid to a final working concentration of approximately 0.5 mg/ml and is stored at 4 °C.

3.3.2 Collagen Gel Preparation

The following steps are carried out in a tissue culture hood under sterile conditions:

1. Pre-chill the 10× MEM, 0.22 M NaOH, and the bottle which will be used to mix the solutions at 4 °C. Keep the collagen solution on ice. Pre-chilling reagents is essential to prevent premature polymerization of the gel.

2. The gels are prepared by combining the collagen solution with 10× MEM and 0.22 M NaOH in an 8:1:1 ratio. For each T75 flask of primary human skin fibroblasts, mix 25 ml rat tail collagen with 3 ml of 10× MEM and 3 ml of 0.22 M NaOH. Finely adjust the solution to pH 7.2 by adding more 0.22 M NaOH in a dropwise manner until the collagen turns orange in color, but stop before it turns pink (*see* **Notes 3** and **4**).
3. Prepare and label twelve 35 mm petri dishes.
4. Prepare a T75 flask of primary human skin fibroblasts (passages 4–7) by trypsinizing, centrifuging down, and resuspending them in 3 ml of FBS.
5. Immediately add the fibroblasts to the collagen gel mixture, on ice, and mix the gel well by pipetting while carefully avoiding the formation of bubbles.
6. Add 2.5 ml of the cell/gel mixture to each 35 mm petri dish. Incubate the dishes at 37 °C in a humidified incubator with 5 % CO_2 for 15–30 min to allow the gel to polymerize.
7. Add 1 ml fibroblast growth media (DMEM/10 % FBS) to each gel, and then carefully detach and release the gel from the petri dish with a pipette tip, being careful not to tear the gel.
8. Leave the gels in the incubator to contract (for approximately 1 week if using TIFFs/MEFs, 2 weeks for primary fibroblasts), changing the media with 2 ml of DMEM/10 % FBS every second day (so that the gel is completely submerged), until they are approximately 1 cm in size.
9. Transfer the contracted gels with sterile non-toothed forceps to a 24-well dish in preparation for tumor cell seeding.
10. Seed 5×10^4 tumor cells alone to the surface of each gel, in 1 ml of tumor cell growth medium (making sure to have gotten rid of any trypsin used to detach the cells). Allow the tumor cells to grow until confluent over the course of 3 days (Fig. 3; *see* **Note 5**).

3.3.3 Organotypic Chemotactic Invasion Assay

1. Place an autoclaved, elevated stainless steel grid into each 6 cm petri dish (*see* **Note 6**). Add enough of the appropriate complete growth medium to each dish so that it just covers the grid (between 6 and 10 ml).
2. Carefully transfer each gel matrix to the dishes containing the grids, adding up to three gels per grid. Carefully aspirate the media so that the bottom of the gels is contacting the media but not submerged, so that the grids are exposed at the air–liquid interface and a fine meniscal ring is visible around each gel (*see* **Note 7**).
3. Change the media every 2 days for 7–21 days (the length of time varies depending on the cell type used).

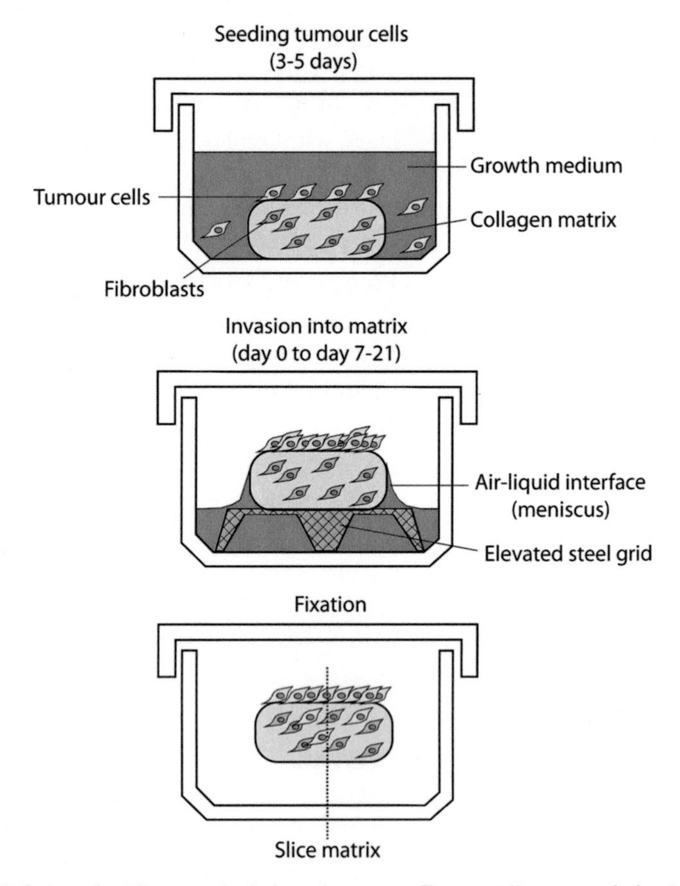

Fig. 3 Setup of a 3D organotypic invasion assay. Tumor cells are seeded onto the matrix and cultured until confluent (*top*). The seeded gel matrix is then transferred to an elevated grid to create an air–liquid interface and media gradient that induces tumor cell invasion (*middle*). Following invasion, the matrix is halved and fixed for analysis (*bottom*). Adapted from [8]

4. Slice the matrix in the middle to form two halves and fix overnight in 4 % paraformaldehyde at 4 °C ready for histological analysis (*see* **Note 8**).

4 Notes

4.1 A

1. A standard microscope stage adaptor will hold a single chamber. To allow multiple experiments to proceed in parallel we use a customized insert consisting of a rectangular frame that can accommodate at least four chambers simultaneously. The XY stage is programmed to photograph each bridge of each chamber in turn. The insert is simply a metal rectangular frame with a thin ledge running lengthwise along the top and bottom of the frame (Fig. 4). The chambers will rest on these

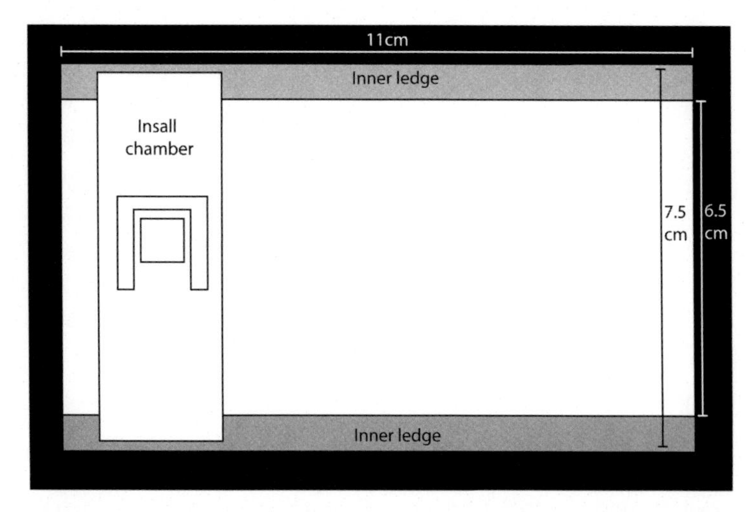

Fig. 4 Dimensions of an Insall chamber insert, to visualize multiple chambers at once

inner ledges, secured by blu-tack. The inner dimensions of the frame are 11 × 7.5 cm, with the ledges being approximately 0.5 cm thick (thus the size of the rectangular space within the frame is 11 × 6.5 cm). The outer perimeter of the frame is 12.8 × 8.7 cm; however these can be of any size as long as they will fit in the plate holder set up on the microscope.

4.2 B

2. When adding pre-warmed media to the CIA dish prior to imaging, using a pipette tip that has been cut to increase the size of the hole allows the media to be pipetted out more diffusely. This reduces the likelihood of disturbing the Matrigel.

4.3 C

3. The careful adjustment of the collagen gel pH to 7.2 should be performed drop by drop, with shaking of the mixture between each pipetted drop because fibroblasts are exquisitely sensitive to alkaline conditions.
4. One T75 flask of sub-confluent fibroblasts contains the correct number of cells for a batch of 12 gels, thereby requiring 25 ml of collagen solution.
5. The precise tumor cell number per gel and duration of the experiment need to be optimized and determined for each cell line (*see* **Note 9**).
6. The stainless steel grids are made by cutting three legs equally spaced around the edge that are then bent to raise the surface of the grid by approximately 0.5–1 cm. The grids are then sterilized by autoclaving overnight. Following their use, they can be washed and cleaned in a detergent before re-autoclaving.

7. When quantifying tumor cell invasion, day 0 is defined as the day the matrices are transferred to the steel grids, as the introduction of the air–liquid interface produces a gradient of cell culture media to trigger invasion.
8. After cutting the organotypic gel matrix in two, they can be transferred to the fixative using a scalpel to prevent the gel from folding over itself.
9. A skin organotypic invasion assay can be constructed by combining the tumor cells with keratinocytes in a ratio of 1:5 [11, 12].

References

1. Insall R (2013) The interaction between pseudopods and extracellular signalling during chemotaxis and directed migration. Curr Opin Cell Biol 25:526–531
2. Roussos ET, Condeelis JS, Patsialou A (2011) Chemotaxis in cancer. Nat Rev Cancer 11:573–587
3. Zigmond SH, Hirsch JG (1973) Leukocyte locomotion and chemotaxis. New methods for evaluation, and demonstration of a cell-derived chemotactic factor. J Exp Med 137:387–410
4. Paguirigan AL, Beebe DJ (2008) Microfluidics meet cell biology: bridging the gap by validation and application of microscale techniques for cell biological assays. Bioessays 30:811–821
5. Zigmond SH (1974) Mechanisms of sensing chemical gradients by polymorphonuclear leukocytes. Nature 249:450–452
6. Zicha D, Dunn GA, Brown AF (1991) A new direct-viewing chemotaxis chamber. J Cell Sci 99:769–775
7. Muinonen-Martin AJ, Knecht DA, Veltman DM et al (2013) Measuring chemotaxis using direct visualization microscope chambers. Methods Mol Biol 1046:307–321
8. Timpson P, McGhee EJ, Erami Z et al (2011) Organotypic collagen I assay: a malleable platform to assess cell behaviour in a 3-dimensional context. J Vis Exp 56:3089
9. Yu X, Machesky LM (2012) Cells assemble invadopodia-like structures and invade into matrigel in a matrix metalloprotease dependent manner in the circular invasion assay. PLoS One 7:e30605
10. Muinonen-Martin AJ, Susanto O, Zhang Q et al (2014) Melanoma cells break down LPA to establish local gradients that drive chemotactic dispersal. PLoS Biol 12:e1001966
11. Santiago-Walker A, Li L, Haass NK, Herlyn M (2009) Melanocytes: from morphology to application. Skin Pharmacol Physiol 22: 114–121
12. Berking C, Herlyn M (2001) Human skin reconstruct models: a new application for studies of melanocyte and melanoma biology. Histol Histopathol 16:669–674

Chapter 18

4D Tumorigenesis Model for Quantitating Coalescence, Directed Cell Motility and Chemotaxis, Identifying Unique Cell Behaviors, and Testing Anticancer Drugs

Spencer Kuhl, Edward Voss, Amanda Scherer, Daniel F. Lusche, Deborah Wessels, and David R. Soll

Abstract

A 4D high-resolution computer-assisted reconstruction and motion analysis system has been developed and applied to the long-term (14–30 days) analysis of cancer cells migrating and aggregating within a 3D matrix. 4D tumorigenesis models more closely approximate the tumor microenvironment than 2D substrates and, therefore, are improved tools for elucidating the interactions within the tumor microenvironment that promote growth and metastasis. The model we describe here can be used to analyze the growth of tumor cells, aggregate coalescence, directed cell motility and chemotaxis, matrix degradation, the effects of anticancer drugs, and the behavior of immune and endothelial cells mixed with cancer cells. The information given in this chapter is also intended to acquaint the reader with computer-assisted methods and algorithms that can be used for high-resolution 3D reconstruction and quantitative motion analysis.

Key words Cell motility, 4D motion analysis, Tumorigenesis model, Tumor microenvironment

1 Introduction

4D models for studying the interactions of cancer cells with each other and their surrounding microenvironments [1] are essential in order to identify the heterogeneous populations of cells that comprise a tumor [2] and their respective contributions to cancer progression, metastasis [3], as well as resistance to treatments [1, 4–6]. Indeed, to that end, numerous in vitro 3D culture systems have been developed that more closely mimic the natural physiological milieu in which tumors grow [7] and in some cases, these models have been used to test the effects of potential anticancer treatments [6, 8]. However, currently available methods are limited in their capacity to perform long-term (14–30 days) dynamic 4D reconstructions of tumorigenesis within a 3D culture and to generate quantitative data from these models.

Tian Jin and Dale Hereld (eds.), *Chemotaxis: Methods and Protocols*, Methods in Molecular Biology, vol. 1407,
DOI 10.1007/978-1-4939-3480-5_18,

Therefore, we developed a 3D culture system that works in conjunction with the 4D motion analysis and reconstruction software, J3D-DIAS 4.1. We initially developed the J3D-DIAS 4.1 prototypes, 2D and 3D-DIAS [9–12], for motion analysis of single cells such as *Dictyostelium discoideum* amoebae [13–21], neutrophils [22–26], and cultured, transfected cells from various cell lines [27–30]. These systems have also been successfully used to analyze movements of *Drosophila* larvae expressing neuronal mutations [31–33], embryogenesis in *C. elegans* [15], embryogenesis in zebrafish [34], and most recently modified specifically for the 4D analysis of tumorigenesis [35]. The major advances of J3D-DIAS 4.1 are that (1) optical sectioning is performed using DIC microscopy with an LED light source, obviating the need for fluorophores and their associated phototoxic effects, while simultaneously circumventing the need for confocal microscopy, which often requires relatively long scan times to generate high resolution images and (2) J3D-DIAS 4.1 generates 4D mathematical models of tumor development that cannot only be viewed in 3D, but can be quantitatively analyzed, for instance, in the presence and absence of experimental anticancer drugs. Hence, the system described here is capable of rapid optical sectioning over long periods of time, yielding high-resolution 4D dynamic reconstructions and quantitative analyses of tumorigenesis in vitro.

The method as described here acquires optical sections at 5–10 μm increments through at least 1 mm of Matrigel matrix in less than a minute, resulting in 100–150 optical sections for each *z*-series, collected, in turn, at 30 min intervals. The process may be repeated indefinitely if a perfusion system is used to refresh media, although typically, experiments are completed within 14–30 days. Between 10 and 25 objects (cells and cell aggregates) in a 446 by 335 μm field can be automatically identified and reconstructed, generating movies comprised of over 800 3D reconstructions equivalent to upwards of 122,000 optical sections per preparation per period of analysis. The optical sections are delivered to the hard drive of the controlling computer as grayscale JPEG images with moderate compression to preserve the grayscale range, and then transcribed into a new file format native to the software program J3D-DIAS 4.1, details of which are presented below.

Application of this system to tumorigenic cells cultured in 3D Matrigel matrices has revealed that cancer cells from both cell lines and human tumors aggregate or coalesce when grown in the 3D culture and furthermore, that this cancer specific, aggregation behavior is mediated by two specialized cell types that we refer to as "facilitators" and "probes." We also identified a third cell type, the "dervish," that forms in aggregates and translocates at a relatively high velocity in a swirling path through the gel, nonadherent to the developing aggregates. In addition, we have shown the efficacy of this model for studying drugs, antibodies and other compounds that effect tumorigenesis. We have so far analyzed the behavior of tumorigenic cell lines in the presence of the monoclonal antibody (mAb) against ß-1-integrin [35–38], the mAb against α-3 integrin [39], and in the presence of Vemurafenib, a drug currently used as a treatment for late stage melanoma, on tumor tissue in comparative studies with susceptible, untreated cells. Finally, the model can be used to study mixed cultures containing tumorigenic and non-tumorigenic cells, including cocultures of normal immune cells, endothelial cells and cancer cells.

2 Materials

2.1 Culturing Cells from Cell Lines

1. All laboratory personnel who handle or otherwise come into contact with cell lines, tissue, and/or blood products should be trained in the use of personal protective equipment (PPE) and the safe handling of biological material (*see* **Notes 1** and **2**).
2. Biological safety cabinet, biohazardous waste decontamination, and biohazardous waste disposal systems.
3. Dulbecco's modified Eagle's medium (DMEM), RPMI 1640, (Life Technologies, Carlsbad, CA) or supplier's recommendation (*see* **Note 3**).
4. Fetal bovine serum (FBS, Atlanta Biologicals) or supplier's recommendation (*see* **Note 4**).
5. Penicillin–streptomycin mixture (Life Technologies).
6. Dulbecco's phosphate buffered saline (DPBS, Life Technologies) without calcium and without magnesium: 2.67 mM potassium chloride (KCl), 1.47 mM potassium phosphate monobasic (KH_2PO_4), sodium chloride (NaCl), and sodium phosphate dibasic ($Na_2HPO_4 \cdot 7H_2O$).
7. Trypsin–0.25 % EDTA with phenol red or enzyme-free cell dissociation buffer for weakly adherent cells (Life Technologies).
8. CO_2 incubator (Thermo Fisher) at 37 °C and 5 % CO_2.
9. T-25 and/or T-75 tissue culture flasks (USA Scientific, Minneapolis, MN) with vented caps, or other tissue culture treated sterile labware for cell culturing.
10. 10 % Bleach for decontamination of biohazardous waste and 70 % ethanol for sterilization of surfaces.

2.2 Culturing Cells from Tissue

1. DMEM/F12 (Life Technologies).
2. M199 or RPMI 1640 (Life Technologies).
3. Fetal calf serum and horse serum.
4. Epidermal growth factor human (EGF) (Sigma-Aldrich, St. Louis, MO).

5. Hydrocortisone suitable for cell culture (Sigma-Aldrich).
6. Insulin (Sigma-Aldrich).
7. Cholera toxin (Sigma-Aldrich).
8. Penicillin–streptomycin (Life Technologies).
9. 6-well plates (Fisher Scientific).
10. Sterile scissors, scalpels, and forceps (Fisher Scientific).
11. CO_2 incubator at 37 °C and 5 % CO_2.

2.3 Isolation of Peripheral Blood Mononuclear Cells (PBMCs) and Macrophage Differentiation

1. Blood collection tubes coated with anticoagulant, such as green-capped heparin-coated Vacutainer® tubes (Becton Dickinson).
2. Histopaque-1077 (Sigma-Aldrich) for separation of PBMCs from whole blood (*see* **Note 6**).
3. Dulbecco's phosphate buffered saline (DPBS, Life Technologies) without calcium and without magnesium for cell dissociation: 2.67 mM potassium chloride (KCl), 1.47 mM potassium phosphate monobasic (KH_2PO_4), sodium chloride (NaCl), and sodium phosphate dibasic.
4. PBS supplemented with 3 mM EDTA (PBS/EDTA) (Life Technologies).
5. RPMI 1640 (Life Technologies) supplemented with penicillin–streptomycin and 5, 10, and 20 % FBS (Atlanta Biologicals).
6. Centrifuge that can operate at room temperature, a force of 400–900 × *g* and no braking during deceleration.
7. Tissue culture plates, 6-well (Fisher Scientific).
8. Recombinant human recombinant M-CSF (R&D Systems) reconstituted to 50 ng/mL in sterile PBS plus 0.1 % BSA (*see* **Note 7**).

2.4 Preparation of 3D Matrigel Cultures in Petri Dishes

1. Matrigel matrix (Becton Dickinson) aliquoted into 500 μL quantities stored at −20 °C (*see* **Note 8**).
2. Chilled pipette tips.
3. Sharp scissors sterilized in 100 % ethanol.
4. 35 mm and/or 65 mm plastic petri dishes with glass insert in dish bottom (InVitro Scientific) (*see* **Note 9**).
5. Petri dish lid with center hole cut out and replaced with a 24 × 30 mm glass coverslip window sealed with Vaseline (*see* **Note 10**).
6. Cell lines at about 70–80 % confluency or cell preparations of fresh biopsy tissue.

2.5 Preparation of 3D Matrigel Cultures in Sykes-Moore Perfusion Chamber

1. Matrigel matrix as described in Subheading 2.4, **item 1**.
2. Sykes-Moore perfusion chamber (Bellco Glass, Inc.) with 2.5 mm gasket, holder, and 25 mm #2 coverslips sterilized in ethanol or autoclaved.
3. Two pieces of Tygon tubing (1/16 in. ID) sterilized by thorough rinsing with ethanol, each connected to a 21-gauge needle via a luer lock for inlet and outlet ports.
4. Two NE-1002X Programmable Microfluidics Syringe Pumps (New Era Pump Systems, Farmingdale, NY) each equipped with a 60 mL BD syringe infusing and withdrawing at a constant rate of 40 μL/min (25 h run time) or each equipped with 60 mL BD syringe set to simultaneously withdraw and infuse 2.5 mL of media every 72 h at a rate of 500 μL/min (72 days runtime).

2.6 Optical Sectioning

1. Microscope equipped with differential interference contrast (DIC) optics, programmable motorized stage, 20× or 40× long-working distance objectives and LED light engine contained within a 37 °C, 5 % CO_2 incubator or other means of regulating temperature and pH for 30 days or more of continual imaging.
2. Cell cultures in modified petri dishes or perfusion chamber as described in Subheadings 2.4 and 2.5.
3. Digital camera controlled by video acquisition software such as the Zeiss AxioCamMRc5 and Laser Vision 6 software, synchronized with the LED light engine and motorized stage.
4. A computer capable of running Microsoft Windows XP, equipped with serial ports (RS-232) to control the microscope *z*-axis stepper motor, a firewire port (IEEE 1394) to capture images from the AxioCam MRc5 camera, and a storage disk capable of storing hundreds of gigabytes of data.

3 Methods

3.1 Passaging Cells

1. Follow supplier's recommendation for thawing cells and for media requirements.
2. Passage cells when they reach 70–80 % confluency by inspection.
3. Working in a biosafety cabinet, remove used media from flask or plate and dispose into a container of 10 % bleach.
4. Rinse the flask with 2 or 5 mL of sterile DPBS for a T-25 or T-75 flask, respectively, remove DPBS and repeat.
5. Add 1–3 mL of trypsin–EDTA for a T-25 or T-75 flask, respectively, and rotate the flask to coat the surface.
6. Remove excess trypsin–EDTA.

7. Observe cells with an inverted microscope. When cells begin to round up and float when gently tapped, then it is time to stop the trypsinization. This process usually takes between 2 and 5 min, depending on the cell type (*see* **Note 11**).
8. Resuspend detached cells in 2–5 mL of fresh media, pipetting gently up and down to break up clumps and transfer 0.5–2.5 mL to a fresh flask.
9. Bring volume of media up to 5 mL for a T-25 or 15 mL for a T-75 flask.

3.2 Establishing Cell Culture from Tissue

1. Working in a biosafety cabinet, mechanically dissociate tissue by maceration with sterile scissors and forceps into approximately 1 mm pieces in a sterile petri dish containing a few milliliters of M199, RPMI 1640, or DMEM media (*see* **Note 12**) supplemented with 10 % FCS, penicillin–streptomycin, and 2.5 Units of Fungizone. Transfer tissue pieces into individual wells of a six-well plate and add 3 mL media.
2. Monitor cell growth. After 2–3 weeks, adapt growing cells to DMEM/F12 medium supplemented with 20 ng/mL EGF, 0.5 μg/mL hydrocortisone, 10 μg/mL insulin, 0.1 μg/mL cholera toxin, 5 % horse serum, penicillin–streptomycin.
3. Transfer to T-25 tissue culture flask when cells are at 70–80 % confluent and passage as needed as described above.

3.3 Isolation of Peripheral Blood Monocytes (PBMCs) and Differentiation into Macrophages and Harvesting Macrophages

1. Collect 40–50 mL of whole venous blood from healthy donors using IRB-approved informed consent protocols (*see* **Note 5**) in anticoagulant-coated blood collection tubes.
2. Bring the Histopaque-1077 and buffers to room temperature. Bring the whole blood in the anticoagulant-coated tube to room temperature with gentle rocking for 30 min.
3. Mix equal volumes of whole blood with PBS/EDTA at room temperature.
4. Carefully layer diluted blood over Histopaque-1077.
5. Centrifuge at room temperature 900 × *g* for 30 min, no brake.
6. Using sterile pipets, remove plasma layer, discard and carefully collect the opaque PBMC layer.
7. Wash monocytes with PBS/EDTA and centrifuge 400 × *g* for 20 min with brake.
8. Wash four times in PBS/EDTA by centrifugation at 200 × *g* for 10 min.
9. Resuspend pellet in 80–100 mL in RPMI without serum and plate 2 mL in each well of a six-well plate.
10. Allow cells to adhere for 1.5 h at 37 °C, 5 % CO_2.

11. Remove non-adherent cells and wash adherent cells three times with PBS.
12. Culture overnight in RPMI supplemented with 5 % FBS.
13. To differentiate monocytes into macrophages, add RPMI 20 % FBS with 20 ng/mL M-CSF for 6 days with a media change at 3–4 days.
14. Dissociate macrophages by rinsing three times with PBS. After the last wash, add 2 mL of ice cold PBS supplemented with 3 mM EDTA to each well and incubate at 4 °C for 20 min.
15. Remove cells by scraping with a sterile scraper and count.
16. Seed appropriate number of cells in 3D Matrigel culture as described in Subheadings 3.4 and 3.5 below.

3.4 Preparation of 3D Culture in Matrigel in Modified Petri Dish with Glass Bottom Insert

1. Thaw desired quantity of previously aliquoted Matrigel matrix on ice.
2. Using a prechilled pipet tip, carefully coat the glass insert of petri dish with 100 μL of Matrigel matrix and incubate for 20 min at 37 °C to allow gelation.
3. Dissociate cells from growth flasks using either trypsin–EDTA or dissociation buffer (*see* Subheading 3.1 and/or Subheading 3.3), count cells, dilute to 5×10^5 cells per mL in 250 μL growth medium, and chill on ice.
4. Mix the entire volume of chilled cell suspension with 500 μL of liquid, chilled Matrigel and gently pipet up and down to mix (Fig. 1a).
5. Transfer the entire volume of 750 μL to the pre-coated coverslip and incubate for 30 min at 37 °C, 5 % CO_2 to allow gelation.
6. Add media to the dish, replace lid, and place in 37 °C, 5 % CO_2 incubator until ready to image.
7. Refresh media every 72 h.
8. To image on a 3D scope, it may be necessary to reduce the height of the bottom of the petri dish in order not to exceed the allowable working distance of the microscope. If so, working in a biosafety cabinet, remove media from the dish and trim about 5 mm from the rim of the dish using a pair of sterile scissors.
9. Cover with a lid that has a glass window (*see* Subheading 2.4, **item 5**, **Note 10** and Fig. 1a).
10. Place dish on the stage of a microscope equipped with a camera, a motor driven stage, and either an environmental chamber or housed in a 37 °C, 5 % CO_2 incubator (Fig. 1c).
11. Monoclonal antibodies or drugs may be added to the cell–Matrigel mixture (*see* **Note 13**).

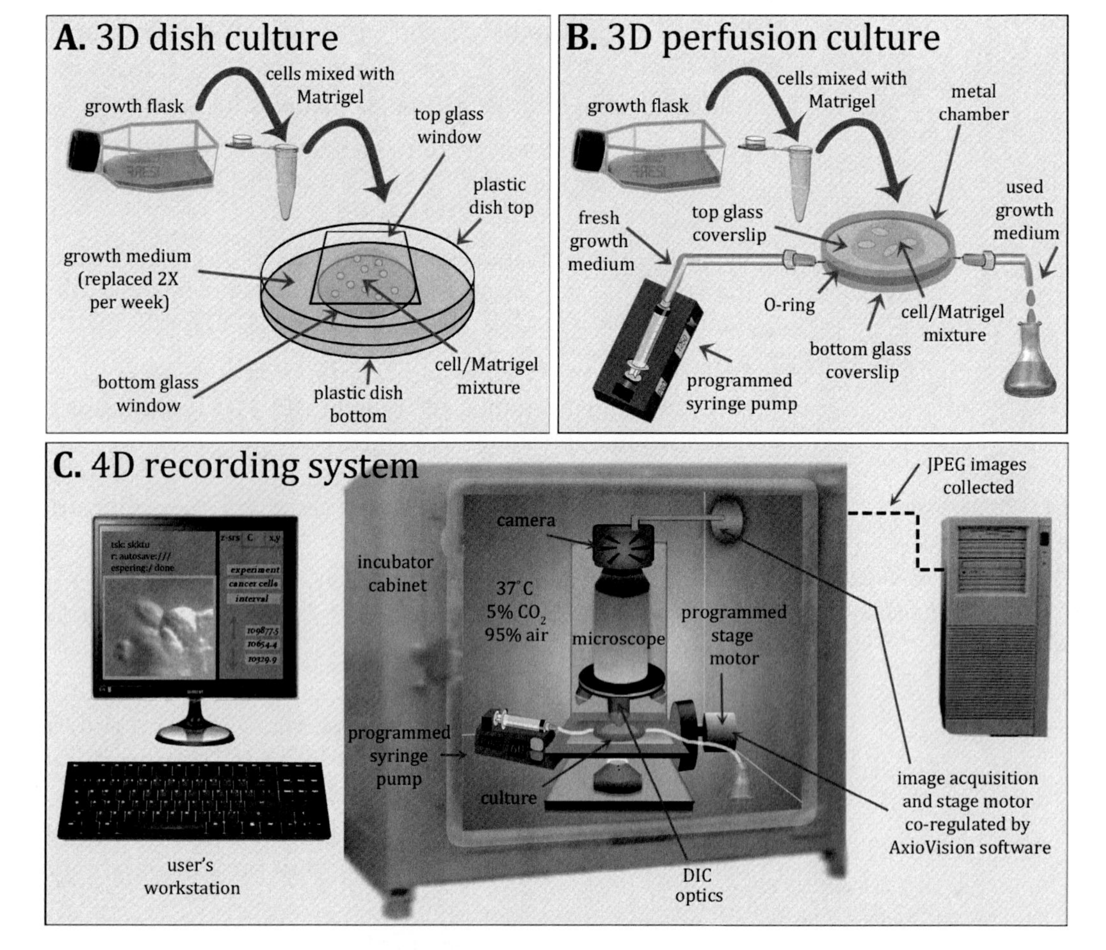

Fig. 1 Methods for 3D culturing and 4D image acquisition. (**a**) Cells are mixed with Matrigel matrix and plated onto a pre-coated glass insert of a petri dish with the lid modified for DIC optics. (**b**) Cells are mixed with Matrigel matrix and plated in a Sykes-Moore perfusion chamber for continuous perfusion with media or with media containing test drugs. (**c**) For long term optical sectioning of 3D cultures, a microscope with a *z*-axis motorized stage synchronized with a digital camera and equipped with an LED light engine as well as DIC optics can be placed inside a 37 °C, 5 % CO_2 incubator

3.5 Preparation of 3D Culture in Matrigel in Perfusion Chamber

1. Place sterile coverslip and sterile gasket inside chamber.
2. Using a prechilled pipet tip, coat the coverslip with 100 μL Matrigel and place in 37 °C incubator to allow gelation (*see* **Note 14**).
3. Dilute cells to 5×10^5 cells per mL in 250 μL growth medium and chill on ice.
4. Mix the entire volume of chilled cell suspension with 500 μL of chilled Matrigel and gently pipet up and down to mix.
5. Transfer the entire volume of 750 μL to the pre-coated coverslip and incubate for 30 min at 37 °C, 5 % CO_2 to allow gelation.

6. Apply the top coverslip and assemble the chamber.
7. Load tubing, needle, and syringe with media or media containing drug or chemokine. Place syringe in holder of pump programmed to deliver media and insert needle into inlet port.
8. Connect second syringe to tubing and needle and place syringe in pump programmed to withdraw media.
9. Turn on pumps and allow pressure within chamber to stabilize.
10. Begin image acquisition.

3.6 Mixing Different Cell Types in 3D Matrigel Culture

1. Coat the glass insert of a petri dish with Matrigel as described in Subheading 3.4.
2. Dissociate cells from growth flasks using either trypsin–EDTA or dissociation buffer (*see* Subheading 3.1 and/or Subheading 3.3), and count.
3. Mix the appropriate volume of cells from each population to achieve 5×10^5 cells per mL in 250 μL growth medium and chill on ice.
4. Proceed with **steps 4–7** as described in Subheading 3.4.

3.7 Optical Sectioning

1. Imaging can be performed through a 20× or 40× long working distance DIC objective.
2. Using Zeiss AxioVision or other controlling software, set the *z* range, *z* increment, the time interval between acquisition of each *z*-series and the duration of the experiment (Fig. 2a-b).
3. The acquisition software generates a sequentially numbered JPEG image stack that can be opened and saved as a movie.
4. If the *x*, *y* scale factor is not automatically determined by the acquisition software, it should be determined prior to analysis by recording a stage micrometer with microscope and camera settings identical to those used for the experiment. The user then demarcates a known distance (for example 10 μm) from the image of the stage micrometer and J3D-DIAS 4.1 converts that distance to μm/pixel.
5. The *z* scale factor is the μm/level that the user determines before starting the acquisition of optical sections.

3.8 Object Detection

1. J3D-DIAS4.1 opens large image stacks, then bundles and indexes them to create a new internal file format native to J3D-DIAS4.1 (*see* **Note 15**).
2. The object of choice can be automatically identified in each optical section, either as an outline or as a pixel bitmap area (Fig. 2c). Both methods identify an object (cell or aggregate in this case) based on the assumption that an object will have a

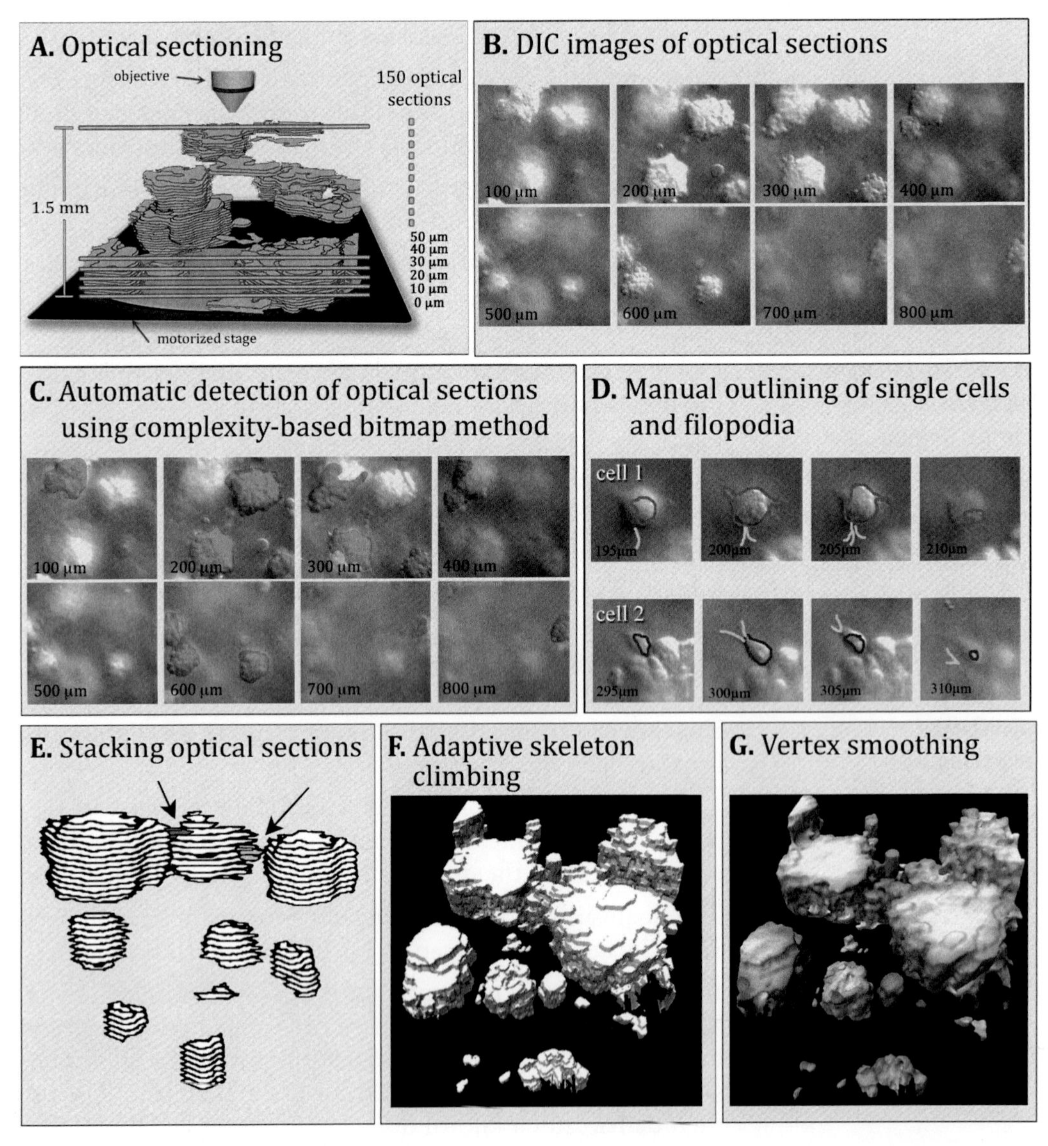

Fig. 2 Summary of 4D reconstruction methods. (**a**) Optical sections acquired using DIC optics through 1.5 mm of Matrigel-embedded cells at 10 µm intervals every 2 min resulting in 150 optical sections per *z*-series for 17 days. (**b**) Representative optical sections acquired by DIC microscopy. (**c**) The automatic complexity-based bitmap traces are *dark gray*. (**d**) Manually traced cells (*dark gray* outlines) and filopodia (*white*) of two representative cells. (**e**) Optical sections of aggregates and single cells (*arrows*) identified by automatic bitmap traces and manual traces, respectively, are combined and overlapping sections are stacked. (**f**) Stacks are rendered in 3D by the adaptive skeleton climbing algorithm. (**g**) Results of vertex smoothing of 3D objects. The height of each slice is given in the lower left of each panel in **b**, **c,** and **d**

greater variation and range of gray scales than the background. A 0-256 gray scale value is therefore assigned to individual pixels within a kernel (pixel matrix). The user defines the size of this kernel, usually 3×3, 5×5, or 7×7. The standard deviation (sd) of the averaged gray scale value of the pixels within the kernel is computed and assigned to the reference pixel within the kernel. If that sd is above the user-determined threshold, then the reference pixel is considered part of the object and is therefore retained. The kernel moves by one pixel, a new reference pixel is established and the sd of the average gray scale within the new kernel is calculated. The process is repeated until every pixel within the field is scanned. Thus, in-focus objects (i.e., objects exhibiting a high degree of gray scale complexity) will be identified while background or out of focus objects will have more uniform gray scale values and be discarded. Finally, the edge detection method discards the interior information while it is retained in bitmap object detection. Thus, the advantages of bitmap tracing over outlining are primarily that less information is lost and that holes or gaps present within the real image are maintained.

3. Manual outlining can be used for single cells, organelles, and/or fine structures such as filopodia. In manual outlining, the gray scale threshold between an object and its background (i.e., the edge of the object) is determined by the user, who uses the mouse to trace and enter the outline of the object, as previously described for 3D-DIAS [9] (Fig. 2d).

3.9 Stacking Optical Sections in J3D-DIAS4.1

1. Upon completion of object detection, the tracings that overlap in the *z*-axis at each time point in the movie are stacked into a 3D *z*-series (Fig. 2e), either as a complexity stack in the case of bitmap tracings or, in the case of outlines, as a series of beta-spline replacements. Beta-splines replace the pixels in the outline with curves that approximate the original pixels [40]. In essence, beta-splines smooth the outline by applying bias, tension, and resolution [41] and generate a mathematical model from which data can then be calculated [9, 41–43].
2. Bitmap pixels are expanded into voxels (3D pixels) and outlines are filled with voxels. In both cases, the *x*, *y*, and *z* scales determine the voxel dimensions and the output is a 3D pixel map (raw voxel block).

3.10 3D Reconstruction Using Adaptive Skeleton Climbing and Vertex Smoothing Algorithms in J3D-DIAS4.1

1. The raw voxel blocks are wrapped to generate a continuous surface. To perform this function, a variation of the "marching cubes" algorithm [44] known as the "adaptive skeleton climbing isoform extraction" algorithm [45] was introduced into J3D-DIAS4.1 (Fig. 2f) and implemented using Java OpenGL (JOGL). Conceptually, adaptive skeleton climbing attaches a triangular facet of a user-determined size to the raw voxel block in the most stable possible configuration. The faceted surface of the entire object is then sequentially built upon this initial facet. Each added facet is placed in the most stable possible configuration. Cracks between facets are filled with irregularly shaped facets.

2. The stacked optical sections can be smoothed using a vertex smoothing algorithm [46]. In vertex smoothing, the *x*, *y*, and *z* coordinates of a given vertex are averaged with neighboring *x*, *y*, and *z* coordinates, respectively, to generate a new vertex that is usually slightly shifted in position, relative to the original one. Three rounds of vertex smoothing are performed. In this manner, irregularities or sharp edges in the surface of the object are attenuated (Fig. 2g).
3. Manually traced and automatically traced objects can be reconstructed as solid, shaded nontransparent objects, such as the cells shown in Fig. 3a or transparent, caged images, such as the aggregates in Fig. 3b.
4. Manually traced and automatically traced objects can be merged, rotated, and viewed from different angles (Fig. 3b), and in movie format.
5. 4D reconstructions of non-tumorigenic cells from normal cell lines or tissues (Fig. 4a) and tumorigenic cells (Fig. 4b) can be viewed at different rotations over time to evaluate the positions of aggregates within the matrix and the process of coalescence in tumorigenic cells.
6. Mechanisms of coalescence and the presence of different cell types can be analyzed in 4D. In Fig. 5a a facilitator cell (color-coded green) that emerged from aggregate 1 of a tumorigenic cell line contacts a probe cell (color-coded red) that emerged from aggregate 2. The facilitator and probe then contact, pulling the two aggregates together, resulting in coalescence. Cables of cells that form between aggregates of tumorigenic cells and pull the aggregates together can also be identified in 4D (Fig. 5b). The movements of the "dervish" (color-coded

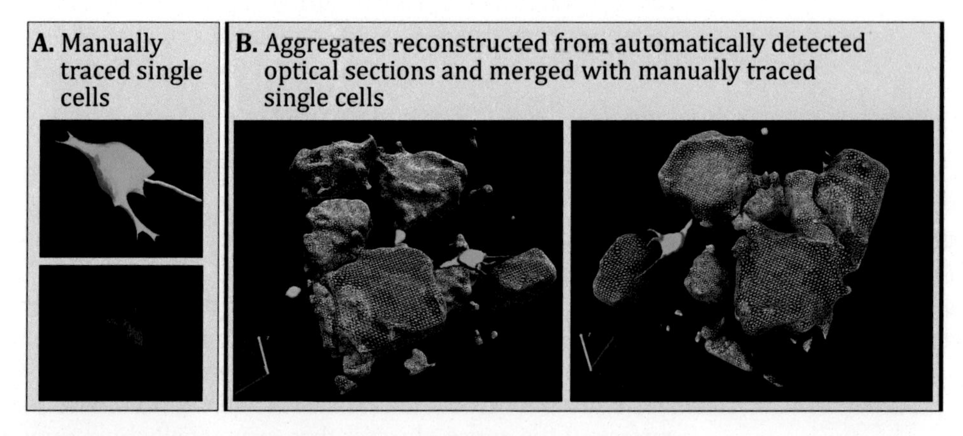

Fig. 3 3D reconstructions of manually outlined single cells (**a**) are merged into a field of automatically bitmap traced aggregates (**b**) and can be rotated and viewed from any angle. Single cells are colored and reconstructed in this example as filled objects while reconstructed aggregates are *gray* and caged, making them transparent. The *xyz* axes in the lower left corner of the panels in (**b**) indicate rotation angle

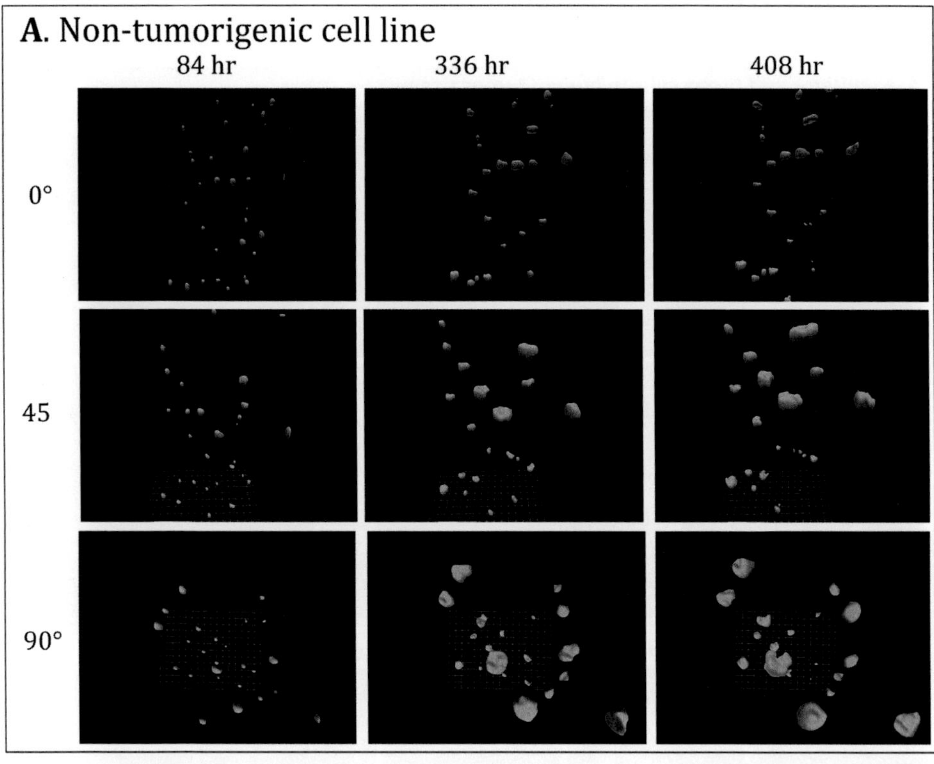

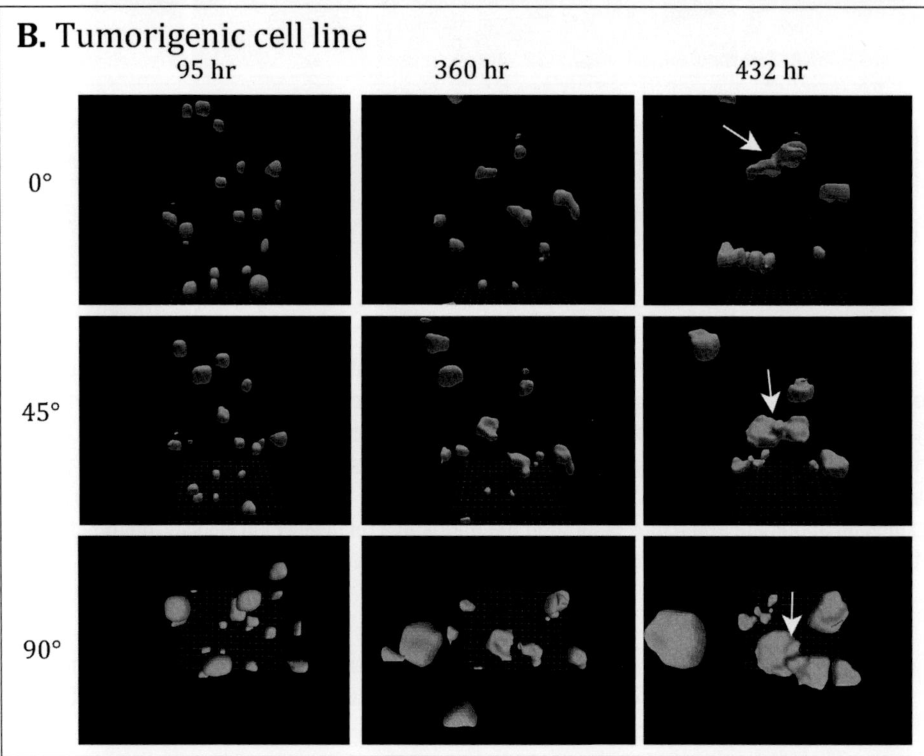

Fig. 4 4D analysis of non-tumorigenic (**a**) and tumorigenic cell lines (**b**) cultured in 3D Matrigel matrix reveals that non-tumorigenic cells divide to form cell islands but remain positionally fixed within the Matrigel while tumorigenic cell islands coalesce into large aggregates. Time in culture is given in hours (hr) across the *top* and angle of rotation in degrees along the side. The merged aggregates indicated by *arrows* in (**b**)

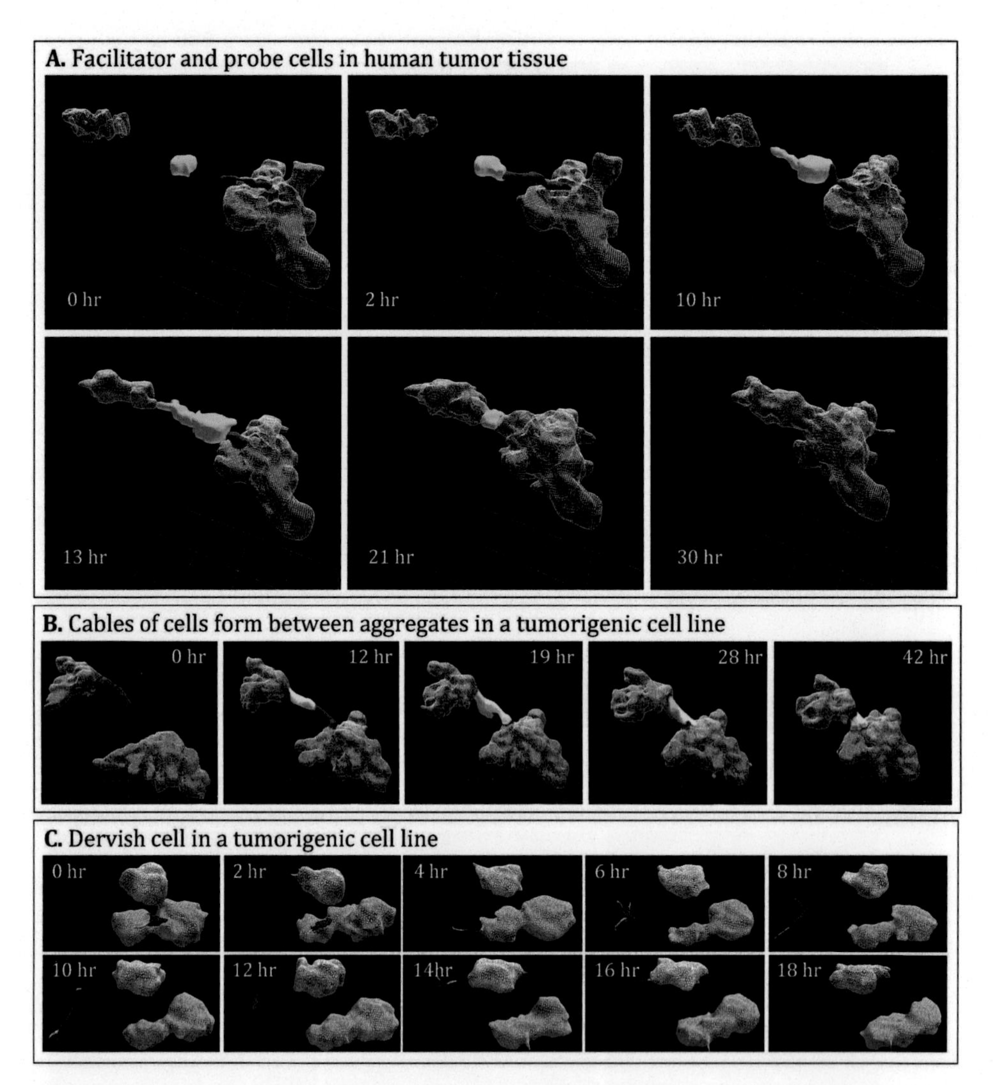

Fig. 5 4D analysis of coalescing tumorigenic cells cultured in Matrigel matrix reveals specific cell types and their role, or lack thereof, in mediating coalescence. (**a**) Facilitator cell (color-coded *green*) that has exited aggregate 1 contacts a probe cell (color-coded *red*) emerging from aggregate 2. The two cell types contact each other at 2 h and pull the aggregates together by 30 h. (**b**) Cables of cells formed between aggregates in a 3D culture of tumorigenic cells mediate their fusion. Cells within the cable are color-coded. (**c**) A dervish cell (*red*) exiting an aggregate at 0 h moves through the matrix at a relatively high velocity, unimpeded by the gel, and subsequently enters a second aggregate (*gray*) at 18 h. hr, hours relative to 0 h time point

red) (Fig. 5c) can also be tracked in 4D as it exits an aggregate (color-coded blue), moves rapidly through the Matrigel and penetrates a second aggregate (gray).

7. The effects of monoclonal antibodies such as anti-β1 integrin (Fig. 6b) and anticancer drugs such as Vemurafenib (Fig. 6c) on emergence of cell types and coalescence in tumorigenic cell lines as well as in human tumor tissue (Fig. 6c) can be evaluated in 4D.

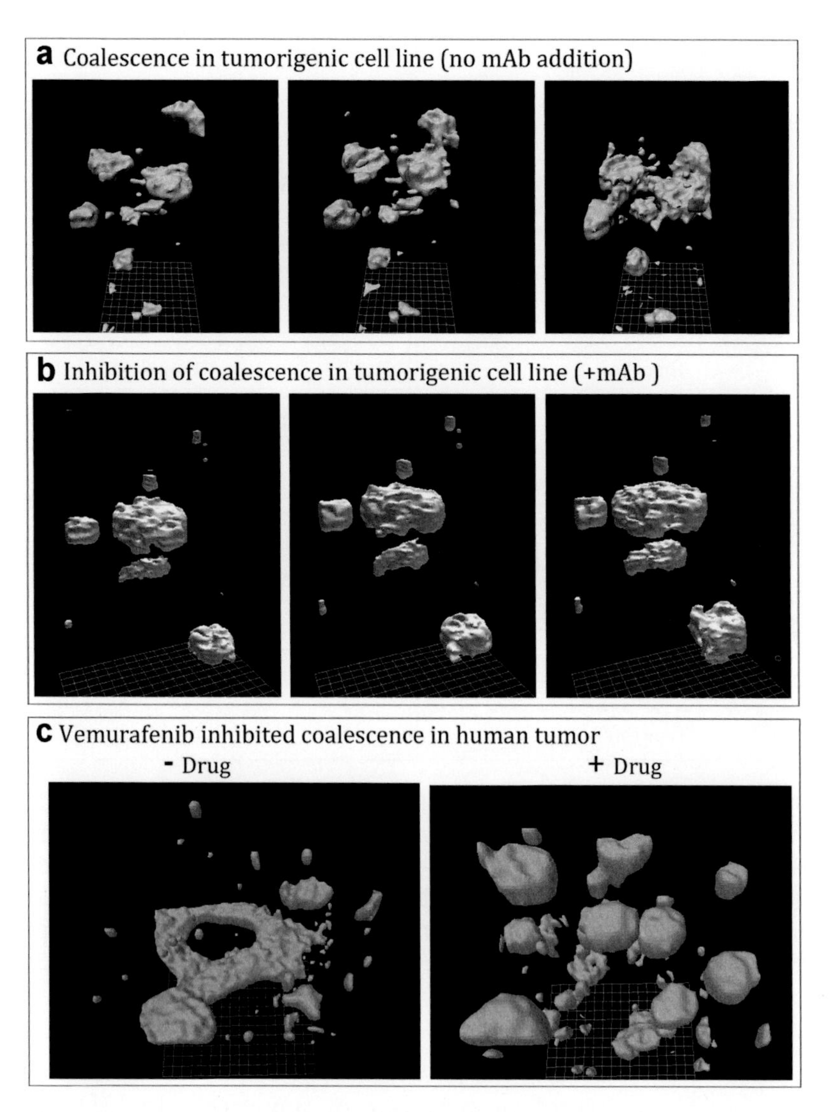

Fig. 6 Effects of drugs on coalescence and cell behavior can be evaluated in the 4D model. (**a**) Coalescence in an untreated tumorigenic cell line is inhibited (**b**) by the addition of anti-β1 integrin monoclonal antibody. *Arrows* indicate aggregates that have coalesced (**c**) Coalescence of cancer cells derived from a human gastric tumor is inhibited by addition of the anticancer drug Vemurafenib

3.11 Motility and Dynamic Morphology Parameters

1. The 3D position of the centroid (center) at each time point can be determined by averaging the *x*-, *y*-, and *z*-coordinates of all interior points of the faceted object.
2. Paths for the objects and the number of paths can be used to determine the number of aggregates (Fig. 7a).
3. The faceted 3D reconstructions are mathematical models of the object [9] and can therefore be used to calculate volume (Fig. 7b). The triangularized surface of the faceted object is a polyhedron. The calculus used to determine the volume of a polyhedron can be found at http:wwwf.imperial.ac.uk/~rn/centroid.pdf.

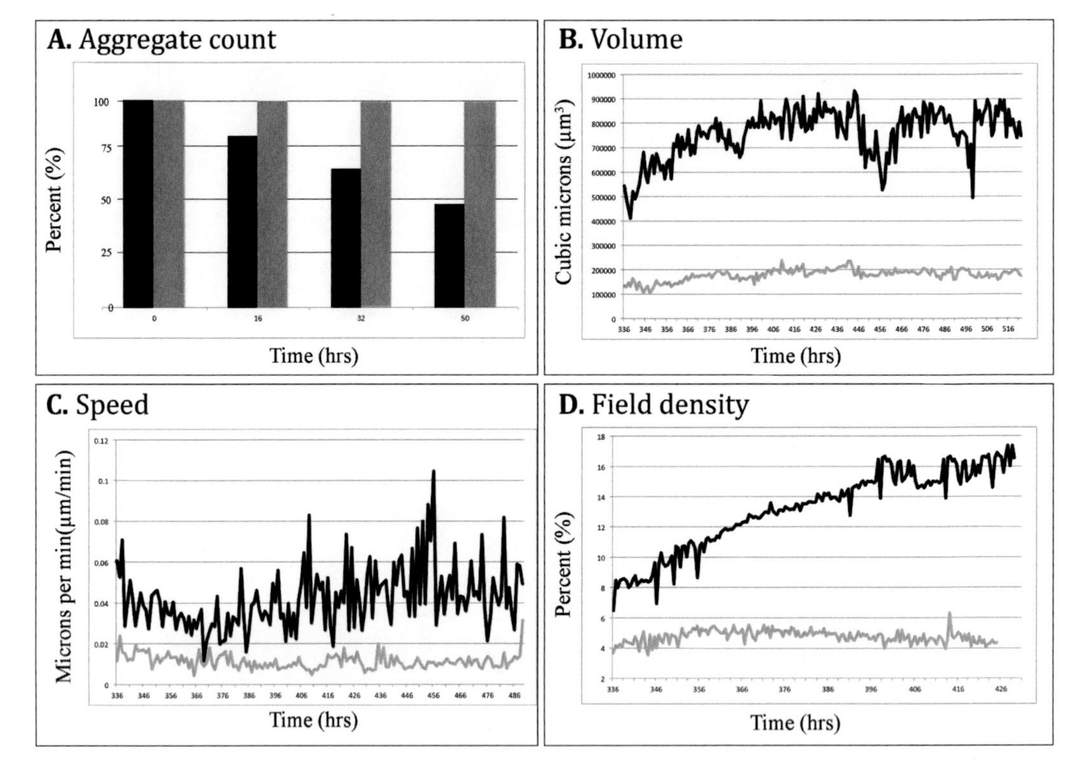

Fig. 7 4D quantitative data comparing tumorigenic and non-tumorigenic cells cultured in a 3D matrix reveal significant differences between the two populations. (**a**) Number of aggregates decreases over time due to coalescence in a tumorigenic cell line but not in non-tumorigenic cells during the 50 h period of analysis. (**b**) Mean volume (**c**) Mean speed and (**d**) Field density. In **b–d**, each point is the average value of all objects, generally between 15 and 30 cells and aggregates, in the area of analysis at each time point over a period of 7 days. *Black*, tumorigenic cells; *gray*, non-tumorigenic cells. Parameters are defined in Subheading 3.11

4. Surface complexity is a function of activity around the periphery of an aggregate or single cell. In J3D-DIAS4.1, it is calculated from volume (described above) and surface area. Surface area is the sum of the areas of all facets. The surface complexity parameter is a derivation of 3D roundness. The formula for the latter is $\mathbf{100} \times 6$ x $\mathbf{sqrt}\boldsymbol{\pi}$ x Vol/(surface area$^{3/2}$), where *sqrt* is the square root. The result is a percentage relative to a perfect sphere, which gives a 3D roundness of 100 %. Thus, the more complex the surface contour of an object is, the lower the 3D roundness parameter will be. However, to make the surface complexity measurement more intuitive, 3D roundness was expressed as the inverse rather than as a percentage of a perfect sphere. That is, surface complexity is computed as 1/(6 x $\mathbf{sqrt}\boldsymbol{\pi}$ x Vol/(surface area$^{3/2}$)). This yields a value for surface complexity that increases, rather than decreases, with increasing complexity.

5. The speed of cell and aggregate translocation (instantaneous velocity) in 3D is calculated from the position of the 3D centroid by the central difference method [47] using the same formulae applied to 2D objects [41] as follows:

$$\mathbf{Speed}[\mathrm{f}] = (\mathbf{scale}x\mathbf{frate})\,x\mathbf{sqrt}\left(\left(\mathbf{x}[\mathrm{f}+\mathbf{I}]-\mathbf{x}[\mathrm{f}-\mathbf{I}]\right)/\,\mathbf{2I}\right)^2 + \left(\left(\mathbf{y}[\mathrm{f}+\mathbf{I}]-\mathbf{y}[\mathrm{f}-\mathbf{I}]\right)/\,\mathbf{2I}\right)^2 + \left(\left(\mathbf{z}\mathrm{f}+\mathbf{I}]-\mathbf{z}[\mathrm{f}-\mathbf{I}]\right)/\,\mathbf{2I}\right)^2 \text{ when } 1 <= \mathrm{f}-\mathbf{I} \text{ and } \mathrm{f}+\mathbf{I} <= \mathbf{F}$$

$$\mathbf{Speed}[\mathrm{f}] = (\mathbf{scale}x\mathbf{frate})\,x\mathbf{sqrt}\left(\left(\mathbf{x}[\mathrm{f}+\mathbf{I}]-\mathbf{x}[\mathrm{f}]\right)/\,\mathbf{I}\right)^2 + \left(\left(\mathbf{y}[\mathrm{f}+\mathbf{I}]-\mathbf{y}[\mathrm{f}]\right)/\,\mathbf{I}\right)^2 + \left(\left(\mathbf{z}\mathrm{f}+\mathbf{I}]-\mathbf{z}[\mathrm{f}-\mathbf{I}]\right)/\,\mathbf{2I}\right)^2) \text{ when } \mathrm{f}-\mathbf{I} < 1 \text{ and } \mathrm{f}+\mathbf{I} <= \mathbf{F}(\text{first frame})$$

$$\mathbf{Speed}[\mathrm{f}] = (\mathbf{scale}x\mathbf{frate})\,x\mathbf{sqrt}\left(\left(\mathbf{x}[\mathrm{f}]-\mathbf{x}[\mathrm{f}-\mathrm{I}]\right)/\,\mathbf{I}\right)^2 + \left(\left(\mathbf{y}[\mathrm{f}]-\mathbf{y}[\mathrm{f}-\mathbf{I}]\right)/\,\mathbf{I}\right)^2 + \left(\left([\mathbf{z}\mathrm{f}+\mathbf{I}]-\mathbf{z}[\mathrm{f}-\mathbf{I}]\right)/\,2\mathbf{I}\right)^2 \text{ when } 1 <= \mathrm{f}-\mathbf{I} \text{ and } \mathrm{f}+\mathbf{I} > \mathbf{F}(\text{last frame})$$

$$\mathbf{Speed}[\mathrm{f}] = 0 \text{ otherwise}$$

where **F** is the total number of frames, f is the "current" frame, (**x**[f],**y**[f]) are the coordinates of the centroid of an object in frame f, 1 <=f<=**F**, **I** is the centroid increment, **frate** is the frame rate in # of frames per unit time, **scale** is the scale factor in distance units per pixel, **sqrt** is the square root function, and x denotes multiplication.

Aggregate speed in cultures of tumorigenic and non-tumorigenic cells is plotted in Fig. 7c.

6. The rate and extent of coalescence can be evaluated with the field density parameter. This parameter draws the smallest possible bounding cube that can encompass all objects in the field in every frame. The volume of the box and the volume of the objects contained within it are determined in each frame. The field density parameter is then computed for each frame as the ratio of the volume of objects over the volume of the bounding cube and multiplied by 100. Therefore, as cells coalesce into aggregates, the volume of the objects will either remain the same or increase while the volume of the bounding box will decrease. A sample data plot of this parameter in tumorigenic and non-tumorigenic cultures is presented in Fig. 7d.

7. Matrix degradation and remodeling [48–50] occur in 3D cultures of tumorigenic cells and can be quantitated in J3D-DIAS4.1 by plotting the *z* position, measured as the high point of an object, over time (Fig. 8a, c) and by stacking perimeter plots of aggregates (Fig. 8b, d).

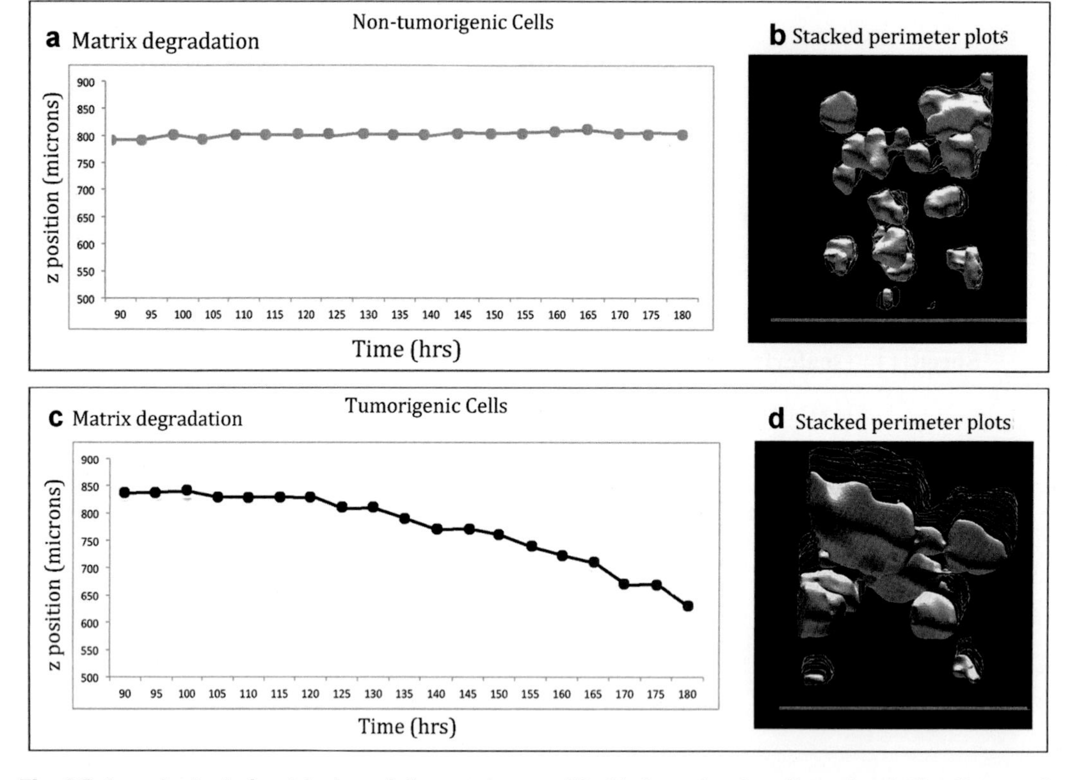

Fig. 8 Rate and extent of matrix degradation can be quantified in tumorigenic cultures by plotting the *z* positions of aggregates and cells over time. (**a**) *z* Height of a non-tumorigenic cell island remains stable over time. (**b**) Stacked perimeter plots confirm positional stability of all cell islands in non-tumorigenic populations. (**c**) *z* height of tumorigenic aggregate declines over time, indicative of matrix degradation. (**d**) Stacked perimeter plots of all cells and aggregates within the tumorigenic culture confirm consistent downward trend in *z* position, indicative of matrix degradation. *White lines* are outlines of objects in consecutive frames. The objects are filled in their final positions

4 Notes

1. Protocols can be found in *Biosafety in Microbiological and Biomedical Laboratories*, 5th ed. HHS Publication No. (CDC) 93-8395. U.S. Department of Health and Human Services Centers for Disease Control and Prevention. Washington, DC: U.S. Government Printing Office or at http://www.cdc.gov/biosafety/publications/bmbl5/index.htm.
2. All human tissues and/or blood products must be obtained under informed consent using protocols approved by the institution's IRB.
3. Specific media requirements are typically provided with commercially available cell lines such as those obtained from ATCC (http://www.atcc.org). Use of the supplier's recommenda-

tions is required for the warranty to be valid. In many cases, however, once a commercial cell line is growing and stocks have been frozen according to the supplier's instructions, cells can be adapted to another type of media, if desired. To do this, at the time of passaging, transfer cells into media consisting of 50 % of the initial media and 50 % of the desired media. At the next passaging, cells can be transferred into 100 % of the desired media. Maintain a back-up flask in the original media in case the adaptation fails.

4. Routine heat inactivation of serum, although once widely practiced, is no longer considered necessary for the majority of applications and if done improperly, may actually reduce the efficacy of cell culturing (http://www.sigmaaldrich.com/technical-documents/protocols/biology/the-cell-environment.html).
5. Peripheral blood mononuclear cells are available commercially, but it may be more cost-effective to isolate them as needed from whole blood drawn from volunteers using IRB approved informed consent protocols.
6. Other density gradient centrifugation media such as Ficoll-Paque (GE Healthcare http://www.gelifesciences.com) are available and should be used according to the manufacturer's directions. The protocol given here is specifically for Histopaque-1077 from Sigma-Aldrich.
7. M-CSF induces monocyte differentiation into macrophages. Macrophages can then be induced to differentiate into different subtypes; e.g., inflammatory or non-inflammatory, using specific growth factors and cytokines [51].
8. Matrigel matrix gels quickly, so care must be taken to keep it chilled along with the surfaces and media that it contacts.
9. If the height of the dish exceeds the working distance of the optics, then approximately 5 mm can be trimmed from the dish bottom with sterile scissors before media is added.
10. To achieve DIC optics for optical sectioning, a glass window should replace the plastic in the light path through the lid of the dish. An area with a diameter of 20–30 mm can be removed from the center of the plastic lid with a drill bit or laser cutter. A glass coverslip of larger dimensions can then be placed over the hole and sealed with Vaseline. The lid with the glass window is sterilized with ethanol and/or by UV exposure within the biosafety cabinet.
11. Closely monitor cell detachment as over-exposure to trypsin can damage cell adhesion sights. Use enzyme-free dissociation buffer to minimize damage to weakly adherent cells.
12. It is not always possible to predict what type of media will support growth of primary cultures. Therefore, it is advisable to

initially set up several tissue samples in different types of media. Once the culture is established, cells can be adapted to different types of media as described in **Note 3** above.

13. We have performed experiments using monoclonal antibodies at 100 μg/mL. Appropriate concentrations of drugs must be determined empirically. We used 20 μM Vemurafenib (http://www.selleckchem.com).
14. To avoid leaks or introduction of air bubbles during perfusion, make certain the Matrigel contacts the edge of the gasket to form a complete, unbroken seal around its entire circumference.
15. J3D-DIAS4.1 can be accessed by collaboration at the W.M. Keck Dynamic Image Analysis Facility, David R. Soll, director, Biology Dept., University of Iowa, Iowa City, IA 52245 USA.

Acknowledgements

This work was supported by the Developmental Studies Hybridoma Bank (DSHB), a National Resource created by the NICHD of the NIH and maintained at The University of Iowa, Department of Biology, Iowa City, IA 52242. The anti-β1 integrin monoclonal antibody, developed by C.H. Damsky was obtained from the DSHB. We thank Brett Hanson, Joseph Ambrose, Kanoe Russell, Emma Buchele, Brian Kroll, Michele Livitz, Benjamin Soll, and Nicole Richardson for technical assistance. We are grateful to Dr. M. Milhem, University of Iowa Hospitals and Clinics, Iowa City, IA 52242 and to Emily Fletcher and Dr. Charles Goldman, Mercy Hospital System of Des Moines, Des Moines, IA for assistance in tissue acquisition.

References

1. Herrmann D, Conway JR, Vennin C et al (2014) Three-dimensional cancer models mimic cell-matrix interactions in the tumour microenvironment. Carcinogenesis 35:1671–1679
2. Godugu C, Patel AR, Desai U et al (2013) AlgiMatrix™ based 3D cell culture system as an in-vitro tumor model for anticancer studies. PLoS One 8:e53708
3. Roh-Johnson M, Bravo-Cordero JJ, Patsialou A et al (2014) Macrophage contact induces RhoA GTPase signaling to trigger tumor cell intravasation. Oncogene 33:4203–4212
4. Heid PJ, Voss E, Soll DR (2002) 3D-DIASemb: a computer-assisted system for reconstructing and motion analyzing in 4D every cell and nucleus in a developing embryo. Dev Biol 245:329–347
5. Kimlin LC, Casagrande G, Virador VM (2013) In vitro three-dimensional (3D) models in cancer research: an update. Mol Carcinog 52:167–182
6. Pampaloni F, Berge U, Marmaras A et al (2014) Tissue-culture light sheet fluorescence microscopy (TC-LSFM) allows long-term imaging of three-dimensional cell cultures under controlled conditions. Integr Biol (Camb) 6:988–998
7. Weigelt B, Ghajar CM, Bissell MJ (2014) The need for complex 3D culture models to unravel novel pathways and identify accurate biomarkers in breast cancer. Adv Drug Deliv Rev 69–70:42–51
8. Nam JM, Onodera Y, Bissell MJ, Park CC (2010) Breast cancer cells in three-dimensional culture display an enhanced radioresponse after

coordinate targeting of integrin alpha5beta1 and fibronectin. Cancer Res 70:5238–5248
9. Soll DR, Voss E (1998) Two and three dimensional computer systems for analyzing how cells crawl. In: Wessels D, Soll DR (eds) Motion analysis of living cells. Wiley, New York, pp 25–52
10. Soll DR, Wessels D, Voss E, Johnson O (2001) Computer-assisted systems for the analysis of amoeboid cell motility. Methods Mol Biol 161:45–58
11. Wessels D, Kuhl S, Soll DR (2009) 2D and 3D quantitative analysis of cell motility and cytoskeletal dynamics. Methods Mol Biol 586:315–335
12. Wessels D, Soll DR (1998) Computer-assisted characterization of the behavioral defects of cytoskeletal mutants of *Dictyostelium discoideum*. In: Soll DR, Wessels D (eds) Motion analysis of living cells. Wiley, New York, pp 101–140
13. Bosgraaf L, Waijer A, Engel R et al (2005) RasGEF-containing proteins GbpC and GbpD have differential effects on cell polarity and chemotaxis in *Dictyostelium*. J Cell Sci 118:1899–1910
14. Breshears LM, Wessels D, Soll DR, Titus MA (2010) An unconventional myosin required for cell polarization and chemotaxis. Proc Natl Acad Sci U S A 107:6918–6923
15. Heid PJ, Wessels D, Daniels KJ et al (2004) The role of myosin heavy chain phosphorylation in Dictyostelium motility, chemotaxis and F-actin localization. J Cell Sci 117:4819–4835
16. Lusche DF, Wessels D, Ryerson DE, Soll DR (2011) Nhe1 is essential for potassium but not calcium facilitation of cell motility and the monovalent cation requirement for chemotactic orientation in *Dictyostelium discoideum*. Eukaryot Cell 10:320–331
17. Lusche DF, Wessels D, Scherer A et al (2012) The IplA Ca^{2+} channel of *Dictyostelium discoideum* is necessary for chemotaxis mediated through Ca^{2+}, but not through cAMP, and has a fundamental role in natural aggregation. J Cell Sci 125:1770–1783
18. Scherer A, Kuhl S, Wessels D et al (2010) Ca2+ chemotaxis in *Dictyostelium discoideum*. J Cell Sci 123:3756–3767
19. Stepanovic V, Wessels D, Daniels K et al (2005) Intracellular role of adenylyl cyclase in regulation of lateral pseudopod formation during *Dictyostelium* chemotaxis. Eukaryot Cell 4:775–786
20. Wessels D, Lusche DF, Kuhl S et al (2007) PTEN plays a role in the suppression of lateral pseudopod formation during *Dictyostelium* motility and chemotaxis. J Cell Sci 120: 2517–2531
21. Zhang H, Heid PJ, Wessels D et al (2003) Constitutively active protein kinase A disrupts motility and chemotaxis in *Dictyostelium discoideum*. Eukaryot Cell 2:62–75
22. Geiger J, Wessels D, Lockhart SR, Soll DR (2004) Release of a potent polymorphonuclear leukocyte chemoattractant is regulated by white-opaque switching in Candida albicans. Infect Immun 72:667–677
23. Gligorijevic B, Bergman A, Condeelis J (2014) Multiparametric classification links tumor microenvironments with tumor cell phenotype. PLoS Biol 12:e1001995
24. Stepanovic V, Wessels D, Goldman FD et al (2004) The chemotaxis defect of Shwachman-Diamond Syndrome leukocytes. Cell Motil Cytoskeleton 57:158–174
25. Wessels D, Srikantha T, Yi S et al (2006) The Shwachman-Bodian-Diamond syndrome gene encodes an RNA-binding protein that localizes to the pseudopod of *Dictyostelium* amoebae during chemotaxis. J Cell Sci 119: 370–379
26. Volk AP, Heise CK, Hougen JL et al (2008) ClC-3 and IClswell are required for normal neutrophil chemotaxis and shape change. J Biol Chem 283:34315–34326
27. Biggs LC, Naridze RL, DeMali KA et al (2014) Interferon regulatory factor 6 regulates keratinocyte migration. J Cell Sci 127:2840–2848
28. Li Y, Wessels D, Wang T et al (2003) Regulation of caldesmon activity by Cdc2 kinase plays an important role in maintaining membrane cortex integrity during cell division. Cell Mol Life Sci 60:198–211
29. Sidani M, Wessels D, Mouneimne G et al (2007) Cofilin determines the migration behavior and turning frequency of metastatic cancer cells. J Cell Biol 179:777–791
30. Wong K, Wessels D, Krob SL et al (2000) Forced expression of a dominant-negative chimeric tropomyosin causes abnormal motile behavior during cell division. Cell Motil Cytoskeleton 45:121–132
31. Song W, Onishi M, Jan LY, Jan YN (2007) Peripheral multidendritic sensory neurons are necessary for rhythmic locomotion behavior in Drosophila larvae. Proc Natl Acad Sci U S A 104:5199–5204
32. Suster ML, Karunanithi S, Atwood HL, Sokolowski MB (2004) Turning behavior in Drosophila larvae: a role for the small scribbler transcript. Genes Brain Behav 3:273–286
33. Wang P, Saraswati S, Guan Z et al (2004) A Drosophila temperature-sensitive seizure mutant in phosphoglycerate kinase disrupts ATP generation and alters synaptic function. J Neurosci 24: 4518–4529

34. Ulrich F, Concha ML, Heid PJ et al (2003) Slb/Wnt11 controls hypoblast cell migration and morphogenesis at the onset of zebrafish gastrulation. Development 130:5375–5384
35. Scherer A, Kuhl S, Wessels D et al (2015) A computer-assisted 3D model for analyzing the aggregation of tumorigenic cells reveals specialized behaviors and unique cell types that facilitate aggregate coalescence. PLoS One 10:e0118628
36. Park CC, Zhang H, Pallavicini M et al (2006) Beta1 integrin inhibitory antibody induces apoptosis of breast cancer cells, inhibits growth, and distinguishes malignant from normal phenotype in three dimensional cultures and in vivo. Cancer Res 66:1526–1535
37. Park CC, Zhang HJ, Yao ES et al (2008) Beta1 integrin inhibition dramatically enhances radiotherapy efficacy in human breast cancer xenografts. Cancer Res 68:4398–4405
38. Weaver VM, Petersen OW, Wang F et al (1997) Reversion of the malignant phenotype of human breast cells in three-dimensional culture and in vivo by integrin blocking antibodies. J Cell Biol 137:231–245
39. Varzavand A, Drake JM, Svensson RU et al (2013) Integrin α3β1 regulates tumor cell responses to stromal cells and can function to suppress prostate cancer metastatic colonization. Clin Exp Metastasis 30:541–552
40. Barsky BA (1988) Computer graphics and geometric modeling using beta-splines. Springer, New York
41. Soll DR (1995) The use of computers in understanding how animal cells crawl. Int Rev Cytol 163:43–104
42. Soll DR (1999) Computer-assisted three-dimensional reconstruction and motion analysis of living, crawling cells. Comput Med Imaging Graph 23:3–14
43. Soll DR, Voss E, Johnson O, Wessels D (2000) Three-dimensional reconstruction and motion analysis of living, crawling cells. Scanning 22: 249–257
44. Lorensen WE, Cline HE (1987) Marching cubes: a high resolution 3D surface construction algorithm. Comput Graph 21:163–169
45. Poston T, Wong TT, Heng PA (1998) Multiresolution isosurface extraction with adaptive skeleton climbing. Comput Graph Forum 17(3):137–147
46. Hermann L (1976) Laplacian-isoparametric grid generation scheme. J Eng Mech Div 102: 749–756
47. Maron MJ (1982) Numerical analysis – a practical approach. Macmillan, New York
48. Li XY, Ota I, Yana I et al (2008) Molecular dissection of the structural machinery underlying the tissue-invasive activity of membrane type-1 matrix metalloproteinase. Mol Biol Cell 19:3221–3233
49. Wang Y, McNiven MA (2012) Invasive matrix degradation at focal adhesions occurs via protease recruitment by a FAK-p130Cas complex. J Cell Biol 196:375–385
50. Wolf K, Müller R, Borgmann S et al (2003) Amoeboid shape change and contact guidance: T-lymphocyte crawling through fibrillar collagen is independent of matrix remodeling by MMPs and other proteases. Blood 102:3262–3269
51. Engström A, Erlandsson A, Delbro D, Wijkander J (2014) Conditioned media from macrophages of M1, but not M2 phenotype, inhibit the proliferation of the colon cancer cell lines HT-29 and CACO-2. Int J Oncol 44:385–392

Chapter 19

An Experimental Model for Simultaneous Study of Migration of Cell Fragments, Single Cells, and Cell Sheets

Yao-Hui Sun, Yuxin Sun, Kan Zhu, Bruce W. Draper, Qunli Zeng, Alex Mogilner, and Min Zhao

Abstract

Recent studies have demonstrated distinctive motility and responses to extracellular cues of cells in isolation, cells collectively in groups, and cell fragments. Here we provide a protocol for generating cell sheets, isolated cells, and cell fragments of keratocytes from zebrafish scales. The protocol starts with a comprehensive fish preparation, followed by critical steps for scale processing and subsequent cell sheet generation, single cell isolation, and cell fragment induction, which can be accomplished in just 3 days including a 36–48 h incubation time. Compared to other approaches that usually produce single cells only or together with either fragments or cell groups, this facile and reliable methodology allows generation of all three motile forms simultaneously. With the powerful genetics in zebrafish our model system offers a useful tool for comparison of the mechanisms by which cell sheets, single cells, and cell fragments respond to extracellular stimuli.

Key words Cell migration, Collective cell migration, Cell fragment, Electric fields, Galvanotaxis, Electrotaxis, Zebrafish

1 Introduction

Cell migration is important in embryonic development, wound healing, and tumor metastasis [1–3], and is a critical process during immune response [4]. Cell motility may be roughly categorized in three forms scaled from cell fragments, as evidenced by cytokineplasts penetrating interendothelial cell junctions in response to a chemoattractant [5], to single cells (cells in isolation), to cohesive cell groups and cell sheets.

The migration of both single cells and groups of cells has been extensively studied in tissue culture and are known to contribute to many physiological motility processes in vivo during embryogenesis and wound healing [6, 7]. Similar migratory behavior is also displayed by many invasive tumor types [8, 9].

Tian Jin and Dale Hereld (eds.), *Chemotaxis: Methods and Protocols*, Methods in Molecular Biology, vol. 1407, DOI 10.1007/978-1-4939-3480-5_19, © Springer Science+Business Media New York 2016

Compared to single cells and cell sheets, migration of cell fragments has been under investigated. In vivo, viable cell fragments that pinch off directly from the cell body have been characterized and linked to various disease states [10, 11]. These anucleated fragments remain functional and migratory and can survive for hours or even days. For instance, fragments can function as intercellular "ferries" to transfer bioactive molecules [12], as marker "beacons" for tumor cells to navigate microenvironments [13] and as phagocytic "defenders" to ingest and kill bacteria [14].

Evaluating models of directional cell migration can provide more comprehensive insight into the complex mechanisms and signaling pathways involved in cellular motility [15]. Although studies of single cell migration, some in conjunction with either cell fragments or cell groups, have been reported [16, 17], a system containing all three motile units has not been reported before. A detailed comparison of the cells, and cell sheets or cell fragments, and their applications of spontaneous and directed migration are summarized in Table 1.

Fish keratocytes provide an excellent locomotion model for better understanding polarization and migration [18–20]. An added advantage to using fish keratocytes from the zebrafish species

Table 1
Comparison of cell types and their use in generating single cells, cell sheets, or cell fragment for migration study

		System of			
Study	**Cell type**	**Cell sheets**	**Single cells**	**Cell fragments**	**Purpose**
Albrecht-Buehler [32]	Human skin fibroblast	No	No	Yes	Spontaneous migration
Malawista et al. [33]	Human PMN	No	No	Yes	Chemotaxis
Euteneuer et al. [16]	Fish keratocyte	No	Yes	Yes	Spontaneous migration
Cooper et al. [34]	Fish keratocyte	Yes	Yes	No	Electrotaxis
Verkhovsky et al. [27]	Fish keratocyte	No	Yes	Yes	Spontaneous and directed migration
Yount et al. [13]	Human glioblastoma	No	Yes	Yes	Spontaneous migration
Li et al. [35]	Many different kinds	Yes	Yes	No	Electrotaxis
Sun et al. [25]	Fish keratocyte	No	Yes	Yes	Electrotaxis
Cohen et al. [36]	HDCK	Yes	No	No	Electrotaxis

[21, 22] is the availability of powerful molecular tools for genetic manipulation. Keratocyte motility is dependent on a synergistic actin treadmill of the self-organizing lamellipodia, which combines the protrusion of growing actin networks in the front and the retraction of the actomyosin contractile network in the rear. As a result, these cells move persistently with steady speeds, shape, and behavior and are thus ideal for in-depth analysis of directional response to extracellular cues, such as electric field [23, 24]. Interestingly, portions of keratocytes can spontaneously detach and form cytoplasmic fragments that lack major cell organelles and microtubules [16], yet still exhibit migrational movement similar to that of their whole cell counterparts. Thus, these keratocyte fragments represent an even simpler model of cell motility and are suitable for experimental cell migration studies [25]. In contrast to single cells or cell fragments, the collective response of cohorts of cells or cell sheets is much more complicated. Collective cell migration requires that all cells or otherwise selected cells sense a guidance cue and interpret it individually and/or cooperatively [26].

Here we describe an optimized protocol for preparing a unique system containing cell sheets, isolated cells, and cell fragments for use in spontaneous and directed migration studies. The protocol relies on the tissue culture of fish scales, and contains detailed step-by-step procedures for cell sheet generation, single cell isolation, and cell fragment induction. The quantitative analysis of these three motile units as a model system enables better insight into the cellular and molecular mechanisms that contribute to the intricate and coherent steps of polarization and directional migration as demonstrated in electrotaxis.

1.1 Overview of the Procedure

Essentially, the procedure described here consists of two parts: (1) isolating fish scales, (2) generating epidermal keratocyte sheets, isolating single keratocytes, and inducing cytoplasmic fragments.

An outline of our protocol is shown in Fig. 1. A detailed illustration of scale preparation and seeding to generate the three motile units of fish keratocytes is shown in Fig. 2. The morphology and characteristics of epithelial cell sheets, isolated cells, and cell fragments are provided in Fig. 3. Quantification of spontaneous migration of these three motile units and applications of these three motile units in electrotaxis are displayed in Figs. 4 and 5.

Fish preparation is critical for scale processing and isolation. Beakers and containers are pre-autoclaved. Fish net is sterilized by 70 % ethanol spray. Water needs to be at room temperature although autoclaving is not necessary. Tap water contains chlorine or chloramines, which can stress and possibly kill the fish. It is important to always use dechlorinated water, which can be achieved by exposing a container of water to air for at least 24 h or by running the water through a carbon filter. The fish must be washed thoroughly to avoid any bacterial or fungal contamination. Care

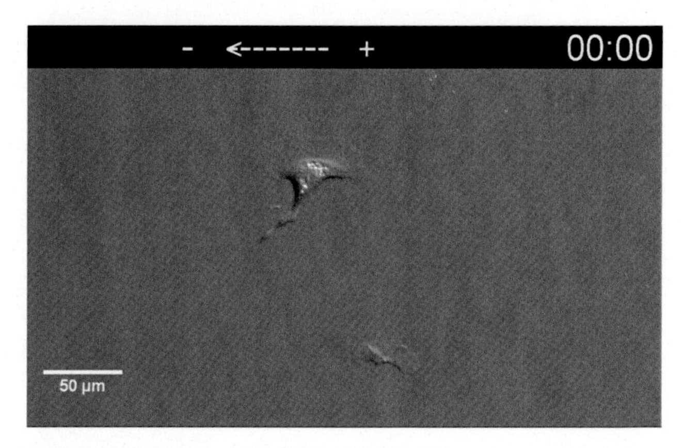

Supplementary movie S3 Electrotaxis of fish keratocyte and fragment. Access this movie at: https://youtu.be/KI1D_t0KU9E

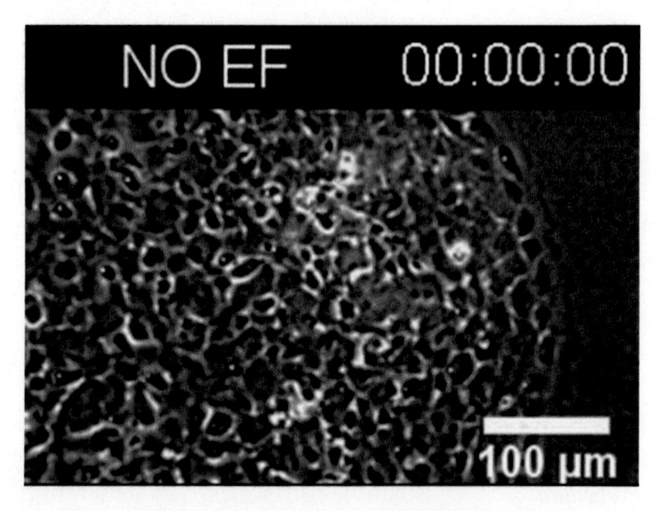

Supplementary movie S4 Collective cell migration in epidermal keratocyte sheet. Access this movie at: https://youtu.be/_TAZKCY2lf8

should be taken when handling the fish to minimize stress and harm. Scales are taken from the fish flanks and laid in culture dish, covered with a 22 × 22 mm glass coverslip. A sterilized stainless steel hex nut is placed on top to hold the scales in place. Detailed illustration of these critical steps is shown in Fig. 2.

Epithelial cells are typically packed together with very little intercellular material between them. An extremely tight bond exists between adjacent cells such that dissociation of epithelium is a difficult process. Methods for isolating primary culture cells have been extensively developed. The culture systems presently used in most laboratories are based on mechanical disaggregation and/or enzymatic digestion of animal tissue into single cells. Currently, several commercialized products such as Gibco cell culture systems (Life Technologies), cell isolation optimizing systems (Worthington

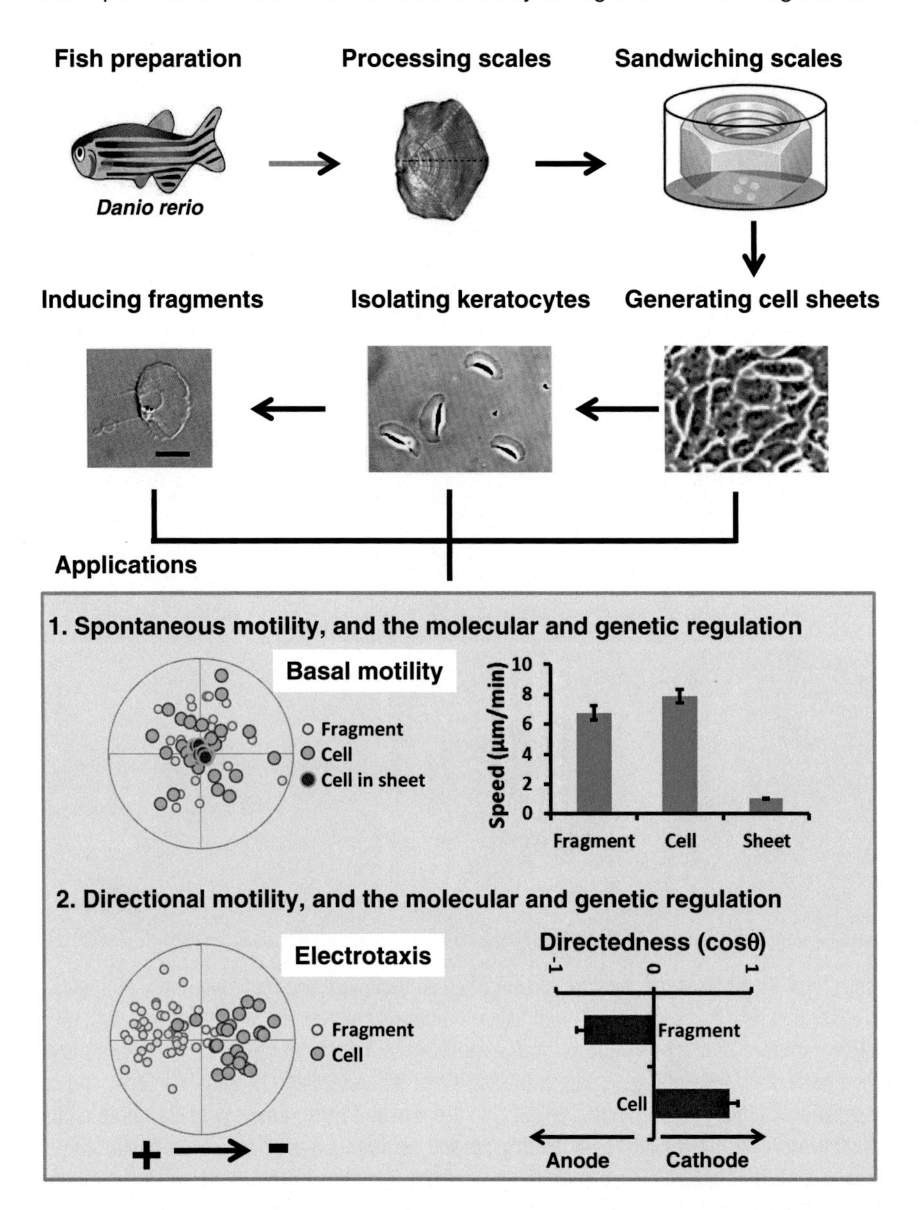

Fig. 1 Schematic outline of production of cell sheets, isolating keratocytes, and cell fragments and examples of application. Fish scales produce keratocyte cell sheets, cells in isolation, and cell fragments. All maintain good motility, and respond to small applied electric fields by directional migration (electrotaxis/galvanotaxis). This provides a unique system to study molecular and genetic control of cell movement in collection and in isolation, and a powerful tool to study directional migration and the mechanisms

Biochemical Corporation), or PrimaCell (Chi Scientific) are available. However, these techniques are time-consuming, require optimization depending on tissue type and the procedures are varied in different laboratories. The technique we developed for generating epithelial cell sheets, isolating keratocytes, and inducing cytoplasmic fragments from the same fish scale samples requires neither mechanical tissue dissociation nor enzymatic digestion.

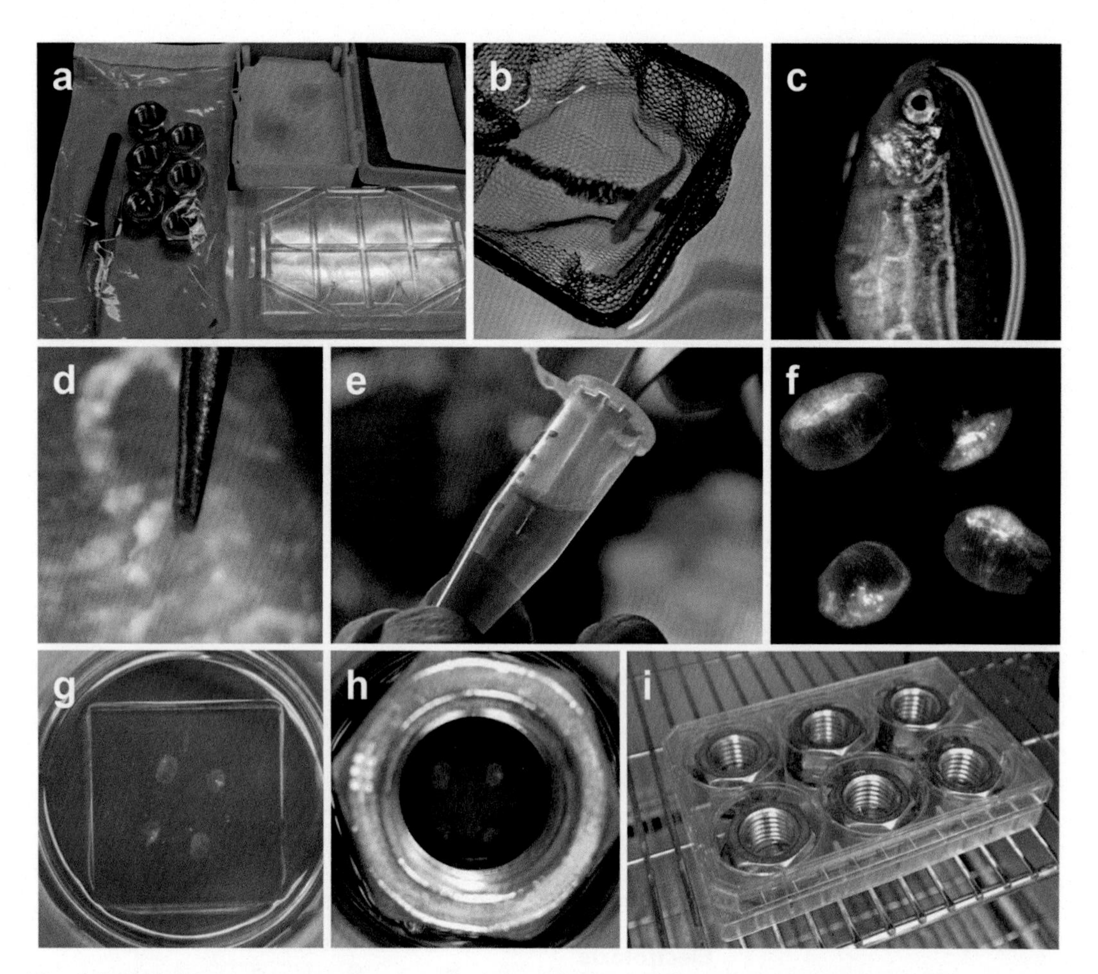

Fig. 2 Critical steps to generate three motile units. (**a**) Autoclaved surgical tweezers, stainless steel nuts, 22 × 22 mm coverslips and a disposable 6-well tissue culture plate. (**b**) Thorough washes with plenty distilled water are key to minimize bacterial or fungal contamination in subsequent steps. (**c**) An anesthetized zebrafish in an operation petri dish. (**d**) Pulling a scale from fish flank with sterilized surgical forceps. The exposed fish flank was sterilized with alcohol prep pads. (**e**) Scales are washed three times with complete culture medium containing antibiotics and antimycotics. (**f**) Evenly spread scales in a well of 6-well tissue culture plate. To prevent drying out of the scales, 20 μl of medium was pre-dropped in the center of the culture area. (**g**) A sterile 22 × 22 mm coverslip is laid over the wet scales. Care was taken to avoid bubble trapped between the coverslip and culture dish. (**h**) An autoclaved stainless steel nut is placed on the coverslip to hold scales in position. Culture medium (1.5 ml) is added to seal coverslip edges. (**i**) The whole plate is placed inside an incubator at room temperature. Sheets of keratocytes migrate off scales normally within 36 h

Traditional cell digestion with excessive volumes of trypsin–EDTA, followed by wash and centrifugation, typically in 15 ml Falcon tubes, may not be efficient for small amount of cells. In addition, inexperienced researchers may face technical difficulties when working with small cell pellets, thereby resulting in variable cell recovery. Thus, the use of small volume on-site trypsinization and subsequent wash and spin in Eppendorf tubes not only simplifies the cell digestion process, but also ensures maximal cell recovery.

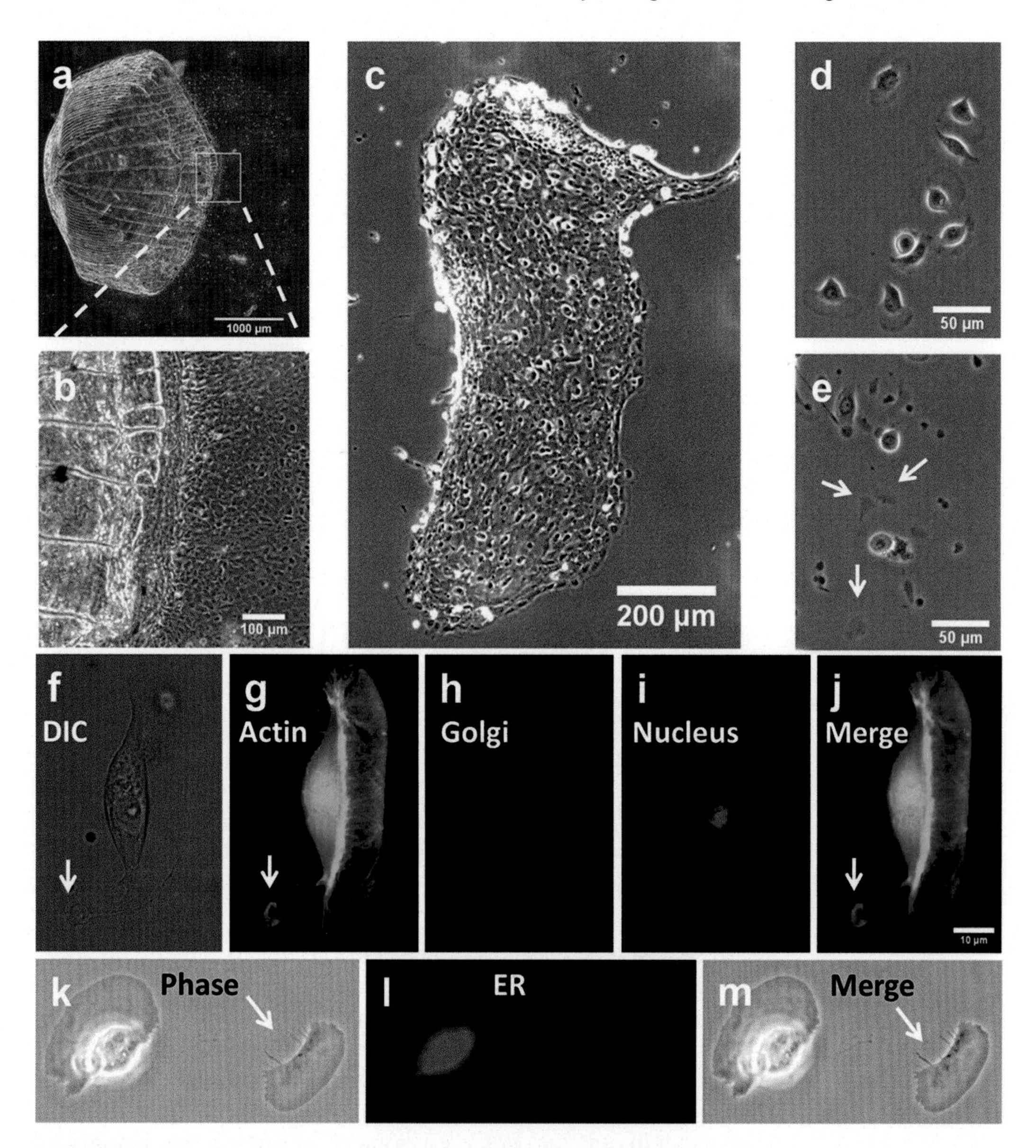

Fig. 3 Micrographs of cell sheets, isolated keratocytes, and cell fragments. (**a**, **b**) Sheets of epidermal keratocyte migrate out of fish scale. (**c**) A typical epidermal keratocyte sheet. (**d**) Isolated keratocytes. Most of the keratocytes have a typical canoe appearance. (**e**) Induced cell fragments (*white arrows*). (**f–j**) Cell fragment (*white arrow*) lacked nucleus (*blue*) as revealed by DAPI staining, and of Golgi body (*red*), as revealed by antibody staining. Actin (*green*) as revealed by FITC-phalloidin presents in both keratocyte and fragment. (**k–m**) Compared to the whole cell, a cell fragment (indicated by *white arrow*) contains much less ER (*cyan*) as revealed by antibody staining (modified from [24])

The following procedures are based on our own experiences or are otherwise adapted from previous publications [27]. We integrated the procedures into a highly successful pipeline for easy and simultaneous generation of epithelial cell sheets, isolated keratocytes, and induced cytoplasmic fragments from the same sample of

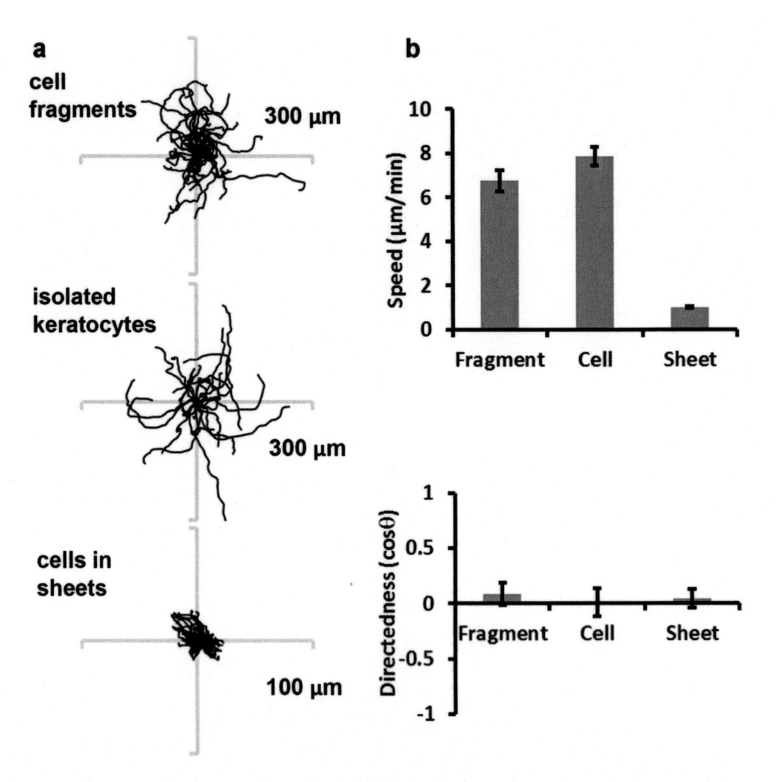

Fig. 4 Migration of cell sheets, single cells, and cell fragments. Single cells and cell fragments migrate much fast, whereas cells in sheets showed very little movement. (**a**) Migration trajectories of cell fragments, cells in isolation, and cells in sheets. Duration: 30 min. Note the scales are different. (**b**) Quantification of migration

fish scales in just 1 h. Once the keratocyte sheets migrate off a scale and attach to the tissue culture treated plate surface (usually after 36–48 h incubation), the scale "sandwiches" are disassembled by carefully removing the weighing steel nuts and the coverslips. The cell sheets formed this way can be either directly used in experiment if they are generated over an appropriate carrier (an electrotaxis chamber in our case) or serve as sources for single cell isolation. Cells in six-well plates are washed and trypsinized by the small volume on-site digestion technique detailed in the procedures. In brief, 0.5 ml trypsin–EDTA is added in each well and cell detachment is constantly monitored under microscope. Once cells are completely dislodged, 0.5 ml complete culture medium containing bovine serum is added to halt and prevent over-digestion. The cells are then transferred into Eppendorf tubes for subsequent wash and spin at room temperature. Cell collection using Falcon tubes and centrifugation with cooling are not necessary as recommended. Concentrated single cells are then seeded in experiment carrier (electrotaxis chambers in our case) and cell density is appropriately adjusted by adding more complete culture medium. As soon as cells attach to the chamber surface, they can be directly assayed or further subject to fragment

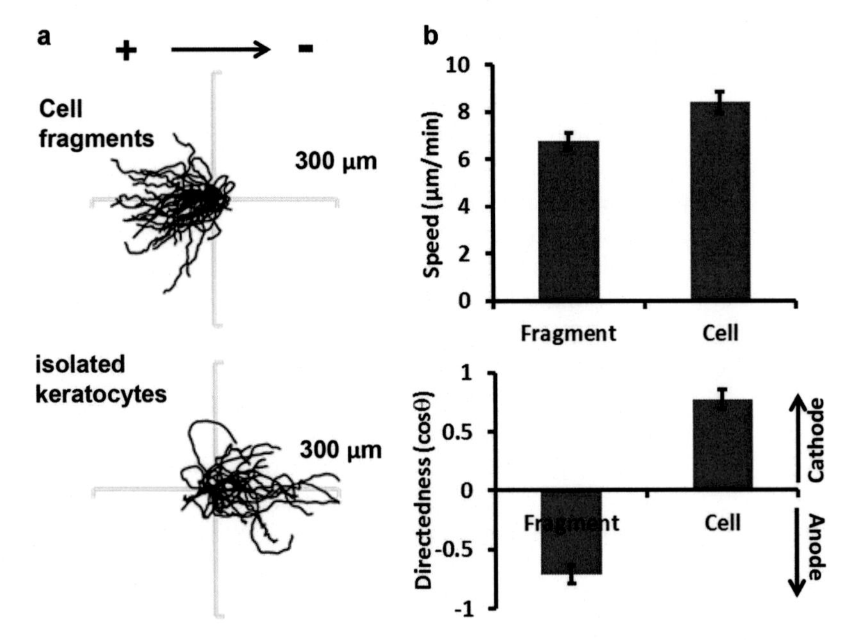

Fig. 5 Contrasting difference in directional migration in response to electrical cues. (**a**) Migration trajectories of cell fragments, single cells (cells in sheet are not shown) in the presence of an electric field with polarity as shown. Duration: 30 min. Note the scales are different. (**b**) Quantification of migration speed and directionality. Upon electric field application single cells migrate toward cathode. Note the cell fragments migrate in the opposite direction of their parental cells to anode. Cells in sheet are not shown

induction. Although cytoplasmic fragments pinch off spontaneously from keratocyte cells, the fastest and most efficient way to generate fragments is to treat with staurosporine, an alkaloid isolated from the bacterium *Streptomyces staurosporeus* [28]. Fragment production is facilitated by increasing the temperature to 35 °C without causing any cell death (*see* Supplementary movie S1). After complete removal of residual staurosporine, the resulting cytoplasmic fragments lack nuclei and other major organelles (Fig. 2), yet are still migratory and can be used for many downstream applications and side-by-side comparison with their parental cells.

In summary, there are many significant modifications and improvements compared to those previously reported in the literature. These are as follows: (1) use of zebrafish scales as keratocyte source; (2) wash fish multiple times (at least 5×) in beakers filled with clean room temperature distilled water; (3) use of stainless steel nuts to hold sandwiched scales in place; (4) use of a vacuum system and sterile pipettes to remove scales efficiently without dislodging attached cell sheets; (5) small volume on-site trypsinization and subsequent wash and centrifugation in Eppendorf tubes to minimize cell loss; (6) resuspending isolated cells in small volume and on-site dilution to achieve optimal cell density; (7) adaptation of treatment with staurosporine in culture media to induce fragment formation and subsequent incubation at 35 °C to facilitate fragment formation.

1.2 Advantages and Limitations of the Method

The present protocol has several limitations: (1) This protocol allows simultaneous generation of epithelial cell sheets and fragments from fish keratoctye model, limiting its use in mammalians cells. There are a number of publications that have demonstrated production of cell fragments in mammalian cells. Our attempt to induce cell fragment from neutrophils, human fibroblasts and glioma cells were unsuccessful. (2) Depending on species and age, not all fishes have equal ability to produce cell fragments, and their corresponding migratory behaviors may be different. (3) The use of staurosporine to induce fragments raises concerns that the treatment may cause cell stress or worse apoptosis, despite being the faster, more convenient and reliable method. It is recommended to monitor fragment formation under microscope, in addition to incubation at 35 °C to reduce drug exposure time. Nonetheless, thorough washes with excessive culture media to remove residual staurosporine results in negligible effect of apoptosis. (4) Our attempt to collect purified cell fragments alone has been unsuccessful. Hence, our electrotactic assay involving cell fragments are performed using a mixture of keratocyte cells and fragments, in some cases with the parent and fragment still tethered together. However, the latter may prove to be useful as an excellent model to study membrane tension in maintaining and regulating polarity and directional migration as demonstrated by Weiner's group [29].

Our protocol has several unique advantages as follows: (1) This protocol is practical in that it allows for simultaneous generation of epithelial cell sheets, isolated keratocytes, and cytoplasmic fragments from the same fish scale samples within 2–3 days. (2) The method allows for the possibility of performing reliable integrative investigation of cell migratory behaviors using three different motile models from the same tissue samples (Figs. 4 and 5). (3) Researchers now take full advantage of the unique genetic manipulability of the zebrafish model organism since well-characterized mutant strains are readily available.

1.3 Applications of the Protocol

The protocol described here has been optimized for the generation of three motile units including epithelial cell sheets, single cells, and cell fragments derived from zebrafish scales. Downstream applications can be performed with this unique system by taking advantage of zebrafish as a powerful model organism and can be integrated into many studies such as basal motility, pharmacological perturbation and manipulation of cellular signaling networks, directional migration such as galvanotaxis/electrotaxis and mechanotaxis. We recently applied this approach to gain insights into the control of cell migration under a guidance cue of applied electric field [25]. In this current manuscript, electrotaxis experiment setup is not discussed, as it has been thoroughly described in detail elsewhere [30, 31]. We expect the new model system will facilitate studies on the mechanisms by which cells detect and respond to external signals.

2 Materials

2.1 Reagents

1. Zebrafish (*Danio rerio*), adult male and female, aged between 3 and 12 months. The line we used is wild type (AB), which is available from Zebrafish International Resource Center (ZIRC) (*see* **Notes 1** and **2**).
2. Distilled water.
3. 70 % ethanol for sterilization.
4. Leibovitz's L-15 medium (Life Technologies).
5. 100× antibiotic–antimycotic (Life Technologies).
6. Fetal bovine serum (Life Technologies).
7. 1 M HEPES pH 7.4 (Life Technologies).
8. 0.25 % Trypsin–0.02 EDTA solution (Life Technologies).
9. Dulbecco's phosphate-buffered saline (Life Technologies).
10. MS-222 (Sigma-Aldrich).
11. Sodium bicarbonate.
12. Dimethyl sulfoxide.
13. Staurosporine (Sigma-Aldrich).
14. Sodium chloride.
15. Potassium chloride.
16. Calcium chloride.

2.2 Equipment

1. Fish net.
2. Beakers, 4 l, plastic (Nalgene).
3. Solvent-resistant marker (Fisher).
4. Stainless steel hex nuts (Bolt Depot, size ¾).
5. No. 1 glass coverslip, size 22 × 22 mm (Corning Life Sciences).
6. Biohazard waste and sharps disposal container.
7. Forceps (Dumont FST, no. 5).
8. Sterile alcohol prep pads.
9. Pipettes for volumes 0.5–1000 μl.
10. Pipette filter tips for volumes 0.5–1000 μl.
11. Sterile plastic Pasteur pipettes (Fisherbrand).
12. Eppendorf tubes, 1.5 ml (Fisherbrand).
13. Falcon conical tubes, 15 ml (Corning Life Sciences).
14. Falcon conical tubes, 50 ml (Corning Life Sciences).
15. Six-well tissue culture treated plates (Corning Life Sciences).
16. Tissue culture petri dishes with 10 cm diameter (Corning Life Sciences).

17. Dissecting microscope.
18. Culture incubator (e.g., Quincy lab, model 12-140).
19. Benchtop centrifuges: non-refrigerated and refrigerated (e.g., Eppendorf 5415D and 5415R, respectively).
20. Autoclave.
21. Cell culture hood with laminar flow and UV light (The Baker Company, Bio-II-A).
22. Time-lapse imaging system, ideally with functions of *X*/*Y*/*Z* multiple position recording and multiple wavelength recording, as well as a CO_2-supplied temperature control chamber incorporated onto the microscope. We currently use MetaMorph imaging software (Molecular Devices).

2.3 Reagent Setup

1. *Complete culture medium*: Leibovitz's L-15 medium + 10 % FBS + 1× antibiotic–antimycotic + 14.2 mM HEPES pH 7.4. The complete medium is used as both cell culture and electrotaxis running buffer. For convenience, aliquot freshly made medium into 50 ml Falcon tubes and stored at 4 °C. Allow medium to reach room temperature before use.
2. *MS-222 stock solution*: Prepare a 0.4 % stock solution in ddH_2O. Adjust pH to 7.2 using sodium bicarbonate as needed. Aliquots in 5 ml can be stored at −20 °C for a few months. To make euthanasia buffer thaw one vial of about 5 ml MS222 stock solution and add 15 ml of fish water (end concentration: 1 g/l).
3. *Fish Ringer's solution*: 116 mM NaCl, 2.9 mM KCl, 1.8 mM $CaCl_2$, 5 mM HEPES, pH 7.2. Fish Ringer's solution can be prepared and sterilized by autoclaving in advance and stored at 4 °C for up to a month.
4. *Staurosporine stock solution*: For convenience it is recommended to make a 1 mM stock solution in DMSO and aliquot in tightly sealed vials at−20 °C. Prior to opening the vial allow it to equilibrate to room temperature for at least 20 min.

2.4 Equipment Setup

1. *Preparation for fish sterilization*: Clean work bench with 70 % ethanol. Fill an autoclaved plastic bucket with plenty of dechlorinated or distilled water that is enough to wash the fish and fish net as described in Subheading 3. Fill two autoclaved 5-l plastic beakers with distilled water. Make sure the water temperature is between 20 and 25 °C.
2. *Preparation for scale processing*: Thaw one vial of about 5 ml MS222 stock solution and mix into ~100 ml clean distilled water to make 200 mg/l buffer for anesthesia. Prepare two disposable tissue culture petri dishes and fill one with 100 ml freshly made 200 mg/l MS-222 in complete culture medium.

Autoclave two surgical forceps, six stainless steel hex nuts, and a box of 22 × 22 mm coverslips. Prepare a stack of sterile alcohol prep pads. Make enough electrotaxis chambers using 10 cm tissue culture dishes if you plan to do experiment with cell sheets. Otherwise, label a six-well plate if you plan to isolate single cells and subsequently induce fragments. Drop 20 μl complete culture medium in the center of each chamber or well. Label three Eppendorf tubes on the side with a solvent-resistant marker and fill each tube with 1 ml complete medium.

3. *Preparation for cell sheet generation*: Set up a vacuum system. Set up a "sharps" container for properly disposing coverslips.
4. *Preparation for single cell isolation*: Set up a vacuum system. Arrange a benchtop centrifuge and a liquid aspiration system. It is recommended to label the tubes on the lid and on the side with a solvent-resistant marker.
5. *Preparation for cell fragment induction*: Make enough electrotaxis chambers using 10 cm tissue culture dishes. Turn on an incubator, set temperature to 35 °C, and allow to equilibrate. Set up a liquid aspiration system. Prepare 1 ml 100 nM staurosporine in complete culture medium. It is recommended to label the tubes on the lid and on the side with a solvent-resistant marker.
6. *Liquid aspiration system setup*: Connect the end of a 1-l vacuum flask to one end of a 0.45-μm membrane filter using a 20-cm rubber tube (in-house or water assisted). Connect the other end of the membrane filter to a vacuum unit using a rubber tube. Close the vacuum flask with a drilled rubber stopper that has a glass Pasteur pipette inserted in the hole. Connect the Pasteur pipette with a 30-cm rubber tube to another glass Pasteur pipette, which will be used to aspirate liquids using a 200-μl pipette tip.

2.5 Sample Handling Recommendations

1. Before starting to work with tissue culture, ensure that appropriate amounts of the required medium, buffers and enzymes have been pre-warmed to room temperature. It is recommended to aliquot complete culture medium into small quantities for single use only. Do not leave solutions open if they are not in use.
2. Like other primary cell cultures, tissue cultures from fish scales are very sensitive to contamination. Clean the area you will process the tissues including pipettes and centrifuges by spraying with 70 % (vol/vol) ethanol.
3. Always wear disposable gloves and replace them regularly.
4. Reusable stainless steel nuts, glassware and plasticware should be autoclaved to ensure sterility.
5. Use sterile instruments, aseptic techniques, and perform work in a laminar flow hoods to maintain sterility.

3 Methods

3.1 Fish Sterilization (Allow 25–30 min per Fish) (See Note 3)

1. Transfer fish into procedure room according to your institution's regulations, and let it sit in a holding tank for at least 20 min.
2. Put a fish in an autoclaved 4-l plastic beaker filled with clean distilled or dechlorinated water and allow the fish to acclimate inside the fish net for 30 s. Water used for washing purposes must be kept at room temperature (20–25 °C) and dechlorinated to avoid stress caused by temperature fluctuations and chlorine.
3. Use the fish net to carefully transfer fish into another 5-l plastic beaker filled with clean distilled water. Repeat wash steps for total of five times. The fish net should be sanitized with 70 % ethanol and rinsed in clean distilled water before and after each use.

3.2 Scale Processing and Assembling (Allow up to 30 min per Fish)

1. Pour 100 ml anesthesia buffer containing 200 mg/l MS-222 into a clean petri dish. The 200 mg/l MS-222 buffer in distilled water must be freshly made for effective anesthesia.
2. Use a pipette to drop 20 μl complete culture medium in the center of each well.
3. Use a pipette to aliquot of 1 ml complete cell culture medium into three Eppendorf tubes.
4. Immerse the fish in anesthesia buffer containing 200 mg/l MS-222. During induction make sure the depth of anesthesia is appropriate by closely monitoring gill movement as an indicator.
5. Transfer anesthetized fish into a new clean sterile petri dish.
6. Gently hold the fish sideways by pressing its head and tail against the petri dish using your left thumb and middle finger.
7. Sterilize the fish flank where you plan to take scales by using alcohol prep pads.
8. With a pair of sterile surgical forceps, gently pull a scale off fish flank from where you just sterilized, and rinse it three times, 1 s each time by sequentially dipping the scale in the three Eppendorf tubes filled with complete cell culture medium (*see* **Note 4**).
9. Carefully lay individual scale into the medium at the center of the well. Use 10-cm tissue culture dishes if you plan to carry out an experiment with cell sheets right after they are formed. Otherwise use six-well tissue culture plates. A drop of complete culture medium helps to prevent the scale from drying out and to spread scales evenly.
10. Repeat **steps 8** and **9** until there are 3–5 scales in each well.
11. Return fish into clean distilled water to facilitate recovery (*see* **Note 5**).

12. Use the same surgical forceps to spread scales evenly in the middle of each well (*see* **Note 6**).
13. Use new sterile surgical forceps to pick up a clean 22 × 22 mm glass coverslip. Tilt one side into the well such that it touches the medium, then slowly lower it to cover the evenly spread scales without trapping any bubbles in between. Care must be taken to avoid bubble as a gas–liquid interface interferes with keratocyte sheet formation.
14. Gently lay a sterile stainless steel hex nut on the top of the square coverslip. Make sure no scales float out from beneath the coverslip.
15. Add 1.5 ml complete culture medium surrounding the nut to immerse the coverslip. Don't add too much medium. Excessive medium is easy to reach inner side of plate cover and to splash over during transportation, therefore increasing chance of contamination.
16. Repeat **steps 12–15** until all the wells are processed.
17. Place covered culture plate in an incubator at room temperature and incubate for 36–48 h (*see* **Note 7**).

3.3 Cell Sheet Generation (Allow ~30 min)

1. Take the culture plate containing the scale assemblies from incubator.
2. Remove the steel nuts using sterile forceps and set them to the side.
3. Using sterile forceps with a sharp tip, carefully remove the coverslips that cover scales in each well. Some of the coverslips might be tightly attached to culture surface. In that case be patient and try to lift from one side (*see* **Note 8**).
4. Carefully aspirate culture medium, making sure not to disrupt the scales.
5. Use the vacuum aspiration system and sterile 200 μl pipette tips to remove scales from each well. Use just enough vacuum power to pick up a scale. Wipe off the suctioned scale carefully with alcohol prep pads (*see* **Note 9**).
6. In each well, wash cells twice with 5 ml Fish Ringer's Solution to remove any debris.
7. Add 3 ml complete culture medium to each well and locate the cell sheets under microscope. At this point the cell sheets can be kept in the dish and assayed later or used directly in the next step for single cell preparation.

3.4 Isolating Single Cells (Allow ~30 min)

1. Aspirate culture medium.
2. Wash cells once with PBS.
3. Add 0.5 ml 0.25 % trypsin–0.02 EDTA solution in each well. Continuously monitor cell detachment under microscope.

Gently tilt well plate to ensure adequate coverage and to speed up digestion. Trypsinization is usually complete within 5 min at room temperature; confirm under microscope.

4. Add 0.5 ml complete culture medium to stop further digestion as soon as all cells are detached.
5. Transfer cells into 1.5 ml Eppendorf tubes, one well per tube.
6. Collect cells by centrifugation at 150 × *g* for 5 min in a benchtop centrifuge at room temperature.
7. Carefully aspirate supernatant using vacuum system without causing significant cell loss (*see* **Note 10**).
8. Combine cells from all wells into a new 1.5 ml Eppendorf tube, and add 1 ml complete culture medium.
9. Repeat **steps 7** and **8** to wash and spin cells twice with complete culture medium.
10. Re-suspend cells in 0.2 ml complete culture medium. If needed single cell yield can be calculated by counting using a hemocytometer. The isolated cells can be assayed right away or proceed to next step for fragment induction.

3.5 Inducing Cell Fragments (Allow ~1 h 15 min)

11. Add 20 μl cell suspension in each experiment carrier (in our case an electrotaxis chamber).
12. Check under inverted microscope and add necessary amount of complete culture medium to achieve optimal cell density.
13. Let cells adhere to chamber surface at room temperature for 30 min.
14. Wash once with complete culture medium to remove unattached cells.
15. Add 0.5 ml complete culture medium containing 100 nM staurosporine.
16. Place the dish in an incubator with its temperature preset at 35 °C.
17. Incubate up to 30 min to induce fragments (*see* **Note 11**).
18. As soon as ideal fragments are formed, bring dish to room temperature and wash twice with complete culture medium.
19. Add enough complete culture medium and let cells and fragments recover for 10 min. Fragments made this way are ready for testing and can survive for a couple of hours (at least 3 h).

3.6 Troubleshooting

Troubleshooting advice can be found in Table 2.

3.7 Anticipated Results

This protocol provides detailed steps for the generation of epithelial cell sheets, single cells, and cell fragments derived from zebrafish keratocytes. The overall procedures, highlights of spontaneous migration and subsequent applications using electrotaxis as an

Table 2
Troubleshooting

Methods section	Problem	Possible reason	Solution
Fish sterilization Subheading 3.1, **steps 2** and **3**	Debris appears in the water during fish cleaning	Water temperature is too low or too high	Leave distill water in containers for longer time or measure water temperature and ensure it's between 22 and 25 °C
Scale processing Subheading 3.2, **step 4**	Fish moves and is difficult to hold	MS-222 is expired	Make new MS-222 stock solution. The 200 mg/l MS-222 in complete culture medium must be made freshly
		Not properly anesthetized	Return fish back to anesthetic buffer for further induction
Scale processing Subheading 3.2, **step 6**	Scale is difficult to pull or multiple scales are pulled	Scales are too tiny	It fish is small and their scales are too tiny to see manipulation under a dissecting microscope is recommended
Scale assembling Subheading 3.2, **step 10**	Scales are difficulty to spread	Scales are dried out	Add medium to pre-wet surface area of each culture well
Scale assembling Subheading 3.2, **step 11**	Scales float away when coverslip is laid	Too much medium around	Add just enough medium (20 μl) when pre-wetting the surface area
Scale assembling Subheading 3.2, **step 15**	Bacterial and fungus in the culture dish	Contamination	The most common contamination lies on the shedding by fish. Therefore thorough wash with excessive clean water is the key to prevent bacterial and fungus contamination
Cell sheet generation Subheading 3.3, **step 1**	Cell sheets disappear	Cell sheets adhere to coverslip	Use untreated clean coverslips to minimize undesired adherence. Otherwise, digest/recover cells from coverslips
Isolating single cells Subheading 3.4, **step 8**	Low isolated cell yield	Room temperature is too low	Increase incubator temperature to 28 °C
		Scales are small	If this is turned out the case try to use more scales in each assembly. Up to 9 scales can be easily accommodated and covered by each 22 × 22 mm coverslip
		Bubbles present during scale assembly	Make sure the coverslips used are clean and dust free
		Significant cell loss during trypsinization and subsequent wash steps	Avoid dislodging cell pellet. Always leave a small amount of solution in the bottom when aspirating supernatant

(continued)

Table 2
(continued)

Methods section	Problem	Possible reason	Solution
Inducing fragments Subheading 3.5, **step 1**	No or few cells attach	Bad tissue culture treatment	Change dishes or re-coat culture dishes with fibronectin
Inducing fragments Subheading 3.5, **step 15**	Too many tiny fragments	Overtreatment of staurosporine	Either decrease staurosporine concentration to 50 nM or induce fragment at room temperature. In either case the induction process must be closely monitored under an inverted microscope

example are summarized in Fig. 1. Figure 2 illustrates the key steps in processing fish scales. Supplementary movie S1 demonstrates formation of cytoplasmic fragment induced by staurosporine. The protocol allows for generation of a unique system of three motile units: a part of cell, a whole cell and a group of cells, from same tissue origin. The morphological characteristics of these three units are shown in Fig. 3. Also in Fig. 3, we confirm that the resulting cytoplasmic fragments lack nuclei and Golgi body, and contains little or no endoplasmic reticulum (ER) as we revealed previously [25].

The model system enabled us to yield some new insights that were not possible using previous model systems. First we compared the basal motility of the cell sheets, single cells, and cell fragments. Under no obvious directed cues cells in isolation migrate fast in a speed about 8 μm/min. While cell fragments migrate a little slower (6.76 ± 0.49 μm/min) the cells in sheets are much less migratory (1.01 ± 0.04 μm/min). Figure 4 shows the distinctive difference of spontaneous migration of the three motile units.

We also applied this protocol to investigate the migratory behaviors of cell fragments as well as cohesive cell sheets in comparison with isolated keratocytes under electrical stimulation. We found that the fragments, devoid of nuclei and major organelles, are still capable of sensing electrical field signals. The fragment, therefore, has the EF sensing mechanism and relevant signal transduction pathways leading to migration. Surprisingly, we observe that under exogenous EF guidance, fragments migrate in opposite directions compared to their intact mother cells (*see* Supplementary movies S2 and S3) [25].

The collective migration of keratocytes in cell sheets is also an interesting phenomenon. Compared to both isolated single cells and cell fragments, cell sheets are less migratory. However, upon electric field application, the cells in sheets immediately respond and migrate, resulting in a significant increase in migration speed (1.99 ± 0.05 μm/min), nearly doubled compared to no EF control

(1.01 ± 0.04 μm/min), and in directionality (mean cosθ = 0.98) that is much improved compared to that of single cells (mean cosθ = 0.77) (*see* Supplementary movie S4). The quantification of the electrotactic migratory behaviors of the three motile units is summarized in Fig. 5.

There are many additional potential downstream applications. These include but not limited to pharmacological perturbation and manipulation of cellular signaling networks, especially when combined with the powerful zebrafish genetic tools, as well as other directed migration such as mechanotaxis. Our protocol provides a valuable system for investigating cell migratory behaviors and may help in dissecting mechanisms of cells in response to external signals.

4 Notes

1. All experiments that use animal tissues should comply with all relevant institutional and governmental guidelines and regulations. Animal protocol must be developed and approved by relevant institute committees/authorities and be active when such experiments is implemented.
2. Animal tissues may contain human pathogens including viruses, bacterium and fungus, which may infect the researcher. Therefore, the use of protective equipment such as gloves, a lab coat, and goggles is recommended.
3. You must gently handle the fish throughout the procedure as detailed in sample handling recommendations in the instruction section to avoid stressing the animal.
4. Forceps are very sharp and there is a risk of laceration when using them, which is a potential source of infection by viruses.
5. After the recovery (usually about 10 min) fish will be returned back to standing water tank and can be reused for future experiment. In case of any sign of suffering immediately immerse fish in euthanasia buffer (distilled water containing 1 g/l MS222). Fish should be dead within minutes and must be disposed properly.
6. Although up to 9 scales can be easily accommodated in a well too many scales cause stack-over problem. Therefore, 3–5 scales in one well are recommended.
7. Occasionally, we see cell sheet formation after overnight incubation. However, incubation for 36–48 h gives us the most stable and maximal cell production. Increased incubation time does not produce more cells but increases chances of contamination.
8. Try to avoid scale shifting when lifting coverslips. Shifting scale may destroy a cell sheet where it migrated from. Dispose used coverslips in a sharps container to avoid potential injury.

9. Try to avoid scale shifting when picking up scales. Shifting scale may destroy a cell sheet where it migrated from.
10. Cell pellet could be very tiny. Try not to touch cell pellet when removing supernatant in this step and subsequent wash steps.
11. Fragmentation process can be monitored on a heated stage under inverted microscope. Do not leave the dish at 35 °C for extended periods of time. Overtreatment in staurosporine at high temperature produces many tiny fragments with decreased mobility and viability, compromising downstream applications such as electrotaxis and other motility assays.

Acknowledgements

We thank Y. Li for help with photography, B. Reid and T. Pfluger for critical reading of the manuscript, and K. Nakajima and Y. Shen for technical assistance. Y.H.S. was supported by NIH grant GM068952 to A.M. Work in the laboratory of M.Z. was supported by NIH grant 1R01EY019101 to M.Z., as well as by California Institute of Regenerative Medicine Research grant RB1-01417 and National Science Foundation Grant MCB-0951199 to M.Z. and P. Devreotes, whose support we gratefully acknowledge.

5 Supplementary Materials

Supplementary movie S1 Cytoplasmic fragment formation induced by incubation with 100 nM staurosporine at 35 °C. Access this movie at: https://youtu.be/_H06Z6klrdQ

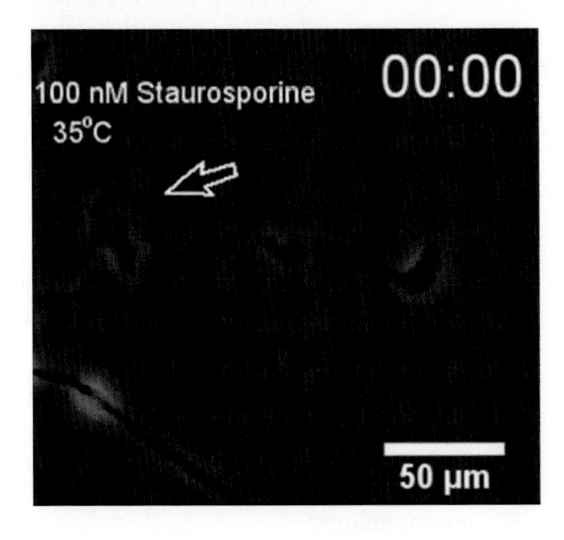

Supplementary movie S2 Random migration of fish keratocyte and fragment. Access this movie at: https://youtu.be/sFf-9k3TF-k

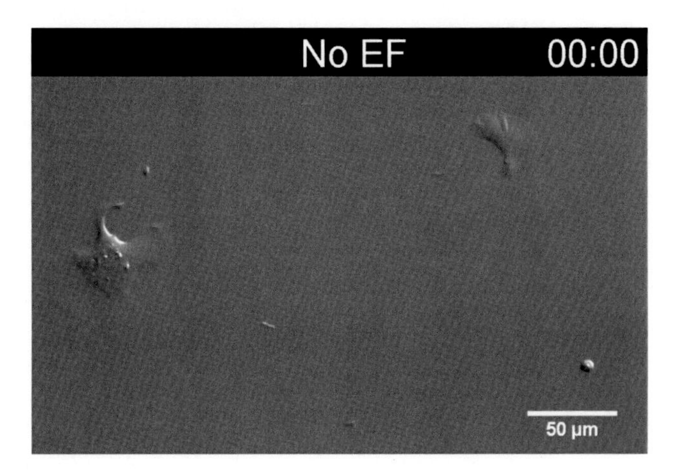

References

1. Aman A, Piotrowski T (2010) Cell migration during morphogenesis. Dev Biol 341:20–33
2. Friedl P, Gilmour D (2009) Collective cell migration in morphogenesis, regeneration and cancer. Nat Rev Mol Cell Biol 10:445–457
3. Calve S, Simon HG (2012) Biochemical and mechanical environment cooperatively regulate skeletal muscle regeneration. FASEB J 26:2538–2545
4. Luster AD, Alon R, von Andrian UH (2005) Immune cell migration in inflammation: present and future therapeutic targets. Nat Immunol 6:1182–1190
5. Malawista SE, Smith EO, Seibyl JP (2006) Cryopreservable neutrophil surrogates: granule-poor, motile cytoplasts from polymorphonuclear leukocytes home to inflammatory lesions in vivo. Cell Motil Cytoskeleton 63:254–257
6. Locascio A, Nieto MA (2001) Cell movements during vertebrate development: integrated tissue behaviour versus individual cell migration. Curr Opin Genet Dev 11:464–469
7. Keller R (2005) Cell migration during gastrulation. Curr Opin Cell Biol 17:533–541
8. Yamaguchi H, Wyckoff J, Condeelis J (2005) Cell migration in tumors. Curr Opin Cell Biol 17:559–564
9. Wang W, Goswami S, Sahai E et al (2005) Tumor cells caught in the act of invading: their strategy for enhanced cell motility. Trends Cell Biol 15:138–145
10. Chironi GN, Boulanger CM, Simon A et al (2009) Endothelial microparticles in diseases. Cell Tissue Res 335:143–151
11. Markiewicz M, Richard E, Marks N, Ludwicka-Bradley A (2013) Impact of endothelial microparticles on coagulation, inflammation, and angiogenesis in age-related vascular diseases. J Aging Res 2013:734509
12. Barry OP, Praticò D, Savani RC, FitzGerald GA (1998) Modulation of monocyte-endothelial cell interactions by platelet microparticles. J Clin Invest 102:136–144
13. Yount G, Taft RJ, Luu T et al (2007) Independent motile microplast formation correlates with glioma cell invasiveness. J Neurooncol 81:113–121
14. Malawista SE, Van Blaricom G (1987) Cytoplasts made from human blood polymorphonuclear leukocytes with or without heat: preservation of both motile function and respiratory burst oxidase activity. Proc Natl Acad Sci U S A 84:454–458
15. Mogilner A, Allard J, Wollman R (2012) Cell polarity: quantitative modeling as a tool in cell biology. Science 336:175–179
16. Euteneuer U, Schliwa M (1984) Persistent, directional motility of cells and cytoplasmic fragments in the absence of microtubules. Nature 310:58–61
17. Zouani OF, Gocheva V, Durrieu MC (2014) Membrane nanowaves in single and collective cell migration. PLoS One 9, e97855
18. Radice GP (1980) Locomotion and cell-substratum contacts of *Xenopus* epidermal cells in vitro and in situ. J Cell Sci 44:201–223
19. Keren K, Pincus Z, Allen GM et al (2008) Mechanism of shape determination in motile cells. Nature 453:475–480

20. Lee J, Ishihara A, Theriot JA, Jacobson K (1993) Principles of locomotion for simple-shaped cells. Nature 362:167–171
21. Howe DG, Bradford YM, Conlin T et al (2013) ZFIN, the Zebrafish Model Organism Database: increased support for mutants and transgenics. Nucleic Acids Res 41:D854–D860
22. Sprague J, Bayraktaroglu L, Bradford Y et al (2008) The Zebrafish Information Network: the zebrafish model organism database provides expanded support for genotypes and phenotypes. Nucleic Acids Res 36:D768–D772
23. Huang L, Cormie P, Messerli MA, Robinson KR (2009) The involvement of Ca^{2+} and integrins in directional responses of zebrafish keratocytes to electric fields. J Cell Physiol 219:162–172
24. Allen GM, Mogilner A, Theriot JA (2013) Electrophoresis of cellular membrane components creates the directional cue guiding keratocyte galvanotaxis. Curr Biol 23:560–568
25. Sun Y, Do H, Gao J et al (2013) Keratocyte fragments and cells utilize competing pathways to move in opposite directions in an electric field. Curr Biol 23:569–574
26. Weijer CJ (2009) Collective cell migration in development. J Cell Sci 122:3215–3223
27. Verkhovsky AB, Svitkina TM, Borisy GG (1999) Self-polarization and directional motility of cytoplasm. Curr Biol 9:11–20
28. Omura S, Iwai Y, Hirano A et al (1977) A new alkaloid AM-2282 of Streptomyces origin. Taxonomy, fermentation, isolation and preliminary characterization. J Antibiot (Tokyo) 30:275–282
29. Houk AR, Jilkine A, Mejean CO et al (2012) Membrane tension maintains cell polarity by confining signals to the leading edge during neutrophil migration. Cell 148:175–188
30. Song B, Gu Y, Pu J et al (2007) Application of direct current electric fields to cells and tissues in vitro and modulation of wound electric field in vivo. Nat Protoc 2:1479–1489
31. Zhao M, Song B, Pu J et al (2006) Electrical signals control wound healing through phosphatidylinositol-3-OH kinase-gamma and PTEN. Nature 442:457–460
32. Albrecht-Buehler G (1980) Autonomous movements of cytoplasmic fragments. Proc Natl Acad Sci U S A 77:6639–6643
33. Malawista SE, De Boisfleury Chevance A (1982) The cytokineplast: purified, stable, and functional motile machinery from human blood polymorphonuclear leukocytes. J Cell Biol 95:960–973
34. Cooper MS, Schliwa M (1986) Motility of cultured fish epidermal cells in the presence and absence of direct current electric fields. J Cell Biol 102:1384–1399
35. Li L, Hartley R, Reiss B et al (2012) E-cadherin plays an essential role in collective directional migration of large epithelial sheets. Cell Mol Life Sci 69:2779–2789
36. Cohen DJ, Nelson WJ, Maharbiz MM (2014) Galvanotactic control of collective cell migration in epithelial monolayers. Nat Mater 13:409–417

Chapter 20

Axon Guidance Studies Using a Microfluidics-Based Chemotropic Gradient Generator

Zac Pujic, Huyen Nguyen, Nick Glass, Justin Cooper-White, and Geoffrey J. Goodhill

Abstract

Microfluidics can be used to generate flow-driven gradients of chemotropic guidance cues with precisely controlled steepnesses for indefinite lengths of time. Neuronal cells grown in the presence of these gradients can be studied for their response to the effects exerted by the cues. Here we describe a polydimethylsiloxane (PDMS) microfluidics chamber capable of producing linear gradients of soluble factors, stable for at least 18 h, suitable for axon guidance studies. Using this device we demonstrate turning of superior cervical ganglion axons by gradients of nerve growth factor (NGF). The chamber produces robust gradients, is inexpensive to mass produce, can be mounted on a tissue culture dish or glass coverslip for long term time-lapse microscopy imaging, and is suitable for immunostaining.

Key words Gradient, Chemotropic cue, Microfluidic, Axon guidance, Chemotaxis, Nerve growth factor, Superior cervical ganglion

1 Introduction

For the brain to develop correctly it must be wired up correctly. To achieve this, growing axons must navigate reliably to their targets and make synaptic connections. Understanding how this navigation occurs is important since axon miswiring may underlie many mental disorders [1], and it is also critical for axons to be able to reform appropriate connections after injury. Due to the enormous complexity of the developing brain, it is often desirable to study these mechanisms in vitro since it is possible to exclude confounding factors.

An important cellular process often required to accomplish correct guidance is the detection, by the growth cones at the tips of developing axons, of gradients of diffusible chemotropic cues within the developing tissues. The in vitro study of axon guidance by diffusible gradient cues ideally requires the generation of temporally and spatially stable gradients with precisely controllable characteristics, that can be applied to a large number of individual axons, and allow

Tian Jin and Dale Hereld (eds.), *Chemotaxis: Methods and Protocols*, Methods in Molecular Biology, vol. 1407,
DOI 10.1007/978-1-4939-3480-5_20, © Springer Science+Business Media New York 2016

time-lapse imaging. However few assays used for axon guidance studies currently achieve these goals. Collagen gel explant coculture assays [2] produce gradients which are poorly characterized and decay with time, while more sophisticated efforts to produce gradients with known steepnesses in collagen gels [3] are expensive to set up and limited to shallow gradients. The widely used "growth cone turning" or "pipette" assay [4, 5] has a low throughput, limited gradient stability, and little control over gradient steepness. The Dunn chamber [6] can be effective for axon guidance studies, but as a passive device suffers from transients and gradient decay [6].

Many of these limitations can be overcome by using microfluidic technologies [7, 8]. Gradients of diffusible factors can be generated dynamically and therefore sustained at a particular steepness and concentration regime almost indefinitely. The number of isolated cells exposed to the gradient is generally greater than that for other assays, and the gradient can be defined with greater precision. Recent advances in microfluidics chamber design have employed innovative approaches which demonstrate that the approach is both powerful and versatile. For instance, microfluidically generated gradients of diffusible Slit-1 or Netrin-1 were able to elicit turning in hippocampal or dorsal root ganglion neurons [9]. Flow-based approaches can cause shear stresses which are damaging for growth cones, which are less robust than cell bodies. Various methods have been used to minimize this problem, such as culturing cells in a 3D hydrogel [9], using micro-well structures [10], or using a permeable membrane separated the fluid-flow driven gradient from the cells [11]. Here we describe a simple and easy to produce flow-based microfluidics chamber which can generate stable linear gradients despite using a flow rate low enough to be suitable for axon guidance studies.

2 Materials

To minimize the blockage of PDMS chambers by dust and other particulates, prepare all materials and solutions in a clean, dust-free environment. Where possible, work in laminar flow hoods to ensure that a minimal amount of dust and particulates are present. Use ultrapure water (deionized to attain 18 MΩ cm at 25 °C) and filter all aqueous solutions with 0.2 μm filters.

2.1 Microfluidics

1. AutoCad software (Autodesk, Australia).
2. Photoplate (Konica, Minolta, New South Wales, Australia).
3. Silicon wafers (M.M.R.C. Pty Ltd, Malvern, Vic, Australia).
4. Photolithography: A clean room with a spin coater, level hotplates, mask writer or photoplotter, mask aligner (EVG, St. Florian, Austria) or UV flood source and a fume hood.

5. Photolithography chemicals: Ti Prime (MicroChemicals, Ulm, Germany), SU-8 2050 and SU-8 2100 (MicroChem, Westborough, MA), Propylene glycol monomethyl ether acetate (PGMEA;Sigma-Aldrich,Australia)andTrichloro(1*H*,1*H*,2*H*,2*H*-perfluorooctyl)silane (Sigma-Aldrich) (*see* **Note 1**).
6. Chamber testing: Epifluorescence microscope or confocal microscope and 10–63× objectives (*see* **Note 2**).
7. Optical profiler (Wyko NT1100, Veeco, Plainview, NY).
8. PDMS mixture: Combine polydimethylsiloxane (PDMS) base elastomer (Sylgard 184, Dow Corning, Midland, MI) and silicon elastomer curing agent in a 10:1 (m/m) ratio in a 50 mL plastic tube. Mix for 1 h either with a wooden tongue depressor or a rotary mixer.
9. Fluorescent dextran: Aqueous 10 mg/mL of 40 kDa dextran conjugated to tetramethylrhodamine (Life Technologies, Australia). Store 10 μL aliquots at −80 °C (*see* **Note 2**).
10. Tubing: Polyethylene tubing I.D. 0.58 mm, O.D. 0.965 mm (Intramedic Clay Adams Brand, Becton Dickinson Co.). Cut to length as required. Store in 70 % ethanol. Air-dry in laminar flow hood and rinse with filtered PBS prior to use.
11. Metal connectors: These can be made by cutting the metal needle from a 23 gauge syringe and smoothing down any rough metal burrs on a wetstone.
12. Syringe connectors: 23 Gauge syringes (Terumo Medical Corp., NSW Australia) with the bevelled-tip cut off with a pair of metal snips. The rough edges are smoothened on a wetstone.
13. Syringes: 100, 250, or 500 μL glass syringes (SGE Analytical Science, NSW Australia) or 1 mL plastic syringes (Terumo Medical Corp.).
14. Microfluidics pump (e.g., Harvard Apparatus Ultra, SGE Analytical Science).
15. 0.75 mm corer (Harris Uni-Core, Ted Pella, CA, USA).
16. Plasma cleaner (e.g., PDC-002, Harrick Plasma, NY, USA).
17. (3-Aminopropyl)triethoxysilane (APTES; Sigma-Aldrich).

2.2 Tissue Culture

1. Laminar flow hood and tissue culture incubator with 5 % CO_2 at 37 °C.
2. Leibovitz's L-15 medium: Add 5 mL of 45 % glucose to 500 mL L-15 (Life Technologies). Store at 4 °C.
3. Petri dishes: 35 mm wide, tissue-culture treated petri dishes (Sigma-Aldrich).

4. Trypsin solution: Add 0.5 mL 2.5 % trypsin to 2.5 mL calcium- and magnesium-free Hanks balanced salt solution. Prepare immediately before use.
5. Superior cervical ganglion (SCG) Growth Medium (SGM): 1× Opti-MEM-1 (Life Technologies) containing 1× penicillin/streptomycin, 10 μg/mL mouse laminin, 4 % (v/v) fetal calf serum, 2 % B-27 supplement (Life Technologies) (*see* **Note 3**).
6. Fixative: 4 % paraformaldehyde in 1× phosphate buffered saline (PBS) (*see* **Note 4**).
7. H Solution: 6 nM NGF in SGM, equivalent to $20 \times K_d$ (*see* **Note 5**).
8. Blocking solution: 4 % normal goat serum in PBS.
9. Primary antibody: 1:1000 mouse anti-neuron specific β-tubulin class III antibody (BD Biosciences, Australia) in blocking solution. Add Triton X-100 to 0.05 %.
10. Secondary antibody: 1:1000 goat anti-mouse Alexa 488 conjugated IgG (BD Biosciences, Australia) in blocking solution.

3 Methods

3.1 Shear Stress Determination

An important consideration when designing any microfluidics chamber with liquid flow is the potential for significant shear stress on the cells in the growing chamber. Various methods have been used to minimize this problem, such as culturing cells in a 3D hydrogel [9], using micro-well structures [10], or using a permeable membrane separating the fluid-flow driven gradient from the cells [11]. Using the chamber design in Fig. 1a, we found no correlation between the final direction of axons and the fluid flow direction (i.e., no bias in neurite growth) due to liquid flow rates up to 200 μL/h (data not shown). Assuming the Poiseuille model [12], the shear stress τ is calculated as follows:

$$\tau = -\frac{12\mu Q}{wh^2} \quad \text{for} \quad w \gg h$$

where Q is the flow rate (m^3/s), μ is the fluid viscosity (Pa s), and h (m) and w (m) are the channel height and width, respectively. According to this equation, a flow rate of 200 μL/h caused shear of 1.7 N/m^2. Since shear is not dependent on position within the growth chamber, and since the cells are all in contact with the growth chamber floor, the shear is uniform across all cells. Little is currently known about how much shear stress can be tolerated by neurons of different type or on different substrates. Morel et al. [11] found that at 5×10^{-2} N/m^2, the growth cones of rat DRG neurons displayed damage, but that at 5×10^{-4} N/m^2, damage due to the shear stress was undetectable. Wang et al. [10] found that at

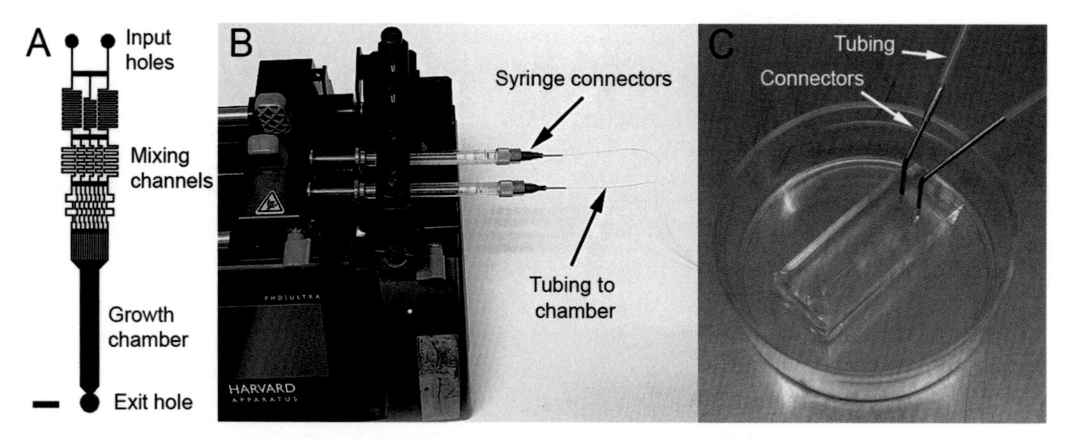

Fig. 1 **(a)** A schematic of the microfluidics chamber channels. The two input holes separately receive SGM containing NGF (H solution) or SGM only (L solution). The media are combined in the mixing channels and enter the growth chamber and eventually leave through the exit hole. The heights of the growth chamber and exit hole are 150 μm, while all other regions of the device are 50 μm. Scale bar = 1 mm. **(b)** Two syringes mounted on a Harvard Apparatus infusion pump via syringe connectors. The polyethylene tubing enters the PDMS chamber (see panel c). **(c)** A completed PDMS chamber bonded to a 35 mm tissue culture grade petri dish. Metal connector tubes are used to connect the polyethylene tubing to the chamber

0.72×10^{-2} N/m^2, *Xenopus* spinal neurons undergo collapse, but that at 4.5×10^{-4} N/m^2, they did not. Consequently, the shear stress in our device at 200 μL/h falls well below that of the detrimental value of 5×10^{-2} N/m^2 used by Morel et al., and is only about twofold above that found to be damaging for *Xenopus* spinal neurons by Wang et al. This suggests that the shear stress in our device is in a domain which, at least for rat SCG cells grown on plastic, is still conducive to growth. However, we note that the no-slip boundary condition for viscous fluid flow states that the velocity profile in a rectangular channel is parabolic [13], implying that the shear stress in the thin layer in which axons are growing may be significantly less than the value calculated above.

3.2 Microfluidics Chambers

3.2.1 Microfluidics Chamber Design

Draft the chamber design using AutoCad software and plot it onto the photoplate. Figure 1a illustrates the design used for the photomask for the first layer of lithography. In order to reduce the shear stress on neurons growing in the growth chamber, the design shown in Fig. 1a is two-layered. The first layer consists of the entire pattern shown in Fig. 1a with a height of 50 μm. The second layer is 100 μm thick and deposited over only the growing region. This means that although the fluid mixing occurs in channel heights of 50 μm, the cells are growing in a region with a 150 μm height. Since shear goes as $1/h^2$, the reduction in shear stress to the cells is considerable. Therefore, a relatively fast flow rate in the mixing channels using only a small expenditure of fluid can be used to establish the gradient quickly.

3.2.2 Microfluidics Chamber Fabrication

Fabricate chamber molds using standard SU-8 multilayer photolithography techniques as follows:

1. Clean silicon wafers using a plasma cleaner with a power of 200 W at a pressure of 200 mTorr for 5 min.
2. Immediately after cleaning the wafers, coat them with Ti Prime by spin-coating at 3000 rpm for 30 s, followed by a 2 min bake at 110 °C. Then allow the wafers to cool to room temperature.
3. Next, deposit a layer of SU-8 2050 on the wafers by spin-coating, according to the manufacturer's recommendations. To achieve an approximately 50 μm thickness, spin the SU-8 at 500 rpm followed by 1500 rpm for 10 and 30 s, respectively.
4. The wafers are then soft baked at 65 °C and 95 °C for 3 min and 9 min, respectively.
5. After allowing the wafers to cool to room temperature, expose them in the mask aligner to a dose of 175 mJ/cm^2. Figure 1a shows the pattern that was used.
6. Perform a post-exposure bake at 65 °C and 95 °C for 2 min and 7 min, respectively, and then allow the wafers to cool to room temperature.
7. Apply a second layer of photoresist, this time SU-8 2100, by spin-coating at 500 rpm followed by 3000 rpm for 10 and 30 s, respectively. This allows for an approximate additional height of 100 μm.
8. The wafers are then soft baked at 65 °C and 95 °C for 5 min and 30 min, respectively.
9. Next align and expose a second mask using the mask aligner at a dose of 250 mJ/cm^2. The second mask is designed to increase the height of the growth chamber and exit hole regions of the devices.
10. When the wafers return to room temperature, develop them in PEGMA for approximately 15 min.
11. Verify the height of the mold using the optical profiler. The lowest and highest features should measure around 50 μm and 150 μm, respectively.
12. Next silanize the molds to prevent PDMS adhesion. Clean the freshly processed masters using an oxygen plasma cleaner and then place them in a vacuum desiccator with several drops of Trichloro(1*H*,1*H*,2*H*,2*H*-perfluorooctyl)silane for approximately 20 min.
13. Pour enough PDMS mixture onto the silicon master to cover it to a depth of about 4 mm.
14. Place the mold with PDMS into a vacuum chamber and apply vacuum for 2 h to degas the PDMS (*see* **Note 6**). During degassing, minute bubbles will form in the low air pressure and rise to the top of the liquid PDMS. After 2–3 h, the PDMS will appear completely transparent.

15. Bake (cure) the mold for at least 2 h at 80 °C. Baking at higher temperatures for slightly shorter times will also lead to curing.
16. Using a scalpel, carefully cut around the chamber, taking care to avoid damaging the lithographed pattern on the silicon wafer. Gently pull the chamber out of the mold.
17. Cover the channel side of the chamber with Scotch magic tape to protect the channels from airborne dust which can lead to clogging during later stages.
18. Using a 0.75 mm corer, core holes into the PDMS where fluid or cells can be introduced. The corer should have an external diameter slightly smaller than that of the metal connecting tubes. Make sure the PDMS "noodle" is pushed out of the holes and the corer goes all the way through the PDMS.
19. To bond the PDMS chamber to a plastic tissue-culture petri dish, plasma treat the petri dish (using 100 W at a pressure of 380 mTorr for 30 s) and then pour enough APTES solution (5 % APTES in 70 % ethanol) into the dish to cover the bottom surface and leave for 5 min. Meanwhile, plasma treat the PDMS chamber with high power for 40 s. Make sure the PDMS chamber is placed into the plasma cleaner so the channels side is face-up, otherwise the plasma will not properly treat the surface. Discard the APTES solution from the petri dish, wash thoroughly with water, and allow it to air-dry or blow-dry. Press the PDMS chamber onto the APTES-treated petri dish. Make sure to press the channel-side surface of the PDMS onto the dish. If bonding of PDMS chambers onto glass is required, plasma treat the glass and the chamber at the same time and gently press the chamber on glass (*see* **Note 7**).
20. Bake the dish for 30 min at 65 °C. Although the bond forms within a few seconds, baking will increase bond strength. After baking, the chamber is ready to be used for tissue culture, otherwise chambers can be prepared ahead of time and stored at room temperature (*see* **Note 8**).
21. Fill the plate with filtered PBS and penicillin/streptomycin and degas in the vacuum chamber for 5 min. Take out the chambers at least 15 min before injecting cells into them to allow the solution to fill up all the channels to avoid air bubbles.

3.3 Tissue Dissociation

1. Cut out the superior cervical ganglia (SCG) from P0–P3 rat pups into about 2 mL Leibovitz medium in a petri dish kept on ice. Leave in trypsin for 30 min at 37 °C in a 15 mL tube. Use a sterilized flame polished glass Pasteur pipette to gently triturate the cells by aspirating up and down slowly (*see* **Note 9**).
2. Stop the trypsin by filling the tube with Leibovitz medium to 15 mL. Centrifuge at 190 rcf for 5 min at 4 °C. Discard the supernatant carefully and then repeat with 15 mL Leibovitz, then 15 mL Opti-MEM and then with 0.5 mL filtered SGM

containing 0.3 nM NGF. Suck out most of the solution to leave 50 μL of solution per SCG. Very gently resuspend the cells. Using the microfluidics pump or pumping by hand with a short length of polyethylene tubing connected to a metal connector, aspirate the solution at 1 μL/s and then inject into a test chamber through the outlet with the chambers fully immersed in PBS/PS. Make sure there is no air bubble in the injected solution or in the chambers. Adjust the cell density so that 20–100 cells enter the growth chamber. Very low density often leads to poor growth and very high density makes imaging difficult.

3.4 Growth in the Gradient

1. Leave the cells in the incubator for at least 1 h to allow cells to adhere to the substrate before setting up the flow.
2. Cut two lengths of polyethylene tubing approximately 60 cm (*see* **Note 10**). Connect a metal connector to one end of each tube (Fig. 1b). Connect the other end of each tube to a syringe connector and connect, via a Luer lock, to a 250, 500, or 1000 μL glass syringe or a plastic 1 mL syringe (the size of the syringe limits the duration of the experiment). Insert the syringes into the microfluidics pump. The syringes contain H solution (SGM with a high concentration of guidance cue (in our case NGF), equivalent to $20 \times K_d$) and L solution (SGM with no guidance cue). Make sure there are no air bubbles in the tubing and that there is no air in the syringes. To generate a linear gradient, the flow rates of both syringes have to be the same (5 μL/h each). To increase the throughput, four chambers can be run in parallel, with eight syringes on two PhD Harvard pumps.
3. Pump out a small amount of solution then quickly put the metal pins of the H and L tubes into the appropriate inlets (Fig. 1c; *see* **Note 11**). Turn on the flow to 5 μL/h. Gradients should establish within 5 min.
4. Place the chambers into a tissue culture incubator (with the pump outside). After 2–4 h, neurites will start to grow from the cell bodies. At this point, the cells may be used for live imaging by moving the chambers and pump(s) to an incubated inverted microscope. The tubing can be attached to the incubator door higher than the plate so that air bubbles will rise and not enter the chambers. Otherwise incubate the chambers overnight. Make sure there is enough solution in the syringes for the desired duration of the experiment.

3.5 Quantification of Guidance

1. The chambers can be used for in vitro imaging of growth cone guidance using time-lapse microscopy. The chambers should be housed in an inverted microscope with environmental con-

trol including temperature (37 °C) and a 5 % CO_2 atmosphere. Images can be obtained using phase contrast microscopy at 1 min intervals for many hours. The degree of guidance can be estimated for each growth cone from the time-lapse data using the definition of the turning angle shown in Fig. 3b [6]; however, other definitions of turning can also be used [10, 14].

2. If, following growth of neurites in the growing chamber, the axons need to be immunostained, then gently remove the tubing from the microfluidics chamber and examine the growing chamber. Neurites should be visible under phase contrast microscopy (*see* **Note 12** and Fig. 3a). Fill a 1 mL plastic syringe with fixative and, with a short length of polyethylene tubing (and using the appropriate connectors) pump through the fixative into the exit hole at 1 μL/s for 60 s. Replace the fixative in the syringe with PBS and pump through at the same flow rate for 5 min to remove all fixative. Pump through blocking solution and then primary antibody at the same flow rate for 60 s. Leave the primary antibody solution in for 15 min without flow. Pump through the appropriate fluorophore labeled secondary antibody at the same flow rate for 60 s and leave without flow for 15 min. Finally, wash with PBS for 5 min (*see* **Notes 13** and **14**).
3. Photograph neurites at 20× using fluorescence microscopy for the appropriate fluorophore.

4 Notes

1. Chamber design: A comprehensive description of the concepts used in the design of our chamber can be found in Campbell et al. [15]. The dimensions of the design were modified to suit the diffusion constant of nerve growth factor.
2. To test whether the chamber is capable of generating a gradient, a fluorescently labeled dextran, of a molecular weight roughly equivalent to the guidance cue being studied, can be used to visualize the gradient. 40 kDa-dextran fluorescently labeled with tetramethylrhodamine has a molecular weight similar to that of nerve growth factor. To visualize the gradient, set up a gradient as in Subheading 3.4, **step 2**; however, exclude the guidance cue and instead, include the fluorophore in the H solution. L solution should contain no fluorophore. The exact concentration of the fluorophore in the H solution is not important as long as it is high enough to provide an image of the gradient using a relatively short exposure (e.g., 1 s) (Fig. 2a, b). Time-lapse imaging of the chamber can be used to assess the stability of the gradient.

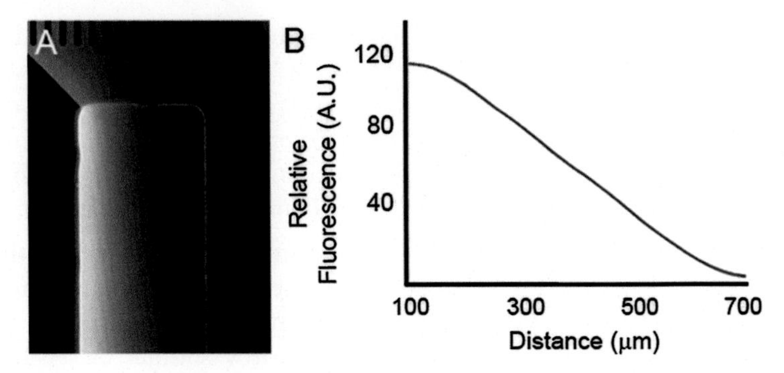

Fig. 2 (a) A gradient of 40 kDa dextran-tetramethylrhodamine in the growth chamber of the PDMS chamber shown in **(b)**. **(b)** The relative fluorescence of 40 kDa dextran-tetramethylrhodamine in a transverse section through the gradient in **(a)**. (A.U. = arbitrary units)

3. This is also suitable for the growth of dorsal root ganglion neurons from newborn rat pups up to about postnatal day 7 of age.
4. 4 % paraformaldehyde in PBS can be stored at −20 °C for several months and should be thawed to room temperature before use. It produces sufficient fixation for the visualization of axons, however we find it unsuitable for the fixation of finer cellular detail such as filopodia.
5. The dissociation constant (K_d) of NGF for its high affinity receptor, TrkA, is about 0.3 nM [16].
6. When PDMS is mixed or poured, tiny bubbles will be generated which, if not removed, will become incorporated into the microfluidics chamber during curing.
7. If PDMS chambers do not adhere to glass or plastic substrates, the plasma oxidation conditions should be optimized. Determine optimal oxidation time, ionization strength and O_2 pressure within the plasma cleaner. For our Harrick Expanded Plasma Cleaner, typical values are 380–410 mTorr O_2 pressure, 30 W power, and 30–50 s ionization time.
8. Contamination by bacterial and fungal cells may be reduced by spraying the chambers with 70 % ethanol following by irradiation with short-wavelength UV for several hours. Once dried, the chambers may be used for tissue culture.
9. Poor cell growth can also be due to poor trituration. Optimize the trypsin concentration and incubation time. Optimize the size of the flame-polished pipette bore. Holes which are too large will result in poor dissociation. Holes which are too small will result in high cell death. Ganglia should dissociate into cells within 2 min of trituration.

10. The length of both tubes has to be enough so that the microfluidics chamber can be placed in a tissue culture incubator with the microfluidics pump outside the incubator.
11. A major problem which will be encountered when using microfluidics is the accidental introduction of air bubbles into the chamber. This can occur mostly as a result of poor connections where the metal connector tubes are inserted into the chambers. The metal connectors can be sealed by making a small amount of PDMS with curing agent and spreading about 10 μL around the insertion site followed by curing at 65 °C. Then fill the plate with filtered PBS to immerse both the chambers and connector pins in solution and degas. Use the tubing to suck solution through the connector pin to fill up all the channels and pins with PBS. However we find that the best solution is to ensure that coring achieves clean defect-free holes which are less likely to leak and inject into the chambers while they are immersed in solution.
12. If neuronal cell growth is poor, it may be necessary to perform PDMS extraction prior to bonding to the substrate. This is done in order to remove unpolymerized PDMS monomers from the PDMS chamber which may cause cell toxicity. Numerous PDMS extraction techniques exist. We have found the following to improve neuronal cell growth: Following removal of the PDMS from the mold, immerse chambers for ~200 mL for 1 h into each of the following; 100 % pentane, 100 % acetone and then 100 % ethanol. PDMS will swell significantly while in the pentane and acetone, and care should be taken to ensure the liquid volume is significantly larger than the PDMS volume. Do not delay in transferring chambers from one solution to the next otherwise cracking of the PDMS will occur. After PDMS have soaked in the ethanol, bake at 65 °C for 2 h and proceed with bonding to the substrate. Discard the used solutions according to institutional guidelines.
13. Image quality of the immunostained cells will be better with a glass substrate.
14. We find that a high solution (H) of $20–40 \times K_d$ and a low solution (L) of $0 \times K_d$ elicited the strongest turning responses in SCG axons whereas a very high solution of $H = 200 \times K_d$ abolishes the turning (Fig. 3c shows turning of growth cones in a gradient using nerve growth factor with $L = 0$ nM, $H = 20 \times K_d$ nM). If no turning is observed, these parameters should be optimized.

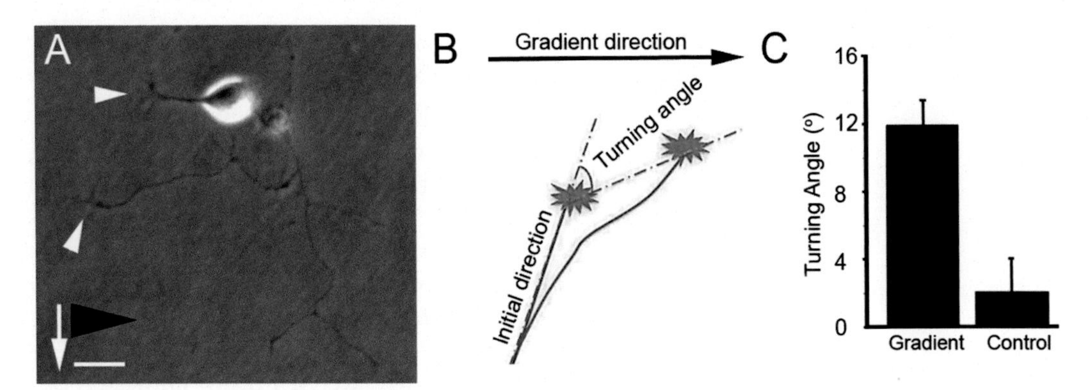

Fig. 3 (a) Phase contrast image, obtained with a 20× objective, of an SCG neuron grown in a microfluidics chamber with an NGF gradient (gradient direction is shown by the *black triangle*). The *white arrow* indicates direction of fluid flow. *White arrowheads* indicate growth cones with filopodia. Scale bar = 10 μm. **(b)** Definition of initial angle and turning angle in short-term turning assay. The reference axis is perpendicular to the direction of flow, which is the gradient direction. The initial direction is defined as the tangential direction of the 15 μm neurite segment. The initial angle is defined as the angle between initial direction and the reference axis. The turning angle is defined as the angle between the line connecting the growth cone before and after the assay and the initial direction. **(c)** Turning angles of axons grown either in a gradient of NGF ($n = 190$ axons) compared to those grown in an NGF plateau control ($n = 110$ axons) using a flow rate of 10 μL/h. Axons grown in the gradient display a positive turning response of 12° ± 1.4° compared to 2.1° ± 2.0° (mean ± S.E.M.) for those grown in a plateau of NGF ($p = 0.033$, *t*-test)

Acknowledgements

We thank Jiajia Yuan for help with an earlier version of the microfluidics chamber. We gratefully acknowledge support from the NHMRC (project grant 1083707). This work was performed in part at the Queensland node of the Australian National Fabrication Facility (ANFF), a company established under the National Collaborative Research Infrastructure Strategy to provide nanofabrication and microfabrication facilities for Australia's researchers.

References

1. Engle EC (2010) Human genetic disorders of axon guidance. Cold Spring Harb Perspect Biol 2:a001784
2. Tessier-Lavigne M, Placzek M, Lumsden AG et al (1988) Chemotropic guidance of developing axons in the mammalian central nervous system. Nature 336:775–778
3. Rosoff WJ, Urbach JS, Esrick MA et al (2004) A new chemotaxis assay shows the extreme sensitivity of axons to molecular gradients. Nat Neurosci 7:678–682
4. Pujic Z, Giacomantonio CE, Unni D et al (2008) Analysis of the growth cone turning assay for studying axon guidance. J Neurosci Methods 170:220–228
5. Lohof AM, Quillan M, Dan Y, Poo MM (1992) Asymmetric modulation of cytosolic cAMP activity induces growth cone turning. J Neurosci 12:1253–1261
6. Yam PT, Langlois SD, Morin S, Charron F (2009) Sonic hedgehog guides axons through a noncanonical, Src-family-kinase-dependent signaling pathway. Neuron 62:349–362
7. Dupin I, Dahan M, Studer V (2013) Investigating axonal guidance with microdevice-based approaches. J Neurosci 33:17647–17655

8. Whitesides GM (2006) The origins and the future of microfluidics. Nature 442:368–373
9. Kothapalli CR, van Veen E, de Valence S et al (2011) A high-throughput microfluidic assay to study neurite response to growth factor gradients. Lab Chip 11:497–507
10. Wang JC, Li X, Lin B et al (2008) A microfluidics-based turning assay reveals complex growth cone responses to integrated gradients of substrate-bound ECM molecules and diffusible guidance cues. Lab Chip 8:227–237
11. Morel M, Shynkar V, Galas JC et al (2012) Amplification and temporal filtering during gradient sensing by nerve growth cones probed with a microfluidic assay. Biophys J 103:1648–1656
12. Lu H, Koo LY, Wang WM et al (2004) Microfluidic shear devices for quantitative analysis of cell adhesion. Anal Chem 76:5257–5264
13. Day MA (1990) The no-slip condition of fluid dynamics. Erkenntnis 33:285–296
14. Bhattacharjee N, Li N, Keenan TM, Folch A (2010) A neuron-benign microfluidic gradient generator for studying the response of mammalian neurons towards axon guidance factors. Integr Biol (Camb) 2:669–679
15. Campbell K, Groisman A (2007) Generation of complex concentration profiles in microchannels in a logarithmically small number of steps. Lab Chip 7:264–272
16. Mortimer D, Feldner J, Vaughan T et al (2009) Bayesian model predicts the response of axons to molecular gradients. Proc Natl Acad Sci U S A 106:10296–10301

Chapter 21

Visualization of Actin Assembly and Filament Turnover by In Vitro Multicolor TIRF Microscopy

Moritz Winterhoff, Stefan Brühmann, Christof Franke, Dennis Breitsprecher, and Jan Faix

Abstract

In response to chemotactic signals, motile cells develop a single protruding front to persistently migrate in direction of the chemotactic gradient. The highly dynamic reorganization of the actin cytoskeleton is an essential part during this process and requires the precise interplay of various actin filament assembly factors and actin-binding proteins (ABPs). Although many ABPs have been implicated in cell migration, as yet only a few of them have been well characterized concerning their specific functions during actin network assembly and disassembly. In this chapter, we describe a versatile method that allows the direct visualization of the assembly of single actin filaments and higher structures in real time by in vitro total internal reflection fluorescence microscopy (TIRF-M) using purified and fluorescently labeled actin and ABPs.

Key words Actin, Actin-binding proteins, Fluorescence labeling, In vitro analysis, SNAP tag, In vitro TIRF microscopy

1 Introduction

Non-muscle cells contain a high concentration of globular monomeric actin (G-actin) which would spontaneously polymerize into filamentous actin (F-actin) if this process was not precisely controlled by a vast pool of ABPs. The formation of F-actin structures decisively contributes to various cellular processes such as endocytosis, cytokinesis, cell migration, and maintenance of cell morphology. The migration of cells is exceptionally complex and the result of the spatial and temporal coordination of four distinct processes: extension of protrusions such as lamellipodia or pseudopodia, stabilization of these structures by attachment to the substrate, translocation of the cell body, and detachment of the rear part of the cell upon disassembly of cell adhesions in this region. Assembly and disassembly of actin filaments are therefore tightly regulated by proteins that effect actin filament nucleation, elongation, branching, capping, and severing as well as actin monomer sequestration and bundling or cross-linking of actin filaments [1].

Tian Jin and Dale Hereld (eds.), *Chemotaxis: Methods and Protocols*, Methods in Molecular Biology, vol. 1407, DOI 10.1007/978-1-4939-3480-5_21, © Springer Science+Business Media New York 2016

To determine the function and the biochemical activities of a distinct ABP, various in vitro assays have been developed in the past and many of them are still valuable tools. The rather simple low- and high-speed sedimentation assays are employed to analyze binding to or bundling of actin filaments [2]. Electron microscopic analyses provide an insight into the effect of ABPs on filament architecture [2–4], whereas the versatile spectroscopic pyrene-actin assay is used to measure actin polymerization kinetics [5]. The disadvantage of this otherwise effective high throughput assay is the difficulty to discriminate between an increased nucleation or elongation rate of filaments during polymerization. In other words, many nucleating filaments that elongate slowly may give the same fluorescence signal as a few rapidly elongating filaments. In 2001, Amann and Pollard overcame these drawbacks by developing the TIRF-M assay [6]. This experimental approach allows to visualize the formation and elongation of fluorescently labeled single actin filaments in presence of nucleation- and elongation-factors such as formins [7], VASP [8, 9] and other accessory proteins [10] in real time (Fig. 1). Since the TIRF-M assay has been continuously improved over the last decade, concerning sensitivity and background reduction, it is now even possible to directly observe interactions of actin filaments with their binding proteins on the single molecule level. By using up to three-color TIRF-M, activation of the Arp2/3 complex [11], formation and subsequent elongation of nucleation seeds by APC/mDia1 [12] and regulation of actin dynamics by Ena/Diaphanous [13] have been visualized and analyzed in detail.

1.1 The Principle of TIRF Microscopy

TIRF-M is an optical technique that allows the selective excitation of fluorophores within a thin region above a glass coverslip and therefore interfering background fluorescence is largely eliminated.

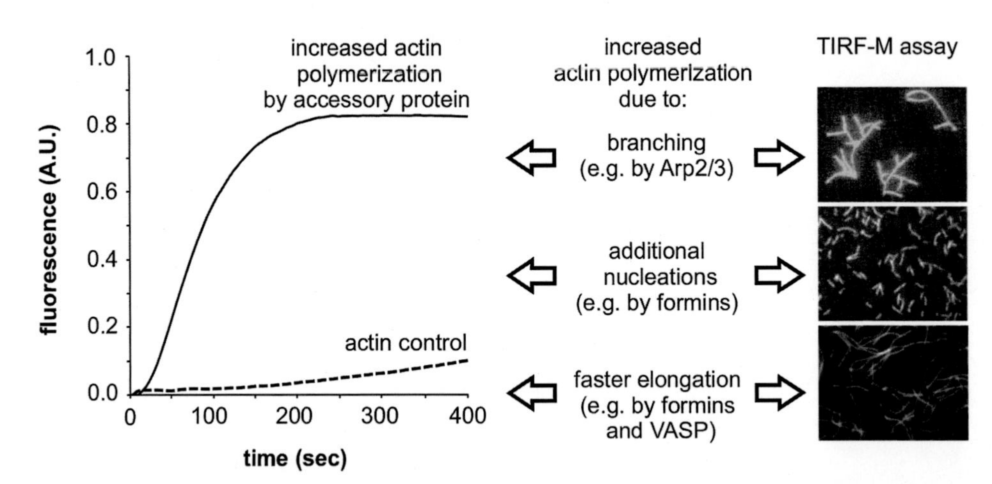

Fig. 1 Comparison of the pyrene-actin assay and TIRF-M as tools for the analysis of in vitro actin assembly. ABPs have various effects on actin polymerization. Whereas the pyrene-actin assay does not allow to clearly distinguish between de novo nucleation (e.g., by formins), nucleation through branching on preexisting filaments (by Arp2/3-complex) or enhanced filament elongation (e.g., by Ena/VASP or formins), these events can be unambiguously discriminated by in vitro TIRF-M

In a TIRF microscope, a laser beam passes a high refractive index medium (immersion oil/glass coverslip) to a lower refractive index medium (aqueous sample) (Fig. 2a). This laser beam is directed to the coverslip liquid interface at an angle which is greater than the critical angle of reflection. Since light cannot be reflected at an infinite thin layer, an electromagnetic "evanescent" wave penetrates into the lower refractive index medium [14]. It propagates parallel to the surface with an exponentially decaying intensity, exciting only fluorophores that are within a range of less than 200 nm above the surface of the reflecting coverslip (Fig. 2b) [15, 16].

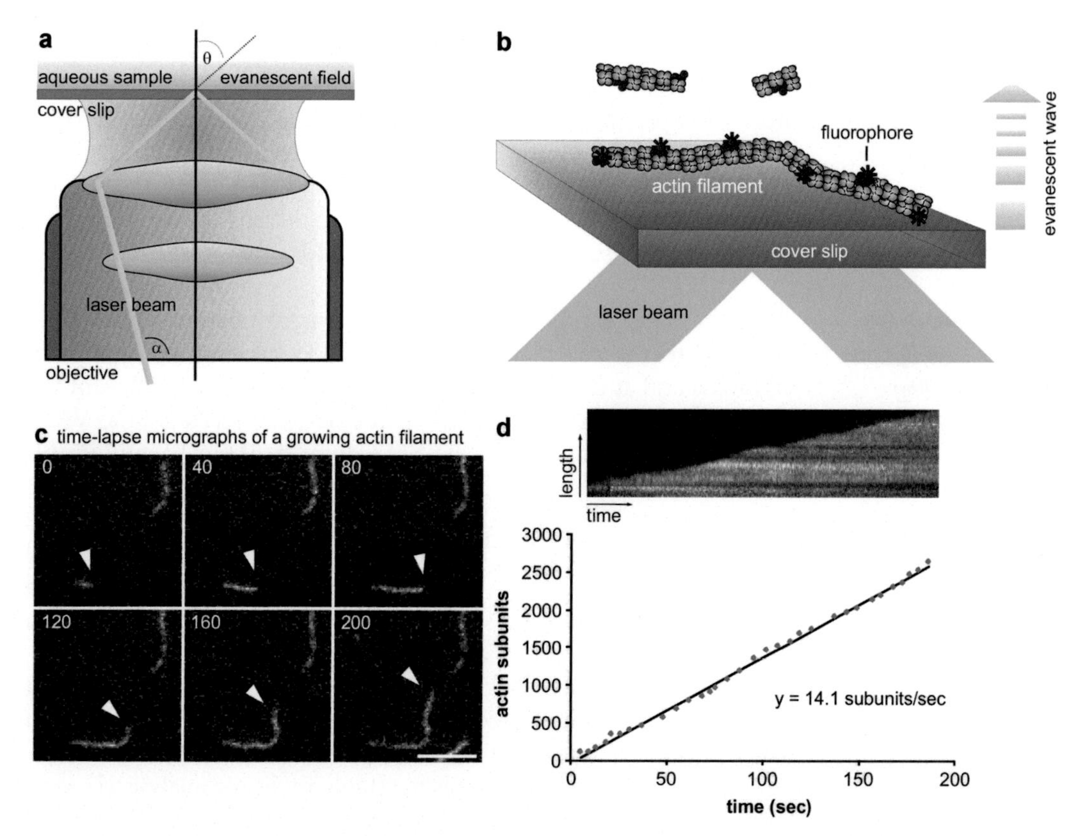

Fig. 2 In vitro TIRF-M of actin assembly. (**a**) Principle of objective-based TIRF-M. A laser beam passes through an objective with high numerical aperture and is reflected from the glass–aqueous solution boundary at an angle that is greater than the critical angle of reflection θ, resulting in an evanescent field above the surface of the coverslip with a penetration depth of less than 200 nm. (**b**) Fluorophores as tracers in an actin filament, floating within the evanescent field, are excited and emit secondary fluorescence, whereas fluorophores more than 200 nm above the surface of the coverslip are not excited. This way, single actin filaments can be visualized at a high signal to noise ratio even in the presence of high amounts of fluorescently labelled G-actin. (**c**) Time-lapse micrographs of actin filament assembly (1 μM G-actin, 10 % labeled with ATTO-488 at Cys374, 0.2 % biotinylated) on an mPEG-coated coverslip (containing 0.1 % biotin-PEG saturated with streptavidin). The fast growing barbed end is marked with an *arrowhead*. Time is given in seconds. Scale bar represents 5 μm. (**d**) Determination of the elongation rate of a single filament. The filament depicted in (**c**) was manually tracked using ImageJ and the Kymograph Plugin. The kymograph (*upper right*) of the tracked filament reveals constant growth and the resulting overall-lengths were plotted over time (*lower right*). The elongation rate in subunits per second can be calculated from the slope of a simple linear regression assuming that 1 μm filament contains 360 actin subunits

1.2 Using In Vitro TIRF Microscopy to Visualize Actin Polymerization

By adding partially labeled monomeric actin to a viscous polymerization buffer, the nucleation and subsequent elongation of single actin filaments can be monitored and analyzed by time-lapse TIRF-M (Fig. 2c). The growing ends of filaments can be manually or automatically tracked by image analysis software such as ImageJ (http://rsb.info.nih.gov/ij/). A plot of the length of a growing filament versus time and subsequent linear regression of the slope yields the elongation rate of a particular actin filament (Fig. 2d) [17]. *Multicolor* TIRF-M assays allow for a much more detailed view concerning localization and function as well as determination of residence times of actin-binding proteins. The labeling of ABPs with fluorescent probes may reveal direct interactions with actin filaments. This way the formation of nucleation complexes by mDia1/APC and subsequent elongation of the formed filaments by mDia1 could be visualized in detail [12]. Furthermore the labeling of nucleation promotion factors, such as the VCA domain of Scar/Wave proteins for Arp2/3 complex activation (Fig. 3c–d), unraveled the chronology of filament branch formation [11]. Finally, the specific bleaching of the fluorescent probe can be utilized to investigate the oligomerization state of a respective ABP [12].

1.3 The SNAP-Tag Provides New Options for Actin Binding Protein Analysis

Several ABPs, like Ena/VASP, are known to operate while being tethered to the cell membrane. Ena/VASP accumulates in tips of filopodia and at the leading edge of migrating cells [20]. Recently it has been shown that the I-BAR protein IRSp53 recruits Ena/VASP proteins to membranes via its SH3-domain [21]. To mimic this directional anchoring for TIRF microscopy studies of clustered Ena/VASP we covalently linked purified SNAP-tagged Ena/VASP proteins to the surface of SNAP-Capture magnetic beads (New England Biolabs) (*see* **Note 1**, Fig. 4c, d). These magnetic beads are coated with O^6-benzylguanine, a synthetic derivative of guanine, which is a substrate of the mammalian enzyme O^6-alkylguanine DNA alkyltransferase [22]. The SNAP-tag itself is an engineered derivative of this enzyme. Furthermore the SNAP-tag offers the opportunity to label purified proteins with various commercially available fluorescence dyes, a fundamental basis for single molecule analysis via *multicolor* TIRF-M (Fig. 4b). Therefore, we developed a vector for recombinant expression of GST-SNAP-tag fusion proteins [23] based on the commercial pGEX-6P-1 vector (GE Healthcare Life Sciences). The fusion proteins contain a cleavage site for PreScission Protease (GE Healthcare Life Sciences) to remove the GST-tag, thus avoiding artificial dimerization of the proteins (Fig. 4a).

Fig. 3 (Continued) SNAP549-VCA dissociated from the generated nucleation seed. (5) A daughter filament started to grow from the mother filament in position of the previous VCA-binding site. (**d**) In time-lapse micrographs, *red arrowhead* marks transient binding of SNAP549-VCA to a mother filament. After dissociation of this Arp2/3-activator a daughter filament started to grow and is marked by a *white arrowhead*. Time is given in seconds. Scale bar represents 5 μm

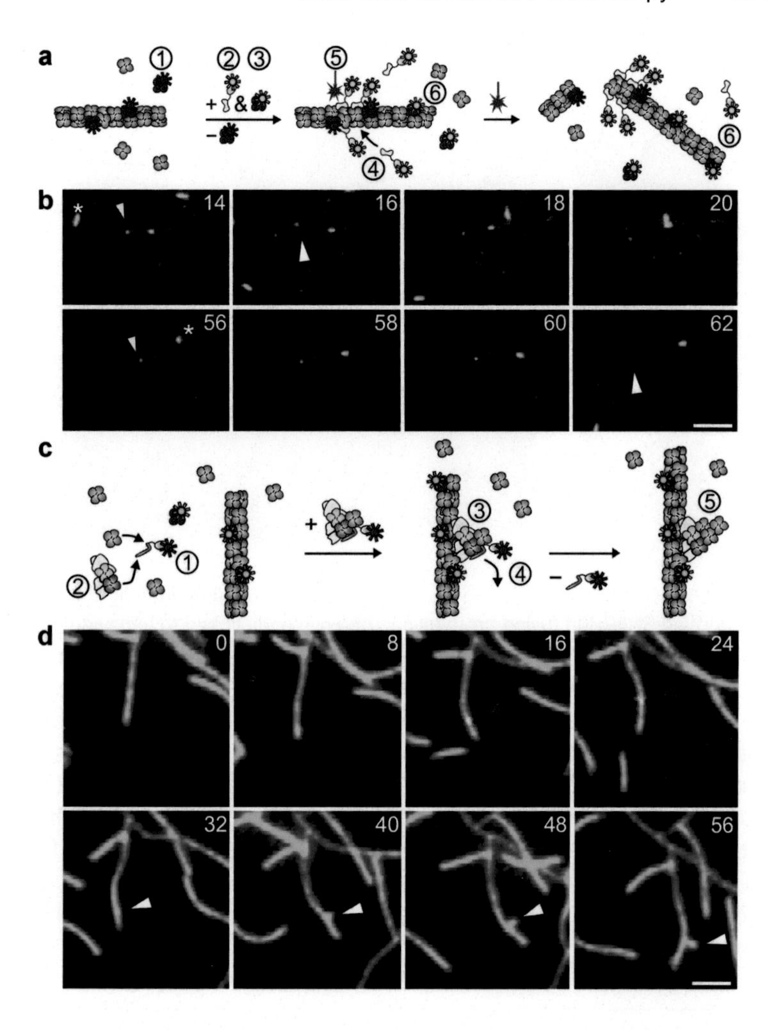

Fig. 3 Examples of in vitro multicolor TIRF-M. (**a**, **b**) Actin filament severing by GFP-Cofilin. (**a**) Schematic description and interpretation of the experiment shown in (**b**). (*1*) ATTO550-actin (1 μM, 20 % labeled) was polymerized for 10 min. Subsequently, (*2*) 100 nM GFP-Cofilin and (*3*) ATTO488-actin (1 μM, 20 % labeled) were flushed into the flow cell while remaining ATTO550-actin monomers were removed simultaneously. (*4*) GFP-cofilin bound to aged (ADP + Pi-actin) ATTO550-labeled actin filaments in a cooperative manner and is known to change the twist of the filaments [18, 19]. (*5*) As these filament regions are more flexible than non-cofilin-decorated segments, shear stress lead to filament severing at the boundaries. (*6*) Barbed ends continued to elongate and were labeled by ATTO488-actin to distinguish "old" and "young" actin filament segments. (**b**) In time-lapse micrographs, GFP-cofilin decorated regions are marked by *yellow arrowheads* and *white arrowheads* mark severing events. *White asterisks* denote freshly growing barbed ends. Time is given in seconds. Scale bar represents 5 μm. (**c**, **d**) Formation of a branched actin network by Arp2/3-complex and its activator SNAP549-VCA. (**c**) Schematic description and interpretation of the experiment shown in (**d**). ATTO488-actin (1 μM, 20 % labeled) was polymerized in presence of (*1*) 5 nM SNAP549-VCA and (*2*) 2 nM Arp2/3-complex. (*3*) SNAP549-VCA activated the Arp2/3-complex by recruiting it along with an actin monomer to a preexisting mother filament. (*4*) Subsequently,

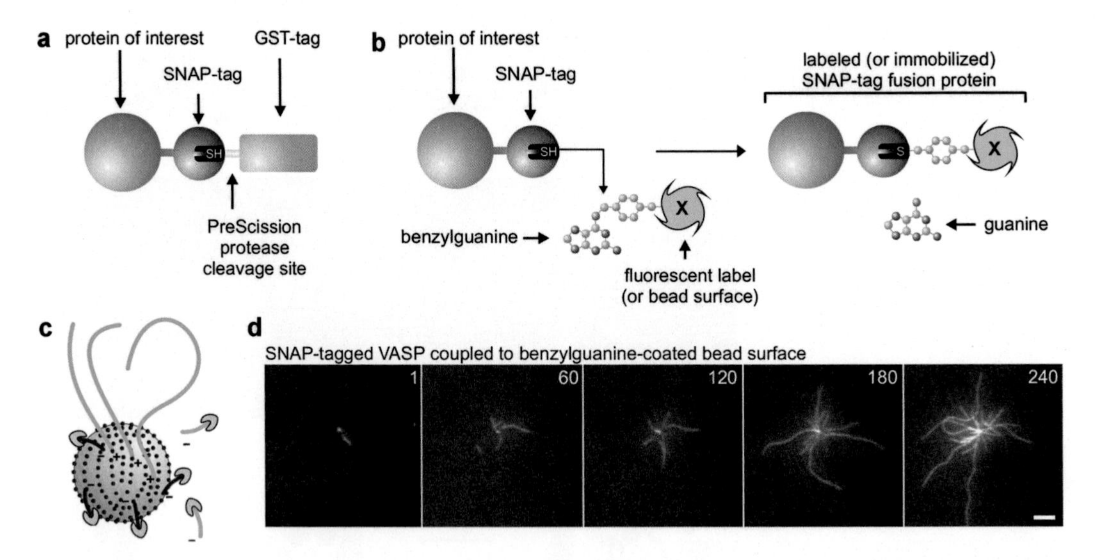

Fig. 4 (**a**) Scheme of purified GST-SNAP-tag fusion protein containing the PreScission Protease cleavage site. (**b**) Purified SNAP-tag fusion proteins can be covalently linked to benzylguanine-activated fluorescent dyes (or a suitable bead surface) by nucleophilic attack of a reactive thiole group (SH) of a cysteine residue on the marked carbon atom of the benzylguanine compound. Following this scheme, labeled ABPs can be analyzed by in vitro *multicolor* TIRF-M to observe complex interactions of actin and proteins of interest even at single molecule level. (**c**) Model of tethered actin filaments elongation onto a bead surface by filament elongators such as Ena/VASP proteins or formins. Growth of filaments that are not attached to the bead surface by their barbed ends as well as of actin filaments growing in the solution around the beads is inhibited by addition of capping protein (*yellow crescents*). (**d**) Time-lapse micrographs of actin filaments (1 μM, 20 % labeled with ATTO488) growing from beads conjugated with SNAP-tagged human VASP. Note that bead-attached filaments were elongated processively in presence of 50 nM capping protein, while filament growth in solution was abolished. Time is given in seconds. Scale bar, 5 μm

2 Materials

2.1 Preparation and Modification of Actin from Rabbit Skeletal Muscle

2.1.1 Preparation of Actin Acetone Powder from Rabbit Skeletal Muscle

Keep all instruments, buffers, and materials at 4 °C.

1. 1.5 kg freshly prepared rabbit muscle from two animals.
2. Sharp knife or scalpel.
3. Meat grinder.
4. 5-L glass or plastic beaker.
5. 5 L high-salt buffer: 0.5 M KCl, 0.1 M K_2HPO_4, 2 mM benzamidine, pH 6.4.
6. 5 L ddH_2O.
7. 10 mL 1 M Na_2CO_3.
8. 5 L acetone at 4 °C.
9. Aluminum foil.
10. Centrifuge (for volumes up to 6 L at 8000 × *g*).

2.1.2 Preparation of G-Actin from Acetone Powder

1. Gauze.
2. Rubber band.
3. 250-mL glass beaker.
4. 5 L G-actin buffer (G-buffer): 2 mM Tris–HCl (pH 8.0 at 4 °C), 0.5 mM DTT, 0.2 mM $CaCl_2$, 0.2 mM ATP, 0.01 % (w/v) NaN_3.
5. 25× actin polymerization buffer: 1.25 M KCl, 50 mM $MgCl_2$, 25 mM ATP (adjust to pH 7.2 before addition of 50 mM $MgCl_2$; *see* **Note 2**).
6. Centrifuge (for centrifugation at 30,000 × *g*) and ultracentrifuge (for centrifugation at 150,000 × *g*).
7. HiLoad 26/600 Superdex 75 gel filtration column (GE Healthcare).
8. Dialysis tubing: e.g., VISKING cellulose, MWCO 12,000–14,000, type 27/32 in. Clean the dialysis tubing thoroughly by washing in distilled water before use.
9. UV/Vis spectrophotometer and quartz cuvette to determine actin concentration.

2.1.3 Labeling of G-Actin with Maleimide-Activated Dyes

1. Alexa Fluor 488 C_5 maleimide (Life Technologies), ATTO-488-maleimide (ATTO-TEC) or other suitable thiol-reactive fluorophores for labeling actin at Cys374.
2. 5 L G-buffer: 2 mM Tris–HCl, pH 8.0, 0.2 mM $CaCl_2$, 0.2 mM ATP, 0.5 mM DTT, 0.01 % (w/v) NaN_3.
3. 3 L Labeling buffer (P-buffer): 10 mM $NaHCO_3$, pH 7 (for maleimide conjugates), 0.2 mM $CaCl_2$, 0.2 mM ATP.
4. Anhydrous dimethyl sulfoxide (DMSO) or dimethyl formamide (DMF).
5. 1 mL 1 M DTT in ddH_2O.
6. Ultracentrifuge (for centrifugation at 150,000 × *g*).
7. HiLoad 26/600 Superdex 75 gel filtration column (GE Healthcare).
8. Dialysis tubing: e.g., VISKING cellulose, MWCO 12,000–14,000, type 27/32 in. or 8/32 for small volumes. Clean the dialysis tubing thoroughly by washing with distilled water before use.
9. UV/Vis spectrophotometer and quartz cuvette to determine actin concentration and the degree of labeling.

2.1.4 Labeling of G-Actin with Biotin

1. NHS-LC-LC-Biotin (Thermo Scientific).
2. 5 L G-buffer: 2 mM Tris–HCl, pH 8.0, 0.2 mM $CaCl_2$, 0.2 mM ATP, 0.5 mM DTT, 0.01 % (w/v) NaN_3.
3. 5 L Biotinylation buffer: 5 mM Hepes pH 7.3, 0.2 mM $CaCl_2$, 0.5 mM DTT and 0.2 mM ATP.

4. Anhydrous dimethyl sulfoxide (DMSO) or dimethyl formamide (DMF).
5. Ultracentrifuge (for centrifugation at 150,000 × *g*).
6. Dialysis tubing: e.g., VISKING cellulose, MWCO 12,000–14,000, type 27/32 in. or 8/32 for small volumes. Clean the dialysis tubing thoroughly by washing with distilled water before use.
7. UV/Vis spectrophotometer and quartz cuvette to determine actin concentration.

2.2 Purification and Modification of GST-SNAP Fusion Proteins

2.2.1 Purification of GST-SNAP Fusion Proteins

1. LB-rich growth medium: 5 g/L NaCl, 8 g/L yeast extract, 16 g/L tryptone/peptone, 10 mM potassium phosphate buffer, pH 7.3.
2. Temperature-controlled shaker for incubation of bacterial cultures (e.g., New Brunswick Innova 40R).
3. Antibiotics stock solution (1000×), sterile filtered. Store at 4 °C.
4. 1 M IPTG stock solution in ddH_2O, sterile filtered. Store at –20 °C.
5. Centrifuges (for culture volumes up to 6 L at 8000 × *g*; for clearing of lysate (100–150 mL) at 60,000 × *g*).
6. 1 L Lysis buffer: 30 mM Tris–HCl, pH 7–8, 200 mM NaCl, containing 5 mM benzamidine, 2 mM EDTA, 3 mM β-mercaptoethanol or 1 mM DTT, 5 mM PMSF.
7. 100 mL Elution buffer: add 30 mM glutathione to lysis buffer and readjust pH with 1 M NaOH to pH 7–8.
8. 1 L Cleavage buffer: 30 mM Tris–HCl, pH 7–8, 150 mM NaCl, 2 mM EDTA, 1 mM DTT.
9. 5–15 mL Glutathione sepharose 4B (GE Healthcare) or Glutathione agarose (Sigma).
10. 2 L Protein filtration buffer: 20 mM HEPES, pH 7–8, 150 mM KCl, 1 mM EDTA, 1 mM DTT.
11. Superdex or Superose gel filtration column (GE Healthcare) depending on size and oligomerization properties of target protein.
12. Dialysis tubing: e.g., VISKING cellulose, MWCO 12,000–14,000, type 27/32 in. or 8/32 for small volumes. Clean the dialysis tubing thoroughly before use.
13. 1 L Protein storage buffer: 20 mM HEPES, pH 7–8, 150 mM KCl, 1 mM DTT, 60 % (v/v) glycerol (for long-term storage at –20 °C).

2.2.2 Coating of SNAP-Capture Magnetic Beads

1. SNAP-Capture magnetic beads (New England Biolabs): 2, 4, or 10 μm in diameter.
2. Coating buffer: 20 mM HEPES pH 7.4, 150 mM KCl, 1 mM EDTA, 1 mM DTT.

3. Refrigerated microcentrifuge for reaction tubes.
4. Temperature-controlled shaker (e.g., Eppendorf Thermomixer).

2.2.3 Labeling of SNAP-Tag Fusion Proteins

1. SNAP-Surface substrates (e.g., 488 or 549, New England Biolabs) or other fluorescent markers coupled to benzylguanine.
2. Labeling buffer: 20 mM HEPES pH 7–8, 100–250 mM KCl, 1 mM EDTA, 1 mM DTT, 0–10 % (v/v) glycerol.
3. Dialysis tubing: e.g., VISKING cellulose, MWCO 12,000–14,000, type 8/32. Clean the dialysis tubing thoroughly by washing with distilled water before use.
4. Anhydrous dimethyl sulfoxide (DMSO).
5. Temperature-controlled shaker or rotator at 4 °C.
6. Refrigerated microcentrifuge for reaction tubes.
7. Zeba Spin desalting columns (0.5 or 2 mL; Thermo Scientific).

2.3 TIRF Microscopy

2.3.1 Microscope Setup

Over the last decade TIRF microscopes have become a common tool and can be purchased from numerous manufacturers such as Olympus, Nikon, or Leica. We currently operate a Nikon Eclipse TI-E inverted microscope equipped with a TIRF Apo 60× and a 100× objective and three Ixon3 897 EMCCD high speed cameras (Andor). The advantage of this setup is the potential to simultaneously acquire three channels (laser lines 488, 561, and 642 nm) at very high sensitivity.

2.3.2 Preparation of Flow Cells

1. 24×24 mm glass coverslips (e.g., Menzel #1 (0,13-0,16 mm)).
2. 76×26 mm microscope slides (e.g., Menzel).
3. Polyvinyl chloride or glass rack for cleaning of glass coverslips.
4. Double faced adhesive tape.
5. Ultrasonic bath (e.g., Bandelin Sonorex).
6. 2 % (v/v) Hellmanex III in ddH_2O (alkaline concentrate for cuvette cleaning, Hellma).
7. 70 % (v/v) ethanol.
8. 1 M HCl.
9. 1 M KOH.
10. mPEG solution: 1 mg/mL mPEG-Silane MW 2000 (Laysan Bio, Inc.) dissolved in 80 % (v/v) EtOH adjusted to pH 2 with HCl.
11. Biotin-PEG-Silane MW 3400 (Laysan Bio, Inc.) for partial biotinylation of coverslips. Prepare a 1 mg/mL stock solution in 80 % (v/v) EtOH adjusted to pH 2 with HCl.

2.3.3 Actin Polymerization Assays

1. Appropriate flow cells (Subheading 3.3.1).
2. 2× TIRF buffer: 40 mM imidazole, pH 7.4, 100 mM KCl, 2 mM $MgCl_2$, 2 mM EGTA, 0.4 mM ATP, 30 mM glucose, 40 mM β-mercaptoethanol and 1 % (w/v) methylcellulose (400 or 4000 cP, Sigma-Aldrich) (*see* **Note 3**).
3. 10× KMEI buffer: 500 mM KCl, 10 mM $MgCl_2$, 10 mM EGTA, 100 mM imidazole, pH 7.4.
4. BSA-buffer: 10 mg/mL bovine serum albumin (BSA, Fraction V) in 1× KMEI buffer.
5. Streptavidin (Rockland Immunochemicals, Biomol) for immobilization of actin filaments. Prepare a 40 μM stock solution in 10 mM Tris–HCl, pH 8.
6. Streptavidin/BSA-buffer: Dilute Streptavidin stock solution in BSA-buffer to obtain a final concentration of 50 nM.
7. 100× O_2 scavenger enzyme stocks: 2 mg/mL catalase, 10 mg/mL glucose oxidase (Sigma-Aldrich) in 1× KMEI with 60 % (v/v) glycerol.
8. Labeled and unlabeled G-actin from Subheadings 3.1.2 and 3.1.3.
9. Particle free ddH_2O.

3 Methods

3.1 Preparation and Labeling of Actin from Rabbit Skeletal Muscle

3.1.1 Preparation of Acetone Powder from Rabbit Skeletal Muscle

1. Prepare everything at 4 °C.
2. Cut 1.5 kg of fresh rabbit muscle meat into small pieces, remove fat generously, and mince the meat twice using a meat grinder.
3. Stir the minced meat for 15 min after addition of 2.5 L high-salt buffer with a solid glass rod and subsequently centrifuge the suspension at 4000 ×*g* for 7 min.
4. Resuspend the pellet with 2.5 L high-salt buffer, stir the suspension again for 15 min, and centrifuge at 4000 ×*g* for 7 min.
5. Resuspend the pellet in 2.5 L of cold ddH_2O, adjust the pH to 8.3 with 1 M Na_2CO_3, and stir for another 10 min. Subsequently, pellet the suspension at 4000 ×*g* for 7 min.
6. Repeat this step two to three times until a recognizable swelling of the pellet occurs (the suspension will also become transparent). This step is critical since washing too long will result in considerable loss of G-actin.
7. Subsequently, resuspend the pellet in 4 °C cold acetone, stir the suspension for 15 min, and centrifuge again for 7 min at 4000 ×*g*.
8. Discard the supernatant and repeat this step one more time.

9. Discard the supernatant, break up the pellet into small pieces and distribute them over a sufficiently large piece of aluminum foil and allow the acetone powder to dry under a fume hood at room temperature overnight.
10. Fill the desiccated acetone powder in 50 mL reaction tubes, close the caps tightly, and store the tubes below −20 °C. 1.5 kg of rabbit skeletal muscle will yield approximately 100 g of acetone powder.

3.1.2 Preparation of G-Actin from Acetone Powder

1. Prepare everything at 4 °C.
2. Stir 5 g of the acetone powder in 100 mL cold G-buffer for 15 min.
3. Filter the suspension through double-folded gauze strained over a 250 mL glass beaker by a rubber band.
4. Repeat the extraction up to five times with 50–75 mL of G-buffer and control the amount of extracted actin by SDS-PAGE later on.
5. Discard the first fraction as it contains most of the impurities.
6. Pool the remaining fractions that contain most of the actin, and subsequently clear spin the suspension at 30,000 × *g* for 30 min.
7. Polymerize actin in the supernatant by addition of 25× actin polymerization buffer to a final concentration of 1× for at least 2 h at room temperature or at 4 °C overnight.
8. Subsequently, increase KCl concentration of the viscous solution to 0.8 M.
9. Spin the F-actin in an ultracentrifuge for 3 h at 150,000 × *g*.
10. Discard the supernatant and wash the pellet carefully several times with 10–20 mL of G-buffer. Resuspend and homogenize the hyaline pellet in 15 mL G-buffer using a Dounce homogenizer. Depolymerize F-actin by dialysis against G-buffer at 4 °C with several buffer changes for at least 3 days.
11. Spin the actin solution in an ultracentrifuge for 3 h at 150,000 × *g*.
12. Take the upper two thirds of the supernatant and separate monomers from dimers and multimers by gel filtration using a Superdex 75 column equilibrated in G-buffer.
13. Determine the actin concentration (c_{actin}) by measuring the absorbance (A) at 290 nm (($A = \varepsilon \cdot c \cdot l$; extinction coefficient of actin: $\varepsilon_{290} = 26{,}600\ M^{-1}cm^{-1}$; l: path length): $c_{actin} = A_{290\ nm} / (26{,}600\ M^{-1}cm^{-1} \cdot l)$.

3.1.3 Labeling of G-Actin with Maleimide-Activated Dyes

1. Prepare everything at 4 °C and protect the fluorescent dye from light (*see* **Notes 4** and **5**).
2. Dialyze purified G-actin against P-buffer with three buffer changes.

3. Dissolve 1 mg of maleimide-activated dye in 50 μL of anhydrous DMSO or DMF.
4. Label G-actin by adding a 1.5–2-fold molar excess of fluorescent dye and stir the solution vigorously while adding the dye (*see* **Note 6**).
5. Stir slowly for up to 6 h at 4 °C in the dark.
6. Add tenfold molar excess of DTT while stirring to consume residual thiol-reactive maleimide.
7. Dialyze the solution against G-buffer to remove unbound fluorescent dye. Change the buffer after a few hours and continue to dialyze the labeled actin overnight.
8. Perform a gel filtration on a 26/60 Superdex 75 column equilibrated with G-buffer to remove residual unbound dye.
9. Dialyze the actin-peak fractions against G-buffer containing 60 % (v/v) glycerol. This will concentrate your protein and allow you to store labeled actin at –20 °C for about 3 month.
10. Before use, dialyze an aliquot against at least 1 L of G-buffer overnight.
11. Perform a gel filtration on a 10/300 Superdex 75 column to separate actin monomers from oligomers.
12. The degree of labeling and concentration of labeled actin can be determined by spectrophotometry using the following equations (exemplified for Alexa Fluor 488):

    $A = \varepsilon \cdot c \cdot l$ (A: absorbance, ε: extinction coefficient, l: path length)

    $c_{\text{actin}} = (A_{290\text{ nm}} - f_{\text{Alexa-488}} \cdot A_{496\text{ nm}}) / (26{,}600\ \text{M}^{-1}\text{cm}^{-1} \cdot l)$

    $f_{\text{Alexa-488}} = 0.13$ is the correction factor for Alexa Fluor 488 at 290 nm (*see* **Note 7**)

    $c_{\text{Alexa-488}} = A_{496\text{ nm}} / (71{,}000\ \text{M}^{-1}\text{cm}^{-1} \cdot l)$

    Degree of labeling $= c_{\text{actin}} / c_{\text{Alexa-488}}$

3.1.4 Labeling of G-Actin with Biotin

1. Prepare everything at 4 °C.
2. Dialyze G-actin against biotinylation-buffer at 4 °C overnight.
3. Rapidly mix 1 mL of actin monomers (150 μM) with 20 μL of a 70 mM NHS-LC-LC-Biotin (Thermo Scientific) solution in DMF.
4. Incubate the reaction for 30 min at RT.
5. Polymerize actin by addition of 25× actin polymerization buffer to a final concentration of 1× for at least 2 h at room temperature or at 4 °C overnight.
6. Spin F-actin in an ultracentrifuge for 3 h at 150,000 × *g*.

7. Discard the supernatant and wash the pellet carefully several times with 5–10 mL of G-buffer. Resuspend and homogenize the hyaline pellet using a Dounce homogenizer with 1 mL G-buffer. Depolymerize F-actin by dialysis against G-buffer at 4 °C with several buffer changes for at least 3 days.
8. Spin the biotin-actin solution in an ultracentrifuge for 3 h at 150,000 × *g*.
9. Take the upper two thirds of the supernatant and determine the actin concentration by a spectrophotometer: $c_{\text{actin}} = A_{290\text{ nm}}/(26{,}600\ \text{M}^{-1}\text{cm}^{-1}\cdot 1)$.
10. The degree of biotinylation can be determined using biotin quantification kits (e.g., from Thermo Scientific) and should be 1–3 biotin molecules per actin monomer.
11. The biotinylated G-actin can be flash-frozen as small aliquots (<5 μL) in thin-walled PCR tubes using liquid nitrogen and stored at −80 °C.

3.2 Purification and Modification of GST-SNAP Fusion Proteins

3.2.1 Purification of GST-SNAP Fusion Proteins

1. As an example, the purification of GST-SNAP-VCA is described here. The purification of other ABPs may be performed accordingly.
2. For expression of GST-SNAP-VCA, the VCA subdomains of human WAVE1 [24] were subcloned into the expression vector pGEX-6P1-SNAP.
3. Transform the expression plasmid in competent *E. coli* BL21 Rosetta2 cells or another suitable expression strain. Verify protein synthesis and optimize expression temperature of GST-SNAP-VCA by small scale induction of individual clones.
4. Inoculate 200 mL of LB-rich medium supplemented with 50 μg/mL of ampicillin and 10 μg/mL chloramphenicol with an expressing clone and grow the preculture overnight at 37 °C while shaking at 150–200 rpm to stationary phase.
5. Dilute 20 mL of the preculture into 1 L of LB-rich medium containing 50 μg/mL of ampicillin and 10 μg/mL chloramphenicol in a 2-L Erlenmeyer flask with baffles and grow the cells at 30–37 °C while shaking at 150 rpm until an $OD_{600nm} > 0.5$ is reached.
6. You may add antifoam reagent Extran AP33 (Merck Millipore) to a final concentration of 0.01–0.02 % (v/v) if too much foam interferes with appropriate aeration.
7. To induce expression of GST-SNAP-VCA, add IPTG to a final concentration of 0.75 mM. Remember to reduce the temperature before induction.
8. Continue to grow the culture while shaking overnight at 22–26 °C at 150 rpm.

9. Harvest the cells by centrifugation in a prechilled centrifuge for 20 min at 4 °C and 5000 × *g*. Discard the supernatant and weigh the pellet. All following steps should be carried out on ice or below 4 °C to prevent degradation of the fusion protein.
10. Resuspend the pellet in ~2 mL/g of lysis buffer, add lysozyme to a final concentration of 2 mg/mL and incubate for 15 min while stirring the suspension.
11. Sonicate the resuspended cells in an ice-water bath with a Branson sonifier 250 (or similar, 2–5 min intervals, 50 % duty cycle, power output of 5–8).
12. Add Benzonase Nuclease (Sigma) according to the manufacturer's protocol to digest the viscous chromosomal *E. coli* DNA.
13. Spin the lysate for 60 min at 60,000 × *g* in a centrifuge at 4 °C and keep the supernatant.
14. While centrifuging, add 10 mL of glutathione sepharose suspension to a Bio-Rad Econo column (or similar) and equilibrate the column with at least two column volumes of lysis buffer.
15. Load the supernatant on the column, using a peristaltic pump at a flow rate of ~0.5 mL/min and pass the flow-through onto the column at least two more times.
16. After loading, wash the column with at least 100 mL of lysis buffer.
17. Elute bound protein with 10–50 mL of elution buffer at a maximum flow rate of 1 mL/min and collect 4 mL fractions using a fraction collector or manually (*see* **Note 8**).
18. To determine the protein concentration and purity, check a sample of each fraction in a Bradford assay and by SDS-PAGE followed by Coomassie blue staining.
19. Pool the fractions containing GST-SNAP-VCA and add PreScission protease according to the manufacturer's protocol (0.1 % (v/v)).
20. Dialyze against 1 L cleavage buffer overnight at 4 °C.
21. Further purify the SNAP-VCA by performing a gel filtration using a Superdex 75 column equilibrated with protein filtration buffer.
22. Dialyze combined pure fractions against protein storage buffer and store them at −20 °C.

3.2.2 Coating of SNAP-Capture Magnetic Beads

1. Prepare everything at 4 °C.
2. Add 80 μL of bead stock solution to 1 mL coating buffer (*see* **Note 9**), centrifuge for 10 min at 6000 × *g*, and carefully discard supernatant. Repeat the washing step twice.

3. Add 120 μL protein solution containing up to 1 mg/mL of purified target SNAP-tag fusion protein.
4. Incubate overnight at 4 °C while mixing in a temperature-controlled shaker.
5. To wash away unbound SNAP-tag fusion protein add 1 mL coating buffer, centrifuge for 10 min at 6000 × *g*, and carefully discard supernatant. Repeat this step twice.
6. Store the bead suspension with the immobilized fusion protein at 4 °C for a maximum of 5 days.

3.2.3 Labeling of SNAP-Tag Fusion Proteins

1. Prepare everything at 4 °C.
2. Dialyze your SNAP-tag fusion protein stock solution against Labeling-buffer overnight.
3. Dissolve SNAP-surface substrate in DMSO to prepare a 1 mM stock solution.
4. Dilute the SNAP-tag fusion protein with Labeling-buffer to 5–10 μM.
5. Add twofold molar excess of SNAP-surface substrate solution to the protein solution and mix well by pipetting up and down.
6. Incubate overnight at 4 °C while mixing in a temperature-controlled shaker at 700–1000 rpm.
7. Equilibrate desalting columns (2 mL resin) with Labeling-Buffer (or your special storage buffer if you want to perform a buffer exchange) by pipetting 1 mL onto the resin, centrifuge at 1000 × *g* for 2 min, and discard the flow-through afterwards.
8. Repeat this step at least three times.
9. Slowly pipette the protein solution onto the center of the column (200–700 μL volume) and let it flow into the resin.
10. Centrifuge at 1000 × *g* for 2 min.
11. Collect the flow-through that is clear of residual SNAP-surface substrate.

3.3 TIRF Microscopy

3.3.1 Preparation of Flow Cells

1. Place coverslips into a suitable polyvinyl chloride or glass rack.
2. Add 60 °C warm ddH_2O containing 2 % (v/v) Hellmanex III.
3. Sonicate for 45 min and rinse thoroughly with ddH_2O.
4. Add 60 °C warm 1 M KOH.
5. Sonicate for 45 min and rinse thoroughly with ddH_2O.
6. Add 60 °C 1 M HCl.
7. Sonicate for 45 min and rinse thoroughly with ddH_2O.
8. Sonicate subsequently in dust free 70 % (v/v) ethanol.
9. Rinse thoroughly with dust free 60 °C ddH_2O and immediately dry the coverslips thoroughly in nitrogen gas stream.

10. Incubate the coverslips with 1 mg/mL mPEG solution at 60 °C overnight. Add Biotin-PEG stock solution to obtain a final concentration of 1 μg/mL (1/1000) if you intend to immobilize actin filaments on the coverslip surface.
11. Rinse thoroughly with dust free 60 °C ddH_2O and immediately dry the coverslips completely under a nitrogen stream.
12. You may store the coverslips in a Parafilm-sealed box at 4 °C for about 1 week.
13. Prepare glass slides with two small stripes of double-faced adhesive tape or Parafilm. The spacing between the stripes should be between 5 and 8 mm.
14. Use forceps to carefully place the clean coverslips onto the stripes. Compress the stripes and the coverslips firmly to seal the sides of the flow cell (Fig. 5).
15. Store the flow cells dry and dust free at room temperature.

3.3.2 Actin Polymerization Assays

It is absolutely crucial to use monomeric actin in the polymerization assay. Even the presence of very small actin multimers will drastically increase the number of growing filaments and therefore impair the analysis. We highly recommend gel filtration of actin using a Superdex 75 column (GE Healthcare) just prior to the experiment. Gel filtered

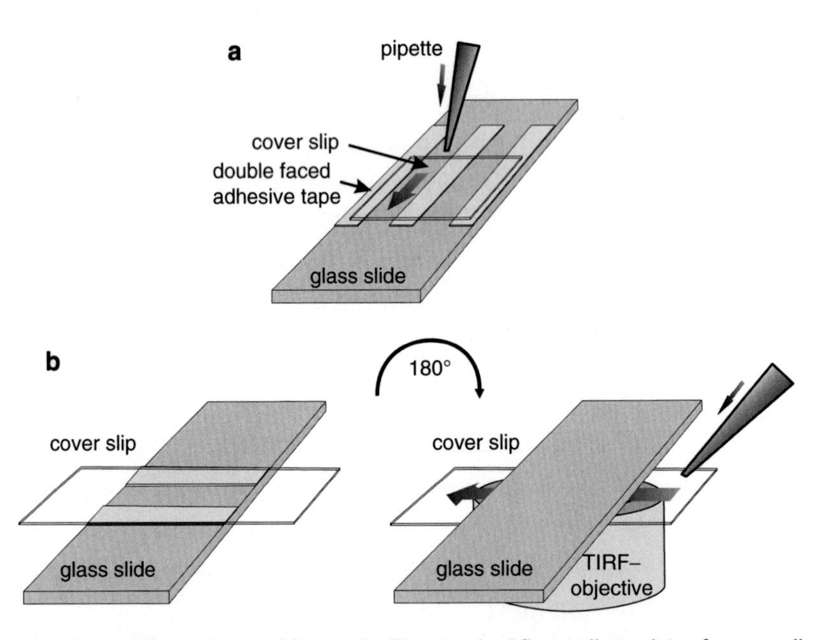

Fig. 5 (**a**) Scheme of two different types of flow cells. The standard flow cell consists of a coverslip attached in parallel orientation to the glass slide. This cell contains two channels to reduce consumption of mPEG-coated coverslips. Due to the restricted spatial access to the channels of the flow cell, this type of flow cell is not well suited for exchange of reaction mixtures during the experiments. (**b**) Attachment of the coverslip perpendicular to the axis of the glass slide allows the infusion and replacement of compounds during the measurement

actin can be used for TIRF polymerization assays for at least 1 week when stored at 4 °C in G-buffer.

1. Mix labeled and unlabeled G-actin to obtain the desired ratio (usually 10–20 % labeling). Add biotinylated actin to obtain a final ratio of 0.1–0.2 % if you intend to immobilize actin filaments on the coverslip surface. Keep this actin mixture on ice.
2. Prepare the reaction mixture (usually 100 μL) in the following order: 49 μL − *x* μL (of accessory protein) − *y* μL (of actin mixture) ddH_2O (depending on the concentrations of actin and the accessory proteins used), 50 μL 2× TIRF buffer, 1 μL O_2 scavenger enzyme solution, and *x*μL of accessory proteins. Incubate the mixture at room temperature for at least 5 min (*see* **Notes 10–12**).
3. In the meantime, wash the flow cell by applying 15–30 μL BSA-buffer to one channel and concomitantly use low-lint tissue to suck it through the cell. Try to avoid air bubbles and incubate the flow cell for 5 min at room temperature.
4. For immobilization of actin filaments on the coverslip surface, load 15–30 μL Streptavidin/BSA-buffer into the pre-blocked flow cell and incubate it at room temperature for 2 min.
5. Add the desired volume of the G-actin mixture (*y*) to the reaction mixture to obtain a final concentration of 1 μM. Mix thoroughly, but prevent formation of air bubbles in the solution. Load 50 μL of the reaction mixture into the flow cell. Add a droplet of immersion oil on the TIRF-objective and place the chamber onto the microscope stage with the coverslip facing down and start recording actin assembly. We recommend exposure times <100 ms to avoid diffusion related blurring and photo bleaching.

4 Notes

1. Currently, the SNAP-Capture magnetic beads are custom made products, which are available at special request from New England Biolabs. We do not recommend the beads listed in the catalog, because they are 75–150 μm in diameter which is excessively too large for use in vitro TIRF-M.
2. If you add hydroxide-ions after addition $MgCl_2$, magnesium hydroxide will be formed, which has a very low solubility in water. Thus, to prevent precipitation of solid magnesium hydroxide adjust to pH 7.2 before adding $MgCl_2$.
3. We also use 100 mM DTT or 40 mM cysteamine hydrochloride as reducing agents instead of β-mercaptoethanol, since we observed variations in stability of the analyzed ABPs depending on the reducing agent.

4. The choice of the fluorescent dye used for labeling actin may affect the readout obtained from the polymerization experiments. Labeling of Cys374 inhibits the interaction of actin with profilin or other G-actin binding proteins and therefore the fluorescent intensities of the filaments may significantly change when assembled for instance by formins. This effect can be utilized to discriminate between spontaneously growing and formin-elongated filaments [7, 23].
5. Alternatively, it may be suitable to label the lysine residues of actin. Currently, NHS-esters variants of the respective fluorophores are the most common reagents. You have to consider that more than one lysine residue is accessible in the actin molecule. Therefore, it is suitable to label F-actin to ensure that your labeled actin is still polymerizable. The amount of fluorescent dye should not be higher than fivefold molar excess to avoid coupling more than one fluorophore to a single actin molecule [25]. Lysine-labeled actin has a considerably higher affinity to profilin as compared to cysteine-labeled actin.
6. Since fluorescent dyes are costly reagents, it is useful to test different conditions such as temperature, incubation time and buffer composition for the labeling reaction in small scale pilot experiments. In any circumstance, for labeling of cysteine residues DTT and azide must be avoided. For the labeling of lysine residues primary amines must be avoided.
7. The correction factor of a fluorescent dye is crucial for accurate calculation of protein concentrations and can be determined by measuring the spectrum of a fluorophore in the respective protein storage buffer alone. The ratio of A_{max}/A_{290} results in the correction factor.
8. The addition of 30 mM of reduced glutathione drastically lowers the pH of the elution buffer. Since these conditions will denature most proteins, it is very important to readjust the pH of the buffer.
9. BSA might be added to a concentration of 1 mg/mL to minimize unspecific binding.
10. For detailed analyses of polymerization kinetics, PBS buffer should be avoided or at least kept at a constant concentration in the experiments, as phosphate alters the polymerization- and ATP-hydrolysis kinetics of actin [26].
11. The fluorescence-quenching O_2-molecules and glucose are converted by glucose oxidase and catalase to gluconic acid and water. Keep in mind that this reaction will lower the pH over time, which in turn might affect protein stability and actin assembly. For long-term experiments >10 min one should control the pH-value of the remaining reaction mixture. Alternatively, one can increase the concentration of the buffer substance.

12. Since the reaction mixture is depleted of oxygen, which in turn is also an effective triplet-state quencher, these unwanted long-lived quantum states have to be quenched by additional reagents to prevent so-called blinking of fluorophores [27, 28]. We are currently optimizing the buffer conditions for methylene blue (Sigma-Aldrich) because it proved to be an efficient anti-blinking reagent. It should be added to a final concentration of ≤5 μM [27]. Alternatively, one may add Trolox (6-hydroxy-2,5,7,8-tetramethylchroman-2-carboxylic acid, Sigma-Aldrich) to a final concentration of ≤1 mM [28].

Acknowledgements

This work was supported by the Deutsche Forschungsgemeinschaft (DFG) by grants to J.F (FA-330/9-1 and FA-330/10-1). We additionally thank Laurent Blanchoin for fruitful discussions.

References

1. Blanchoin L, Boujemaa-Paterski R, Sykes C, Plastino J (2014) Actin dynamics, architecture, and mechanics in cell motility. Physiol Rev 94:235–263
2. Faix J, Steinmetz M, Boves H et al (1996) Cortexillins, major determinants of cell shape and size, are actin-bundling proteins with a parallel coiled-coil tail. Cell 86:631–642
3. Kondo H, Ishiwata S (1976) Uni-directional growth of F-actin. J Biochem 79:159–171
4. Pollard TD (1986) Rate constants for the reactions of ATP- and ADP-actin with the ends of actin filaments. J Cell Biol 103:2747–2754
5. Kouyama T, Mihashi K (1981) Fluorimetry study of N-(1-pyrenyl)iodoacetamide-labelled F-actin. Local structural change of actin protomer both on polymerization and on binding of heavy meromyosin. Eur J Biochem 114:33–38
6. Amann KJ, Pollard TD (2001) Direct real-time observation of actin filament branching mediated by Arp2/3 complex using total internal reflection fluorescence microscopy. Proc Natl Acad Sci U S A 98:15009–15013
7. Kovar DR, Harris ES, Mahaffy R et al (2006) Control of the assembly of ATP- and ADP-actin by formins and profilin. Cell 124:423–435
8. Breitsprecher D, Kiesewetter AK, Linkner J et al (2008) Clustering of VASP actively drives processive, WH2 domain-mediated actin filament elongation. EMBO J 27:2943–2954
9. Breitsprecher D, Kiesewetter AK, Linkner J et al (2011) Molecular mechanism of Ena/VASP-mediated actin-filament elongation. EMBO J 30:456–467
10. Breitsprecher D, Koestler SA, Chizhov I et al (2011) Cofilin cooperates with fascin to disassemble filopodial actin filaments. J Cell Sci 124:3305–3318
11. Smith BA, Padrick SB, Doolittle LK et al (2013) Three-color single molecule imaging shows WASP detachment from Arp2/3 complex triggers actin filament branch formation. Elife 2:e01008
12. Breitsprecher D, Jaiswal R, Bombardier JP et al (2012) Rocket launcher mechanism of collaborative actin assembly defined by single-molecule imaging. Science 336:1164–1168
13. Bilancia CG, Winkelman JD, Tsygankov D et al (2014) Enabled negatively regulates diaphanous-driven actin dynamics in vitro and in vivo. Dev Cell 28:394–408
14. Axelrod D (1981) Cell-substrate contacts illuminated by total internal reflection fluorescence. J Cell Biol 89:141–145
15. Axelrod D (2001) Total internal reflection fluorescence microscopy in cell biology. Traffic 2:764–774
16. Axelrod D (2013) Evanescent excitation and emission in fluorescence microscopy. Biophys J 104:1401–1409
17. Kuhn JR, Pollard TD (2005) Real-time measurements of actin filament polymerization by total internal reflection fluorescence microscopy. Biophys J 88:1387–1402
18. McCullough BR, Blanchoin L, Martiel JL, De la Cruz EM (2008) Cofilin increases the bending flexibility of actin filaments: implications for severing and cell mechanics. J Mol Biol 381:550–558

19. Suarez C, Roland J, Boujemaa-Paterski R et al (2011) Cofilin tunes the nucleotide state of actin filaments and severs at bare and decorated segment boundaries. Curr Biol 21:862–868
20. Rottner K, Behrendt B, Small JV, Wehland J (1999) VASP dynamics during lamellipodia protrusion. Nat Cell Biol 1:321–322
21. Disanza A, Bisi S, Winterhoff M et al (2013) CDC42 switches IRSp53 from inhibition of actin growth to elongation by clustering of VASP. EMBO J 32:2735–2750
22. Keppler A, Gendreizig S, Gronemeyer T et al (2003) A general method for the covalent labeling of fusion proteins with small molecules in vivo. Nat Biotechnol 21:86–89
23. Winterhoff M, Junemann A, Nordholz B et al (2014) The Diaphanous-related formin dDia1 is required for highly directional phototaxis and formation of properly sized fruiting bodies in *Dictyostelium*. Eur J Cell Biol 93:212–224
24. Block J, Breitsprecher D, Kühn S et al (2012) FMNL2 drives actin-based protrusion and migration downstream of Cdc42. Curr Biol 22:1005–1012
25. Isambert H, Venier P, Maggs AC et al (1995) Flexibility of actin filaments derived from thermal fluctuations. Effect of bound nucleotide, phalloidin, and muscle regulatory proteins. J Biol Chem 270:11437–11444
26. Fujiwara I, Vavylonis D, Pollard TD (2007) Polymerization kinetics of ADP- and ADP-Pi-actin determined by fluorescence microscopy. Proc Natl Acad Sci U S A 104:8827–8832
27. Schäfer P, van de Linde S, Lehmann J et al (2013) Methylene blue- and thiol-based oxygen depletion for super-resolution imaging. Anal Chem 85:3393–3400
28. Cordes T, Vogelsang J, Tinnefeld P (2009) On the mechanism of Trolox as antiblinking and antibleaching reagent. J Am Chem Soc 131:5018–5019

Chapter 22

Quantitative Monitoring Spatiotemporal Activation of Ras and PKD1 Using Confocal Fluorescent Microscopy

Xuehua Xu, Michelle Yun, Xi Wen, Joseph Brzostowski, Wei Quan, Q. Jane Wang, and Tian Jin

Abstract

Receptor activation upon ligand binding induces activation of multiple signaling pathways. To fully understand how these signaling pathways coordinate, it is essential to determine the dynamic nature of the spatiotemporal activation profile of signaling components at the level of single living cells. Here, we outline a detailed methodology for visualizing and quantitatively measuring the spatiotemporal activation of Ras and PKD1 by applying advanced fluorescence imaging techniques, including multichannel, simultaneous imaging and Förster resonance energy transfer (FRET).

Key words Confocal fluorescence microscopy, Ras activation, Protein kinase D 1 (PKD1), Phosphorylation of PKD1, Förster resonance energy transfer (FRET)

1 Introduction

Chemotaxis, a chemoattractant gradient-directed cell migration, plays critical roles in many physiological processes such as recruitment of neutrophil to the sites of inflammation, neuron patterning, immune response, angiogenesis, and metastasis of cancer cells. Eukaryotic cells detect many chemoattractants by different receptors, including G protein-coupled receptors (GPCRs). Ras GTPases are one of the most evolutionarily conserved proteins among eukaryotes and play essential roles in many signaling pathways. By coupling with many different receptors, Ras GTPases are universal molecular switches that act as kinetic timers of signal transduction events upon extracellular stimulation. The engagements of various chemoattractants to their GPCRs generally trigger Ras activation in neutrophil and neutrophil-like HL60 cells [1]. It is important to measure the spatiotemporal dynamics of Ras activation in response to various fields of chemoattractant stimuli in neutrophils.

Tian Jin and Dale Hereld (eds.), *Chemotaxis: Methods and Protocols*, Methods in Molecular Biology, vol. 1407, DOI 10.1007/978-1-4939-3480-5_22,

The protein kinase D family of serine/threonine kinases plays critical roles in many physiological processes, including cell growth, protein trafficking, and lymphocyte biology [2]. A recent study has shown that PKD1 phosphorylates SSH1 and regulates cofilin to mediate F-actin depolymerization in breast cancer cells [3]. PKD isoforms are also known to be highly expressed in immune cells, including neutrophils [4]. We have shown that PKD and its spatiotemporal activation play essential roles in neutrophil chemotaxis [5]. Though several methods had been developed to measure the activation profile of PKDs [6–8], it is becoming increasingly important to measure the spatiotemporal activation of PKD levels in single live cells from different tissue origins.

Here, we report two imaging methods that have been successfully applied to measure spatiotemporal changes of signaling events in single live cells in real time. First, we describe how to simultaneously visualize multiple signaling events upon application of different fields of chemoattractant stimulation. Next, we describe how to apply the Förster resonance energy transfer (FRET) imaging technique to monitor and measure PKD1 activation in single live cells using two different intramolecular FRET probes, DKAR and CFP-PDK1-YFP (hereafter referred to as “PKD1-CY”).

2 Materials

2.1 HL60 Cell Culture

1. RPMI 1640 medium GlutaMAX™ Supplement (Life Technologies, Grand Island, NY, #61870-036).
2. Fetal bovine serum (FBS) (Gemini Bio-Products, West Sacramento, CA).
3. HEPES 1 M sterile solution, pH 7.3 (Quality Biological Inc. Gaithersburg, MD).
4. Penicillin (10,000 U/ml) and streptomycin (10 mg/ml) solution (Gibco, Invitrogen Life Science).
5. RPMI 1640 complete culture media (RPMI 1640 medium GlutaMAX™ Supplement with 20 % FBS, 25 mM HEPES, and 1 % Penicillin/Streptomycin).
6. RPMI 1640 starving media (RPMI 1640 medium GlutaMAX™ Supplement with 25 mM HEPES).
7. T75 flask (BD Biosciences, Falcon).
8. HL-60 cell line (ATCC. Manassas, VA).
9. Cell culture incubator, 37 °C, 5 % CO_2.
10. Cell culture hood (NUNR Class II biological safety cabinet typeA/B3, Plymouth, MN).

11. *N*-Formyl-Met-Leu-Phe (fMLP) (Sigma, St. Louis, MO). A 1 mM stock is made in DMSO and frozen at −20 °C in 10 μl aliquots. The working concentration of fMLP is 100 nM–1 μM.
12. Alexa 594 (Molecular Probes, Eugene, OR). A 1 mg/ml stock is made in water and frozen at −20 °C in 10 μl aliquots. The working concentration is 0.1 mg/ml.

2.2 Plasmids for Ras Activation and PKD1 FRET Probe

1. Plasmid encoding an active Ras binding domain (RBD) tagged with GFP (Addgene, plasmid # 18664).
2. Plasmid encoding Ras tagged with RFP (John F. Hancock, University of Texas Medical School at Houston).
3. Plasmids encoding CFP-CXCR1 (single donor control) and CXCR1-YFP (single acceptor control) (T. Jin, NIAID/NIH).
4. Plasmid encoding the FRET probe PKD1-CY (Q. Jane Wang, University of Pittsburg).

2.3 Transfection of HL60 Cells

1. Cell Line Nucleofector® Kit V (Lonza, Basel, Switzerland).
2. Plastic bulb sterile pipette included in Nucleofector® Kit.
3. 1.6 ml tube (Neptune, #4445.S.X).

2.4 Coating Cover-Glass Surface of 4-Well Chamber

1. 10 % Gelatin solution (Sigma-Aldrich, St. Louis, MO, #G1393).
2. HBSS (Hanks' Balanced Salt Solution, Life Technologies, Grand Island, NY).
3. Single- and four-well Lab-Tek II cover-glass chambers (Nalge Nunc International, Naperville, IL).

2.5 MDA-MB-231 Cell Culture

1. DMEM Medium with high glucose, l-glutamine, and phenol red (Gibco, Invitrogen).
2. Fetal bovine serum (FBS) (Gemini Bio-Products, West Sacramento, CA).
3. Penicillin/streptomycin stock solution (Gibco, Invitrogen Life Science, #15140). Penicillin (10,000 U/ml) and streptomycin (10 mg/ml) solution is frozen at −20 °C in 5 ml aliquots.
4. DMEM complete culture medium. 500 ml DMEM Medium with high glucose, l-glutamine, and phenol red containing DMEM Medium with 10 % FBS and 5 ml penicillin/streptomycin stock solution.
5. Trypsin–EDTA (0.05 %) (Gibco, Life Technologies, Cat# 25300-054).
6. PBS 10× (KD Medical, Cat# RGF-3210).
7. DEPC Treated Water (Quality Biological, Cat# 351-068-131).
8. 1× PBS. Combine 50 ml of 10× PBS with 450 ml of DEPC-treated water.

9. T75 flask (BD Biosciences, Falcon).
10. Cell lines: MDA-MB-231 cells.
11. Cell culture incubator, 37 °C, 5 % CO_2.
12. Cell culture hood (NUNR Class II biological safety cabinet type A/B3, Plymouth, MN).

2.6 Transfection of MDA-MB-231 Cells by Lipofectamine 2000

1. Lipofectamine 2000 Reagent (Invitrogen, Cat# 11668-027).
2. Opti-MEM (Invitrogen, Cat#51985091).
3. Plasmids: CFP, YFP, CFP-PKD1-YFP (Q. Jane Wang, University of Pittsburg).
4. DMEM transfection medium (DMEM medium with 10 % FBS).
5. DMEM complete culture medium.
6. 6-Well cell culture plate (Costar, Corning Incorporated, Cat# 07-200-80).
7. Trypsin–EDTA (0.05 %) (Gibco, Life Technologies, Cat# 25300-054).
8. 1× PBS. Combine 50 ml of 10× PBS with 450 ml of DEPC-treated water.
9. 4-well chamber with cover glass (Lab-Tek, Nunc, Cat# 155383).
10. Dimethyl Sulfoxide (Sigma-Aldrich).
11. Phorbol 12-Myristate 13-Acetate (Sigma-Aldrich).

2.7 Imaging with Carl Zriss Zen710 Microscope

1. Four-well Lab-Tek II coverglass chambers (Nalge Nunc International, Naperville, IL).
2. FemtoJet microcapillary pressure supply (Eppendorf, Germany).
3. TransferMan NK2 micromanipulator (Eppendorf, Germany).
4. Zen 710 or equivalent with 40× 1.3 NA objective lens.
5. Other microscope accessories:
   (a) Heating Insert P Lab-Tek S1 (Pecon, Part# 131-800 029).
   (b) TempModule S1 (Pecon, Part# 800-800 000).
   (c) Zen 2010 Software, Macro FRET 55.
   (d) Carl Zeiss Immersol 518 F (Zeiss).
   (e) Lasos Ar-lon Laser Series (Zeiss, Model LGN 4000, Remote Control RMC 7812 Z2).
   (f) Controller pad (Wienecke & Sinske GmbH, Serial# 12-0902305, Gleichen, Germany).

3 Methods

3.1 Simultaneous Monitoring of Ras Activation in Response to Various Visible Fields of Chemoattractant Stimulation

3.1.1 Cell Culture

1. Culture HL60 cells in RPMI 1640 medium with 20 % FBS, 1 % Penicillin and streptomycin containing 25 mM HEPES and passage them every 2 or 3 days (*see* **Note 1**).

3.1.2 Coating the Cover-Glass Surface of the 4-well Chamber

1. Dilute 10 % Gelatin Stock Solution with 1× HBSS to final concentration of 0.2 % gelatin.
2. Add 1 ml 0.2 % gelatin in HBSS to the two middle wells of a 4-well chamber bottomed with cover glass. Incubate the chamber at 37 °C for 1 h.
3. Just before seeding cells, remove the gelatin solution from 4-well chamber (*see* **Note 2**).

3.1.3 Transfection of HL60 Cells

1. Differentiate HL60 cells in 1.3 % DMSO for 5 days. Cell density is measured before transfection (*see* **Note 3**).
2. Centrifuge the cells at 200 × *g* for 5 min.
3. After removing the supernatant, suspend the cells in transfection reagent according to the manufacturer's recommendation (usually two million cells in 100 μl of reagent per transfection) and add the mixture to electroporation cuvette.
4. Use the manufacturer's program T-019 to perform the electroporation.
5. Without delay, gently add 500 μl RPMI 6140 complete culture medium to the electroporated cells in the cuvette.
6. Gently transfer the cells from the cuvette to a 1.6 ml tube with the provided plastic pipettes.
7. Place the tube into a cell culture incubator for 30 min.
8. Seed a 100 or 500 μl of cells into pre-coated wells of 4-well or 1-well chamber, respectively (*see* **Note 4**).
9. Add culture medium up to a final volume of 1 or 4 ml to one well of a four-well chamber or a one-well chamber, respectively.
10. After allowing the cells to recover for 3 h in the incubator, gently remove all of the culture medium from the well of chamber and replace it with RPMI 1640 starving medium (*see* **Note 5**).

11. One hour later, the cells are ready for imaging experiments.
12. Meanwhile, prepare 100 nM or 1 μM fMLP stimuli (*see* **Note 6**).

3.1.4 Simultaneous Monitoring Application of Chemoattractant Stimuli and Ras Activation in Response to Chemoattractant Stimulation

Chemotactic cells are able to respond to small spatiotemporal differences in concentrations of chemoattractants. We have developed a method that allows us to indirectly visualize the spatiotemporal changes of an applied cAMP stimulus around cells by mixing the chemoattractant with a fluorescent dye and measuring the changes in fluorescence intensity [9]. This method also allows us to measure the difference of a chemoattractant gradient to which cells are exposed or to measure the spatiotemporal application of stimuli exposed to cells. Moreover, by using cells expressing a different emission fluorescent marker, we are able to simultaneously measure multiple events, including application of chemoattractant stimuli and cellular dynamics in response to the stimuli. Here, we present different procedures to apply several kinds of fMLP stimuli to neutrophil like cells, HL60, using live cell imaging experiments.

1. Sudden exposure of a cell to uniform fMLP stimulation. To generate a uniform stimulation, 200 μl of a mixture of fMLP and Alexa 633 (0.1 μg/μl) is applied to the cells covered with 800 μl buffer and placed in a four-well chamber (Fig. 1a, b). Under this condition, cells are normally exposed to a uniform fMLP stimulation that was about 2.5-fold of the final concentration.
2. Exposure of a cell to a steady fMLP gradient. To establish a fMLP gradient, a micropipette (Eppendorf Femtotips) is filled with 30 μl solution of fMLP and Alexa 633 and then attached to a FemtoJet and micromanipulator 5171 (Eppendorf). FemtoJet is set at Pc = 70 and Pi = 70 to inject a constant and tiny volume of a mixture of fMLP and Alexa 633 into a one-well chamber that is covered with 5 ml buffer. Under this condition, a gradient can be established with 100 μm around the tip of the micropipette and is usually stable for more than 1 h (Fig. 1c; *see* **Note 7**).

We presented three channel setting (DIC, red, and far red) here and observed a clear membrane localization of H-Ras in HL60 cells as previously reported [10]. If necessary, it can be set up to five channels by adding blue and green emission channels [9]. Although these procedures are well established, it is recommended to optimize the channel distribution by the abundance and biological relevance of the proteins, the options and photon yields of the labeling fluorophores, and other necessary considerations for the experiments. For the fluorescent dye, it is has to be

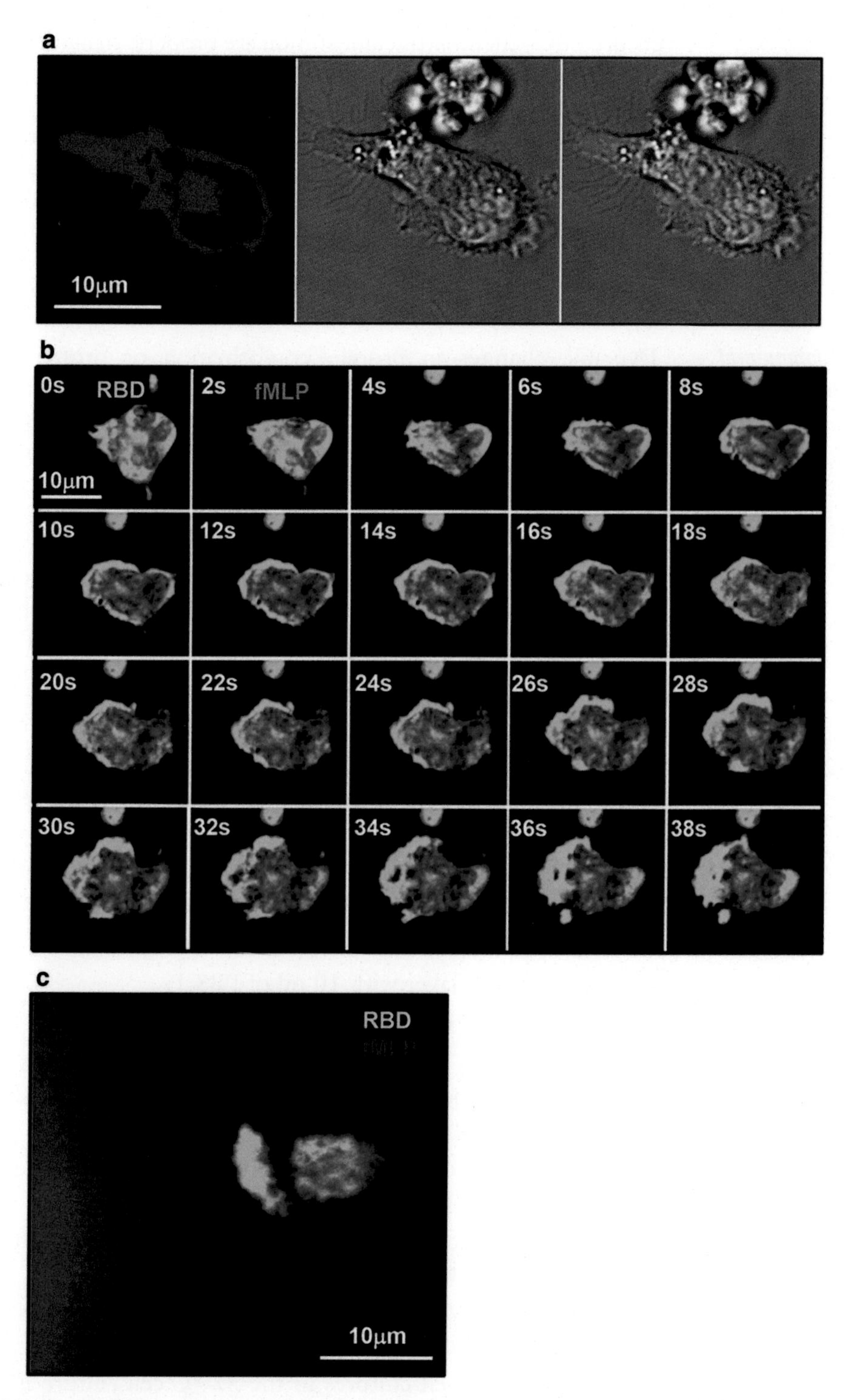

Fig. 1 GPCR-mediated Ras activation in HL60 cells monitored by active Ras probe. (**a**) Cellular localization of H-Ras in HL60 cells. HL60 cells express mRFP-HRas(G12V). (**b**) Activation of Ras in HL60 cells upon chemoattractant fMLP stimulation. HL60 cells expressing 100 nM fMLP is mixed with Alexa 644. (**c**) An enriched activation of Ras in the leading edge of a chemotaxing HL 60 cells. A final concentration of 1 μM fMLP is mixed with Alexa 644 (0.1 μg/ml)

ensured that chemoattractant stimuli are properly delivered in each experiment. It is also important to select an appropriate fluorescent dye to monitor a chemoattractant. Because the delivery procedures depend on the diffusion properties of a chemoattractant, the selected dyes should be compared with the chemoattractants under the experimental conditions. The dye with a diffusion coefficient similar to that of the chemoattractant should be used in order to accurately measure the concentrations of the chemoattractant in live cell experiments.

3.2 Spatiotemporal Monitoring PKD1 Activation by Förster Resonance Energy Transfer

Dynamic changes in protein conformation or protein–protein interaction can be monitored in single living cells in real time using FRET imaging methodologies. FRET detection and its efficiency depend upon the distance between the donor and acceptor fluorophores, which ranges between 10 and 100 Å. A good fret pair would have an efficient overlapping of donor CFP emission spectrum with the excitation spectrum of acceptor YFP (>30 %). Previously, we presented a spectral-resolved FRET measurement of heterotrimeric G protein activation in subcellular level in real time [9, 11]. Below, we describe the two different FRET probes designed by different principles to spatiotemporally measure PKD1 activation upon extracellular stimuli.

3.2.1 MDA-MB-231 Cell Culture

1. Culture the MDA-MB-231 cells in DMEM complete medium containing DMEM medium, 10 % FBS and 1 % Penicillin/ Streptomycin.
2. Monitor the cells daily. When they reach a density of 85–95 % confluency, passage the cells every 2–4 days as follows.
3. Remove culture medium using vacuum.
4. Rinse the cell layer with 10 ml of PBS 1×.
5. Remove the PBS with vacuum.
6. Rinse the cell layer with 2 ml of 0.05 % Trypsin–EDTA. To facilitate dissociation of the cell layer, place the flask in the 37 °C cell culture incubator for 1 min.
7. Lightly tap the flask and confirm that the cells have detached from the flask under the microscope.
8. Add 10 ml of fresh DMEM growth medium and gently pipette up and down to prevent clumping of cells.
9. Aliquot the MDA-MD-231 cells to new T75 flasks at subcultivation ratios of 1:2 to 1:6 as desired.
10. Incubate the cell cultures in the 37 °C cell culture incubator with 5 % CO_2.

3.2.2 Transfection of MDA231 Cells with CFP, YFP, and CFP-PKD1-YFP

1. Cell cultures should be prepared so that a T75 flask of MDA-MB-231 cells is 90 ~ 95 % confluent 3 days before the planned imaging experiment.
2. Remove the culture medium from T75 flask using vacuum and gently rinse the cell layer with 10 ml of 1× PBS and remove the PBS using vacuum.
3. To dissociate the cell layer, transfer 2 ml of 0.05 % Trypsin–EDTA and place the flask in the 37 °C cell culture incubator for 1 min. Lightly tap the flask and confirm that the cells have detached from the flask under the microscope. Then, add 6 ml of fresh DMEM culture medium and gently pipette up and down to prevent clumping of cells.
4. Add 1 ml of the cell suspension to each well of a 6-well cell culture plate and incubate the plated cells in the 37 °C cell culture incubator with 5 % CO_2 overnight.
5. In three 1.5 ml Eppendorf tubes, dilute 25 μl of Lipofectamine 2000 Reagent (room temperature; mix gently before use) in 250 μl of Opti-MEM (room temperature).
6. In another three tubes, dilute 10 μl of CFP, YFP, or PKD1-CY plasmids in 250 μl of Opti-MEM (room temperature). After the 5 min of incubation, combine the diluted Lipofectamine 2000 and diluted DNA. Gently pipette to mix and Incubate for 20 min at room temperature.
7. Add the DNA-Lipofectamine complexes drop by drop into 6-well plates of cells. As the solution is added, gently swirl the plate to evenly distribute to the cells.
8. After 3–4 h of incubation in 37 °C cell culture incubator, change the medium to fresh DMEM growth medium (10 % FBS and 1 % Penicillin/Streptomycin). Continue incubation overnight in 37 °C cell culture incubator.
9. Aliquot the cells obtained from one well of the 6-well plates into 4-well chambers.
10. Gently pipette out the DMEM medium from each well of the 6-well plate.
11. Wash each well with 2 ml of PBS 1× into each well and add 0.5 ml of 0.05 % Trypsin–EDTA into each well and place in 37 °C cell culture incubator for 1 min. Tap gently and view under microscope to confirm all cells have detached and then add 1 ml of culture medium into each well.
12. Centrifuge the cells at 300 rpm for 5 min and discard the supernatant.

13. Suspend the cells with 1.5 ml fresh DMEM culture medium and aliquot 180 μl into the center of each well of the 4-well Lab-Tek Chambers (only the middle two wells of each Lab-Tek Chamber were used for convenience during imaging). Note: eight total wells per plasmid were used. And then incubate overnight in 37 °C cell culture incubator.
14. Four hours before imaging, add 820 μl of fresh DMEM culture medium to each well and prepare 2 μM PMA stimuli for imaging experiment.

3.2.3 Sensitized Emission FRET Imaging Monitors PKD1 Activation Using Intramolecular FRET Probe DKAR

FRET Sensitized Emission is used for measuring the actual FRET efficiency. Same as the spectrally resolved FRET measurement, FRET Sensitized Emission is widely used to measure protein–protein interaction in live cells. Particularly, FRET Sensitized Emission provides a powerful approach to detect protein–protein interaction (FRET measurement) in fixed cells or tissue samples in which a dynamic of FRET is either not possible or necessary.

Same as spectral resolved FRET detection, sensitized emission FRET measurement also requires three types of cells: donor cells containing only the donor fluorophore; receptor cells containing only the acceptor fluorophore; and FRET cells containing both donor and acceptor fluorophores. FRET measurement starts with the acquisition of images in three channels under Channel Mode: (1) Donor channel: Excitation with the donor's excitation wavelength and detection with donor's parameter settings. (2) FRET channel: Excitation with donor's excitation wavelength and detection with acceptor's parameter settings. (3) Acceptor channel: Excitation with acceptor's excitation wavelength and detection with acceptor's parameter settings. The two single labeled samples provide the reference values to correct the optical bleed through of the donor (donor emission bleeds into acceptor emission) and the possible excitation of the acceptor through the wavelength of the donor. The third sample is the actual FRET sample on which the measurement is performed and measured.

3.2.4 Acquisition Configuration of Sensitized Emission FRET Measurement by Applying CFP/YFP FRET Pairs

Sensitized emission FRET measurement calculates the breed through of donor and cross-talk of acceptor which must be obtained in cells expressing single fluorophore, as an example here are cells expressing CFP or YFP, respectively. The key point of Sensitized Emission FRET image acquisition is that all the imaging parameters (laser output, zoom, pinhole, detector gain, and amplifier offset and so on) should be kept constant.

1. Set the system to the Channel acquisition mode and configure Donor CFP channel, FRET channel, and Acceptor YFP channel as follows:

(a) *CFP channel*: Excitation with 458 nm laser and application of NFT 515 nm to split the signal and detection with parameter settings of donor (BP 480–520 nm).

(b) *FRET channel*: Excitation with 458 nm laser and detection with parameter settings of acceptor (BP 500–550 nm).

(c) *YFP channel*: Excitation with 514 nm laser and detection with parameter settings of acceptor (BP 500–550 nm).

2. Adjust the pinhole size for each channel as an optical slice of the specific type of cells. Here, we set it to about 3 μm.
3. To image single CFP expressing cells, adjust the detector gain and amplifier offset of the donor and acceptor channels to obtain a decent signal. Imaging parameters should be kept constant. FRET channel will show the bleed-through from the donor CFP emission channel using donor CFP excitation wavelength (Fig. 2b, upper panel) (*see* **Note 8**).
4. For single YFP expressing cells, the FRET channel will show the cross talk from the acceptor YFP emission channel using donor CFP excitation wavelength (Fig. 2b, second panel) (*see* **Note 9**).
5. For CFP/YFP double expressing cells, select those with both CFP and YFP expression (*see* **Note 10**).
6. Set up time series acquisition as detailed in **step 1**.
7. After two or three scans, add stimuli homogenously to the cells and continue the acquisition (*see* **Notes 11** and **12**).

3.2.5 Quantitative Calculation of Sensitized Emission FRET Using Carl Zeiss Zen FRET plus 55

We followed the data analysis procedures of Zeiss Zen FRET Plus 55. During the data analysis, we found that following key points should be carefully followed for accurate data analysis:

1. Be aware that whole sets of data for sensitized emission FRET data analysis should be acquired with the exact same acquisition parameters.
2. To set up background threshold of donor single, acceptor single, or FRET samples, choose the areas without any signal.
3. To set up bleed through ratio in donor/FRET/acceptor channels, choose the areas without any saturated pixels. We recommend a medium range intensity for this setup.
4. We recommend choosing the ROIs (regions of interest) based on the biology of the interested protein.

3.2.6 FRET Loss of KDAR Indicates PKD1 Activation upon Stimulation

The protein kinase D family of serine/threonine kinases plays critical roles in many physiological processes, including cell growth, protein trafficking, and lymphocyte biology [2]. It is

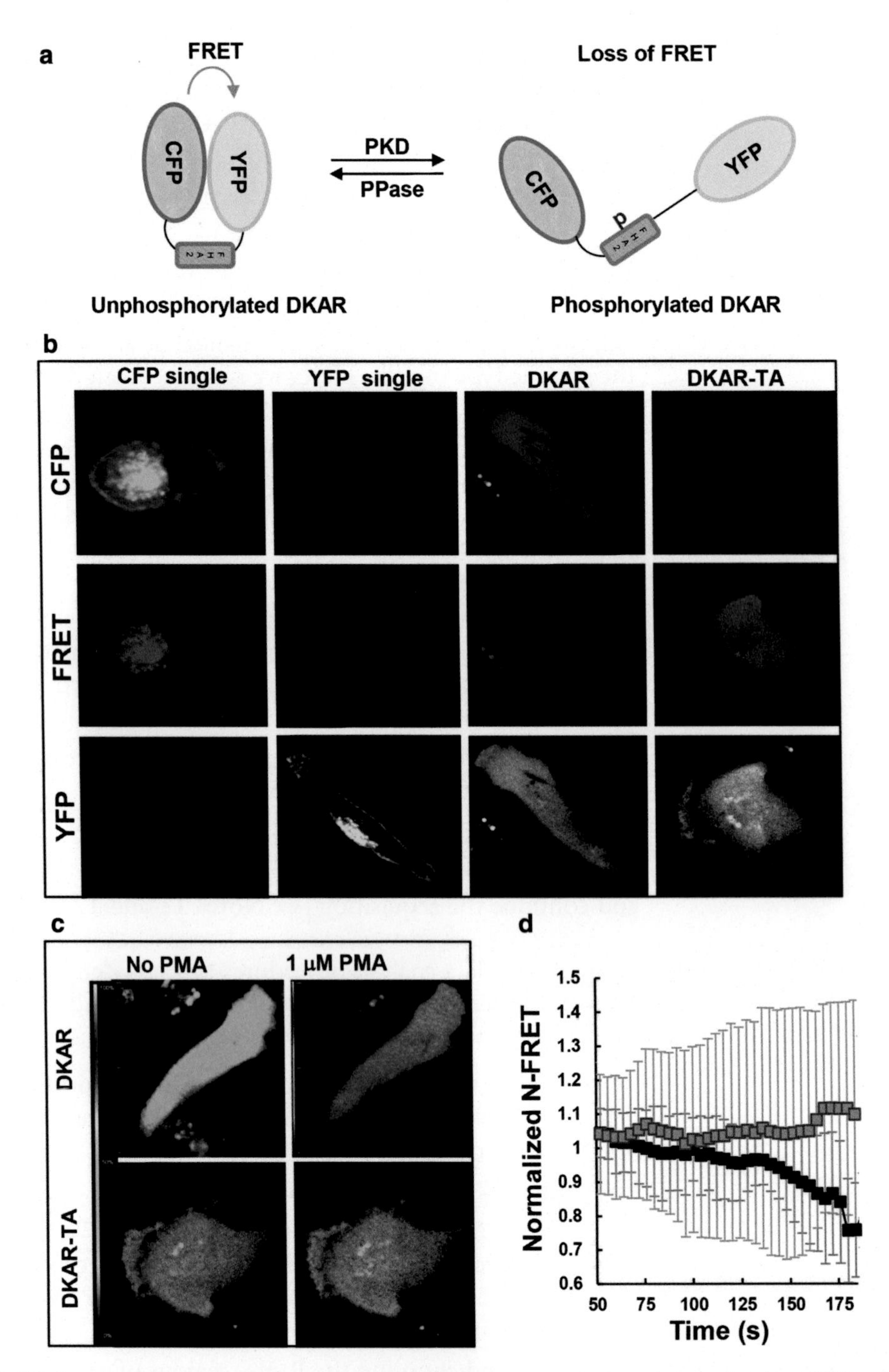

Fig. 2 PKD1 activation monitored by spectrum FRET measurement of an intramolecular FRET probe CFP-PKD1-YFP. (**a**) Scheme shows the principle of the applied FRET probe DKAR for PKD1 activation. The peptide of PKD1 substrates is inserted between FRET pair CFP and YFP. Phosphorylation of the peptide by PKD1 results in loss of FRET between CFP and YFP. DKAR-TA encodes PKD1 substrate peptide with mutation and serves as a negative control for DKAR. (**b**) Configuration and channel settings of CFP, FRET, and YFP for sensitized emission FRET measurement of CFP single control, YFP single control, DKAR, and DKAR-TA using confocal fluorescent microscopy. (**c**) N-FRET image of PKD1 is processed and obtained by Zen710 FRET plus 55. (**d**) The dynamic change of the normalized N-FRET upon stimulation is presented in (**d**). Mean ± SD is shown

important to measure the activation profile of PKDs in different type of mammalian cells. There are several methods had been developed to detect PKD activation [6–8]. Confocal imaging approaches, including applying FRET technique, had been used for PKD activation as well. An intramolecular FRET probe designed by inserting a substrate peptide of PKD1 between CFP/YFP FRET pairs (Fig. 2a) [6, 8]. Using sensitized emission FRET measurement, we also measured a FRET loss of DKAR and confirmed the feasibility of FRET measurement in MDA-MD-231 cells at the level of whole cell (Fig. 2b–d). Consistent with previously reported [6, 8], we measured a FRET loss of DKAT upon PMA treatment while its negative control DKAR-TA, in which serine residue is replace by arginine, did not display a FRET loss (Fig. 2d). The advantage of this FRET probe is able to measure kinase activity of the endogenously expressed PKD1 upon stimulation. However, there are two major disadvantages of DKAR probe: (1) the specificity issue of the substrate peptide in the construct, (2) the lack of spatial information of the measurement due to the diffusion nature of this FRET probe. These two disadvantages do not meet the increasing demands on the understanding of the temporal-spatial activation and regulation of PKDs.

3.2.7 Monitoring Spatiotemporal PKD1 Activation Using Intramolecular FRET Probe PKD1-CY

Phosphorylation mediates activation and conformational change of PKD1 [2, 12, 13]. The underlying mechanisms of PKD1 activation depend on the relief of autoinhibition by the PH domain following activation-loop phosphorylation [12, 14]. Hence, we construct an intramolecular FRET probe by inserting PKD1 protein between CFP/YFP FRET pairs (CFP-PKD1-YFP, also PKD1-CY) to measure the tempo-spatial dynamics of PKD1 activation (Fig. 3a). Upon PMA stimulation PKD1-CY displays a clear membrane translocation (Fig. 3c), confirming the functionality of FRET probe and, more importantly, the subcellular localization of PKD1. Along with membrane translocation, we detected a clear FRET gain of PKD1-CY, which is inhibited by PKD specific inhibitor CID755673 (Fig. 3d; *see* **Note 13**). This result indicates that FRET gain can be used to measure PKD1 activation. More importantly, simultaneous monitoring of the subcellular location of PKD1-CY and FRET change upon PMA stimulation provides an efficient way to visualize and quantitatively measure the spatiotemporal dynamics of PKD1 activation at the subcellular level upon extracellular stimulation.

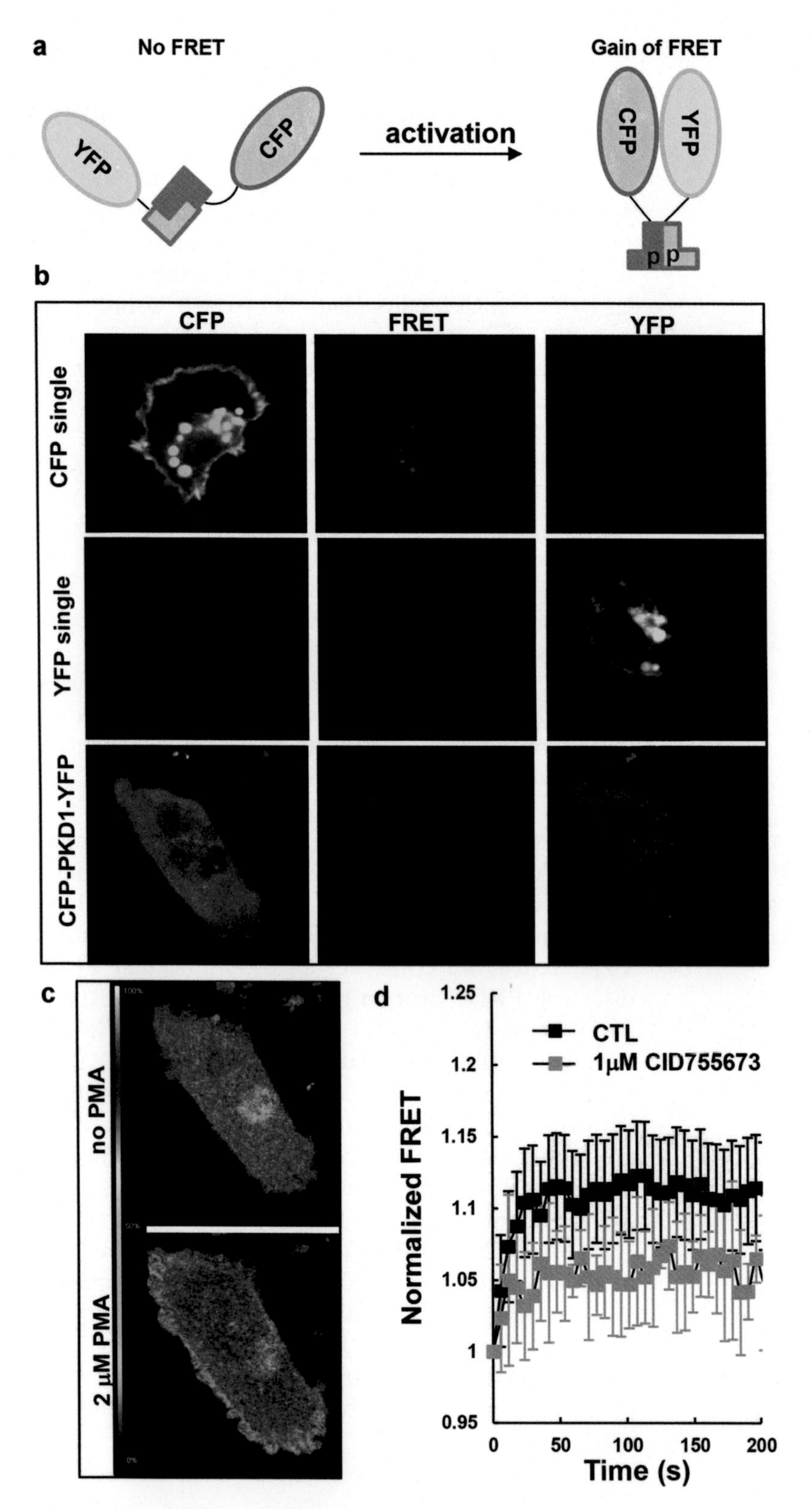

Fig. 3 PKD1 activation monitored by sensitized emission FRET measurement of an intramolecular FRET probe CFP-PKD1-YFP. (**a**) Scheme shows the possible mechanism of the applied FRET probe CFP-PKD1-YFP for PKD1 activation. PKD1 is inserted between FRET pair CFP and YFP and is designated as PKD1-CY. Phosphorylation of PKD1 results in its conformational change which cause the gain of FRET between CFP and YFP.

4 Notes

1. It is not recommended that HL60 cells are passaged more than 30 times.
2. About 5–10 % differentiated cells adhere to gelatin coated cover glass. We suggest a thorough removal of excess gelatin since we often saw less HL60 cells adhere to the cover glass containing floating gelatin.
3. After 5 days of the differentiation, cell densities vary from 1 to 2.5×10^6 cells/ml. The best chemotactic cells might be found at days 5 or 6 after differentiation. Best guess bases on the morphological changes from the round-shape undifferentiated cells to an oval-like polarized, chemotactic cells.
4. $3–5 \times 10^5$ HL60 cells in an appropriate volume are usually seeded into one well of four-well chamber for the experiments. Up to 1 ml of medium is added to one well of 4-well chamber for slightly longer time of incubation or starvation.
5. For stimulation experiment, cells are much more responsive after a short period of starvation. In order to starve cells, RPMI1640 complete culture medium is replaced by RPMI1640 starving medium without FBS. Caution: to remove medium from the cells already seeded in either 4-well or 1-well chamber, we tried several different ways to figure out a best way to exchange the medium without washing away the adhering cells. We found that directly flipping the chamber to dump the excess medium works well.
6. To visualize the application of chemoattractant stimuli to the cells, a final concentration of 100 nM fMLP is mixed with Alexa 633 (final concentration is 0.1 μg/μl). For gradient experiment, a final concentration of 1 μM fMLP is mixed with Alexa 633 (0.1 μg/μl). The selected dyes should be compared under the experimental conditions. It is important to select proper fluorescence dyes to monitor a chemoattractant. Because the delivery procedures depend on the diffusion, diffusion properties of a chemoattractant, the selected dyes should be compared with the chemoattractant under experimental conditions. The dye with similar diffusion efficiency as that of the chemoattractant should be used in order to accurately measure the concentrations of the chemoattractant in live cell experiments.
7. We often experience blockage of the needle of microinjector during chemotaxis assay o fHL60 cells. To avoid this problem, we suggest to remove the non-adhering cells by change the medium prior to the experiments.

Fig. 3 (continued) (**b**) Configuration and channel settings of CFP, FRET, and YFP for sensitized emission FRET measurement of CFP single control, YFP single control, and PKD1-CY using confocal fluorescent microscopy. (**c**) N-FRET image of PKD1 is processed and obtained by Zen710 FRET plus 55. The dynamic change of the normalized FRET efficiency upon stimulation with or without PKD1 inhibitors is presented in (**d**). Mean ± SD is shown

8. While a shorter wavelength laser would provide an excitation line closer to the optimal excitation wavelength of CFP, effective FRET measurements can be made using the 458 nm line from the Argon laser.
9. Choose cells for FRET analyses that have a balanced CFP and/or YFP signal. Make adjustments for relative signal strength of each fluorophore depending on the limitations of the available filter sets on your system. For example, we visually inspect for strong CFP signal with the mercury lamp using a CFP (or DAPI) filter set. We use a GFP filter set to inspect for the cells co-expressing YFP and choose those that have a slightly higher YFP signal relative to the CFP signal for analysis.
10. Due to the very close emission spectrum of GFP and YFP, it is normal to use a GFP filter cube to observe YFP expressing cells.
11. Since whole sets of comparable experiments have to be acquired with exact same acquisition condition, it is recommended to use CFP/YFP double expressing cells to set up acquisition condition instead of using single control cells. We often choose those cells with a similar level of CFP and YFP expression instead of those highly expressing only one fluorescent protein.
12. We noticed that a high dose of PMA treatment often triggers a strong apoptosis of MDA-MB-231 cells. In the current chapter, we use a range of stimuli ranging from 100 nM to 1 μM PMA as indicated in the figures.
13. There might be other possibilities resulting in a FRET increase of PKD1-CY, such as a local accumulation of PKD1-CY. However, we observed that a PKD inhibitor abolished the FRET signal without affecting the membrane accumulation of PKD1 [5], suggesting that the gain of FRET is most likely due to PKD activation resulting to/from the intramolecular conformational changes. The detailed activation mechanisms will require further investigation.

Acknowledgements

The authors would like to thank all members of the Chemotaxis Signal Section. This research was supported by the Intramural Research Program of the NIH, NIAID. We thank Dr. Derek C. Braun at the Gallaudet University, Washington, D.C. for his assistance in construction of PKD1-CY and John F Hancock for sharing the plasmid of Ras-mRFP for this work. We also thank Howard Boudreau and Thomas Leto for providing MDA-MD-231 cells for this work. Q. Jane Wang was supported in part by National Institutes of Health grants R01CA129127 and R01CA142580.

References

1. Williams JM, Savage CO (2005) Characterization of the regulation and functional consequences of p21ras activation in neutrophils by antineutrophil cytoplasm antibodies. J Am Soc Nephrol 16:90–96
2. Wang QJ (2006) PKD at the crossroads of DAG and PKC signaling. Trends Pharmacol Sci 27:317–323
3. Eiseler T, Döppler H, Yan IK et al (2009) Protein kinase D1 regulates cofilin-mediated F-actin reorganization and cell motility through slingshot. Nat Cell Biol 11:545–556
4. Balasubramanian N, Advani SH, Zingde SM (2002) Protein kinase C isoforms in normal and leukemic neutrophils: altered levels in leukemic neutrophils and changes during myeloid maturation in chronic myeloid leukemia. Leuk Res 26:67–81
5. Xu X, Gera N, Li H et al (2015) GPCR-mediated PLCβγ/PKCβ/PKD signaling pathway regulates the cofilin phosphatase slingshot 2 in neutrophil chemotaxis. Mol Biol Cell 26:874–886
6. Bossuyt J, Chang CW, Helmstadter K et al (2011) Spatiotemporally distinct protein kinase D activation in adult cardiomyocytes in response to phenylephrine and endothelin. J Biol Chem 286:33390–33400
7. Jacamo R, Sinnett-Smith J, Rey O et al (2008) Sequential protein kinase C (PKC)-dependent and PKC-independent protein kinase D catalytic activation via Gq-coupled receptors: differential regulation of activation loop Ser(744) and Ser(748) phosphorylation. J Biol Chem 283:12877–12887
8. Kunkel MT, Toker A, Tsien RY, Newton AC (2007) Calcium-dependent regulation of protein kinase D revealed by a genetically encoded kinase activity reporter. J Biol Chem 282:6733–6742
9. Xu X, Brzostowski JA, Jin T (2009) Monitoring dynamic GPCR signaling events using fluorescence microscopy, FRET imaging, and single-molecule imaging. Methods Mol Biol 571:371–383
10. Zhou Y, Liang H, Rodkey T et al (2014) Signal integration by lipid-mediated spatial cross talk between Ras nanoclusters. Mol Cell Biol 34:862–876
11. Xu X, Meier-Schellersheim M, Jiao X et al (2005) Quantitative imaging of single live cells reveals spatiotemporal dynamics of multistep signaling events of chemoattractant gradient sensing in *Dictyostelium*. Mol Biol Cell 16:676–688
12. Storz P, Döppler H, Johannes FJ, Toker A (2003) Tyrosine phosphorylation of protein kinase D in the pleckstrin homology domain leads to activation. J Biol Chem 278:17969–17976
13. Zugaza JL, Sinnett-Smith J, Van Lint J, Rozengurt E (1996) Protein kinase D (PKD) activation in intact cells through a protein kinase C-dependent signal transduction pathway. EMBO J 15:6220–6230
14. Waldron RT, Rozengurt E (2003) Protein kinase C phosphorylates protein kinase D activation loop Ser744 and Ser748 and releases autoinhibition by the pleckstrin homology domain. J Biol Chem 278:154–163

Chapter 23

Fluorescence Readout of a Patch Clamped Membrane by Laser Scanning Microscopy

Matthias Gerhardt, Michael Walz, and Carsten Beta

Abstract

In this chapter, we describe how to shield a patch of a cell membrane against extracellularly applied chemoattractant stimuli. Classical patch clamp methodology is applied to allow for controlled shielding of a membrane patch by measuring the seal resistivity. In *Dictyostelium* cells, a seal resistivity of 50 MΩ proved to be tight enough to exclude molecules from diffusing into the shielded membrane region. This allowed for separating a shielded and a non-shielded region of a cell membrane to study the spatiotemporal dynamics of intracellular chemotactic signaling events at the interface between shielded and non-shielded areas. The spatiotemporal dynamics of signaling events in the membrane was read out by means of appropriate fluorescent markers using laser scanning confocal microscopy.

Key words Patch clamp, Confocal microscopy, Excitable dynamics, *Dictyostelium discoideum*, Chemotaxis, Actin dynamics

1 Introduction

Chemotactic cells rely on an intracellular mechanism to sense chemical gradients to decide into which direction to move [1]. A common assumption to describe the gradient sensing mechanism is the local excitation global inhibition (LEGI) model [2, 3]. The LEGI mechanism is based on the intracellular release of an activator and an inhibitor upon extracellular stimulation with chemoattractant molecules (see for example [4–7]). In the case of *Dictyostelium discoideum*, a common model organism for the study of eukaryotic chemotaxis, cyclic-adenosine-monophosphate (cAMP) molecules are detected by the cAMP receptors car1 or car3 [8]. According to the LEGI assumption, activator and inhibitor have different diffusion coefficients causing the inhibitor to diffuse much quicker through the cytosol than the activator. When the cell is exposed to a chemoattractant gradient, the activator thus remains nonuniformly distributed within the cell, so that its local concentration at the front might surpass the inhibitor concentration and thus

Tian Jin and Dale Hereld (eds.), *Chemotaxis: Methods and Protocols*, Methods in Molecular Biology, vol. 1407,
DOI 10.1007/978-1-4939-3480-5_23,

exceeds the threshold to trigger further downstream signaling and, finally, motility. However, it remains still elusive which molecules could take on the role of the activator and the inhibitor. Also the relation of the gradient sensing mechanism to wave-like spreading of signaling activity and oscillatory actin dynamics remains the subject of current debate [9–13].

In the absence of cAMP stimuli, combined actin/PIP_3-waves spontaneously emerge at the substrate-attached membrane of *Dictyostelium* cells [11, 13, 14]. These traveling waves show the typical hallmarks of an excitable system. Can we initiate such wave patterns by localized cAMP stimuli? Recently, we published results obtained by spatially confined cAMP stimulation of chemotactic *Dictyostelium* cells to address this question [15]. This approach was inspired by localized perturbation experiments in non-biological active media, where spreading waves could be readily initiated [16, 17]. Our results showed that PIP_3-formation and increased actin polymerization remained spatially confined to the membrane region in contact with cAMP. No wave-like spreading of the activity into non-stimulated membrane regions was observed. Thus, a localized receptor stimulus did not initiate traveling actin/PIP_3-waves.

What is suppressing the initiation of waves from a locally stimulated membrane region? Further experiments based on the membrane shielding technique presented here may provide an answer to this question. In particular, the response to localized cAMP stimuli of varying concentration will yield insight into the intracellular decision-making process [2] that allows chemotactic cells to switch between different dynamic states of their cortical machinery.

In the past, controlled cAMP stimulation was mostly based on microfluidic tools [18]. However, while such techniques provide excellent temporal resolution, subcellular spatial localization of the stimulus is difficult to achieve [19]. In this chapter, we present how to combine patch clamp techniques to achieve local shielding of the membrane and laser scanning microscopy for fluorescence readout of the spatiotemporal distribution of intracellular markers while the cell is stimulated extracellularly. A simpler version of this method was given by Luo et al. [20] who focused on laser scanning microscopy of aspirated membrane patches without chemical stimulus. We extended this method by combining fluorescence readout of aspirated membranes with patch clamp to control the sealing between the membrane and the glass pipette. A proper seal allows for shielding experiments while extracellular stimulation with molecules of choice is carried out. A short description of this method was already given in Gerhardt et al. [15]. The method relies on the patch clamp technique, which is the standard approach for studying ligand-gated ion channels in the membrane of eukaryotic cells [21]. To study the signaling of muscles or ganglion cells, the so-called loose patch method [22, 23] allows for extracellular recording without destruction of the membrane of the recorded cell. Herein a patch of a cell membrane is aspirated into the open tip of a glass micropipette by means of negative pressure. Simultaneously, the electrical "seal" resistivity of the cell-to-glass pipette connection is measured based on an electrical current in the picoampere range through the pipette. We implemented this so-called "current-clamp approach" using a patch clamp amplifier, which changes the required voltage to keep the current constant. The cell-to-glass pipette contact can be described by capacitive and resistive components of an electrical circuit. The ohmic resistivity across the membrane is defined by Ohm's law, given by the voltage required to regulate a constant current if all capacitive components have been charged entirely (Fig. 1).

For practical reasons the voltage is measured using an oscilloscope. However, in contrast to classical patch-clamp approaches, the response of a chosen marker molecule is detected using confocal laser scanning microscopy (LSM). This technique allows for

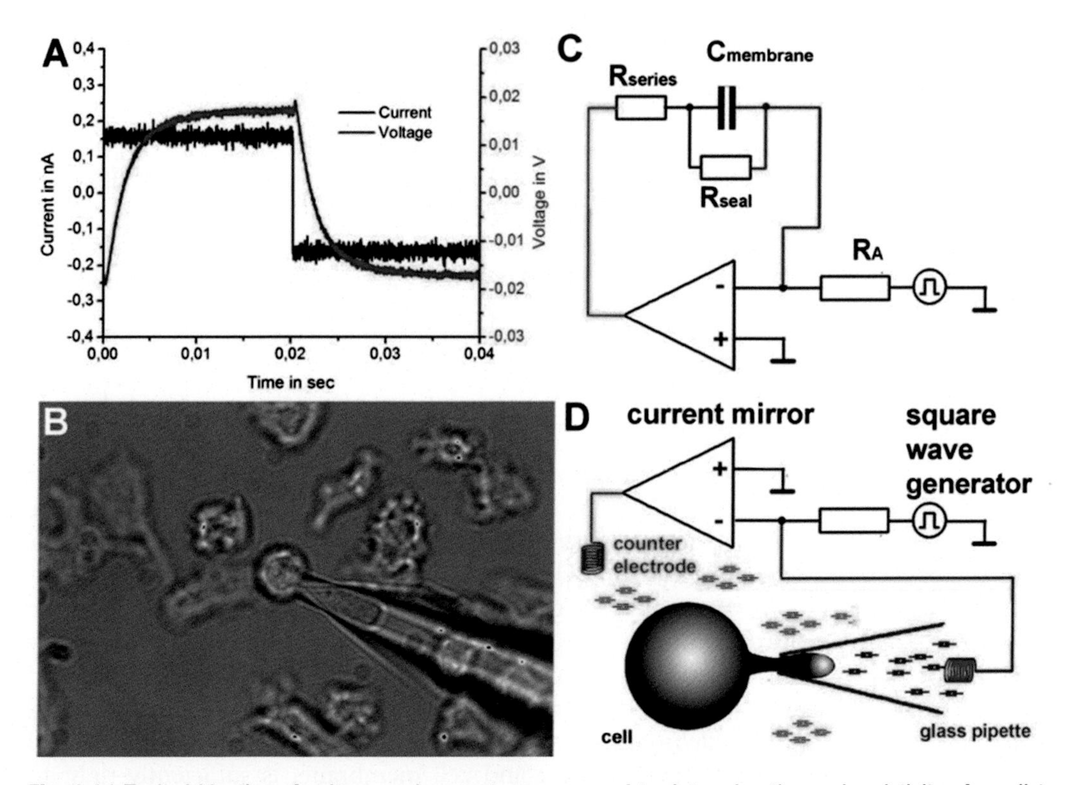

Fig. 1 (**a**) Typical kinetics of voltage and current as measured to determine the seal resistivity of a cell-to-pipette contact which itself is shown in (**b**). To understand how that kinetics occurs, an equivalent circuit of the cell-to-pipette contact connected to a current mirror is shown in (**c**). The operational amplifier in the circuit regulates its output voltage such that the difference between the current across the cell-to-pipette contact equals that of the resistor R_A. A rising edge of a square wave causes charging of the membrane capacity until the selected current can be maintained by the seal resistivity entirely. If the series resistivity was compensated properly the seal resistivity can be calculated by the quotient of voltage and current amplitude accordingly to Ohm's law. Note that "current clamp" is electronically implemented in a professional patch clamp amplifier such as the Axopatch 200B. In panels (**c** and **d**), the equivalent circuit is given for illustration. Reproduced and modified according to Gerhardt et al. [15]

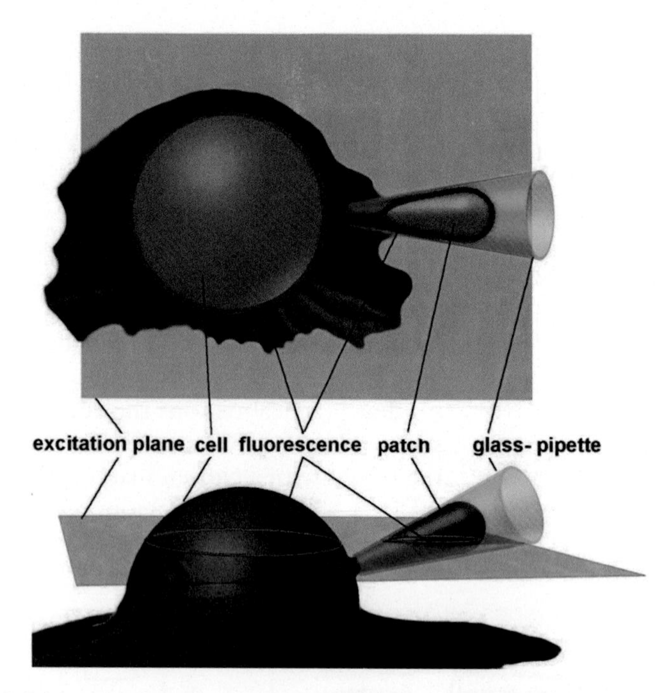

Fig. 2 Schematic view of a cell which is intersected by the excitation plane of a confocal laser scanning microscope. A patch of the cell membrane is aspirated into the opening of a glass pipette. The excitation plane is intersecting the patch inside the glass pipette causing fluorescence of intracellular markers in the patch as well as in the remaining cell outside the pipette. Reproduced and modified according to Gerhardt et al. [15]

excitation of fluorescence in a plane intersecting the shielded membrane patch in the glass pipette as well as the remaining cell outside. Interestingly, neither the excitation focus nor the emission focus is significantly blurred by the presence of the glass wall of the pipette, which allows for proper imaging of the fluorescent marker molecules (*see* Fig. 2).

While increasing the negative pressure, the seal between the patch and the glass pipette is established and the seal resistivity increases simultaneously. By investigating the receptor-induced translocation of intracellular fluorescent markers in a differentiated *Dictyostelium* cell, we find that a seal resistivity of 50 MΩ, the seal between glass pipette and cell membrane, is sufficiently tight to exclude diffusion of cAMP molecules into the sealed patch.

2 Materials

2.1 Differentiated Cells

1. Cell culture equipment such as flow bench, centrifuge, 250 ml and 50 ml Erlenmeyer flasks and shaker, computer controlled pump (syringe or peristaltic), photometer (*see* **Note 1**), and cell counter.

2. Cells that express fluorescent markers amenable for imaging using LSM, such as the following *Dictyostelium discoideum* cell lines:
   (a) DdLimEΔ-RFP and PTEN-GFP in AX2 (described in Ref. 12]).
   (b) PH_{CRAC}-GFP in AX3 (described in Ref. [24]) (*see* **Note 2**).
3. Selection factors to maintain expression of the fluorescent markers, including Blasticidin, Hygromycin, and Geneticin (AppliChem/PanReac, Germany).
4. Cell culture medium: HL5 medium (Foremedium, UK) supplemented with:
   (a) 5 μg/ml Blasticidin, 10 μg/ml Geneticin, and 33 μg/ml Hygromycin for cell line 2a above; or
   (b) 5 μg/ml Geneticin for cell line 2b above.
5. Phosphate buffer at pH 6 (PB): 2.1 mM Na_2HPO_4, 14.6 mM KH_2PO_4, 2 mM $CaCl_2$; prepared in 0.05 μS/cm water.
6. 1 mM cyclic adenosine monophosphate (cAMP; Sigma, USA) in PB.

2.2 Patch Clamp

1. Patch clamp amplifier equipped with head stage (Axopatch 200B, Molecular Devices Inc., USA) (*see* **Note 3**).
2. Micromanipulator (Luigs & Neumman, Scientifica, or Eppendorf).
3. Digital storage oscilloscope with a sampling rate ≥20 MHz (Tektronix, USA).
4. Function generator for generating a square wave output (e.g., HM8030-4, HAMEG, Germany).
5. Pipette puller (Sutter Instruments, USA).
6. Glass pipettes: (GBT150F, Science Products, Germany) (*see* **Note 4**).
7. Pipette-holder compatible with the head stage being used (Molecular Devices, USA).
8. Silicone tubing, syringe, three-way valve, Luer-Lock adaptors and connectors, and microloaders (Eppendorf).
9. Counterelectrode and holder (e.g., silver chloride pellet electrode, Science Products, Germany).

2.3 Microscope

1. Confocal laser scanning microscope (e.g., Zeiss LSM710 or LSM780) mounted on an anti-vibration desk.
2. 488 and 526 nm LASERs.
3. 40× Plan Apo N.A. 1.4 Oil lens.
4. Adaptor for mounting a micromanipulator for manipulating the head stage of the patch clamp amplifier (*see* **Note 5**).
5. Glass bottomed dishes, diameter ~1.5″ (*see* **Note 6**).

3 Methods

To study intracellular responses evoked by extracellular stimulation, the cells need to be initially differentiated to develop their capability to respond on extracellular stimulation. For the case of *Dictyostelium discoideum* cells the process of differentiation is carried out by letting the cells starve under simultaneous exposure to pulses of cAMP (*see* **Note 7**). The point in time when the cells have differentiated sufficiently needs to be determined by fluorescence microscopic observation of intracellular responses of the markers upon extracellular stimulation of the cells with cAMP. Once the intracellular markers have shown a clear response, shielding experiments can be carried out. This requires glass pipettes with a tip opening of 2–4 μm which are ideally manufactured using a pipette puller. If the pipette is filled with PB and its opening is gently manipulated into the vicinity of a cell, the electrical resistance of the pipette needs to be compensated. After compensation of the pipette resistance, the seal between the cell and the micropipette is established by application of a negative pressure to the pipette. While applying negative pressure, the voltage displayed on the oscilloscope increases until the quotient between voltage and current exceeds 50 MΩ (only for pipettes with a tip opening in the range from 2 to 4 μm). This is the moment when the aspirated membrane patch and the remaining cell can be imaged using laser scanning microscopy while at the same time the cell is stimulated with cAMP. The scanned image reveals whether the marker in the aspirated membrane patch reacts upon cAMP stimulation or not.

3.1 Growth and Differentiation of Cells

1. Using sterile technique, inoculate *Dictyostelium discoideum* in HL5 containing appropriate selection factors as specified in Subheading 2.1, **item 4**, at an initial density of 0.5×10^6 cells/ml in the 250 ml Erlenmeyer flask using a shaking incubator at 150 rpm at 22 °C. Large flasks allow for sufficient air supply to the cells in the HL5 medium.
2. Grow the cells until they reach the exponential growth phase, which can be either determined by manually counting the cells or by means of a photometer if available (*see* **Note 1**). The exponential growth phase is usually reached 3–4 days after inoculation.
3. 50×10^6 cells are harvested by centrifugation at a relative centrifugal force (RCF) of $50 \times g$ and washed in PB twice.
4. 25 ml of PB containing 50×10^6 cells are transferred to a 50 ml Erlenmeyer flask and shaken at 150 rpm under periodic pulses of cAMP (50 nM final concentration each 6 min).
5. Typically, after 4–6 h of starvation, the cells are differentiated, which needs to be checked by the intracellular responses of the chosen markers upon extracellular stimulation with 20 μM of cAMP using fluorescence microscopy (*see* **Note 7**).

3.2 Patch Clamp Setup

1. Connect the output of a function generator to both the rear side input "Ext. Command/Front." of the Axopatch 200B amplifier and channel "1" of the oscilloscope using BNC cables and a T-connector. The "scaled output" on the front side of the Axopatch 200B needs to be connected to channel "2" of the oscilloscope. Make sure the headstage is connected with the Axopatch 200B and mounted properly on the micromanipulator (*see* **Note 5**). Connect the pressure input of the pipette holder with one end of a silicone tube. Connect the other end of the tube to a three-way valve and to a syringe (50 ml).
2. Set the following parameters on the Axopatch 200B as follows: PipetteOffset: 2/0, C_fast: 0/0, C_slow: 0/cont., SeriesResistance: 100 %, Prediction: 0 %, Lag: 100 %, WholeCellCap: Off, CommandSetting: 0/Off/1x/ExtSwitch:Off, Meter: Vm, Mode: I_fast, Lowpass: 5 kHz, Outputgain: 10.
3. Set the following Parameters on the function generator (e.g., HAMEG HM8030-4): Mode: square wave, frequency: 2 Hz, amplitude: 150 mV (Amp: min and –80 dB).
4. Mount a glass bottomed dish (1.5″) onto a suitable stage of the LSM and add a drop of 400–500 μl of cell suspension (2×10^6 cells/ml) onto the glass bottom. Make sure that the drop is well balanced due to its surface tension and does not flow over the wall of the glass bottom. Do not forget to add immersion oil to the objective lens to match its refractivity index to that of the glass bottom in the dish.
5. Let the cells sediment for 5 min onto the glass bottom and use the time in between to prepare glass pipettes and the patch clamp setup.
6. Set the following parameters to the Sutter Instrument P97 pipette puller and pull a glass pipette (use GBT150F glass tubing): Heat 384, Pull 0, Velocity: 35, Time 250, Pressure 500, Loops: 3–4.
7. Setup of a counter electrode. Mount a silver oxide pellet onto the stage of the microscope so that it touches the surface of the buffer solution in the petri dish. Connect the pellet with the ground connector of the head stage. Since neither a counter electrode nor appropriate wires come with the Axopatch 200B, ready to use equipment needs either to be purchased or custom machined or improvised by the operator (*see* **Note 8**).
8. After the 5 min for sedimentation of cells have elapsed, add carefully PB to reach a final volume of ~1 ml. Make sure that the silver oxide pellet now touches the surface of the buffer filled into the petri dish.
9. Fill the glass pipette with PB. To avoid air bubbles inside the pipette, aspirate at first a small volume of PB into the pipette tip

by application of negative pressure to the open end of the pipette. Subsequently, fill the remaining part of the pipette from the open end by means of a microloader. Remove the remaining air bubbles between the small amount of buffer at the tip and the large volume of buffer filled into the remaining pipette by gently knocking against the pipette with your fingers.

10. Mount the filled micropipette onto the pipette holder of the headstage of the Axopatch 200B.
11. Activate the manual bright field mode of the LSM microscope and move the cells on the glass bottom into the focus. Subsequently move the focus above the cells so that you can still recognize in which plane the cells can be found.
12. Initiate the function generator to generate square waves and switch on "Ext.Switch" in the commands section of the Axopatch 200B. Move the pipette into the center of the field of view above the buffer using the manual alignment mode of the micromanipulator.
13. Lower the pipette tip into the buffer and observe meanwhile the voltage signal displayed at the oscilloscope. The amplified voltage of the pipette tip should be visible as a square wave with rounded corners, which arise from charging and discharging of stray capacitance associated with the membrane, the pipette, and the remaining circuit.
14. Turn the potentiometer "resistance compensation" until the pipette voltage vanishes entirely. Alternatively, check whether the patch clamp setups works properly by turning the potentiometer "resistance compensation" until the voltage signal approaches zero. If the pipette opening is between 2 and 4 μm, the reading of the potentiometer should be between 3 and 14 MΩ to reach a fully compensated pipette resistance.
15. Under microscopic observation, lower the pipette tip further until it is well visible in the focal plane previously selected.
16. Move the cells into the focal plane and micromanipulate the pipette tip into the vicinity of a cell (*see* Fig. 3a).
17. Check again the compensation of the series resistance. If the series resistivity is not compensated well, then the voltage signal will look similar to the current signal. In this case turn the series resistance compensation until the reading of the oscillograph resembles that shown in Fig. 3d. Increase the pressure inside the pipette by means of a syringe.
18. Observe via the microscope whether the pressure driven fluid flow out of the pipette tip indents the membrane of the selected cell. A small moon-shaped indentation in the cell membrane close to the pipette opening will form due to outflow of buffer from the pipette opening, *see* Fig. 3b. If so, decrease the distance between the opening of the glass pipette and the cell

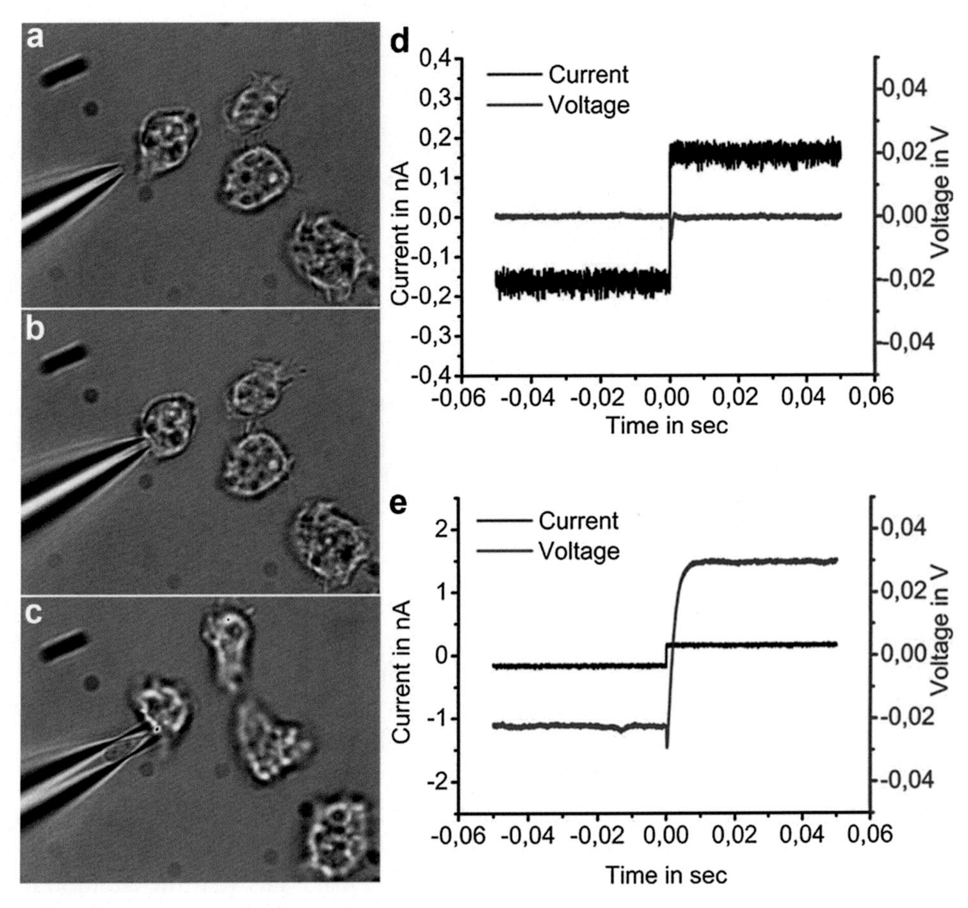

Fig. 3 Step-by-step procedures how to patch clamp a *Dictyostelium* cell. (**a**) The patch pipette initially needs to be micromanipulated into the vicinity of a cell. The oscillograph reading should appear as shown in panel (**d**). (**b**) Increase the pressure of the pipette slightly and micromanipulate its opening closer to cell. The bright field image should look like panel (**b**). (**c**) Decrease the pipette pressure until a patch of the cell membrane aspirates into the pipette. The reading of the oscillograph should look like panel (**e**). In this case the seal resistivity abruptly increased up to 150 MΩ which will be quite enough to seal the patch against molecules in the extracellular space

using the micromanipulator and decrease the pressure by opening the three way valve.

19. Aspirate a patch of the cell membrane into the tip of the glass pipette by reducing the pressure inside the pipette with, for example, a syringe and observe on the oscilloscope how the voltage at the pipette tip increases. Increase this voltage until the Ohmic quotient between the voltage and the current at the pipette tip reaches or surpasses 50 MΩ. Make sure that all correction factors of the Axopatch 200B have been correctly calculated. Note also that the seal resistivity depends on the pipette opening which can be either checked by secondary electron microscopy if available (see Note 9) or by the reading of the series resistance compensation potentiometer of the Axopatch 200B.

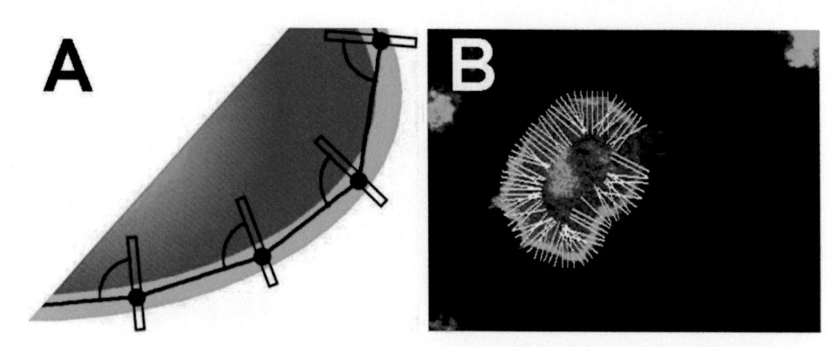

Fig. 4 Procedure to quantitatively measure the intensity of fluorescence across a cell membrane by scanning the intensity profile perpendicularly to the perimeter of the cell as shown schematically in panel (**a**) and as applied to the image of a real cell in panel (**b**), showing scanning paths assembled perpendicularly to the cell's perimeter. To obtain the mean intensity profile all vertices that define the perimeter of the cell should to be averaged. Reproduced and modified according to Gerhardt et al. [15]

20. If 50 MΩ seal resistivity has been surpassed, the seal can be considered to be tight. Switch on the laser scanning mode of the LSM and adjust the image plane such that both the cell and the patch inside the glass pipette are captured. For excitation of both the GFP and RFP markers the wavelengths 488 and 526 nm are required.
21. Once a stable image of the cell and of the patch is established, the extracellular stimulus can be added. Add carefully 50 μl of 1 mM cAMP in PB to the cells. To make sure that the addition of volume does not causes strong convection, which eventually damages the established cell-to-pipette contact, add the 50 μm of cAMP a distance of 1 cm away from the aspirated cell—optimally at the border of the petri dish.
22. By means of the syringe, regulate the pipette pressure such that the seal resistivity and, thus, the seal remain stable. If the seal resistivity increases too much, the patch might be further aspirated into the pipette and might eventually rupture. To avoid this, a slight decrease of the pressure is required and vice versa (It takes some practice to maintain the seal.).
23. To analyze the obtained images, either qualitative interpretation can be used or the intensity change of fluorescence inside the membrane can be quantitatively measured using either ImageJ or custom developed software (Fig. 4).

4 Notes

1. To continuously monitor the optical density of growing cells, we custom developed an in situ photometer (Fig. 5) that relies on a bypass pipe added to an Erlenmeyer flask. The photome-

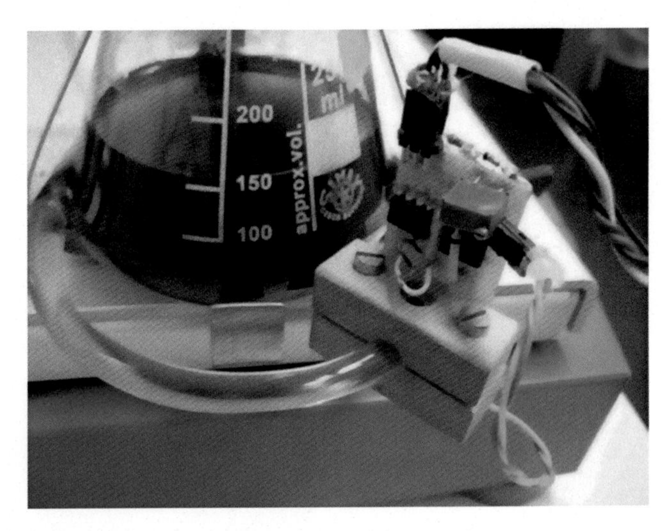

Fig. 5 Custom developed in situ photometer to monitor the optical density of a cell culture inside a shaking flask

ter itself consists of a light emitting 420 nm diode, which emits light into this bypass pipe and a photodiode, which measures the intensity of the transmitted light. The optical density can be calculated by the measured intensity of the transmitted light at the beginning of an incubation period and at a point in time during incubation which is of interest. The photometer allows for exact determination of the point in time when the cells have reached the exponential growth phase.

2. We used cells expressing the PH_{CRAC}-GFP label in an AX3 background. They contain the plasmid pWF38, which can be obtained from DictyBase.org. Although these cells are no longer able to develop spores, they are still able to become chemotactic during starvation.
3. To interconnect the patch clamp amplifier with oscilloscope and function generator, cables equipped with BNC male plugs at both ends will be required. Cables of desired lengths can either be purchased ready-made or they can be custom fabricated using a crimping tool and appropriate BNC connectors to save costs or obtain lengths which are not commercially available.
4. Since custom pulled glass pipettes are very delicate it is recommended to store them in a suitable rack. We custom manufactured such a rack to accommodate 7×7 glass pipettes. The rack should be equipped with a "safety bar" to avoid accidental contact between the very sharp pipette tips and your skin (Fig. 6).
5. To image the cell body and the shielded patch inside the glass pipette simultaneously, it is necessary to mount the micromanipulator such that the glass pipette and the sample layer

Fig. 6 Micropipette rack to accommodate 49 micropipettes. A bar on top makes sure that nobody can be accidently injured by the sharp glass tips

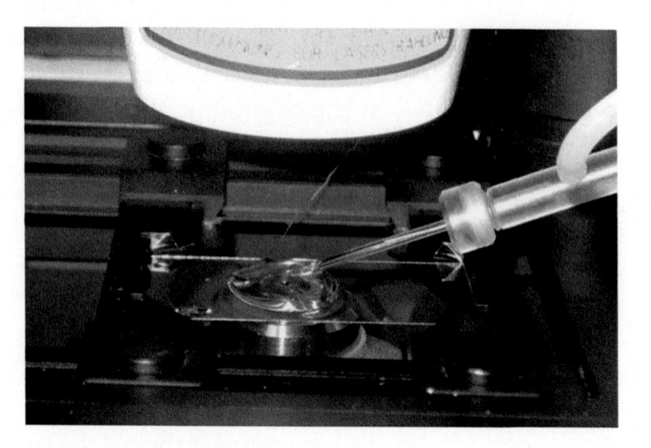

Fig. 7 Assembly of micropipette and sample stage of an LSM. The angle between pipette and sample stage needs to be sharp to allow for proper imaging of cell body and shielded membrane patch

enclose a sharp angle. The sample layer will be a glass bottomed dish or a cover slide (Fig. 7).

6. If glass bottomed dishes are used it is necessary to remove parts of its plastic wall (Fig. 8) to allow for a sharp angle between glass pipette and sample layer such as described in **Note 5**. Pliers with narrow tip are best suited to crack away small pieces of the wall. If the pieces are too large, the entire dish might break.
7. Cells that express the $PH_{CRAC}GFP$ label can easily be checked for their development stage using a LSM. In the absence of a

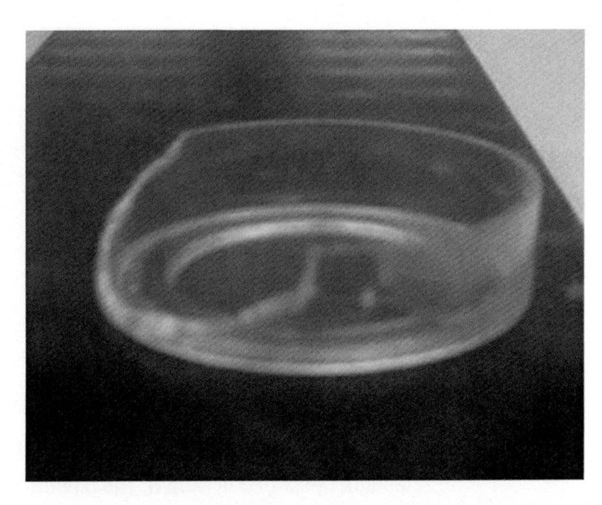

Fig. 8 Custom-modified glass bottomed dish, which allows for a sharp angle between a pipette coming from the left side and the glass bottom

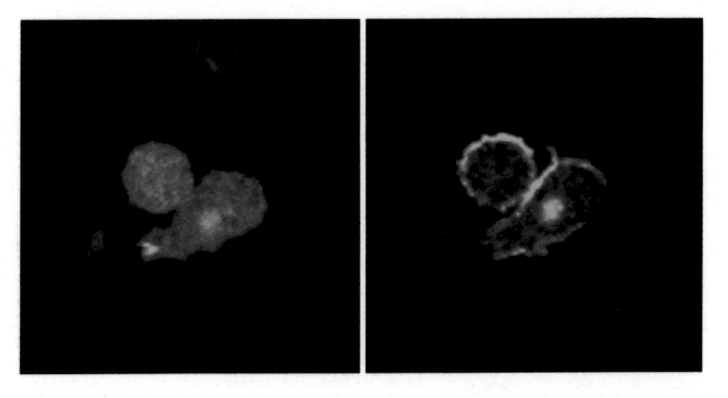

Fig. 9 *Left panel*: The green PH_{CRAC}GFP label is evenly distributed throughout the cytosol. *Right panel*: After the addition of 10 μM cAMP (final concentration), the green PH_{CRAC}GFP label translocates to the inner cell membrane, indicating that the cells are well differentiated

receptor stimulus, the PH_{CRAC}GFP label is evenly distributed inside the cytosol. If cAMP molecules are added to the surrounding medium with a final concentration not higher than 10 μM, a transient translocation of the PH_{CRAC}GFP label from the cytosol to the cell membrane can be observed if the cells are well differentiated (Fig. 9). The translocation reaches a maximum after about 7 s.

8. To mount the counter electrode to the microscope stage as required for patch clamping of Dictyostelium cells, a small aluminum angle to host a banana plug "MST3" female (test socket 2 mm) is suitable. The angle can be reversibly glued to the microscope stage using double sided adhesive tape. A silver oxide pellet usually comes ready equipped with a wire which can be soldered to a banana plug "MST3" male. Now the silver

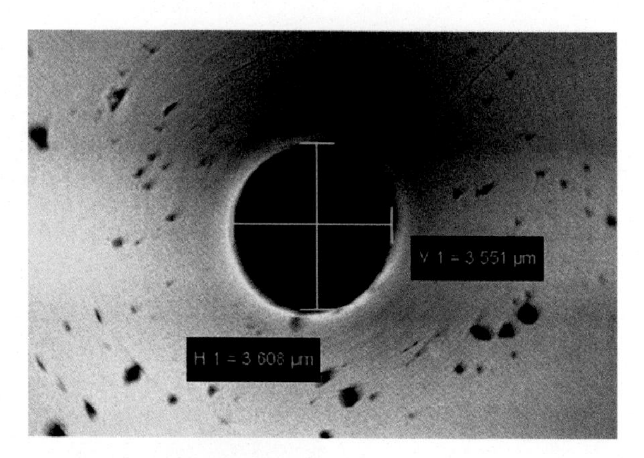

Fig. 10 Secondary electron microscopy image of a pipette pulled using the procedure described in **step 8**. The pipette tip opening shown has a diameter of 3.5 µm. For patch clamping *Dictyostelium* cells, make sure that the diameter is not larger than 4 µm, otherwise given readings of a minimum seal resistivity to seal a patch might not be transferable (*see* **Note 9**)

oxide pellet can easily be plugged into the socked which is hosted by the aluminum angle. The contact of the socket can then be easily connected with the ground connector of the Axopatch 200B headstage. However, such holders for counter electrodes are also available ready to use from World Precision Instruments Inc., USA.

9. To check the tip opening of pulled micropipettes, it is recommended to employ a scanning electron microscope (SEM) to image the tip (in this case, also a sputter coater is needed to add a conductive coating to the surface of the pipettes) (Fig. 10). However, in case such equipment is not available, the ohmic resistance of the tip opening can be measured using the patch amplifier. The ohmic resistance of the tip opening is the reading of the "resistance compensation" potentiometer of the patch amplifier after compensating the series resistance. If the reading is within the range of 3–14 MΩ, the tip opening is suitable to perform shielding experiments.

Acknowledgment

We thank Kirsten Krüger for cell culture support, Otto Baumann for support and access to the LSM710 microscope, and Günther Gerisch for providing *Dictyostelium* cell lines.

References

1. Swaney KF, Huang CH, Devreotes PN (2010) Eukaryotic chemotaxis: a network of signaling pathways controls motility, directional sensing, and polarity. Annu Rev Biophys 39:265–289
2. Parent CA, Devreotes PN (1999) A cell's sense of direction. Science 284:765–770
3. Levchenko A, Iglesias PA (2002) Models of eukaryotic gradient sensing: application to chemotaxis of amoebae and neutrophils. Biophys J 82:50–63
4. Levine H, Kessler DA, Rappel WJ (2006) Directional sensing in eukaryotic chemotaxis: a balanced inactivation model. Proc Natl Acad Sci U S A 103:9761–9766
5. Beta C, Amselem G, Bodenschatz E (2008) A bistable mechanism for directional sensing. New J Phys 10:083015
6. Xiong Y, Huang CH, Iglesias PA, Devreotes PN (2010) Cells navigate with a local-excitation, global-inhibition-biased excitable network. Proc Natl Acad Sci U S A 107:17079–17086
7. Takeda K, Shao D, Adler M et al (2012) Incoherent feedforward control governs adaptation of activated ras in a eukaryotic chemotaxis pathway. Sci Signal 5:ra2
8. Insall RH, Soede RD, Schaap P, Devreotes PN (1994) Two cAMP receptors activate common signaling pathways in *Dictyostelium*. Mol Biol Cell 5:703–711
9. Huang CH, Tang M, Shi C et al (2013) An excitable signal integrator couples to an idling cytoskeletal oscillator to drive cell migration. Nat Cell Biol 15:1307–1316
10. Westendorf C, Negrete J Jr, Bae AJ et al (2013) Actin cytoskeleton of chemotactic amoebae operates close to the onset of oscillations. Proc Natl Acad Sci U S A 110:3853–3858
11. Gerisch G, Bretschneider T, Müller-Taubenberger A et al (2004) Mobile actin clusters and traveling waves in cells recovering from actin depolymerization. Biophys J 87:3493–3503
12. Gerisch G, Schroth-Diez B, Müller-Taubenberger A, Ecke M (2012) PIP3 waves and PTEN dynamics in the emergence of cell polarity. Biophys J 103:1170–1178
13. Gerhardt M, Ecke M, Walz M et al (2014) Actin and PIP_3 waves in giant cells reveal the inherent length scale of an excited state. J Cell Sci 127:4507–4517
14. Bretschneider T, Anderson K, Ecke M et al (2009) The three-dimensional dynamics of actin waves, a model of cytoskeletal self-organization. Biophys J 96:2888–2900
15. Gerhardt M, Walz M, Beta C (2014) Signaling in chemotactic amoebae remains spatially confined to stimulated membrane regions. J Cell Sci 127:5115–5125
16. Wolff J, Stich M, Beta C, Rotermund HH (2004) Laser-induced target patterns in the oscillatory CO oxidation on Pt(110). J Phys Chem B 108:14282–14291
17. Punckt C, Stich M, Beta C, Rotermund HH (2008) Suppression of spatiotemporal chaos in the oscillatory CO oxidation on Pt(110) by focused laser light. Phys Rev E 77:046222
18. Beta C, Bodenschatz E (2011) Microfluidic tools for quantitative studies of eukaryotic chemotaxis. Eur J Cell Biol 90:811–816
19. Bae AJ, Beta C, Bodenschatz E (2009) Rapid switching of chemical signals in microfluidic devices. Lab Chip 9:3059–3065
20. Luo T, Mohan K, Iglesias PA, Robinson DN (2013) Molecular mechanisms of cellular mechanosensing. Nat Mater 12:1064–1071
21. Neher E (1992) Nobel lecture. Ion channels for communication between and within cells. EMBO J 11:1672–1679
22. Almers W, Roberts WM, Ruff RL (1984) Voltage clamp of rat and human skeletal muscle: measurements with an improved loose-patch technique. J Physiol 347:751–768
23. Gerhardt M, Groeger G, Maccarthy N (2011) Monopolar vs. bipolar subretinal stimulation-an in vitro study. J Neurosci Methods 199:26–34
24. Xiao Z, Zhang N, Murphy DB, Devreotes PN (1997) Dynamic distribution of chemoattractant receptors in living cells during chemotaxis and persistent stimulation. J Cell Biol 139:365–374

Chapter 24

Use of Resonance Energy Transfer Techniques for In Vivo Detection of Chemokine Receptor Oligomerization

Laura Martínez-Muñoz, José Miguel Rodríguez-Frade, and Mario Mellado

Abstract

Since the first reports on chemokine function, much information has been generated on the implications of these molecules in numerous physiological and pathological processes, as well as on the signaling events activated through their binding to receptors. As is the case for other G protein-coupled receptors, chemokine receptors are not isolated entities that are activated following ligand binding; rather, they are found as dimers and/or higher order oligomers at the cell surface, even in the absence of ligands. These complexes form platforms that can be modified by receptor expression and ligand levels, indicating that they are dynamic structures. The analysis of the conformations adopted by these receptors at the membrane and their dynamics is thus crucial for a complete understanding of the function of the chemokines. We focus here on the methodology insights of new techniques, such as those based on resonance energy transfer for the analysis of chemokine receptor conformations in living cells.

Key words Chemokines, GPCR, Oligomerization, FRET, BRET, BiFC, SRET

1 Introduction

1.1 Chemokines and Their Receptors: A Dynamic System

Chemokines are a family of structurally related, low-molecular-weight pro-inflammatory cytokines involved in recruitment of specific cell populations to target tissues by interacting with members of the G protein-coupled receptor (GPCR) family [1–4]. Although originally described as specific mediators of leukocyte directional movement, they are now implicated in a much wider variety of physiological and pathological processes including tumor cell growth and metastasis, atherosclerosis, angiogenesis, chronic inflammatory disease, and HIV-1 infection [2, 5–10].

In man, more than 50 chemokines and 20 receptors have been described and classified according to functional criteria as pro-inflammatory or homeostatic chemokines, depending on their role in inflammation or in immune system homeostasis, respectively [1, 11]. Although chemokines and their receptors were once considered

Tian Jin and Dale Hereld (eds.), *Chemotaxis: Methods and Protocols*, Methods in Molecular Biology, vol. 1407,
DOI 10.1007/978-1-4939-3480-5_24, © Springer Science+Business Media New York 2016

independent, isolated entities, the situation is much more complex than initially anticipated. Chemokines can form dimers, tetramers, and even oligomers [12, 13]. They interact not only with chemokine receptors but can also bind cell surface glycosaminoglycans as well as non-signaling scavenger receptors [1, 14, 15]. Chemokine receptors interact with chemokines and also other molecules, such as defensins and virally encoded chemokines [2, 16–19].

Whereas some data indicate that GPCR family members can function as monomers [20], increasing experimental evidence indicates that these receptors form homodimeric and heterodimeric complexes at the cell membrane [21–25] not only with other chemokine receptors, but also with other GPCR (e.g., EBI2, delta opioid receptor) [26, 27] and other membrane proteins (e.g., CD4, CD26, T cell receptor, tetraspanins) [28–32]. These receptor complexes are functional entities that mediate biological responses and are associated with modulation of ligand binding and of G protein associations, and with activation of signaling events distinct from those triggered by individual receptors [33, 34]. Heterodimerization is the mechanism that underlies delayed AIDS progression in HIV-1-infected patients bearing the CCR2V64I polymorphism [35]; CCR5 expression alters CD4/CXCR4 heterodimer conformation, thus blocking M-tropic HIV-1 binding and infection [28]. Heterodimers activate specific signaling cascades that differ from those activated by homodimers [33, 36]. In contrast to homodimer-triggered Gi signaling, CCR2/CCR5 heterodimers activate G_{11}, which alters PI3K induction kinetics [33]. CCR5 and CXCR4 heterodimerization with opioid receptors modulates chemokine responses [27, 37, 38]. Oligomeric complexes can also regulate ligand affinity; for example, CXCR5 and EBI2 heterodimers reduce CXCL13 affinity for CXCR5, as well as subsequent Gi protein activation [26]. In this scenario, chemokine-mediated signaling properties depend on the receptor complex stabilized. This defines a very dynamic universe that offers ample possibilities for regulating chemokine function, and allows the design of innovative drugs to target specific chemokine-mediated functions without altering others.

Although biochemical approaches were initially used to determine chemokine receptor expression at the cell membrane as well as the signaling molecules involved in chemokine function, they render a static view of the system that can lead to misinterpretation of results. For example, western blot and coimmunoprecipitation showed CCR2 dimers only after ligand activation, which led us to describe an active role for ligands in receptor dimerization [23, 39]. Later studies using imaging-based techniques showed that dimers form in the absence of chemokines [21, 40, 41]. Our initial conclusion was thus incorrect; the difficulty in detecting these complexes by immunoprecipitation might be due to dimer conformational instability in the absence of ligand.

1.2 Resonance Energy Transfer Technology

Newer methods now in wide use for evaluating chemokine receptor oligomerization in living cells are based on resonance energy transfer (RET). These techniques are also useful for determining conformation dynamics, identifying the role of ligands and/or receptors in this process, and defining the dimerization site in the cell [42].

RET measurement allows identification of molecular interactions using techniques based on the quantitative theory developed by Förster in the 1940s. RET is a non-radioactive quantum mechanical process that neither requires electron collision nor involves heat production. There are two main types of RET, fluorescence resonance energy transfer (FRET) and bioluminescence resonance energy transfer (BRET); in the former, the donor fluorochrome transfers energy to an acceptor fluorochrome and in the latter, the donor molecule is luminescent [43, 44].

Both methods require generation of fusion proteins between the receptor and the fluorescent/luminescent donor and acceptor proteins, as well as the use of transfected cells [45]. Although BRET has been used for single-cell analysis [46], it is a more appropriate approach for cell suspensions [47]. It allows measurement of energy transfer between receptors, independently of their expression pattern. Dimers at the plasma membrane cannot be distinguished from those being synthesized or trafficking through the endoplasmic reticulum. BRET saturation curves can nonetheless be quantitated, as reported for CCR5/CCR2 heterodimers [48] and CXCR4 homodimers [40]. In contrast, FRET imaging using confocal microscopy allows measurements in single cells and identification of specific cell locations. FRET requires robust controls to discard direct acceptor activation by the light used to excite the donor, to eliminate nonspecific random collisions, and to monitor receptor overexpression [49].

2 Materials

2.1 Materials Common to all Techniques Described

1. HEK-293T cells (ATCC® CRL-11268™, Manassas, VA) (*see* **Note 1**).
2. DMEM culture medium containing 10 % fetal calf serum, 1 mM sodium pyruvate, and 2 mM l-glutamine (complete DMEM).
3. Tissue culture 6-well plates.
4. Cell incubator with 5%CO_2.
5. Polyethylenimine (PEI, Sigma-Aldrich, St Louis, MO).
6. 150 mM NaCl.
7. Hank's balanced salt solution (HBSS) supplemented with 0.1 % glucose.

Table 1
Combinations of fluorescent protein fragments for BiFC

Fusion[a]	Purpose	Excitation filter(s) (nm)	Emission filter(s) (nm)
A-YN155 B-YC155	A–B interaction	480–520	495–565
A-YN173 B-YC173	A–B interaction	480–520	495–565
A-CN155 B-CC155	A–B interaction	426–446	440–500
A-YN155 B-CN155 Z-CC155	Simultaneous visualization of A and B interaction with Z	480–520 and 426–446	505–565 and 440–500

[a]YN155 corresponds to EYFP residues 1–154, YC155 to EYFP 155–238, YN173 to EYFP 1–172, YC173 to EYFP 173–238, CN155 to ECFP 1–154, and CC155 to ECFP 155–238 (Table adapted from Kerppola [56])

8. Bradford protein assay (Bio-Rad, Hercules, CA).
9. Black 96-well microplates (OptiPlate, PerkinElmer, Waltham, MA).
10. Multilabel fluorescent plate reader (EnVision, Perkin Elmer).
11. Statistical analysis software (GraphPad PRISM, GraphPad Software Inc., San Diego, CA and MATLAB, The Mathworks Inc., Natick, MA).

2.2 Additional Material for BRET, BRET-BiFC, and SRET Analysis

1. Flat-bottom white 96-well microplates (Corning 3912, Corning, NY).
2. Coelenterazine H (p.j.k GmbH, Germany).
3. DeepBlueC (Biotium Inc., Hayward, CA).

2.3 Expression Vectors for Fluorescent and Luminescent Proteins

1. pECFP-N1, pEYFP-N1, pEdsRed-N1 (Clontech, Mountain View, CA) (*see* **Notes 2** and **3**).
2. For BRET assays use C-terminal part of the chemokine receptor fused to the *Renilla* luciferase gene using the pRLuc-N1 plasmid (PerkinElmer).
3. For BiFC assays, use N- and C-terminal nonfluorescent fragments of a specific fluorescent protein (*see* Table 1) cloned in pcDNA3.1 (Addgene, Cambridge, MA) (*see* **Note 4**).

3 Methods

3.1 Fluorescence Resonance Energy Transfer

The FRET mechanism involves a donor fluorophore in an electron-excited state that can transfer its excitation energy to a nearby acceptor chromophore in a non-radioactive fashion through

Table 2
Properties of fluorescent protein pairs for FRET

Fluorescence protein pair	Donor excitation maximum (nm)	Acceptor emission maximum (nm)	Donor quantum yield	Förster distance (nm)
EBFP2–mEGFP	383	507	0.56	4.8
ECFP–EYFP	440	527	0.40	4.9
Cerulean–Venus	440	528	0.62	5.4
MICy–mKO	472	559	0.90	5.3
TFP1–mVenus	492	528	0.85	5.1
CyPet–YPet	477	530	0.51	5.1
EGFP–mCherry	507	510	0.60	5.1
Venus–mCherry	528	610	0.57	5.7
Venus–tdTomato	528	581	0.57	5.9
Venus–mPlum	528	649	0.57	5.2

long-range dipole–dipole interactions. This phenomenon is not mediated by photon emission and in many applications, energy transfer results in quenching of donor fluorescence and subsequent reduction of fluorescence lifetime; this process is obviously also accompanied by an increase in acceptor fluorescence emission. The range over which energy transfer can take place is limited to approximately 10 nm (100 Å), a sufficient distance to consider that molecular interactions take place. FRET efficiency is thus extremely sensitive to the distance between donor and acceptor.

Although there are many fluorescent proteins pairs used as FRET donor/acceptors (Table 2), they respond to several common requirements.

1. The donor emission spectrum must overlap the acceptor excitation spectrum. If there is donor–acceptor interaction, donor excitation with the maximum absorbance wavelength must increase the intensity of the maximum emission fluorescence of the acceptor.
2. Energy transfer efficiency depends on the relative orientations of the donor emission dipole and the acceptor absorption dipole.
3. Donor and acceptor molecules must be located within 1–10 nm of one another. Energy transfer efficiency between donor and acceptor molecules decreases as the sixth power of the distance between the two fluorescent molecules as described by the Förster equation:

$$E_t = R_0^{\,6} / R^6 + R_0^{\,6}$$

where E_t represents energy transfer efficiency, R is the distance between the fluorescent proteins, and R_0 is the donor–acceptor distance at which 50 % of the excitation energy is transferred whereas the remaining 50 % decays as non-radioactive or radioactive energy. This efficiency depends on the fluorescent partners used (Table 1).

4. The fluorescence lifetime of the donor molecule (quantum yield of the donor) must be long enough to allow energy transfer to the acceptor.

Several methods are used to determine and quantify FRET efficiency (*see* **Note 5**). To study receptor interactions in dynamic processes the most adequate FRET technique is based in the sensitized emission of the acceptor.

3.1.1 Sensitized Emission of the Acceptor

FRET efficiency can be determined in individual (fixed) cells using confocal microscopy or in living cell populations using a multilabel fluorescent plate reader with specific detectors. The method requires specific excitation of the donor fluorescent protein and use of specific light filters to determine donor and acceptor fluorescent emissions. Specific detectors must be used to collect the maximum peak of fluorescent emission for both donor and acceptor molecules.

As in other FRET methods, the donor emission spectrum must overlap the acceptor excitation spectrum. For many FRET partners, the requirement for spectral proximity can lead to strong fluorophore cross talk and cross-excitation processes that might alter quantitative analysis. To correct for detection of ratiometric signals and sensitized emission, a linear unmixing method is used that is based on the assumption that total detected signal (S) for each channel (λ) can be expressed as a linear combination of the contributing fluorophores (FluoX), according to the formula:

$$S(\lambda) = A_1 \times \text{Fluo1}(\lambda) + A_2 \times \text{Fluo2}(\lambda)$$

where An is the contribution of a specific fluorophore, and fluorophore concentration in the signal observed therefore determines their respective contribution to the total signal [50, 51]. This signal is redistributed into the specific fluorescence channels and can be analyzed quantitatively. To calculate fluorophore contributions, linear unmixing requires knowledge of reference values for samples that contain the fluorophores of interest separately. For successful separation of overlapping signals, the number of detection channels used and the number of fluorophores in the sample must be the same.

For highly efficient FRET partners such as CFP and YFP, which have very close overlapping signals, the FRET pair must be excited by a single excitation wavelength and FRET interactions resolved using linear unmixing (Fig. 1, example spectrum CFP-YFP, see FRET measurement). This is the best way to analyze the contribution of individual signals to the mixture.

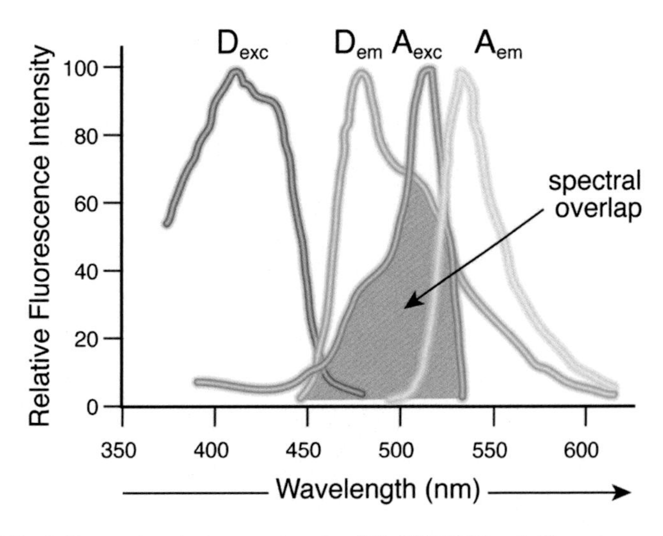

Fig. 1 Excitation and emission spectra of a CFP–YFP FRET pair. The scheme shows absorbance (exc) and emission (em) spectra of CFP (cyan fluorescent protein; donor; D) and YFP (yellow fluorescent protein; acceptor, A). Spectral overlap between CFP emission and YFP excitation (*shaded region*) is a prerequisite for FRET

3.1.2 FRET Saturation Curves by Sensitized Emission

1. Plate HEK293T cells (3.5×10^5 cells/well in 2 ml complete DMEM, using 6-well plates) and culture (24 h, 37 °C, 5 % CO_2).
2. To obtain FRET saturation curves, prepare two transfection mixtures: (*a*) mix 25 μl of 150 mM NaCl/μg of DNA and vortex (10 s); (*b*) at a 4:1 ratio, mix 150 mM NaCl and PEI (5.47 mM in nitrogen residues). Add mixture *b* to mixture *a* at a 1:1 ratio, vortex (10 s) and incubate (15–30 min, room temperature). Add the mixture (*a* + *b*) to the cells in 1 ml serum-free DMEM, incubate (4 h, 37 °C), and replace medium with complete DMEM (*see* **Note 6**).
3. At 48 h post-transfection, wash cells twice in HBSS supplemented with 0.1 % glucose and resuspend in this same buffer. Determine total protein concentration for each cell sample using a Bradford assay kit. Pipette Cell suspensions (20 μg protein/100 μl) into black 96-well microplates and read in a multilabel fluorescent plate reader equipped with a high-energy xenon flash lamp, a short-wavelength filter (8 nm bandwidth, 405 nm) and a long-wavelength emission filter (10 nm bandwidth, 486 nm for CFP channel and 10 nm bandwidth, 530 for YFP channel) (*see* **Note 7**). To determine the spectral signature, HEK293T cells are transiently transfected with the chemokine receptor coupled to CFP or to YFP separately. The contributions of CFP and YFP alone are measured in each detection channel and normalized to the sum of the signal obtained in both channels. Analyze the spectral signatures of CFP or YFP fused to chemokine receptors and check variability

(it must not vary significantly ($p > 0.05$)) from the signatures determined for each fluorescent protein alone. For FRET quantitation and receptor-YFP expression quantitation in FRET saturation curves, the spectral signature is taken into account for linear unmixing in order to separate the two emission spectra. To determine the fluorescence emitted by each fluorophore in FRET experiments, apply the following formulas:

$$\text{CFP} = S/(1+1/R) \text{ and } \text{YFP} = S/1+R$$

where

$$S = \text{ChCFP} + \text{ChYFP},$$

$$R = (\text{YFP}_{530}Q \quad \text{YFP}_{486})/(\text{CFP}_{486} - \text{CFP}_{530}Q), \text{ and}$$

$$Q = \text{ChCFP}/\text{ChYFP}$$

ChCFP and ChYFP represent the signal detected for CFP in the 486 nm and 530 nm detection channels (Ch), respectively; CFP_{486}, CFP_{530}, YFP_{530}, and YFP_{486} are the normalized contributions of CFP and YFP to 486 and 530 nm channels, as determined from their spectral signatures.

4. Sensitized emission FREt allows measurement of the energy transferred relative to the acceptor/donor ratio to generate FRET saturation curves. The curves show FRET efficiency as a function of the acceptor–donor ratio and are characterized by two important parameters; $FRET_{max}$, is the (asymptotic) maximum of the curve and $FRET_{50}$ corresponds to the acceptor–donor ratio that yields half $FRET_{max}$ efficiency (only when energy transfer reaches saturation and the curve is hyperbolic). $FRET_{max}$ is associated with the number of receptor complexes formed and/or changes in complex conformation, and $FRET_{50}$ allows estimation of the apparent affinity between the two partners that form the complex [52, 53]. These two parameters are deduced from data analysis using a nonlinear regression equation applied to a single binding site model (based on Michaelis–Menten model, Fig. 2) (*see* **Note 8**).

3.2 BRET

The BRET technique is based on non-radiative energy transfer between a bioluminescent donor (usually *Renilla* luciferase (Luc)) and a fluorophore acceptor. Like FRET, this method requires the generation of fusion proteins for donors between the receptor and luciferase (Luc) and for acceptors fusion between receptor and YFP, CFP, or GFP^2. When Luc is present, it oxidizes its substrate (coelenterazine or DeepBlueC) and triggers photons release. Close proximity (10 nm) of an appropriate fluorophore ensures its excitation; that is, electron movement to a higher energetic state and thus, photon emission of longer wavelengths (*see* **Note 9**).

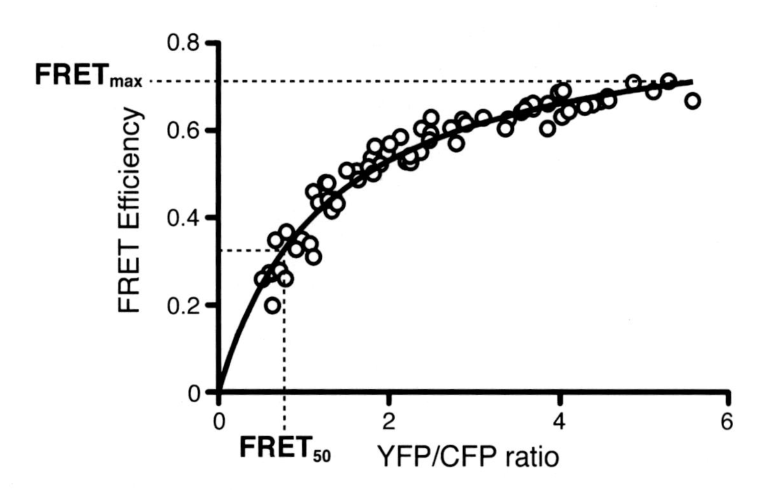

Fig. 2 FRET saturation curve generated by the sensitized emission method. A representative FRET saturation curve for a chemokine receptor pair. The curve reaches FRETmax (maximum FRET efficiency detected) and is hyperbolic. $FRET_{50}$ indicates the FRET efficiency value that corresponds to the acceptor–donor ratio (YFP–CFP ratio) that yields half FRETmax

3.2.1 BRET Titration Assays

1. Plate HEK293T cells (3.5×10^5 cells/well in 2 ml complete DMEM, using 6-well plates) and culture (24 h, 37 °C, 5 % CO_2).
2. Cotransfect the cells using the PEI protocol (*see* Subheading 3.1.1 above) with a constant amount of donor (Luc-fused chemokine receptor) and increasing amounts of acceptor protein (YFP-, CFP-, or GFP^2-fused chemokine receptor, depending on the luciferase substrate; see above) (*see* **Note 10**).
3. At 48 h post-transfection, wash cells twice in HBSS supplemented with 0.1 % glucose and resuspend cells in this buffer. Using a Bradford assay, determine total protein concentration in each well.
4. Quantify the fluorescent protein (20 μg) using a multilabel plate reader equipped with a high-energy xenon flash lamp (for CFP or GFP^2 acceptor, 8 nm bandwidth excitation filter at 405 nm; for YFP, 10 nm bandwidth excitation filter at 510 nm). Receptor fluorescence expression is determined as fluorescence of the sample minus the fluorescence of cells that express donor alone. For $BRET^2$ and $BRET^1$ measurements, the equivalent of 20 μg cell suspension is distributed in 96-well microplates, followed by 5 μM DeepBlueC (for $BRET^2$) or coelenterazine H (for $BRET^1$). For $BRET^2$ experiments, signals are obtained immediately (30 s) after DeepBlueC addition using the multilabel plate reader, which allows integration of signals detected in the short- (8 nm bandwidth, 405 nm) and the long-wavelength filters (10 nm bandwidth, 486 nm). For $BRET^1$, readings are collected 1 min after coelenterazine H addition, as the plate

reader allows integration of signals detected in the short- (10 nm bandwidth, 510 nm) and long-wavelength filters (10 nm bandwidth, 530 nm). Receptor-Luc luminescence signals should be acquired 10 min after coelenterazine H (5 μM) addition. BRET efficiency ($BRET_{eff}$) is defined as:

$$BRET_{eff} = [(\text{long wavelength emission}) / (\text{short wavelength emission})] - C_f$$

where C_f is [(long wavelength emission)/(short wavelength emission)] for the Luc construct expressed alone in the same experiment (control).

5. Statistical analysis (*see* **Note 5** and Subheading 3.1.2 above).

3.3 Bimolecular Fluorescence Complementation (BiFC)

Bimolecular fluorescence complementation (BiFC) technology enables simple, direct visualization of protein–protein interactions in living cells [54–56]. The BiFC assay uses receptors fused to two fluorescent protein fragments that are nonfluorescent individually; fluorescence is recovered only when both fragments interact, that is, when the accompanying receptor form complexes. The approach can be used for analysis of interactions between many types of proteins and does not require information about the structures of the interaction partners (*see* **Note 11**).

Like FRET and BRET, BiFC requires fusion of the fluorescent protein fragments to chemokine receptors, as well as testing that neither protein expression nor receptor function is modified by the fluorescent fragments (*see* **Notes 12** and **13**).

3.3.1 BiFC Measurement

1. Plate HEK293T cells 24 h before transfection (3.5×10^5 cells/2 ml, in 6-well plates) and cultured (37 °C, 5 % CO_2).
2. Cotransfect using PEI or JetPei methods (according to manufacturer's protocol), with chemokine receptors fused to the nonfluorescent fragments at a 1:1 ratio (i.e., 0.7 μg CXCR4-nYFP + 0.7 μg CXCR4-cYFP) and culture (48 h, 37 °C, 5 % CO_2).
3. Wash cells with HBSS supplemented with 0.1 % glucose and resuspend in the same buffer. Determine the protein concentration in the cell pool using a Bradford assay kit (Bio-Rad).
4. Pipette cells into black 96-well plates, at 20 μg of protein in 100 μl (0.2 μg/μl) per well.
5. Quantify the fluorescent protein using a multilabel plate reader equipped with a high-energy xenon flash lamp (for YFP, 10 nm bandwidth excitation filter at 510 nm and 10 nm bandwidth emission filter at 530 nm).

3.4 BRET-BiFC

BRET-BiFC allows identification of heterotrimeric complexes in living cells. This method combines BRET and BiFC techniques sequentially. It is thus necessary take into account all controls and considerations described in Subheadings 2 and 3. In BRET-BiFC, the donor is a Luc-fused chemokine receptor or protein and the acceptor is formed by interaction of the two nonfluorescent fragments fused to the chemokine receptors of interest.

1. Plate HEK293T cells (3.5×10^5 cells/well in 2 ml complete DMEM, use 6-well plates) and culture (24 h, 37 °C, 5 % CO_2).
2. Using the PEI method (*see* Subheading 3.1.2, **step 2** above), cotransfect cells with a constant amount of donor (Luc-fused chemokine receptor) and increasing amounts of a mixture of acceptor proteins at a 1:1 ratio (i.e., nYFP- and cYFP-fused chemokine receptors) (*see* **Note 14**).
3. At 48 h post-transfection, wash cells twice in HBSS supplemented with 0.1 % glucose and resuspend the pellet in the same buffer. Determine protein concentration for the cell pool using a Bradford assay kit.
4. Quantify the fluorescent protein in samples containing 20 μg protein, using a multilabel plate reader equipped with a high-energy xenon flash lamp (for YFP, 10 nm bandwidth excitation filter at 510 nm). Expression of the fluorescent protein-fused receptor is determined as fluorescence of the sample minus fluorescence of cells expressing donor-Luc alone. Distribute the cell suspension (20 μg/well) in 96-well microplates (Corning 3912, flat-bottom white plates). Add coelenterazine H (for $BRET^1$, 5 μM/well; PJK GmbH) and read 1 min later in the plate reader, which allows integration of signals detected in the short- (10 nm bandwidth, 510 nm) and long-wavelength filters (10 nm bandwidth, 530 nm). Receptor-Luc luminescence signals are acquired 10 min after coelenterazine H addition. BRET-BiFC efficiency is calculated using the same formula given for BRET efficiency in Subheading 3.2.1 above.

   In each BRET-BiFC titration curve, the relative amount of acceptor is given by the ratio between acceptor (YFP) formed by the two nonfluorescent proteins and the luciferase activity of the donor (Luc).
5. Statistical analysis as in Subheading 3.1.2 above (*see* **Note 5**).

3.5 Sequential BRET-FRET (SRET)

The SRET method allows detection of complexes of three proteins in living cells. There are several possible combinations depending on the donor and acceptor fusion proteins (Fig. 3).

$SRET^1$ combines $BRET^1$ and FRET. The first donor protein is a Luc-fused chemokine receptor. The first acceptor protein is YFP, which is then used as a second donor to excite the next acceptor, a chemokine receptor fused to dsRed, which is the signal detected. This method uses coelenterazine H as a Luc substrate.

$SRET^2$ combines $BRET^2$ and FRET. The Luc-fused chemokine receptor is the first donor (using DeepBlueC as a specific Luc substrate). T CFP- or GFP^2-fused receptor is then excited and its

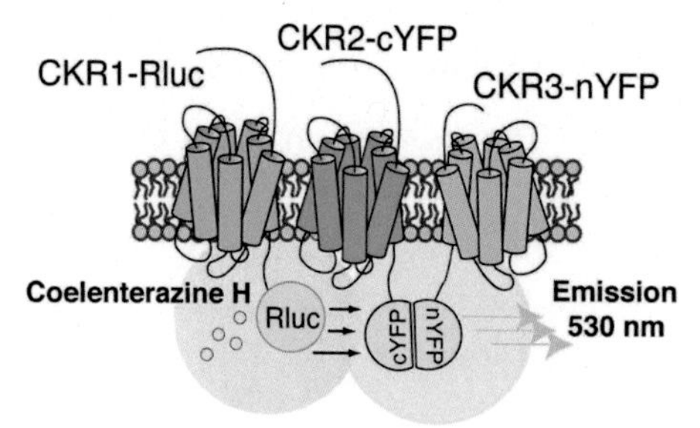

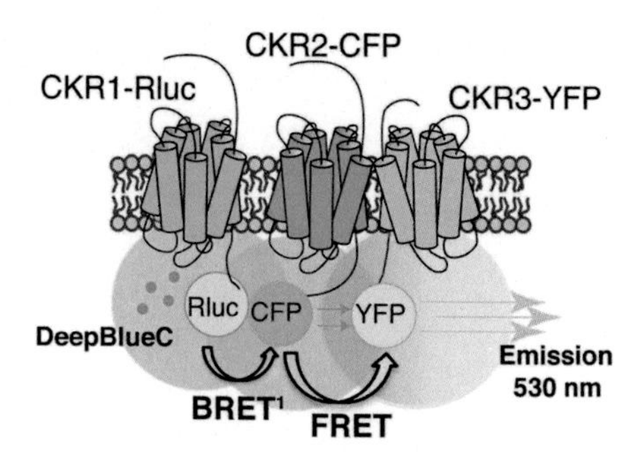

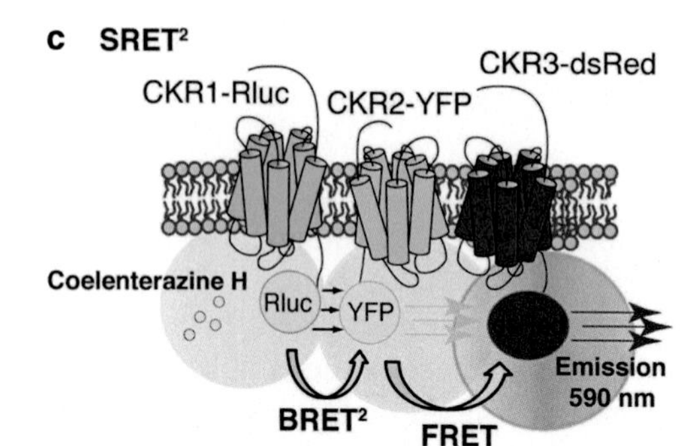

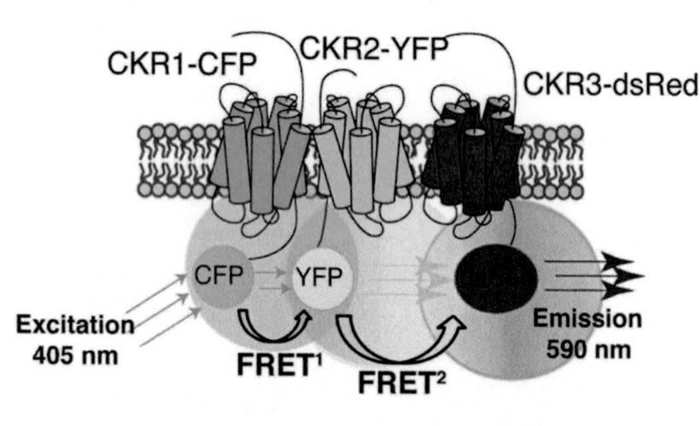

Fig. 3 Scheme of BRET-BiFC, SRET^1, SRET^2, and SRET^3 techniques. (**a**) BRET-BiFC for the chemokine receptors CKR1, CKR2, and CKR3 fused to Rluc, cYFP, and nYFP, respectively. Following interaction between chemokine receptors CKR2 and CKR3 (*colored halos*), YFP is reconstituted and is susceptible to excitation by Rluc activated by its substrate, coelenterazine H (*yellow dots*). EYFP emission is detected at 530 nm. (**b**) SRET^1 for the chemokine receptors CKR1, CKR2, and CKR3 fused to Rluc, CFP, and YFP, respectively. Due to activation by its substrate DeepBlueC (*blue dots*), Rluc excites CFP (BRET), which in turn excites YFP (FRET) that is detected at 530 nm. (**c**) In SRET^2, initial BRET between Luc and YFP triggered by coelenterazine H (*yellow dots*) excites dsRed, the last acceptor fluorescence protein, which then emits light at 590 nm. (**d**) SRET^3 is a FRET^1 and FRET^2 sequence. Due to the interaction between CFP, YFP, and dsRed fused to a chemokine receptor (CKR1, CKR2, CKR3, respectively), CFP excitation (at 405 nm) triggers dsRed light emission at 590 nm

emission energy (BRET^2) excites the YFP-fused receptor, which is then the last SRET acceptor.

SRET^3 combines two sequential FRET methods. The donor protein is the CFP- or GFP^2-fused chemokine, the first acceptor/second donor is the YFP-fused receptor which excites the last acceptor, which is the dsRed-fused receptor.

Here we describe in detail the SRET^2 method, which can easily be adapted for SRET^1 or SRET^3 using the detection filters for each specific fluorophore and the appropriate Luc substrate in the case of SRET^1.

1. Plate HEK293T cells (3.5×10^5 cells/well in 2 ml complete DMEM, use 6-well plates) and culture (24 h, 37 °C, 5 % CO_2).
2. Using the PEI method (Subheading 3.1.2, **step 2** above), cotransfect the cells with a constant amount of Luc-fused

receptor and increasing amounts of receptor fused to acceptors (CFP or GFP^2 and YFP at 1:1 ratio); culture (48 h, 37 °C, 5 % CO_2).

3. Wash cells with HBSS supplemented with 0.1 % glucose and resuspend the pellet in the same buffer. Total protein concentration is determined for the cell pool using a Bradford assay kit (Bio-Rad).
4. Use aliquots of transfected cells (20 μg protein/100 μl in a 96-well microplate) to perform three experiments in parallel.
5. In the first experiment, quantify FRET efficiency for the FRET pairs used (CFP or GFP^2 and YFP) and the amount of acceptor protein–YFP at each ratio. Distribute cells (20 μg) in black 96-well microplates and read in a multilabel plate reader equipped with a high-energy xenon flash lamp, using an excitation filter at 405 ± 8 nm and 10 nm bandwidth emission filters corresponding to 530 nm (channel 1) and 486 nm (channel 2). As in FRET, separate the relative contribution of fluorophores to each detection channel for linear unmixing. Then measure the contribution of CFP or GFP^2 and YFP proteins alone to the two detection channels and normalize to the sum of the total signal obtained in the two detection channels. Quantify the total amount of receptor-YFP at each ratio in the same equipment using a 510 ± 10 nm excitation filter at and a 530 ± 10 nm emission filter.
6. In the second experiment, quantify the receptor-Luc expression by determining its luminescence. Pipette the cells (20 μg) into 96-well microplates (white-bottomed white plates), add the substrate (5 μM coelenterazine H); after 10 min, detect luminescence in a multilabel plate reader.
7. For $SRET^1$ evaluation in the third experiment, distribute cells (20 μg) in 96-well microplates (white-bottomed white plates) and add 5 μM DeepBlueC as Luc substrate; after 30 s, collect the SRET signal using a multilabel plate reader with detection filters for short (486 nm) and long wavelengths (530 nm).
8. By analogy with BRET, net SRET is defined as:

$$\text{netSRET} = ((\text{long wavelength emission}) / (\text{short wavelength emission})) - C_f$$

where

$$C_f = ((\text{long wavelength emission}) / (\text{short wavelength emission})$$

for cells expressing the receptor-Luc, receptor-CFP or -GFP^2 separately. SRET is only detected if all receptors interact and the corresponding pairs (Luc/CFP or Luc/GFP^2 and CFP/YFP or GFP^2/YFP or Luc/YFP) are located at <10 nm distance.

9. For statistical analysis, *see* Subheading 3.1.2 above and **Note 5**.

4 Notes

1. Any other cell line with high transfection efficiency for the receptors of interest can be employed. In this case the transfection method should be optimized to these cell lines. Take into account that most cell lines can express endogenously chemokine receptors (most frequently CXCR4) that may alter chemokine oligomers.
2. Several plasmids are commercially available to fuse fluorescent proteins to a receptor (Clontech).
3. For chemokine receptors and to avoid interference with ligand binding it is best to fuse fluorescent proteins to the C-terminal end of the receptor.
4. All the techniques described in this chapter require fusion of chemokine receptors to fluorescent (FRET) or luminescent proteins (BRET), or N- or C-terminal fragments of fluorescent proteins (BiFC). Insertion of a fluorescent probe in the C-terminal region of the receptor involves eliminating the receptor stop codon, whereas N-terminal insertion requires elimination of the fluorescent protein stop codon. Transfected cells should therefore be analyzed for receptor expression and function.

   The constructs are obtained using standard molecular biology techniques in commercially available vectors that bear the luminescent or fluorescent proteins (Clontech). In the case of BiFC, the fluorescent protein used must be cut into N- and C-terminal fragments, which must then be included in separate expression vectors (i.e., pcDNA3.1; *see* Table 2). Insertion of these fragments in-frame with the chemokine receptor C-terminal region requires elimination of the receptor stop codon. Transfected cells with these chimeric proteins should be tested for receptor expression and function.

   Several experiments are needed to test whether the chimeric proteins maintain expression and function. Using flow cytometry, stain the chemokine receptors expressed at the cell surface with specific antibodies. Chemokine receptor function is usually tested using calcium mobilization assays or migration in transwells. The chemokine receptors fused to fluorescent/luminescent proteins must behave similarly to the wild type receptor.
5. Methods used to determine and quantify FRET efficiency include:

   (a) *Donor quenching or acceptor photobleaching* is a method based on quenching donor fluorescence. As some donor photons excite the acceptor, a decrease is detected in donor emission energy. After acceptor photobleaching, increased donor light emission is detected. Due to cell damage caused by the extended laser exposure needed to photobleach the acceptor, this method cannot be used in live cells.

(b) For *sensitized emission of the acceptor*, the acceptor signal is quantitated after donor excitation. This method is useful to determine FRET in dynamic processes in the cell, such as the consequences of ligand stimulation, the effects triggered by other proteins coexpressed on cell surface, and so on.

(c) *Fluorescence lifetime imaging microscopy (FLIM)* is a method based on measurement of a constant parameter of each fluorescent protein in each experimental condition, termed lifetime. Lifetime is the duration of the excited state of a fluorophore before returning to its ground state. This technique allows spatial resolution of biochemical processes. The fluorescence lifetime of a donor molecule decreases in FRET conditions independently of fluorescent protein concentrations and of excitation intensities.

6. The cells must be transiently cotransfected with a constant amount of donor (chemokine receptor fused to the donor fluorophore, CFP) and increasing amounts of acceptor protein (chemokine receptor fused to the acceptor fluorophore, YFP) using the polyethylenimine method (PEI).

7. Gain settings must be identical for all experiments to maintain a constant relative contribution of each fluorophore to the detection channels for spectral imaging and linear unmixing.

8. Statistical analysis is needed to reduce experimental variability (at least five replications of the saturation curves are usually generated), and will finally determine the homogeneous curves. We currently use three distinct statistical methods (bootstrap, F test, and Akaike information criterion (AIC)), which usually lead to similar conclusions [57]. For example, we can determine the conformational changes in dimers promoted by a given ligand ("the treatment"). Curves for untreated and ligand-treated groups are naturally paired when dimerization is evaluated in the same group of cells before and after "treatment." To determine which model best fits the data for pairs of saturation curves, we use the AIC method corrected for small size samples ($n \leq 10$), in which the null hypothesis is one curve for all data sets (before and after "treatment") and the alternative hypothesis is the existence of different curves for each data set. If the majority of the AICc difference (Δ) is positive, the preferred model is a distinct curve for all data sets, whereas if Δ is negative, the preferred model is a single curve for all data sets [28]. When the number of individual determinations (n) in each curve is >10 observations, a t-test can be used to compare the components of each pair [58–60]. When the p value is <0.05, we can conclude that the changes observed after a specific treatment are statistically significant.

9. Distinct luciferase substrates can be chosen depending on the acceptor fluorophore. Requirements for energy transfer in BRET are the same that those for FRET (*see* Subheading 3.1 above). The most used luciferase substrates are coelenterazine, used in $BRET^1$ assays, whose maximal emission is at 515 nm and which is used in combination with YFP as acceptor. DeepBlueC, used in $BRET^2$ assays, is an analogue of the natural luciferase substrate with maximal emission wavelength at 410 nm; it is used in combination with CFP or GFP^2, which emit at 485 nm and 515 nm respectively, yielding a spectral separation of >100 nm.
10. The cells must also be transfected separately with donor and acceptor. Use untransfected HEK293T cells as a background control for each experiment.
11. One advantage of this technique is that the complex formed has strong intrinsic fluorescence, which allows direct visualization of the protein interaction without exogenous agents. This avoids disturbance of the cells, but also is a clear disadvantage, as the user must know that there is a delay between the time needed by the proteins to interact and the time required for the reconstituted complex to become fluorescent [54].
12. Flexible linker sequences are recommended, to allow maximal mobility of the nonfluorescent fragments after complex formation.
13. To determine BiFC specificity, many controls are included in each experiment, such as nonfluorescent fragment-fused receptors with point mutations in the interaction interface that impede reconstitution of the complete protein [55, 61]. Control proteins with the same cell expression pattern as the chemokine receptors should be included.
14. Controls include cells transfected with the receptor-Luc construct alone and with receptors fused to the acceptor fragments. In addition, untransfected HEK293T cells are needed to determine the background.

Acknowledgments

We specially thank the present and former members of the DIO chemokine group who contributed to some of the work described in this review. We also thank C. Bastos and C. Mark for secretarial support and editorial assistance, respectively. This work was partially supported by grants from the Spanish Ministry of Economy and Competitiveness (SAF-2011-27270), the RETICS Program (RD 12/0009/009 RIER) and the Madrid regional government (S2010/BMD-2350; RAPHYME).

References

1. Griffith JW, Sokol CL, Luster AD (2014) Chemokines and chemokine receptors: positioning cells for host defense and immunity. Annu Rev Immunol 32:659–702
2. Bachelerie F, Ben-Baruch A, Burkhardt AM et al (2013) International Union of Basic and Clinical Pharmacology. [corrected]. LXXXIX. Update on the extended family of chemokine receptors and introducing a new nomenclature for atypical chemokine receptors. Pharmacol Rev 66:1–79
3. Baggiolini M (1998) Chemokines and leukocyte traffic. Nature 392:565–568
4. Rossi D, Zlotnik A (2000) The biology of chemokines and their receptors. Annu Rev Immunol 18:217–242
5. Mackay CR (2001) Chemokines: immunology's high impact factors. Nat Immunol 2:95–101
6. Balkwill F (2004) Cancer and the chemokine network. Nat Rev Cancer 4:540–550
7. Baggiolini M, Dahinden CA (1994) CC chemokines in allergic inflammation. Immunol Today 15:127–133
8. Belperio JA, Keane MP, Arenberg DA et al (2000) CXC chemokines in angiogenesis. J Leukoc Biol 68:1–8
9. Gerard C, Rollins BJ (2001) Chemokines and disease. Nat Immunol 2:108–115
10. Godessart N, Kunkel SL (2001) Chemokines in autoimmune disease. Curr Opin Immunol 13:670–675
11. Proudfoot AE (2002) Chemokine receptors: multifaceted therapeutic targets. Nat Rev Immunol 2:106–115
12. Proudfoot AE, Handel TM, Johnson Z et al (2003) Glycosaminoglycan binding and oligomerization are essential for the in vivo activity of certain chemokines. Proc Natl Acad Sci U S A 100:1885–1890
13. Jansma A, Handel TM, Hamel DJ (2009) Chapter 2. Homo- and hetero-oligomerization of chemokines. Methods Enzymol 461:31–50
14. Hamel DJ, Sielaff I, Proudfoot AE, Handel TM (2009) Chapter 4. Interactions of chemokines with glycosaminoglycans. Methods Enzymol 461:71–102
15. Salanga CL, Handel TM (2011) Chemokine oligomerization and interactions with receptors and glycosaminoglycans: the role of structural dynamics in function. Exp Cell Res 317:590–601
16. Alcami A (2003) Viral mimicry of cytokines, chemokines and their receptors. Nat Rev Immunol 3:36–50
17. Seet BT, McFadden G (2002) Viral chemokine-binding proteins. J Leukoc Biol 72:24–34
18. Murphy PM (2001) Viral exploitation and subversion of the immune system through chemokine mimicry. Nat Immunol 2:116–122
19. De Paula VS, Gomes NS, Lima LG et al (2013) Structural basis for the interaction of human β-defensin 6 and its putative chemokine receptor CCR2 and breast cancer microvesicles. J Mol Biol 425:4479–4495
20. Chabre M, Deterre P, Antonny B (2009) The apparent cooperativity of some GPCRs does not necessarily imply dimerization. Trends Pharmacol Sci 30:182–187
21. Hernanz-Falcón P, Rodríguez-Frade JM, Serrano A et al (2004) Identification of amino acid residues crucial for chemokine receptor dimerization. Nat Immunol 5:216–223
22. Rodríguez-Frade JM, Mellado M, Martínez-A C (2001) Chemokine receptor dimerization: two are better than one. Trends Immunol 22:612–617
23. Rodríguez-Frade JM, Vila-Coro AJ, de Ana AM et al (1999) The chemokine monocyte chemoattractant protein-1 induces functional responses through dimerization of its receptor CCR2. Proc Natl Acad Sci U S A 96: 3628–3633
24. Thelen M, Muñoz LM, Rodríguez-Frade JM, Mellado M (2010) Chemokine receptor oligomerization: functional considerations. Curr Opin Pharmacol 10:38–43
25. Wu B, Chien EY, Mol CD et al (2010) Structures of the CXCR4 chemokine GPCR with small-molecule and cyclic peptide antagonists. Science 330:1066–1071
26. Barroso R, Martínez Muñoz L, Barrondo S et al (2012) EBI2 regulates CXCL13-mediated responses by heterodimerization with CXCR5. FASEB J 26:4841–4854
27. Pello OM, Martínez-Muñoz L, Parrillas V et al (2008) Ligand stabilization of CXCR4/delta-opioid receptor heterodimers reveals a mechanism for immune response regulation. Eur J Immunol 38:537–549
28. Martínez-Muñoz L, Barroso R, Dyrhaug SY et al (2014) CCR5/CD4/CXCR4 oligomerization prevents HIV-1 gp120IIIB binding to the cell surface. Proc Natl Acad Sci U S A 111:E1960–E1969
29. Iizumi M, Bandyopadhyay S, Watabe K (2007) Interaction of Duffy antigen receptor for chemokines and KAI1: a critical step in metastasis suppression. Cancer Res 67:1411–1414

30. Kumar A, Humphreys TD, Kremer KN et al (2006) CXCR4 physically associates with the T cell receptor to signal in T cells. Immunity 25:213–224
31. Herrera C, Morimoto C, Blanco J et al (2001) Comodulation of CXCR4 and CD26 in human lymphocytes. J Biol Chem 276:19532–19539
32. Yoshida T, Ebina H, Koyanagi Y (2009) N-linked glycan-dependent interaction of CD63 with CXCR4 at the Golgi apparatus induces downregulation of CXCR4. Microbiol Immunol 53:629–635
33. Mellado M, Rodríguez-Frade JM, Vila-Coro AJ et al (2001) Chemokine receptor homo- or heterodimerization activates distinct signaling pathways. EMBO J 20:2497–2507
34. Rozenfeld R, Devi LA (2010) Receptor heteromerization and drug discovery. Trends Pharmacol Sci 31:124–130
35. Mellado M, Rodríguez-Frade JM, Vila-Coro AJ et al (1999) Chemokine control of HIV-1 infection. Nature 400:723–724
36. Molon B, Gri G, Bettella M et al (2005) T cell costimulation by chemokine receptors. Nat Immunol 6:465–471
37. Chen C, Li J, Bot G et al (2004) Heterodimerization and cross-desensitization between the mu-opioid receptor and the chemokine CCR5 receptor. Eur J Pharmacol 483:175–186
38. Szabo I, Wetzel MA, Zhang N et al (2003) Selective inactivation of CCR5 and decreased infectivity of R5 HIV-1 strains mediated by opioid-induced heterologous desensitization. J Leukoc Biol 74:1074–1082
39. Rodríguez-Frade JM, Vila-Coro AJ, Martín A et al (1999) Similarities and differences in RANTES- and (AOP)-RANTES-triggered signals: implications for chemotaxis. J Cell Biol 144:755–765
40. Percherancier Y, Berchiche YA, Slight I et al (2005) Bioluminescence resonance energy transfer reveals ligand-induced conformational changes in CXCR4 homo- and heterodimers. J Biol Chem 280:9895–9903
41. Wilson S, Wilkinson G, Milligan G (2005) The CXCRl and CXCR2 receptors form constitutive homo- and heterodimers selectively and with equal apparent affinities. J Biol Chem 280:28663–28674
42. Harrison C, van der Graaf PH (2006) Current methods used to investigate G protein coupled receptor oligomerisation. J Pharmacol Toxicol Methods 54:26–35
43. Pfleger KD, Eidne KA (2006) Illuminating insights into protein-protein interactions using bioluminescence resonance energy transfer (BRET). Nat Methods 3:165–174
44. Cardullo RA (2007) Theoretical principles and practical considerations for fluorescence resonance energy transfer microscopy. Methods Cell Biol 81:479–494
45. Boute N, Jockers R, Issad T (2002) The use of resonance energy transfer in high-throughput screening: BRET versus FRET. Trends Pharmacol Sci 23:351–354
46. Coulon V, Audet M, Homburger V et al (2008) Subcellular imaging of dynamic protein interactions by bioluminescence resonance energy transfer. Biophys J 94:1001–1009
47. Pfleger KD, Seeber RM, Eidne KA (2006) Bioluminescence resonance energy transfer (BRET) for the real-time detection of protein-protein interactions. Nat Protoc 1:337–345
48. El-Asmar L, Springael JY, Ballet S et al (2005) Evidence for negative binding cooperativity within CCR5-CCR2b heterodimers. Mol Pharmacol 67:460–469
49. Marullo S, Bouvier M (2007) Resonance energy transfer approaches in molecular pharmacology and beyond. Trends Pharmacol Sci 28:362–365
50. Dickinson ME, Bearman G, Tille S et al (2001) Multi-spectral imaging and linear unmixing add a whole new dimension to laser scanning fluorescence microscopy. Biotechniques 31:1272, 1274–1276, 1278
51. Zimmermann T, Rietdorf J, Girod A et al (2002) Spectral imaging and linear un-mixing enables improved FRET efficiency with a novel GFP2-YFP FRET pair. FEBS Lett 531: 245–249
52. Mercier JF, Salahpour A, Angers S et al (2002) Quantitative assessment of beta 1- and beta 2-adrenergic receptor homo- and heterodimerization by bioluminescence resonance energy transfer. J Biol Chem 277:44925–44931
53. Fuxe K, Ferré S, Canals M et al (2005) Adenosine A2A and dopamine D2 heteromeric receptor complexes and their function. J Mol Neurosci 26:209–220
54. Hu CD, Chinenov Y, Kerppola TK (2002) Visualization of interactions among bZIP and Rel family proteins in living cells using bimolecular fluorescence complementation. Mol Cell 9:789–798
55. Hu CD, Kerppola TK (2003) Simultaneous visualization of multiple protein interactions in living cells using multicolor fluorescence complementation analysis. Nat Biotechnol 21: 539–545

56. Kerppola TK (2006) Visualization of molecular interactions by fluorescence complementation. Nat Rev Mol Cell Biol 7:449–456
57. Baíllo A, Martínez-Muñoz L, Mellado M (2013) Homogeneity tests for Michaelis-Menten curves with application to fluorescence resonance energy transfer data. J Biol Syst 21:1350017
58. Levoye A, Balabanian K, Baleux F et al (2009) CXCR7 heterodimerizes with CXCR4 and regulates CXCL12-mediated G protein signaling. Blood 113:6085–6093
59. Martínez Muñoz L, Lucas P, Navarro G et al (2009) Dynamic regulation of CXCR1 and CXCR2 homo- and heterodimers. J Immunol 183:7337–7346
60. Motulsky H, Christopoulos A (2004) Fitting models to biological data using linear and nonlinear regression: a practical guide to curve fitting. Oxford University Press, New York
61. Grinberg AV, Hu CD, Kerppola TK (2004) Visualization of Myc/Max/Mad family dimers and the competition for dimerization in living cells. Mol Cell Biol 24:4294–4308

Chapter 25

Multi-State Transition Kinetics of Intracellular Signaling Molecules by Single-Molecule Imaging Analysis

Satomi Matsuoka, Yukihiro Miyanaga, and Masahiro Ueda

Abstract

The chemotactic signaling of eukaryotic cells is based on a chain of interactions between signaling molecules diffusing on the cell membrane and those shuttling between the membrane and cytoplasm. In this chapter, we describe methods to quantify lateral diffusion and reaction kinetics on the cell membrane. By the direct visualization and statistic analyses of molecular Brownian movement achieved by single-molecule imaging techniques, multiple states of membrane-bound molecules are successfully revealed with state transition kinetics. Using PTEN, a phosphatidylinositol-3,4,5-trisphosphate (PI(3,4,5)P_3) 3′-phosphatase, in *Dictyostelium discoideum* undergoing chemotaxis as a model, each process of the analysis is described in detail. The identified multiple state kinetics provides an essential clue to elucidating the molecular mechanism of chemoattractant-induced dynamic redistribution of the signaling molecule asymmetrically on the cell membrane. Quantitative parameters for molecular reactions and diffusion complement a conventional view of the chemotactic signaling system, where changing a static network of molecules connected by causal relationships into a spatiotemporally dynamic one permits a mathematical description of stochastic migration of the cell along a shallow chemoattractant gradient.

Key words Single-molecule imaging, Membrane, Lateral diffusion, Reaction kinetics, Molecular state, Phosphatidylinositol-3,4,5-trisphosphate, PTEN

1 Introduction

Cell movement depends on the spatiotemporal dynamics of signaling molecules that generates anterior-posterior polarity and coordinates pseudopod formation and tail retraction [1, 2]. Pseudopod formation is triggered by molecules that accumulate locally and transiently on the cell membrane in the absence of any extracellular spatial cues [3–6]. The spontaneous generation of signals exhibiting stereotypical spatiotemporal characteristics leads to random cell movement and can be explained by an excitatory mechanism [7]. The same signals are utilized to migrate directionally along a chemoattractant gradient in eukaryotic chemotaxis, with the firing frequency enhanced at the side facing the higher concentration [8]. A central goal of chemotaxis study is to uncover the underlying

Tian Jin and Dale Hereld (eds.), *Chemotaxis: Methods and Protocols*, Methods in Molecular Biology, vol. 1407,
DOI 10.1007/978-1-4939-3480-5_25,

molecular mechanism responsible for both these random and biased movements.

Signaling molecules inherently adopt multiple states. Chemoattractants bind to receptors integrated into the cell membrane, causing a transition of the receptor's state from ligand-unbound to ligand-bound. State transitions are usually coupled to conformational changes, with the ligand-bound state activating effectors in the cytoplasm. Phosphorylation/dephosphorylation and an exchange of GTP/GDP are typically associated with transitions between the active and inactive states of the effectors and other signaling molecules downstream. Thus, stimulation with a chemoattractant triggers a series of equilibrium shifts between states of the signaling molecules. In order to understand the molecular mechanism of the signaling system, it is necessary to reveal the dynamics of molecules that adopt active states in addition to molecular entities that cause state transitions.

Chemotactic signaling is based on stochastic reactions and movements of the signaling molecules [9]. Although the total amount of ligand-bound receptors on the membrane is kept constant under steady state conditions, individual ligand molecules repeatedly bind and unbind to the receptors and the receptor molecules themselves exhibit random Brownian movement on the cell membrane [10]. The density of ligand-bound receptors fluctuates around an average value, and the deviation, or molecular noise, is determined by the reaction kinetics and lateral diffusion mobility of the receptor. Gradient sensing requires overcoming molecular noise to detect shallow spatial gradients [11, 12]. On the other hand, an excitable system utilizes molecular noise to generate stereotypical spatiotemporal responses that finally lead to pseudopod formation [3, 13]. Despite this system being imperative to cell movement, how the signaling molecules in each state are produced, degraded and dispersed within the cell before and after chemoattractant stimulation is poorly understood.

In this chapter, we describe the method of single molecule imaging in living cells to quantify molecular state transition kinetics and lateral diffusion on the cell membrane [14, 15]. We have applied the method to PTEN, a 3′-phosphatase of $PI(3,4,5)P_3$, to reveal the membrane localization mechanism in *Dictyostelium discoideum* cells undergoing chemotaxis [16]. Upon stimulation with a spatial gradient of the chemoattractant cAMP, PTEN accumulates ubiquitously on the cell membrane except for the side facing the higher concentration [2, 17]. The cAMP-dependent exclusion of PTEN from the cell membrane does not require phosphatase activity and is insensitive to the substrate $PI(3,4,5)P_3$ [18, 19], suggesting that binding sites of PTEN on the membrane are regulated by cAMP. Single-molecule imaging and multistate kinetics analyses revealed that state transitions mediate asymmetric membrane localization of PTEN harboring a phosphatase-deficient mutation, G129E, according to the chemoattractant gradient. The key regulation is suggested to be suppression of the most stable of three states that the membrane-bound PTEN adopts. The suppression that takes place locally at the front of the cell leads to an exclusion of PTEN from the cell membrane, illustrating a molecular mechanism that underlies the polarity formation of PTEN in chemotactic *Dictyostelium* cells.

2 Materials

This chapter, which focuses on discerning state transition kinetics using single-molecule image analysis, expands upon a previous chapter of the authors [20]. Therefore, regarding basic procedures of single-molecule imaging, including microscopy, cell culture, and preparation of specimens, the reader is referred to [20].

2.1 Single Molecule Imaging

Components of total internal reflection fluorescence microscopes (TIRFM) for single molecule imaging are detailed in [20].

2.2 Trajectory Data Acquisition

Software for image analyses. Examples include Image J (public domain), Image-Pro (Media Cybernetics, MD), and G-Track (G-Angstrom, Sendai, Japan).

2.3 Lateral Diffusion Analysis

Software for calculations. Examples include MatLab (The MathWorks, MA), Origin (OriginLab, MA), and gnuplot (open source). Maximum likelihood estimation requires software capable of solving optimization problems.

2.4 Reaction Kinetics Analysis

Software for calculations. Examples include MatLab (The MathWorks, MA), Origin (OriginLab, MA), and gnuplot (open source).

3 Methods

3.1 Single Molecule Imaging

Single-molecule imaging in living *Dictyostelium* cells under a total internal reflection fluorescence microscope (TIRFM) is performed according to the methods described in [20]. Briefly, a molecule of interest is labeled with a fluorophore such as tetramethylrhodamine (TMR) (Fig. 1). TMR is excited only when it comes near the basal cell membrane because of an evanescent field that arises within approximately 200 nm from the surface of the coverslip where the cell sits [21]. When the molecule is bound to the membrane, a quantized fluorescence from the conjugated fluorophore is captured via a sensitive camera such as an EM-CCD camera. Typically, single-molecule images in living cells are visualized at a frame rate of 30 fps, or frame interval of $dt = 1/30$ s [10, 22].

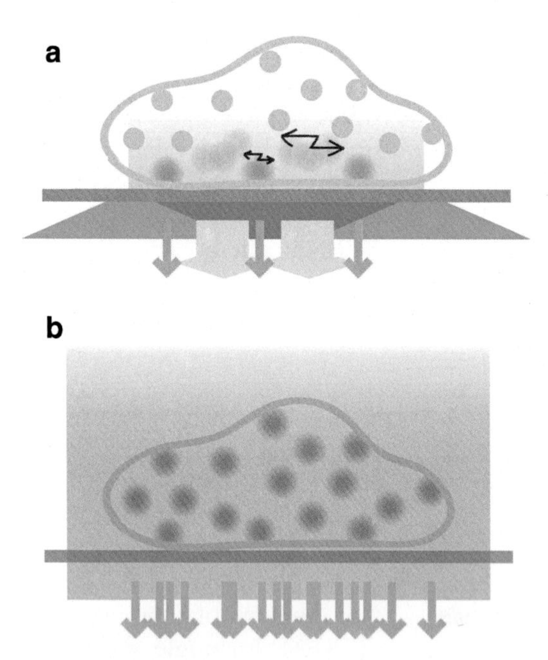

Fig. 1 Single-molecule imaging in living cells. (**a**) The principle of single molecule imaging. An evanescent field made by the total internal reflection of incident laser limits the excitation of fluorophores to those within ~200 nm from the surface of the coverslip. Molecules exhibiting relatively slow diffusion on the cell membrane are detected as fluorescent spots, while those exhibiting relatively fast diffusion in the cytoplasm cause background fluorescence to increase. (**b**) Epi-fluorescent imaging. All fluorophores located in the light path are excited, and thus the fluorescence of membrane-bound molecules is overwhelmed by the background fluorescence

3.2 Trajectory Data Acquisition

Analyses of lateral diffusion and reaction kinetics are based on trajectories of molecules moving on the cell membrane, which are acquired from movies made from the TIRFM imaging as follows (Fig. 2).

1. Select a target from the quantized fluorescent signals.
2. Find a frame in which the target first appears. The time when the signal becomes visible is designated $t = 0$ s, which represents the time when the molecule starts an association with the membrane. t should be distinguished from the time after the movie starts, which is designated as T (s) hereafter.
3. Fit a two-dimensional distribution of the fluorescence intensity to a Gaussian function. The mean of the Gaussian distribution corresponds to the estimated position of the molecule. Obtain x- and y-coordinates (μm) of the mean with the origin set at the bottom left corner of the movie.
4. Feed the frame and repeat the position estimation until the target disappears. A time series of estimated positions corre-

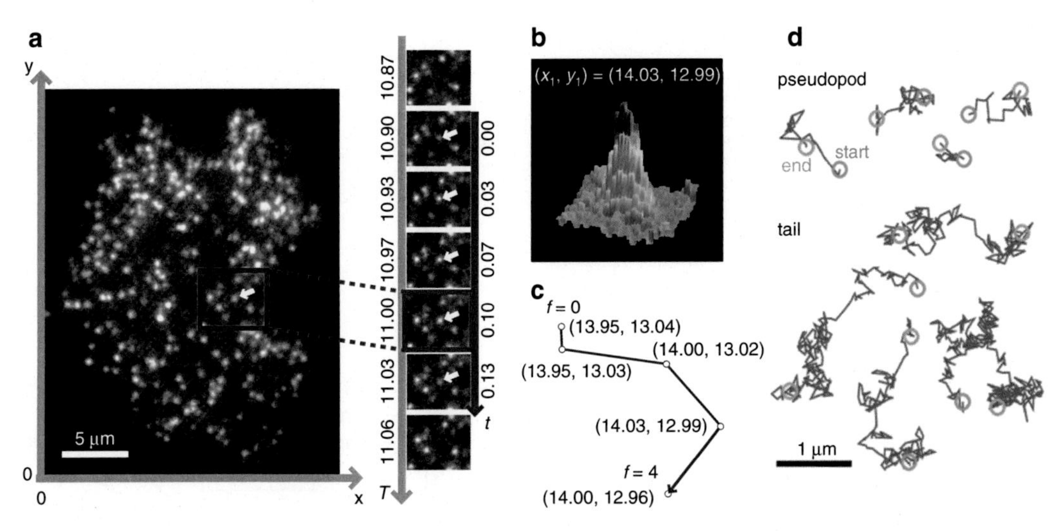

Fig. 2 Tracking single molecules. (**a**) Single-molecule images of PTEN in living *Dictyostelium discoideum* cells. The image is a snapshot from a movie at $T = 11.00$ (s). The fluorescent spot indicated by the *white arrow* appeared on the membrane at $T = 10.90$, exhibited lateral diffusion and disappeared before $T = 11.06$. (**b**) Two-dimensional distribution of fluorescence intensity. By fitting the distribution to a Gaussian function, the position of the molecule is estimated to be $(x, y) = (14.03, 12.99)$ (µm) at $T = 11.00$. The origin is set at the lower-left corner of the movie. (**c**) A trajectory of lateral diffusion. The trajectory of the *i*-th molecule is represented by $(xi(t), yi(t))$. $t = 0$ is the time when the signal first appeared. (**d**) Trajectories of $PTEN_{G129E}$ in cells undergoing chemotaxis. $PTEN_{G129E}$ exhibited longer membrane association and slower lateral diffusion at the tail than the pseudopod (*see* Figs. 5a and 6)

sponds to a trajectory of the molecule, which is written as $(xi(t), yi(t))$ for the *i*-th molecule ($t = fdt = 0, dt, 2dt, \ldots, Fidt$; $f = 0, 1, 2, \ldots, Fi$; $i = 1, 2, \ldots I$).

5. Select other targets and acquire their trajectories. It is recommended to track most of the signals in the movie.
6. Collect trajectories from the movies obtained under the same experimental conditions until the number of tracked molecules, *I*, is sufficient (*see* **Note 1**).

3.3 Lateral Diffusion Analysis

3.3.1 Diffusion Mode

Interactions between membrane proteins and cortical cytoskeletons are decomposed into modes of diffusion: simple, sub-diffusion, and super-diffusion [23, 24]. Confinement by the lattices of filamentous actin anchoring the membrane shifts the mode from simple to sub-diffusion [25]. Directional transport by motor proteins results in super-diffusion [26]. The diffusion modes are mathematically described by the mean squared displacement (MSD), which is calculated as $\text{MSD}(\tau) = \langle (x(t+\tau) - x(t))^2 + (y(t+\tau) - y(t))^2 \rangle$ from the trajectories. τ represents a lag time in measuring the displacement between positions at time t and $t+\tau$.

1. Consider the *i*-th molecule, whose trajectory consists of $Fi+1$ positions (Fig. 3). Calculate the squared displacement using $(xi(fdt+\tau fdt) - xi(fdt))^2 + (yi(fdt+\tau fdt) - yi(fdt))^2$, where τfdt

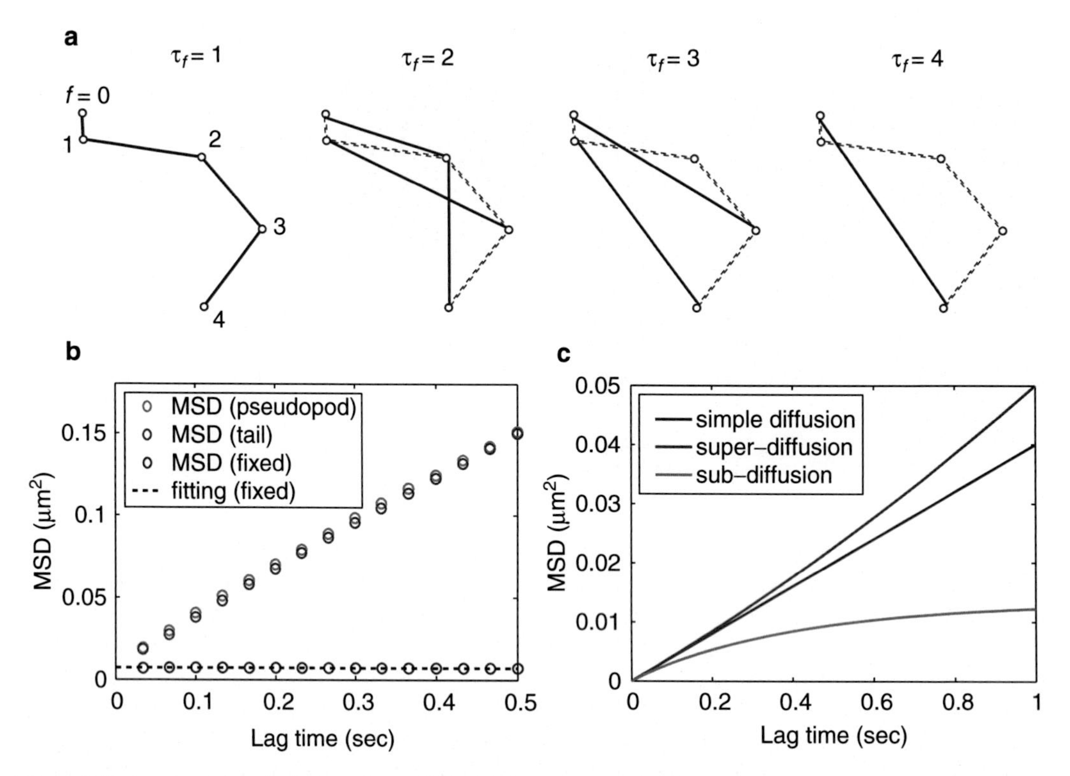

Fig. 3 Diffusion mode analysis based on MSD. (**a**) Sampling of displacement data. The length of the paths shown in *solid lines* is measured. (**b**) MSD calculated from all trajectories of $PTEN_{G129E}$. When observed in chemotactic *Dictyostelium* cells, $PTEN_{G129E}$ exhibited simple diffusion on the membrane of the pseudopod and tail. MSD obtained in living cells and fixed cells converged at the vertical axis. The *y*-intercept corresponds to $4\varepsilon^2$, and ε was estimated to be 0.036 μm. (**c**) Simulation of MSD assuming sub-diffusion and super-diffusion. MSD is calculated assuming the presence of either confinement in a 0.04-μm² area (*green*) or directional flow at a velocity of 0.1 μm/s (*red*) (23). $D = 0.01$ μm²/s

represents the lag time ($\tau f = 1, 2, \ldots, Fi$). Under conditions for constant τf, collect data of all possible calculations with variable f to acquire a set of $Fi - \tau f + 1$ squared displacement data for the lag time. Then, collect Fi sets with variable τf. The number of data included in the set decreases with increasing τf, finally reaching 1 ($\tau f = Fi$).

2. Perform **step 1** for I trajectories.

3. Calculate a mean of squared displacements obtained from all trajectories for each lag time (*see* **Note 2**).

4. Plot the mean squared displacement against lag time. By setting MSD to a function of $\tau\gamma$, a diffusion mode is predicted by the γ value (Fig. 3c). MSD(τ) shows a linear dependence on τ, that is $\gamma = 1$, in the case of simple diffusion. γ is smaller and larger than 1 for sub- and super-diffusion, respectively. Spatial characteristics of the confinement, which may reflect

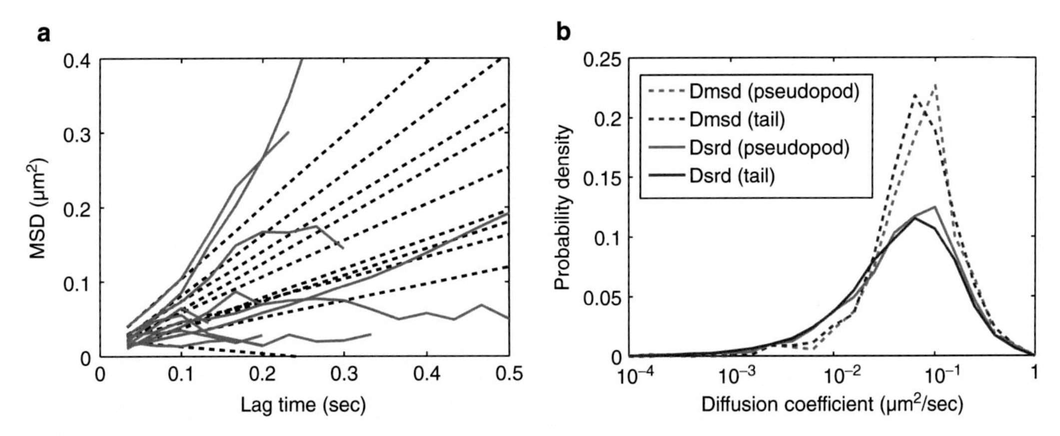

Fig. 4 Diffusion coefficient estimation based on MSD. (**a**) MSD calculated from single trajectories of $PTEN_{G129E}$. Ten typical results of $PTEN_{G129E}$ bound to the pseudopod membrane are shown (*solid lines*), in which data between 0.033 and 0.100 s were fitted to Eq. 1 (*dotted lines*). (**b**) Distribution of the estimated diffusion coefficient. While normal MSD analysis showed single-peak distributions (*dotted lines*), SRD analysis showed broader distributions (*solid lines*), suggesting multiple peaks. Thus, accurate estimation of the diffusion coefficient requires maximum likelihood estimation based on displacement Δr (*see* Fig. 5)

the density of the lattice, for example, and a velocity vector of the directional transport are examined by fitting MSD(τ) to appropriate equations [24].

3.3.2 Diffusion Coefficient

Two methods have been proposed to estimate the diffusion coefficient, D μm²/s, from single-molecule trajectories on the assumption that the molecules exhibit simple diffusion (*see* **Note 3**) [14, 27]. The first method is based on the theory that MSD(τ) can be expressed by the following equation,

$$\mathrm{MSD}(\tau) = 4D\tau + 4\varepsilon^2 \tag{1}$$

where ε represents a standard deviation of the localization error that arises from the Gaussian fitting of the fluorescence distribution. This method is valid for examining a mean diffusion coefficient.

1. Consider the i-th molecule. Estimate a diffusion coefficient by fitting the MSD(τ) – τ plot to Eq. 1 (*see* **Note 4**) (Fig. 4a). Use the ε value obtained in Subheading 3.5.1. Note that the estimate represents a mean even if the diffusivity changed during the membrane association.
2. Estimate the mean diffusion coefficients for all molecules and make a histogram (Fig. 4b). Multiple peaks in the histogram suggest that the molecule changed its diffusion coefficient relatively slowly, once in a trajectory at most, for example. If the histogram has only a single peak, further analysis is required to distinguish two other possibilities: constant diffusivity or

averaged variable diffusivity due to diffusivity transitions that are faster than the temporal resolution of the imaging and/or analysis processes. To improve the resolution of the analysis, short-range diffusion (SRD) analysis, in which trajectories are divided into pieces to avoid the averaging of multiple diffusion coefficients, is effective as described below.

3. Extract all possible short trajectories with an arbitrary duration from each trajectory. In Fig. 4b, short trajectories with 133-ms duration were derived from each trajectory. Successive 42 short trajectories can be derived when a trajectory of 1.5 s was acquired at 30 fps.
4. Make a $\text{MSD}(\tau)-\tau$ plot for each short trajectory and fit the plot to Eq. 1. The slope of the first several data points ($\tau=33$–100 ms in Fig. 4a) is used to obtain an estimate of diffusion coefficient.
5. Create a histogram of the SRD estimates. This histogram shows the distribution of diffusion coefficients. If the molecules have two types of diffusion states, the SRD histogram becomes broader or has multiple peaks (Fig. 4b). This analysis is simple and useful to detect multiple diffusion states, but precise diffusion coefficients cannot be obtained because no appropriate distribution function for fitting the histogram is available.

The second method is based on a probability density function (PDF) of displacement, Δr, for a molecule moving during time interval, Δt,

$$P_J^D(\Delta r, \Delta t) = \sum_{j=1}^{J} q_j^D \frac{\Delta r}{2D_j \Delta t + 2\varepsilon^2} e^{-\frac{\Delta r^2}{4D_j \Delta t + 4\varepsilon^2}}. \tag{2}$$

The PDF is built assuming that a molecule has J states with different diffusion coefficients designated as Dj ($j=1,2,\ldots,J$). The probability of adopting the j-th state is qjD, where

$$\sum_{j=1}^{J} q_j^D = 1.$$

This method provides a quantitative estimation of multiple diffusion coefficients by fitting a statistic distribution of Δr to Eq. 2 (Fig. 5).

1. Measure Δr (μm) for $\Delta t = dt = 1/30$ s. In **step 1** of Subheading 3.3.1, collect squared displacement data of all possible calculations using $\tau f=1$ and then obtain their square root values. Perform the same calculations for all I trajectories to generate N samples of Δrn ($n=1,2,\ldots N$). The best possible temporal resolution, i.e., the frame interval of the movie, is applied to accurately estimate multiple diffusion coefficients (*see* **Note 5**).

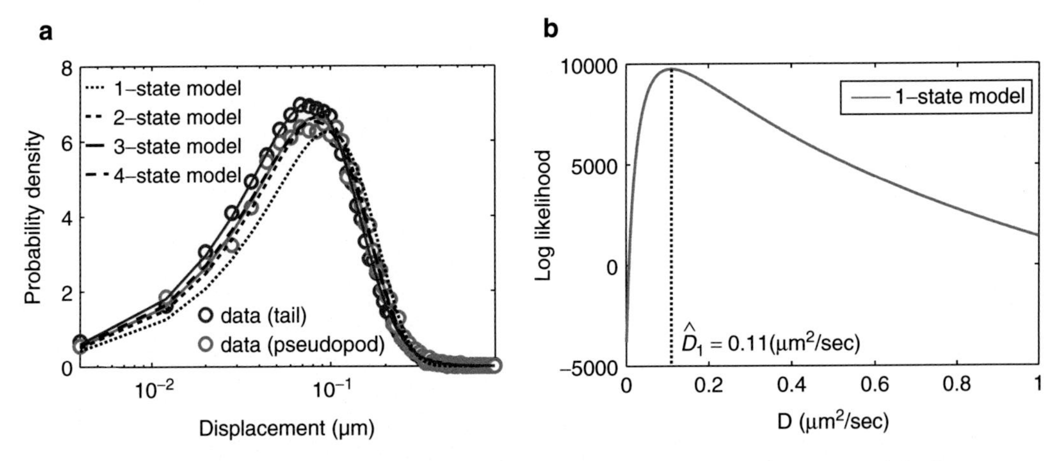

Fig. 5 Diffusion coefficient estimation based on displacement distribution. (**a**) Distribution of displacement Δr ($\Delta t = 0.033$ (s)). $PTEN_{G129E}$ molecules exhibited faster diffusion on the membrane of the pseudopod (*green*) than the tail (*red*). Curves show PDFs (Eq. 2) in which maximum likelihood estimates are introduced. AIC suggested that the 3-state model is most likely for molecules at the pseudopod and tail (*see* Table 1) (*solid lines*). The results of the MLE assuming 1, 2, or 4 diffusion states are shown for molecules on the pseudopod (*dotted lines*). (**b**) A log likelihood function. The result shown is obtained from displacement measured at the pseudopod and PDF ($J = 1$) and returns the MLE $D_1 = 0.11$ (µm²/s)

2. Estimate the diffusion coefficient(s) by the maximum likelihood estimation (MLE). For a given model assuming J states, Dj and qjD are estimated by finding the maximum of a log likelihood function which is computed as,

$$L_J(\theta_J) = \sum_{n=1}^{N} \log P_J^D(\Delta r_n \mid \theta_J).$$

The log likelihood is a function of a vector of the parameters to be estimated, $\theta_J = (D_j, q_j^D)$ ($j = 1,2,\ldots J$). At first, consider a model in which the molecule shows a single diffusion coefficient. That is, $J = 1$, $q_1{}^D = 1$ and D_1 is the parameter to be estimated. If the D_1 value is defined, Eq. 2 returns $P_1{}^D(\Delta rn)$ for each of the samples of Δrn. The sum of all $\log P_1{}^D(\Delta rn)$ values is the log likelihood for the defined D_1 (Fig. 5b). The D_1 value that returns a maximum of the log likelihood function is the maximum likelihood estimate, D_1 (*see* **Note 6**). Use the ε value obtained in Subheading 3.5.1.

3. Check if the single-state model is appropriate. Make a histogram of Δr and superimpose Eq. 2 into which the estimate of D_1 is introduced (*see* **Note 7**) (Fig. 5a). If the curve fits the histogram well, it is concluded that a multiplicity of diffusion coefficients is not required. If not, proceed to **step 4**.

4. Increase the state number J and obtain estimates of Dj and qjD until Eq. 2 fits the histogram well (*see* **Note 6**). In order to avoid excessive complication, J should be kept to a minimum.

Table 1
Maximum likelihood estimation

Model	1 State ($J=1$)	2 States ($J=2$)	3 States ($J=3$)	4 States ($J=4$)
Pseudopod ($I=865$, $N=7423$, $\varepsilon=0.033$)				
D_1 (q_1D)	0.11 (1.00)	0.08 (0.90)	0.04 (0.34)	0.00 (0.06)
D_2 (q_2D)	–	0.41 (0.10)	0.11 (0.61)	0.08 (0.82)
D_3 (q_3D)	–	–	0.56 (0.05)	0.30 (0.11)
D_4 (q_4D)	–	–	–	0.93 (0.01)
LJ	9755	10,050	10,059	10,061
AICJ	-19,509	-20,094	-20,107	-20,107
Tail ($I=1583$, $N=27{,}401$, $\varepsilon=0.037$)				
D_1 (q_1D)	0.09 (1.00)	0.05 (0.88)	0.03 (0.64)	0.03 (0.51)
D_2 (q_2D)	–	0.39 (0.12)	0.15 (0.33)	0.11 (0.42)
D_3 (q_3D)	–	–	0.72 (0.03)	0.41 (0.06)
D_4 (q_4D)	–	–	–	1.23 (0.01)
LJ	35,938	37,746	37,819	37,823
AICJ	-71,874	-75,486	-75,626	-75,630

The diffusion coefficients and probabilities of the j-th state ($j=1,2,3,4$) are estimated assuming 1–4 diffusion states

This minimum can be estimated by criteria for model selection such as Akaike Information Criterion (AIC) (Table 1) [28]. AIC is calculated as,

$$AIC_J = -2L_J(\hat{\theta}_J) + (2J-1)\log(\log(N))$$

for each model with different J. The AIC value of a J-state model is composed of the minimum of $-LJ(\theta J)$ and a penalty for increasing the dimension of the parameter vector. AIC returns the minimum when the model is most likely.

3.4 Reactions Kinetics Analyses

3.4.1 Membrane Dissociation

A reaction rate constant is quantified by directly measuring the time it takes for each molecule to change the state by the reaction,

$$A \xrightarrow{k} B,$$

where k denotes a rate constant (s^{-1}) [29]. Consider a dissociating reaction from the cell membrane: the molecule is bound to the membrane in state A and freely diffusing in the cytoplasm in state B. The time length of a trajectory corresponds to the time it took for the molecule to dissociate from the membrane without photo-bleaching of the conjugated fluorophore (*see* **Note 8**). The time length follows a PDF,

$$p_J^k(t) = \sum_J^{j=1} q_j^k (k_j + k_b) e^{-(k_j + k_b)t}.$$

The PDF is built assuming J states with different rate constants, kj ($j=1,2,\ldots,J$), and no state transitions among them. kb is a rate constant of fluorophore photo-bleaching, which is estimated in Subheading 3.5.2. The probability of adopting the j-th state at $t=0$ is qjk, where

$$\sum_J^{j=1} q_j^k = 1.$$

A cumulative description of the PDF,

$$P_J^k(t) = \sum_{j=1}^{J} q_j^k e^{-(k_j + k_b)t}, \tag{3}$$

is conventionally used to fit the data. The analysis proceeds as follows.

1. Measure the length of a trajectory, which is $Fi+1$ frames for the i-th molecule. Count the number of molecules for which trajectory length is f ($f=1,2,\ldots$) (Fig. 6a). The number of molecules is represented by If.

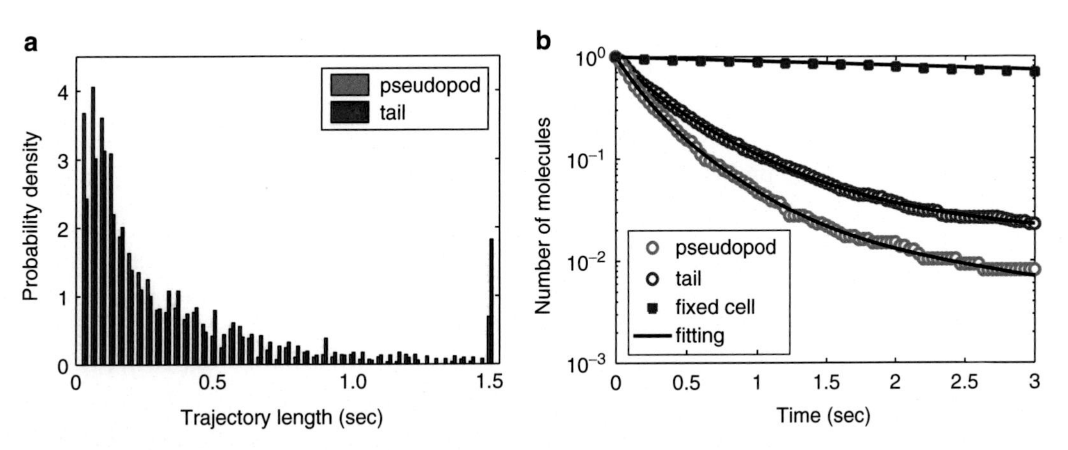

Fig. 6 Analysis of membrane dissociation kinetics. (**a**) Histograms of the length of trajectories. A density calculated by If/Idt is plotted against time, $t=fdt$. (**b**) Dissociation curves. Dissociation from the cell membrane was faster at the pseudopod than the tail. Both dissociation curves were fitted to Eq. 3 with $J=3$, providing the parameter estimates $k_1=0.48$, $k_2=2.75$, and $k_3=7.39$ (s^{-1}) for the dissociation curve at the pseudopod and $k_1=0.22$, $k_2=2.04$, and $k_3=5.99$ (s^{-1}) at the tail. Note that these values do not represent the dissociation rate constants of those molecules adopting the j-th state. TMR photo-bleaching occurred at a sufficiently slower rate than membrane dissociation reactions. Fitting the dissociation curve obtained in the fixed cells to Eq. 3 provided the estimate of the photo-bleaching rate, $kb=0.10$ (s^{-1})

2. Count the number of molecules, $s(t)$, that are bound to the membrane at $t = fdt$ ($f = 0,1,2,\ldots$). It is I at $f = 0$, when all molecules start associating with the membrane. After that, $s(f)$ is calculated by subtracting I_f from $s(f-1)$. Divide these numbers by I to calculate the probability that the molecule remains bound to the membrane after the onset of membrane association,

$$S(f) = \frac{s(f)}{I} = \frac{I - \sum_{f}^{m=0} I_m}{I}.$$

3. Plot the probability, $S(t)$, against t, to produce a dissociation curve (Fig. 6b). A multiplicity of dissociation kinetics is obvious when the vertical axis is scaled logarithmically. The number of slopes is correlated to the number of states with different dissociation rate constants (*see* **Note 9**). Fitting the plot to Eq. 3 provides estimates of k_j and q_j^k. Note that Eq. 3 is built assuming no state transitions. In the case that the molecule exhibits state transitions, a fitting function needs to be built based on an appropriate model in order to quantify the kinetics of both membrane dissociation and state transitions (*see* Subheading 3.4.2).

3.4.2 State Transition

A multiplicity of molecular states suggests that the molecule possibly exhibits state transitions. When each state can be characterized by a specific diffusion coefficient, one can distinguish whether the state transitions occur or not by a time series of the displacement, Δr (Fig. 7a) [14, 16]. Due to alternations in diffusivity, an autocorrelation function of squared displacements shows an exponential decay (Fig. 7b). This function will decay similarly to a delta function in the absence of state transitions. Kinetics of the state transitions is quantified as follows after the number of states, $\hat{J}$, and the specific diffusion coefficients, $\hat{D}_j (j = 1,2,\ldots \hat{J})$, are estimated by MLE (*see* Subheading 3.3.2).

1. Group Δrn ($n = 1,2,\ldots N$) by the time after an onset of membrane association. The samples of displacement that the molecules moved between times fdt and $f'dt$ are included in the f'-th subset, where $f' = f + 1$.
2. Estimate the probability of the j-th state, q_j^D, for each subset by MLE using Eq. 2, which generates a time series of the estimated probability, $q_j^D(f')$, where $\sum_{j=1}^{\hat{J}} q_j^D(f') = 1$. When estimating q_j^D, fix the diffusion coefficients to $\hat{D}_j$ throughout the subsets.
3. Calculate the probability, $Q_j^D(t)$, that the molecule adopts the j-th state on the membrane at time t by multiplying $S(t)$ by $q_j^D(t)$, where $\sum_{j=1}^{\hat{J}} Q_j^D(t) = S(t) \le 1$ (Fig. 7c, d). Because $q_j^D(0)$

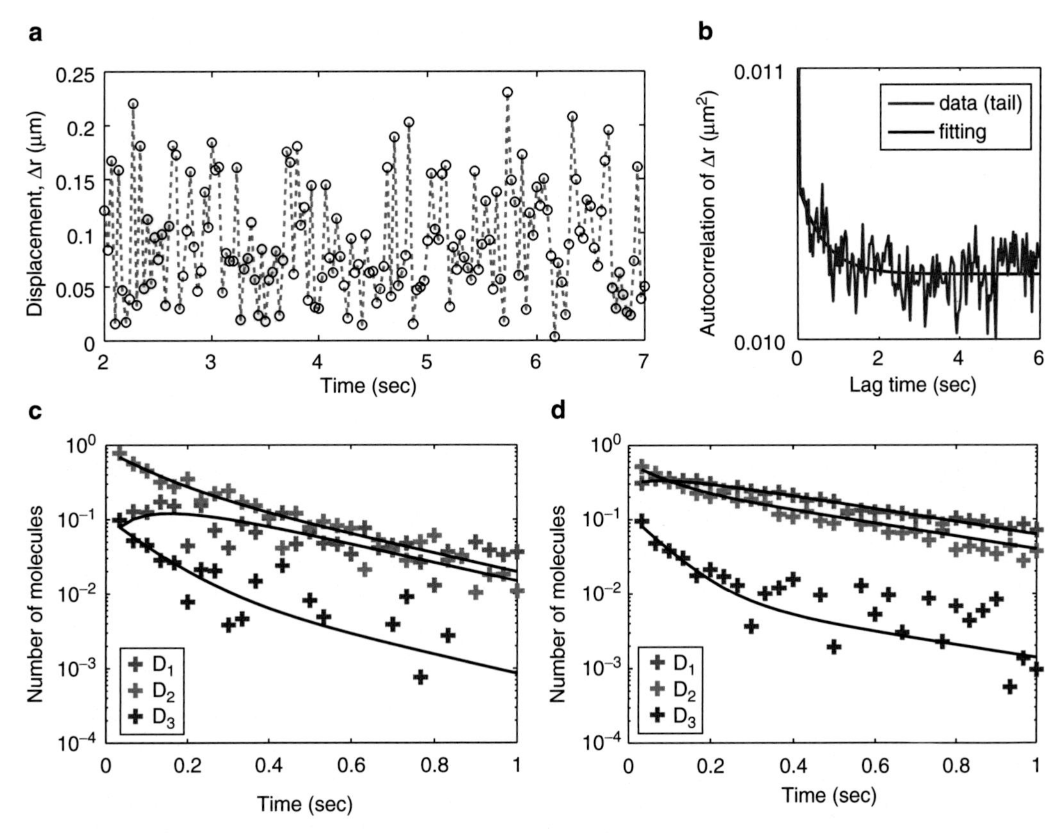

Fig. 7 Analysis of state transitions. (**a**) A time series of displacements ($\Delta t = dt$) calculated from a trajectory observed at the tail membrane. (**b**) An autocorrelation function of displacements. An average obtained from 10 trajectories at the tail membrane is shown. The fitting function is $f(t) = a^*\exp(-Kt) + b$, with $K = 1.64$ (s^{-1}). (**c**, **d**) Probabilities of PTEN$_{G129E}$ adopting three diffusion states at time t on the pseudopod (**c**) or tail (**d**) membrane. The estimates of the parameters are $k_{12} = 5.4$, $k_{21} = 1.9$, $k_{23} = 0.4$, $k_{32} = 0.04$, $k_{off1} = \sim 0$, $k_{off2} = 4.5$, $k_{off3} = 12.8$, $Q_1(0) = 0.04$, $Q_2(0) = 0.85$, and $Q_3(0) = 0.11$ at the pseudopod, and $k_{12} = 4.7$, $k_{21} = 4.2$, $k_{23} = 0.4$, $k_{32} = 0.03$, $k_{off1} = \sim 0$, $k_{off2} = 4.6$ $k_{off3} = 13.8$, $Q_1(0) = 0.30$, $Q_2(0) = 0.58$, and $Q_3(0) = 0.12$ at the tail

does not exist, start the multiplication from $t = fdt$ with $f = 1$. $Q_j^D(t)$ will decay at the same rate for all j in a longer time scale with state transitions. Without state transitions, $Q_j^D(t)$ decays monotonously at specific rate kj.

4. Estimate the rate constants of both state transitions and membrane dissociations by fitting $Q_j^D(t)$ to the functions obtained by assuming a model. Here the simplest model in which a molecule exhibits state transitions mutually between states 1 and 2 and membrane dissociations is considered. The probabilities that the molecule adopts state 1 or 2 at time t follow the simultaneous differential equations,

$$\frac{dQ_1(t)}{dt} = -k_{12}Q_1(t) + k_{21}Q_2(t) - (k_{off1} + k_b)Q_1(t)$$

$$\frac{dQ_2(t)}{dt} = k_{12}Q_1(t) - k_{21}Q_2(t) - (k_{off2} + k_b)Q_2(t),$$

where $k_{12/21}$ and $k_{\text{off1/2}}$ are rate constants of state transitions and membrane dissociation, respectively. By solving these equations, theoretical descriptions for $Q_1D(t)$ and $Q_2D(t)$ are obtained,

$$Q_1(t) = C_1 e^{-\alpha t} + C_2 e^{-\beta t},$$
$$Q_2(t) = C_3 e^{-\alpha t} + C_4 e^{-\beta t},$$

where

$$\alpha = -(A + B)/2$$
$$\beta = -(A - B)/2$$
$$A = -(k_{12} + k_{21} + k_{\text{off1}} + k_b + k_{\text{off2}} + k_b)$$
$$B = \sqrt{A^2 - 4(k_{12}(k_{\text{off2}} + k_b) + k_{21}(k_{\text{off1}} + k_b) + (k_{\text{off1}} + k_b)(k_{\text{off2}} + k_b))}$$
$$C_1 = \{2k_{21} - Q_1(0)(k_{12} + k_{21} + k_{\text{off1}} - k_{\text{off2}} - B)\}/2B$$
$$C_2 = \{-2k_{21} + Q_1(0)(k_{12} + k_{21} + k_{\text{off1}} - k_{\text{off2}} + B)\}/2B$$
$$C_3 = \{2k_{12} + Q_2(0)(-k_{12} - k_{21} + k_{\text{off1}} - k_{\text{off2}} + B)\}/2B$$
$$C_4 = \{-2k_{12} - Q_2(0)(-k_{12} - k_{21} + k_{\text{off1}} - k_{\text{off2}} - B)\}/2B.$$

By fitting $Q_{1/2}{}^D(t)$ to $Q_{1/2}(t)$, the values of α, β, C_1, C_2, C_3, and C_4 are estimated, from which $k_{12/21}$, $k_{\text{off1/2}}$, and $Q_{1/2}(0)$ are calculated.

When each state cannot be characterized by a specific diffusion coefficient, it is useful to fit $S(t)$ to $Q_1(t) + Q_2(t)$.

These analyses have suggested that $\text{PTEN}_{\text{G129E}}$ adopts three states with different diffusivity and exhibits state transitions among them (Figs. 5 and 8) [16]. The difference in the state transition kinetics between pseudopod and tail was revealed by assuming the model shown in Fig. 8. The suppression of the stably binding state, in which PTEN shows the slowest lateral diffusion and membrane dissociation, at the pseudopod is most likely to be a key mechanism to establish the asymmetric distribution on the cell membrane in *Dictyostelium* cells undergoing chemotaxis.

3.5 Calibration of Measurement Errors

3.5.1 Localization Error

A localization error arises due to uncertainty in the position of the molecule that is estimated by fitting a rough fluorescence intensity distribution to a Gaussian (*see* **Note 10**) (Fig. 2b). The standard deviation, ε, of the error is quantified as follows.

1. Prepare a specimen using buffer solution in which fluorophore molecules are dissolved at an appropriate concentration. Use the same fluorophore as that used for labeling target molecules. It is optional to include the cells in the specimen, which makes it easier to focus the microscope on the glass-buffer interface. Fixed cells in which labeled molecules are immobilized on the cell membrane with fixatives such as formaldehyde can be used also as shown in Fig. 3b and 6b.

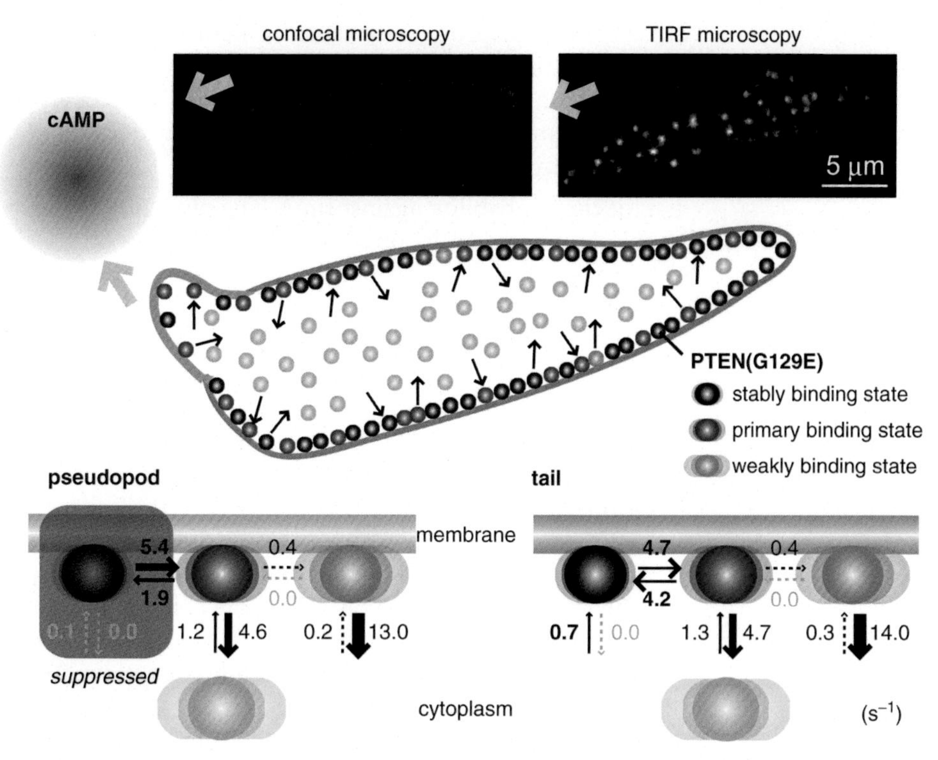

Fig. 8 A model of the multi-state transition kinetics of $PTEN_{G129E}$. Suppression of the stably binding state, which has the slowest membrane dissociation rate and lateral diffusion mobility, leads to an exclusion of PTEN from the cell membrane at the pseudopod

2. Capture immobile fluorescence signals from those adsorbed onto a surface of the coverslip under the same conditions as those used for imaging in living cells.
3. Perform single-molecule tracking of the signals.
4. Calculate MSD according to Subheading 3.3.1.
5. Estimate ε by fitting $\text{MSD}(\tau)-\tau$ plot to Eq. 1. The plot will be constant against τ if $D=0$ (Fig. 3b). It is also effective to estimate ε by extrapolating the $\text{MSD}(\tau)-\tau$ plot of target molecules.

3.5.2 Photo-bleaching Rate Constant

Fluorophores exhibit photo-bleaching during continuous excitation. The trajectory of a molecule is limited to the duration of fluorescence, which ends independently of the state of the molecule, in a stochastic manner (*see* **Note 8**). The rate of photo-bleaching is necessary for accurate quantification of the reaction kinetics.

1. Perform single-molecule imaging of the fluorophores adsorbed onto the glass surface according to Subheading 3.5.1. Track the signals from an initial frame of the movie until they diminish (*see* **Note 11**).
2. Measure the duration of fluorescence. When the trajectory of the i-th molecule consists of $Fi+1$ coordinates, the signal is detected for $Fi+1$ frames and the duration is $(Fi+1)dt$s.

3. Calculate the probability that the fluorophore emits fluorescence at time T after the movie starts in the same manner as Subheading 3.4.1. Count the number of molecules that are fluorescent at time T. The number divided by I is the probability.
4. Plot the probability against time. The plot shows an exponential decay that can be fitted to $F(t) = e^{-k_b t}$ (Fig. 6b). kb is a rate constant of photo-bleaching and dependent on the excitation intensity (*see* **Note 12**).

4 Notes

1. Parameter estimation becomes more precise by increasing the number of trajectories to be analyzed [16]. Typically, we use ~1000 and ~2000 trajectories when estimating kinetic parameters in the absence and presence of state transitions, respectively. When the molecule is involved in more complicated reactions, a larger number of trajectories will be required. On the other hand, when estimating diffusion coefficients, a smaller number of trajectories provides enough confident estimates, since multiple displacement data are collected from a single trajectory.
2. When a molecule has a relatively long trajectory, as is often the case with membrane-integrated proteins, it is effective to analyze the $\mathrm{MSD}(\tau) - \tau$ plot for each molecule. On the other hand, the plot obtained from a single short trajectory is usually stochastic, and it is recommended to analyze the average of all trajectories.
3. Equations 1 and 2 are derived from the same diffusion equation,

$$\frac{\partial P(x,y,t)}{\partial t} = D\left(\frac{\partial^2}{\partial x^2} + \frac{\partial^2}{\partial y^2}\right)P(x,y,t).$$

The diffusion equation statistically explains how the position of a diffusing molecule changes as time passes. By solving this equation, the probability density function of the position of a diffusing molecule at time t is obtained as,

$$P(x,y,t) = \frac{1}{4\pi Dt + 4\pi\varepsilon^2} e^{-\frac{x^2+y^2}{4Dt+4\varepsilon^2}}.$$

The PDF shows the same profile as a Gaussian distribution with mean $(x, y) = (0, 0)$ and variance $4Dt + 4\varepsilon^2$, which is dependent on t. Equation 2 is obtained by transforming variables x and y into distance r. Therefore, while only the variance is utilized in the method based on Eq. 1, the distribution itself is utilized in the method based on Eq. 2. That is why the latter can dissect multiple diffusion coefficients.

4. The $\text{MSD}(\tau)-\tau$ plot calculated for a single trajectory shows stochastic deviations at larger τ due to a lack of squared displacement data. When estimating a mean diffusion coefficient, it is recommended to limit fitting to a part of the plot showing linear dependence on τ. We usually use $\text{MSD}(\tau)$ data for $\tau <= 0.2$ s.
5. The detection of multiple diffusion coefficients depends on the temporal resolution of both imaging and analysis. For the analysis, the method based on Eq. 2 provides the best temporal resolution when displacement during a frame interval of the movie is used (*see also* **Note 3**). If the rates of state transitions are comparable to the frame rate, the accuracy of diffusion coefficients estimated will be improved by modifying Eq. 2 and introducing an autocorrelation function of squared displacements (for details, *see* ref. [14]). If the rates are much faster, diffusivity changes within an exposure of 1/30 s such that it cannot be resolved into separate states by the analysis. In that case, higher temporal resolution of imaging is required to analyze the states.
6. It is more effective to search a minimum of $-L_J(\boldsymbol{\theta}_J)$ than a maximum of $L_J(\boldsymbol{\theta}_J)$. This is an optimization problem that can be solved by using a solver, e.g., fmincon from MatLab Optimization toolbox.
7. The PDF is comparable to the histogram in which the number of samples of displacement between Δr and $\Delta r+\delta r$ is divided by $N\delta r$. In order to incorporate the constraint, $\sum_{j=1}^{\hat{J}} q_j^D = 1$, use the equation,

$$q_j^D = \frac{q_j^D}{\sum_{J}^{j=1} q_j^D} \quad (j = 1,2,\ldots J),$$

and set the lower bounds to q_j^D at 0. The equation,

$$q_J^D = 1 - \sum_{J-1}^{j=1} q_j^D,$$

does not exclude the possibility of q_J^D being negative.
8. The reactions are rewritten after a photo-bleaching process is taken into account as follows:

$$C \overset{k_b}{\leftarrow} A \overset{k}{\rightarrow} B.$$

The molecule is not fluorescent in state C. The two reactions of membrane dissociation and fluorophore photo-bleaching are independent of each other: the fluorescent state of the fluorophore does not affect the membrane dissociation kinetics of the conjugated molecule. Thus, the probability that the molecule adopts state A decreases with time at the rate constant $k + kb$.

9. In order to reveal relatively slow kinetics of membrane dissociation, it is necessary to use a fluorophore that exhibits photobleaching at a rate slower than the reactions.
10. The localization error is critical in the case of slow diffusion, where typical displacements during $\Delta t = 1/30$ s are comparable to the standard deviation of the error. The error reduction is achieved by using more intense excitation laser to increase the signal-to-noise ratio of the images. On the other hand, the excitation laser should be kept as a low as possible so not to harm the cells. One must find a compromise between accuracy and physiological significance of the quantification.
11. Signals become detectable in the course of the movies when fluorophores dissolved in the buffer adhere to the glass surface and are excluded from the analysis. Freely diffusing fluorophores can be removed by washing and submerging the coverslip with buffer on which the fluorophore molecules are adsorbed by air-drying an aliquot of the solution.
12. The intensity of the evanescent field is dependent on both the intensity and angle of the incident laser. When the microscope configuration is changed, *kb* should be measured accordingly.

Acknowledgement

The authors thank Peter Karagiannis for critical reading of the manuscript.

References

1. Ridley AJ, Schwartz MA, Burridge K et al (2003) Cell migration: integrating signals from front to back. Science 302:1704–1709
2. Swaney KF, Huang CH, Devreotes PN (2010) Eukaryotic chemotaxis: a network of signaling pathways controls motility, directional sensing, and polarity. Annu Rev Biophys 39:265–289
3. Arai Y, Shibata T, Matsuoka S et al (2010) Self-organization of the phosphatidylinositol lipids signaling system for random cell migration. Proc Natl Acad Sci U S A 107:12399–12404
4. Funamoto S, Meili R, Lee S et al (2002) Spatial and temporal regulation of 3-phosphoinositides by PI 3-kinase and PTEN mediates chemotaxis. Cell 109:611–623
5. Postma M, Roelofs J, Goedhart J et al (2004) Sensitization of *Dictyostelium* chemotaxis by phosphoinositide-3-kinase-mediated self-organizing signalling patches. J Cell Sci 117: 2925–2935
6. Gerisch G, Schroth-Diez B, Müller-Taubenberger A, Ecke M (2012) PIP_3 waves and PTEN dynamics in the emergence of cell polarity. Biophys J 103:1170–1178
7. Nishikawa M, Hörning M, Ueda M, Shibata T (2014) Excitable signal transduction induces both spontaneous and directional cell asymmetries in the phosphatidylinositol lipid signaling system for eukaryotic chemotaxis. Biophys J 106:723–734
8. Tang M, Wang M, Shi C et al (2014) Evolutionarily conserved coupling of adaptive and excitable networks mediates eukaryotic chemotaxis. Nat Commun 5:5175

9. Miyanaga Y, Matsuoka S, Yanagida T, Ueda M (2007) Stochastic signal inputs for chemotactic response in *Dictyostelium* cells revealed by single molecule imaging techniques. Biosystems 88:251–260
10. Ueda M, Sako Y, Tanaka T et al (2001) Single-molecule analysis of chemotactic signaling in *Dictyostelium* cells. Science 294:864–867
11. Ueda M, Shibata T (2007) Stochastic signal processing and transduction in chemotactic response of eukaryotic cells. Biophys J 93:11–20
12. Amselem G, Theves M, Bae A et al (2012) Control parameter description of eukaryotic chemotaxis. Phys Rev Lett 109:108103
13. Xiong Y, Huang CH, Iglesias PA, Devreotes PN (2010) Cells navigate with a local-excitation, global-inhibition-biased excitable network. Proc Natl Acad Sci U S A 107:17079–17086
14. Matsuoka S, Shibata T, Ueda M (2009) Statistical analysis of lateral diffusion and multistate kinetics in single-molecule imaging. Biophys J 97:1115–1124
15. Matsuoka S (2011) Statistical analysis of lateral diffusion and reaction kinetics of single molecules on membranes of living cells. In: Sako Y, Ueda M (eds) Cell signaling reactions: single-molecule kinetic analysis, 1st edn. Springer, New York, pp 265–296
16. Matsuoka S, Shibata T, Ueda M (2013) Asymmetric PTEN distribution regulated by spatial heterogeneity in membrane-binding state transitions. PLoS Comput Biol 9:e1002862
17. Iijima M, Devreotes P (2002) Tumor suppressor PTEN mediates sensing of chemoattractant gradients. Cell 109:599–610
18. Iijima M, Huang YE, Luo HR et al (2004) Novel mechanism of PTEN regulation by its phosphatidylinositol 4,5-bisphosphate binding motif is critical for chemotaxis. J Biol Chem 279:16606–16613
19. Hoeller O, Kay RR (2007) Chemotaxis in the absence of PIP_3 gradients. Curr Biol 17: 813–817
20. Miyanaga Y, Matsuoka S, Ueda M (2009) Single-molecule imaging techniques to visualize chemotactic signaling events on the membrane of living *Dictyostelium* cells. Methods Mol Biol 571:417–435
21. Funatsu T, Harada Y, Tokunaga M et al (1995) Imaging of single fluorescent molecules and individual ATP turnovers by single myosin molecules in aqueous solution. Nature 374:555–559
22. Sako Y, Minoghchi S, Yanagida T (2000) Single-molecule imaging of EGFR signalling on the surface of living cells. Nat Cell Biol 2:168–172
23. Qian H, Sheetz MP, Elson EL (1991) Single particle tracking. Analysis of diffusion and flow in two-dimensional systems. Biophys J 60: 910–921
24. Kusumi A, Sako Y, Yamamoto M (1993) Confined lateral diffusion of membrane receptors as studied by single particle tracking (nanovid microscopy). Effects of calcium-induced differentiation in cultured epithelial cells. Biophys J 65:2021–2040
25. Fujiwara T, Ritchie K, Murakoshi H et al (2002) Phospholipids undergo hop diffusion in compartmentalized cell membrane. J Cell Biol 157:1071–1081
26. Tani T, Miyamoto Y, Fujimori KE et al (2005) Trafficking of a ligand-receptor complex on the growth cones as an essential step for the uptake of nerve growth factor at the distal end of the axon: a single-molecule analysis. J Neurosci 25:2181–2191
27. Saxton MJ (1997) Single-particle tracking: the distribution of diffusion coefficients. Biophys J 72:1744–1753
28. Akaike H (1974) A new look at the statistical model identification. IEEE Trans Automat Contr 19:716–723
29. Phillips R, Kondev J, Theriot J et al (2012) Rate equations and dynamics in the cell. In: Physical biology of the cell, 2nd edn., Garland Science, UK, pp 573–621

Chapter 26

Mathematics of Experimentally Generated Chemoattractant Gradients

Marten Postma and Peter J.M. van Haastert

Abstract

Many eukaryotic cells move in the direction of a chemical gradient. Several assays have been developed to measure this chemotactic response, but no complete mathematical models of the spatial and temporal gradients are available to describe the fundamental principles of chemotaxis. Here we provide analytical solutions for the gradients formed by release of chemoattractant from a point source by passive diffusion or forced flow (micropipettes) and gradients formed by laminar diffusion in a Zigmond chamber. The results show that gradients delivered with a micropipette are formed nearly instantaneously, are very steep close to the pipette, and have a steepness that is strongly dependent on the distance from the pipette. In contrast, gradients in a Zigmond chamber are formed more slowly, are nearly independent of the distance from the source, and resemble the temporal and spatial properties of the natural cAMP wave that *Dictyostelium* cells experience during cell aggregation.

Key words Diffusion, Equation, Chemotaxis, Dictyostelium, Point source, Pipette, Zigmond chamber

1 Introduction

Chemotaxis is a vital process in a wide variety of organisms, ranging from bacteria to vertebrates. Prokaryotes use chemotaxis to move towards high nutrient concentrations or away from unfavorable conditions, while in eukaryotes it is also involved in embryogenesis, wound healing, and the immune response. Chemotaxis is achieved by coupling gradient sensing to basic cell movement. Prokaryotes are too small to sense spatial gradients and therefore rely on temporal changes of the chemoattractant concentration to achieve chemotaxis. They do this by adjusting their tumbling frequency in response to temporal changes of the chemoattractant concentration [1]. Eukaryotic cells are typically large enough to be able to measure a spatial gradient. The difference in receptor occupation between each side of the cell leads to an internal polarization. A pseudopod is extended at the side with the highest receptor

Tian Jin and Dale Hereld (eds.), *Chemotaxis: Methods and Protocols*, Methods in Molecular Biology, vol. 1407,
DOI 10.1007/978-1-4939-3480-5_26, © Springer Science+Business Media New York 2016

occupation and at the same time, pseudopod formation at all other sides is repressed, resulting in directional cell migration [2, 3].

Dictyostelium is an eukaryotic organism that is widely used to study chemotaxis [4–6]. Starved *Dictyostelium* cells periodically secrete cAMP. Through relay of the cAMP signal by neighboring cells, concentric cAMP waves are generated. Starved *Dictyostelium* cells are chemotactically sensitive to cAMP and by movement in the direction of the origin of the cAMP waves the cells are able to aggregate into groups of up to 1,000,000 cells. The chemotactic response of *Dictyostelium* is optimized for the dynamic cAMP waves that coordinate both aggregation and multicellular development. Cells show a much stronger chemotactic response to a cAMP wave where the mean concentration increases over time, than to a static spatial gradient. *Dictyostelium* uses both spatial gradient sensing and the "bacterial-like" temporal gradient sensing to respond to these dynamic chemoattractant gradients [7–9].

Several chemotaxis assays have been developed that can be divided into three groups, depending on how the gradient developed. In Zigmond chambers, cells are placed on a bridge separated by a chemoattractant source and a sink reservoir [10]. A gradient will be formed under the bridge, which will be nearly linear when at equilibrium. Depending on the geometry of the bridge (which in most setups is a few mm) half-maximal equilibrium is reached only after several minutes. Dunce [11] and Insall [12] chambers have similar properties: a linear gradient that is formed during several minutes of incubation.

Many experiments are performed with micropipettes, because this allows the precise positional stimulation of the cell [13]. Since the pipette is usually in the field of microscopic observation, distances are relatively small (less than 200 μm) and equilibrium is established very fast on the order of seconds. Since a micropipette behaves like a point source, the gradient will be nonlinear approaching the equation $dC / dx = \nabla C = 1 / x^2$, where x is the distance from the pipette. Thus, close to the pipette, the gradient is very steep (~100 % per cell of 10 μm length at a distance of 10 μm from the pipette) while more shallow at the edge of the field of observation (10 % per cell at a distance of 100 μm).

Microfluidic devices are designed to provide a defined gradient in terms of its spatial and temporal properties (reviewed in Refs. [14–16]). Many chamber have two input lines pumping buffer or chemoattractant solutions into the chamber where cells can freely move. Using pre-chamber mixing devices, simple linear or complex gradients are generated. The gradients can be linear across the width of the channel from concentration X at one side to concentration Y at the other side of the channel, but the gradient can also have a bidirectional shape (maximal concentration in the middle of the channel and lower at both sides) or exponential. These gradients are very stable [17]. Microfluidic devices are also very useful to provided well-controlled temporal modulations of the spatial gradient by changing the chemoattractant concentrations at the inlet lines. This allows to switch between gradients and uniform concentrations, or to suddenly invert the direction of the gradient and thereby determine the response time of cells [18, 19]. Since experiments with microfluidics devices are designed to produce specific spatial and temporal gradients, researchers use dyes to visualize the gradient; therefore, the theory of gradient formation in microfluidic devices is not discussed here.

In this chapter we derive mathematical equations for the temporal and spatial properties of the gradients formed in a Zigmond chamber and delivered from a pipette. We compare the theoretical properties of these gradients with experimental data measuring the gradients using fluorescent dyes. Finally we compare these experimental gradients with those observed during the natural aggregation of *Dictyostelium* cells.

2 Zigmond Chamber-Generated Gradients

2.1 Experimental Setup Zigmond Chamber

Figure 1a shows the experimental setup with our modified Zigmond chamber [10]. On a microscope slide, a glass bridge of ~2 mm wide and 24 mm long was placed on top of two supporting glass strips with thickness 0.15 mm. Cells were placed under the bridge to yield a density of 3×10^4 cells/cm^2. A block of agar containing only buffer was placed at one side of the bridge, while a block of agar containing cAMP and buffer was placed at the other side, which induces the formation of a cAMP gradient under the bridge. Cells were observed by phase contrast microscopy in an area of 350×270 μm or by confocal fluorescence microscopy in an area of 150×150 μm both at a distance of 600–700 μm from the agar block containing cAMP.

2.2 Measurement/Analysis Zigmond Chamber

The formation of the cAMP gradient was deduced by measuring the diffusion across the bridge of the modified Zigmond chamber using bromophenol blue (Mw = 670 Da). This reveals that a gradient developed during the first few minutes after the cAMP containing agar block was placed against the bridge (Fig. 1b). A stable linear spatial gradient was reached at 5–10 min. This spatial gradient remained approximately constant for 30 min, and then slowly diminished due to depletion of the cAMP source and accumulation of cAMP in the buffer sink (Fig. 1c). Thus the gradients in the modified Zigmond chamber have temporal and spatial components during the first 5 min, but stable spatial gradients during the subsequent 30 min of the experiment.

2.3 Diffusion Equations for Zigmond Chamber

When a large reservoir with cAMP is connected to another large empty reservoir through a thin bridge, diffusion will occur that can be modeled essentially with a one-dimensional diffusion equation in Cartesian coordinates:

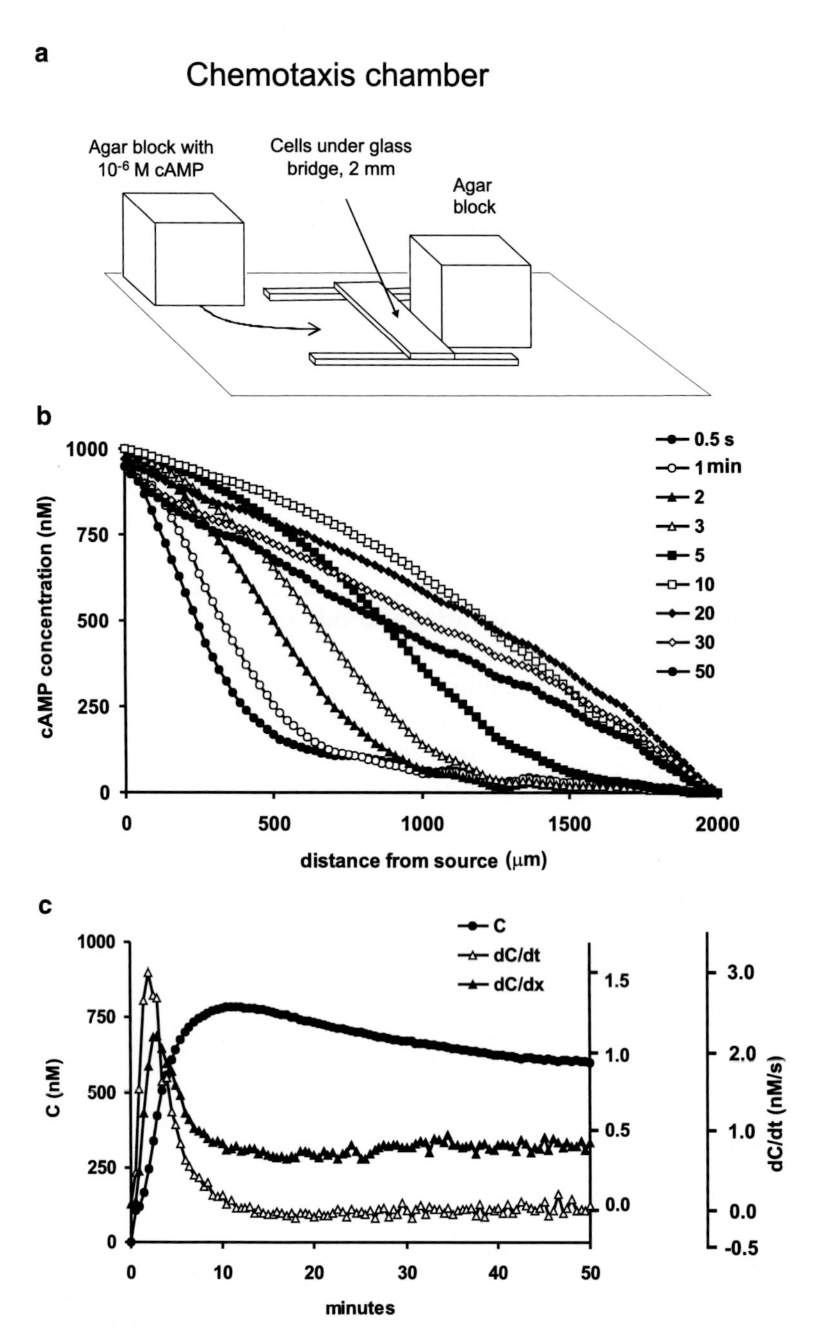

Fig. 1 Observed gradients in the modified Zigmond chamber. (**a**) Setup of the Zigmond chamber. A glass bridge of ~2 mm wide and 24 mm long is placed on top of two supporting glass strips with thickness 0.15 mm. Cells are placed under the bridge. A block of agar containing only buffer is placed at one side of the bridge, while a block of agar containing 1 μM cAMP and buffer was placed at the other side. (**b**) A gradient of cAMP develops under the bridge, visualized by diffusion of a dye added to the agar block containing cAMP. (**c**) Gradients at the position of chemotactic observation (650–750 μm from the source). Using the local concentration of the dye we calculate the cAMP concentration *C*, the spatial gradient *dC*/*dx*, and temporal cAMP gradient *dC*/*dt*

$$\frac{\partial C(x,t)}{\partial t} = D\frac{\partial^2}{\partial x^2}C(x,t) \tag{1}$$

In this equation D ($\mu m^2\ s^{-1}$) denotes the diffusion coefficient of cAMP, $C(x,t)$ denotes the concentration at time t (s) and position x (μm) from the source. This equation is solved with the following boundary conditions: the concentration at $x=0$ (source) is constant with value Cs, the concentration at $x=L$ (sink) is assumed to be constant and zero. L is the length of the bridge. The complete space-time solution of the concentration profile is then given by:

$$C(x,t) = C_S(1-\frac{x}{L}) - C_S\sum_{\infty}^{n=1} 2a_n^{-1}\sin(a_n\frac{x}{L})e^{-\frac{t}{\tau_n}} \tag{2}$$

where $a_n = \pi n$, the time constants $\tau_n = a_n^{-2}\tau_D$ and $\tau_D = L^2 / D$. The slowest time constant is $\tau D/\pi^2$.

After some time equilibrium will appear with a time constant $\tau_D = L^2 / D$. The equilibrium concentration profile is given by the first term of Eq. 2:

$$C(x) = C_s(1-\frac{x}{L}) \tag{3}$$

The gradient and the relative gradient are then given by:

$$\nabla C(x) = \frac{C_s}{L} \tag{4}$$

and

$$\frac{\nabla C(x)}{C(x)} = -\frac{1}{L-x} \tag{5}$$

Figure 2 reveals that cAMP will diffuse from the source block approaching a steady state after about 10–20 min. The spatial gradient is initially very steep close to the source block and very shallow at the sink block (Fig. 2b). Over time the gradient decreases in the half of the chamber closest to the source, but increases at the other half closest to the sink. Finally a steady state is reached at 10–20 min, with a magnitude that depends on the cAMP concentration in the source and the width of the bridge. With $Cs=1$ μM cAMP as source and a bridge of $L=2000$ μm, the steady state spatial gradient is constant at $\nabla C = 5\text{nM} / .\text{m}$. In the field of observation (600–700 μm from source) the absolute concentration C changes from 700 to 650 nM, and the relative spatial gradient from 0.77 to 0.83 % across the cell. The temporal gradient profile is presented in Fig. 2c, showing that it reaches a maximum with magnitude and at a time that depends on the distance from the

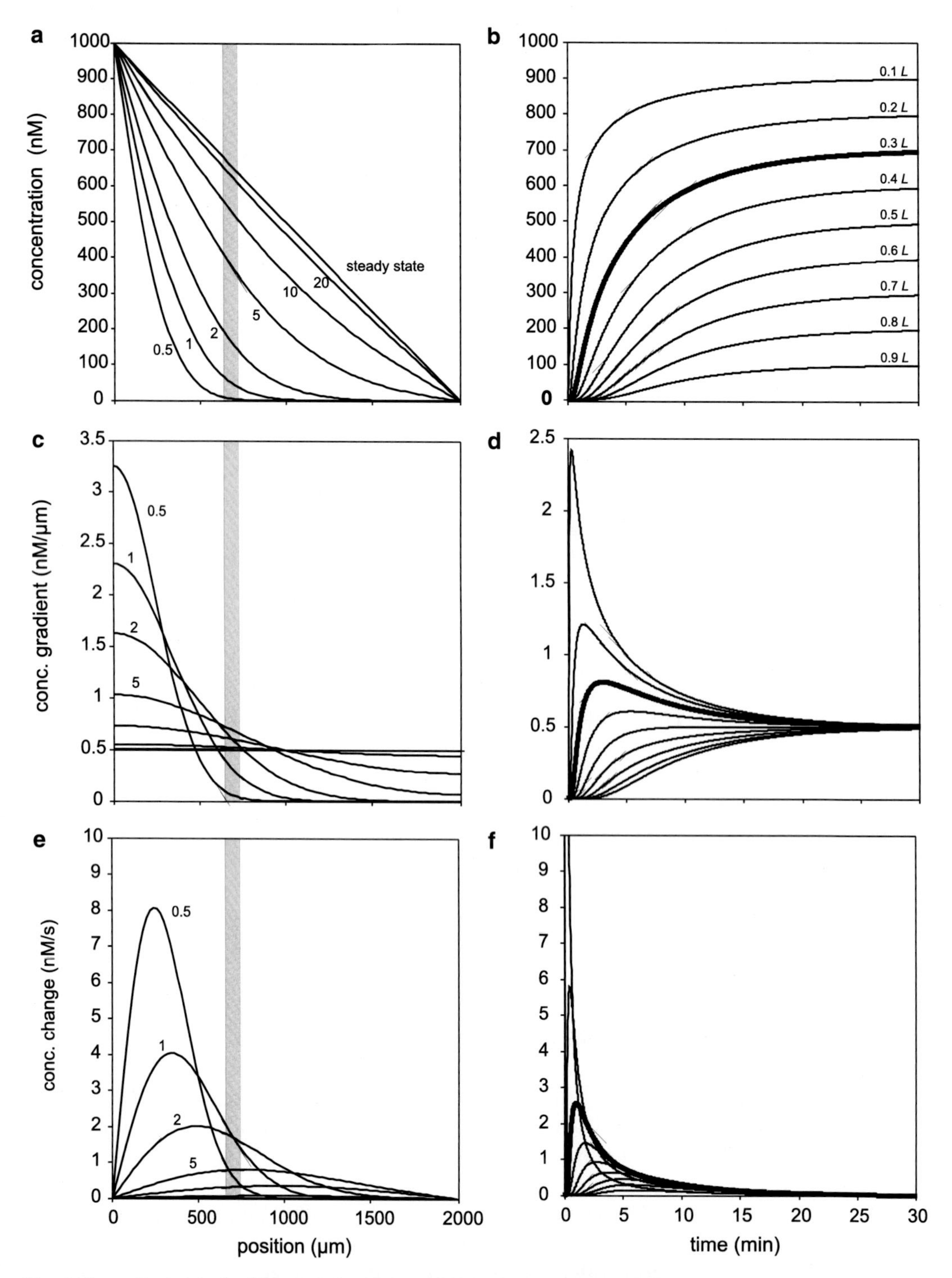

Fig. 2 Theoretical data for Zigmond chamber. Equation (2) is used to calculate the concentration dynamics. Panels (**a**, **d**) present the concentration *C*. The spatial gradient *dC*/*dx* is presented in panels (**b**, **e**) and the temporal gradient *dC*/*dt* in panels (**c**, **e**), all at different positions and times from the source containing 1 μM of cAMP. The *gray bars* in the *left panels* and the *thick line* in the *right panels* indicate the place of chemotactic observation (650–750 μm from the source)

source. Close to the source the maximal temporal gradient is very high and early, and becomes lower and later further away from the source.

The experimental data of bromophenol blue diffusion (Fig. 1b) are in close agreement with the model, except for very long incubation times. Since the source and sink are not indefinitely large, the concentrations in the source and sink slowly decrease and increase, respectively. As a consequence, the gradient will slowly collapse (Fig. 1c). At the position where the cells are observed using the microscope (600–700 μm from the source), this becomes significant only after more than 2 h, much longer than the 30 min to perform the experiments. Closer to the source or the sink the gradient deviates from a linear gradient sooner (Fig. 1b). At short incubation times, the gradient is formed from 10 to 90 % of the equilibrium value in about 4 min, which is only slightly slower than the formation of the cAMP wave during *Dictyostelium* cell aggregation (see below). The model and experimental data imply that the modified Zigmond chamber allows two assay conditions: during the first 5 min, the gradient is formed and therefore cells are exposed to both temporal and spatial gradients. The magnitude of these gradients resembles the temporal and spatial gradient of the natural cAMP wave during *Dictyostelium* chemotaxis (*see* Fig. 5 and Subheading 4). After 10 min a stable gradient is formed without a temporal component.

3 Micropipette-Generated Gradients

3.1 Experimental Setup Pipette Assay

For the micropipette assay, a droplet of a cell suspension was placed on a glass slide yielding a cell density of 5×10^4 cells/cm^2; the droplet has a diameter of about 10 mm and a height of 3 mm. After the cells were allowed to adhere, a pipette filled with 100 μM cAMP is placed just above the glass surface. cAMP was delivered from the femtotip at a pressure ranging from 0 to 100 hPa.

3.2 Measurement/Analysis Pipette Assay With and Without Flow

The formation of the cAMP gradient was deduced by measuring the diffusion of the fluorescent dye lucifer yellow (Mw = 457 Da) using confocal microscopy. The fluorescence intensity at different distances from the pipette was recorded in pixel elements (0.404×0.404 μm) using excitation at 488 nm and a 520–550 band pass filter for the emission. The data were calibrated using the fluorescence intensity of a diluted lucifer yellow added homogeneously to the bath (Fig. 3a). The deduced cAMP concentration profile (Fig. 3b) reveals a steep gradient in the vicinity of the pipette that rapidly declines at longer distances from the pipette. This gradient was stable within 20 s after application of the pipette. The concentration at the tip of the pipette is only 150 nM, compared to the 100 μM cAMP inside the pipette.

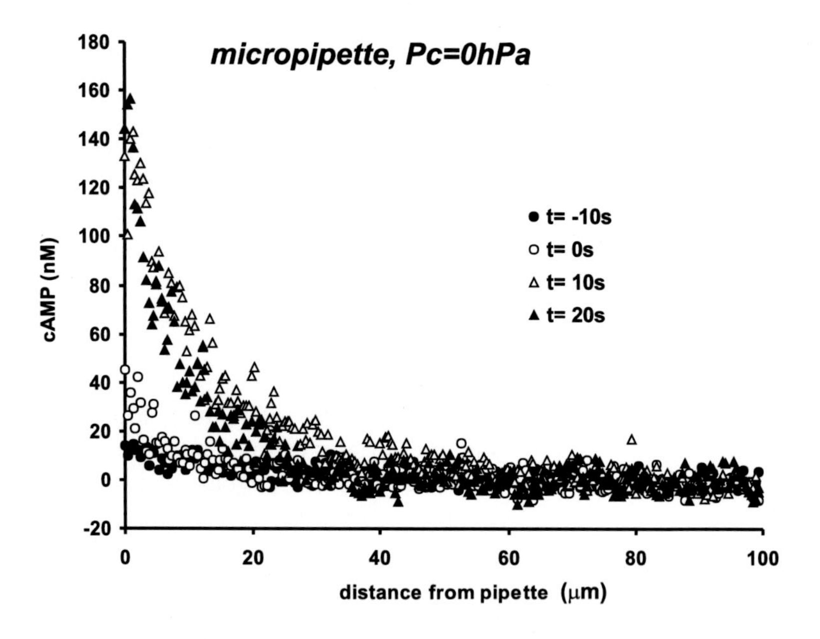

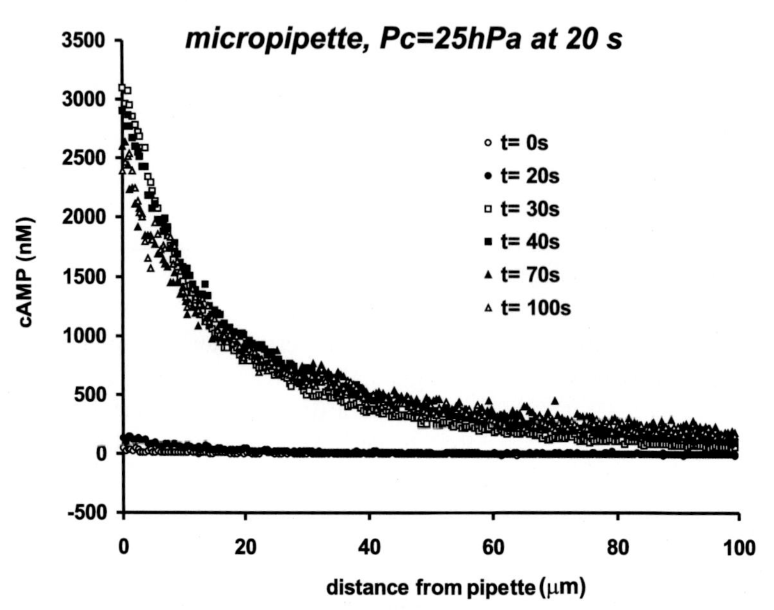

Fig. 3 Observed gradients in the micropipette assay. A micropipette filled with 100 μM cAMP and lucifer yellow is applied just above the glass surface in a droplet of cells. The fluorescence intensity is measured by confocal microscopy at different times after positioning of the pipette in the fluid. Initially no pressure is applied to the pipette (panel **a**), and a pressure of 25 hPa is applied at 20 s (panel **b**). The concentration of cAMP is deduced using a dilution series of lucifer yellow added homogeneously to the bath

The amount of cAMP released from the pipette can be increased by applying pressure to the pipette, which induces a steady flow of cAMP that diffuses away. The cAMP concentration profile was again deduced from the pipette containing lucifer yellow, revealing that the concentration at the tip increases from 150 nM cAMP without pressure to 3000 nM cAMP at a pressure of 25 hPa (Fig. 3c). Similar to the case without applied pressure, the concentration rapidly decreases with distance from the pipette, except close to the pipette (within 2 μm) where the concentration remains relatively uniform.

3.3 Diffusion Equations for Pipette Gradients without Flow

When a pipette filled with cAMP is positioned just above the floor of a chamber, cAMP will diffuse effectively in a half-sphere. The point of the pipette is in the center of the sphere; the opening of the pipette with opening r_0 is regarded as a small sphere from which cAMP diffuses. The diffusion equation is then given in sphere-coordinates by:

$$\frac{\partial C(x,t)}{\partial t} = D\frac{1}{x^2}\frac{\partial}{\partial x}x^2\frac{\partial}{\partial x}C(x,t) \tag{6}$$

In this equation, D (μm^2 s^{-1}) denotes the diffusion coefficient of cAMP, $C(x,t)$ denotes the concentration at time t (s) and distance x (μm) from the pipette. We assume that the gradient in the pipette builds up very fast. Hence, equilibrium in the small sphere at the tip of the pipette will be reached rapidly leading to a constant concentration Cs at the tip. Numerical analysis of diffusion in the pipette and the small sphere suggests that the time constant of this process is less than 1 s. In addition, measurements presented in Fig. 3b reveal that equilibrium at the tip is reached within 10 s. We use the boundary condition that the concentration at the edge of the bath, $x = R$ is constant and zero. The complete space-time solution is then given by:

$$C(x,t) = C_S\frac{r_0}{x}\frac{R-x}{R-r_0} - C_S\frac{r_0}{x}\sum_{\infty}^{n=1}2a_n^{-1}\sin(a_n\frac{x-r_0}{R-r_0})e^{\frac{t}{\tau_n}} \tag{7}$$

where $a_n = \pi n$, the time constants $\tau_n = a_n^{-2}\tau_D$ and $\tau_D = (R-r_0)^2/D$

.

A very good approximation for Eq. 7 is:

$$C(x,t) = C_S\frac{r_0}{x}\operatorname{erfc}(\frac{1}{2}\frac{x-r_0}{\sqrt{Dt}}) \tag{8}$$

The time needed to reach this equilibrium strongly depends on the distance from the pipette $\tau_D = (x-r_0)^2/D$; when t equals this time constant about half-maximal equilibrium is reached.

After some time the gradient reaches equilibrium. The equilibrium concentration profile is given by the first term of Eq. 7, which for a large bath ($R \rightarrow \infty$) and $x > r_0$ is given by:

$$C(x) = C_S \frac{r_0}{x} = \frac{\alpha C_p}{x} \tag{9}$$

where Cp is the cAMP concentration in the pipette. The absolute spatial gradient and the relative spatial gradient are then given by:

$$\nabla C(x) = -C_S \frac{r_0}{x^2} = -\frac{\alpha C_p}{x^2} \tag{10}$$

and

$$\frac{\nabla C(x)}{C(x)} = -\frac{1}{x} \tag{11}$$

The experimental equilibrium data (Fig. 4a) were fitted using Eq. 9 showing that they are in close agreement with the calculated gradient profile, except very close to the pipette. The experiment reveals that the measured cAMP concentration at the tip (i.e., Cs) is only 150 nM compared to 100 μM in the pipette, indicating that a very strong gradient is formed in the pipette.

3.4 Diffusion Equations for Pipette Gradients with Flow

When pressure is applied to the pipette, liquid will flow out of the pipette with flux F (μm^3 s^{-1}). To account for the flow, the diffusion equation has to be extended with convection:

$$\frac{\partial C(x,t)}{\partial t} = D\frac{1}{x^2}\frac{\partial}{\partial x}x^2\frac{\partial}{\partial x}C(x,t) - \frac{F}{2\pi x^2}\frac{\partial}{\partial x}C(x,t) \tag{12}$$

A complete space-time solution of this equation is difficult to obtain. However, the equilibrium solution can be obtained relatively easy. For a large bath the steady state concentration profile is given by:

$$C(x) = C_s \frac{1 - e^{-\frac{F}{2\pi D x}}}{1 - e^{-\frac{F}{2\pi D r_0}}} \tag{13}$$

The absolute spatial gradient and the relative gradient are then given by:

$$\nabla C(x) = -C_s \frac{F}{2\pi D x^2} \frac{e^{-\frac{F}{2\pi D x}}}{1 - e^{-\frac{F}{2\pi D r_0}}} \tag{14}$$

and

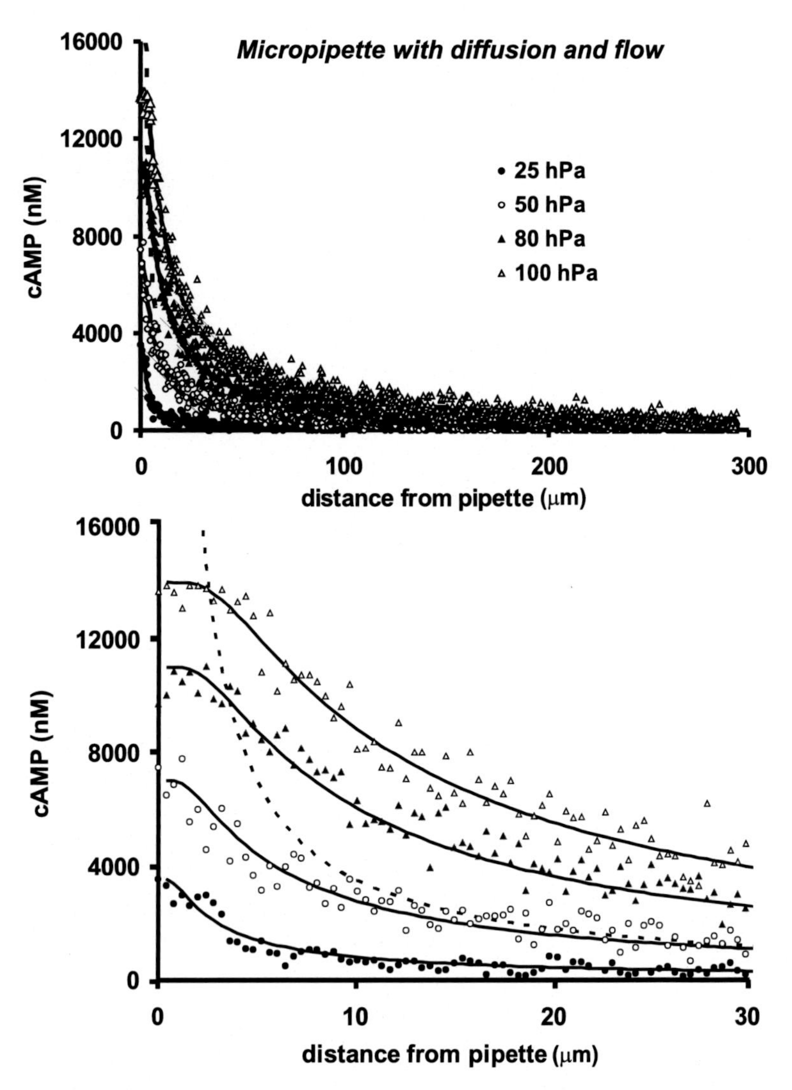

Fig. 4 Experimental equilibrium data of the cAMP gradient with different flow from the micropipettes. A micropipette filled with cAMP and lucifer yellow is applied just above the glass surface in a droplet of cells. The pressure applied is 25, 50, 80, and 100 hPa. The equilibrium fluorescence intensity is measured by confocal microscopy at 30 s after application of the pipette. The lines are the fitted data using Eq. 13 with *Cs* and *F* as indicated in Table 1; the *dashed line* is Eq. 17 for 50 hPa. The *lower panel* shows the same data as *upper panel*, but only at shorter distance from the pipette

$$\frac{\nabla C(x)}{C(x)} = -\frac{F}{2\pi D x^2} \frac{e^{-\frac{F}{2\pi D x}}}{1 - e^{-\frac{F}{2\pi D x}}} \tag{15}$$

The results shown in Figs. 3 and 4 reveal that with an increase of pressure from 0 to 25 hPa, the concentration at the tip of the pipette increases substantially from 150 nM at 0 hPa to 3000 nM

Table 1
Values of *F* and *Cs* obtained by fitting experimental observation in Fig. 4 with Eq. 13

Applied pressure (*Pc*, hPa)	Fitted flow (F, $\mu m^3/s$)	Fitted concentration at tip (*Cs*, nM)	$\alpha = \frac{C_S}{C_0}\frac{F}{2\pi D}(\mu m^{-1})$
25	15,000	3500	0.0875
50	30,000	7000	0.35
80	48,000	11,000	0.88
100	60,000	14,000	1.4

at 25 hPa. In the absence of pressure, the concentration at the tip is very low due to limited diffusion in the narrow tip of the pipette, causing a large concentration gradient inside the pipette. With liquid flow, the liquid at the tip is replaced by interior liquid of higher concentration, resulting in a higher cAMP concentration at the tip.

The gradient profiles at different applied pressures (Fig. 4) were fitted using Eq. 13, yielding the values for the concentration at the tip (*Cs*) and flow (*F*) as presented in the Table 1. The calculated lines are in very good agreement with experimental data, both close to the pipette and at longer distances from the pipette. This suggests that we derived and accurate model for gradient formation from a pipette with diffusion and flow.

Using the observed data and fitted values of *F* and *Cs*, the Eqs. 13–15 can be simplified. The calculated flow is large relative to $2\pi D r_0$, which implies that the denominator in Eqs. 13–15 reduces to 1. Furthermore, at longer distances from the pipette (i.e., at large x), $F/2\pi Dx$ in Eq. 13 becomes very small, and consequently Eq. 13 reduces to the following equation:

$$C(x) = C_S \frac{F}{2\pi Dx} (\text{for large} x) \tag{16}$$

This equation has the same form as Eq. 9:

$$C(x) = \frac{\alpha C_p}{x} \tag{17}$$

where $\alpha = \frac{C_S}{C_p}\frac{F}{2\pi D} \text{in} \mu\text{m}^{-1}$.

In Fig. 4 the dashed line represents Eq. 17 with the experimentally fitted values of *F* and *Cs*, showing that the experimental data are very well described with the simple Eq. 17 at distances beyond about 15 μm from the pipette. Finally, inspecting Table 1, we noted that the fitted values of both *Cs* and *F* increase linearly with the applied pressure Pc, indicating that α, and thus the absolute concentration, depends on Pc^2.

The usual field of observation is 100×100 µm with the pipette placed somewhere in the field. The time constant to reach equilibrium, $\tau_D = (x - r_0)^2 / D$, indicates that in the field of observation (x maximal 100 µm; r_0 = 0.5 µm, D = 1000 µm^2/s) equilibrium is reached within 10 s after application of the pipette, as was observed experimentally (Fig. 3). Thus gradients from pipettes are essentially stable spatial gradients, except at very long distances from the pipette.

4 Gradients Generated by Aggregating *Dictyostelium* Cells

Dictyostelium cells secrete cAMP in a pulsatile manner with a periodicity of about 5 min. The cAMP signal is relayed through the field leading to waves of cAMP that propagate with a velocity of about 300 µm/min. These waves have been visualized by fluorography using competition between secreted cAMP and added [^{3}H] cAMP for binding to a cAMP-binding protein.

We calculated the extracellular cAMP concentration during natural cell aggregation using the original fluorographs made by Tomchik and Devreotes [20] as presented in Ref. [21]. The fluorographs represent the competition of a fixed amount of [^{3}H] cAMP and secreted cAMP for binding to the regulatory subunit of cAMP-dependent protein kinase. In essence, this experiment is an isotope dilution assay in space that follows the equation:

$$A(x) = a[\frac{C_0 - bl}{C_x - bl} - 1] \tag{18}$$

where $A(x)$ is the cAMP concentration at position x, C_0 is the amount of [^{3}H]cAMP-binding in the absence of cAMP, bl is the blank of the assay (i.e., the amount of [^{3}H]cAMP-binding in the presence of excess cAMP), Cx is the amount [^{3}H]cAMP-binding at position x competed by unknown amount of cAMP, and a is a proportionality constant that is determined by measuring the amount [^{3}H]cAMP-binding at known amounts of unlabelled cAMP. For cAMP determinations in a test tube, the measured units are counts per minute (cpm), while in fluorographs the units are in gray scales. We determined the proportionality constant a in a test tube (550 nM), and estimated the values of C_0 and bl from the information provided by Devreotes et al. [21]. C_0 is the gray level in the absence of cAMP), while bl is the gray level in the presence of saturating cAMP. C_0 and bl were 127 and 90 in Fig. 2 of ref. [21], respectively. The cAMP concentration in natural waves was calculated as the average of two successive waves. From the spatial resolution and the speed of wave propagation we calculated the

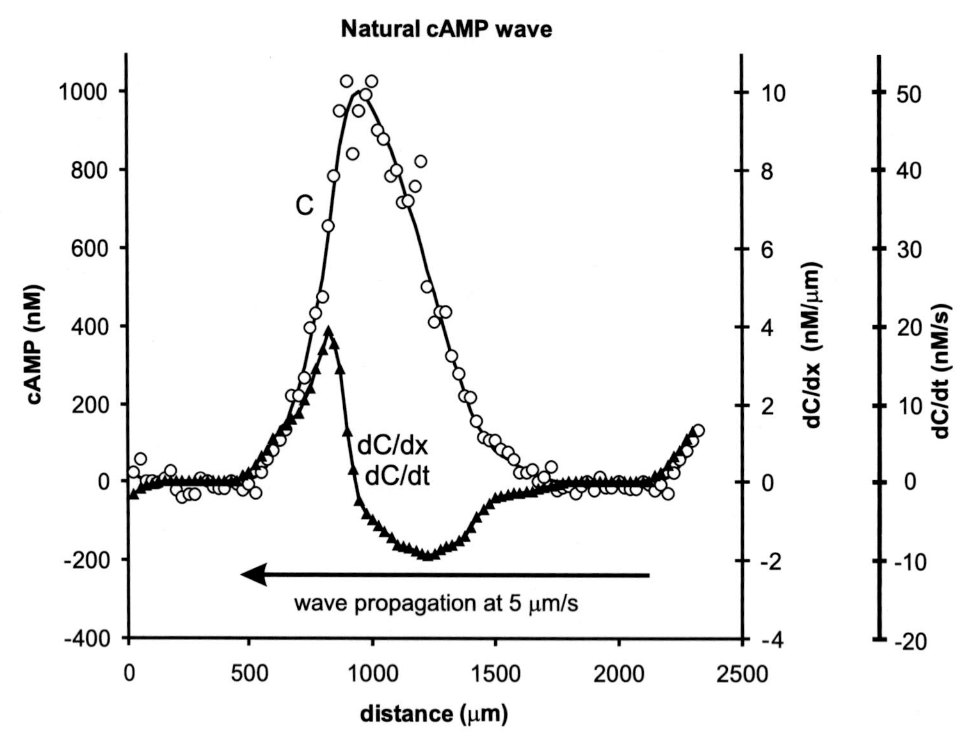

Fig. 5 Natural cAMP wave during cAMP aggregation. The cAMP concentration is deduced from fluorographs of released cAMP measured by [21]. See methods for calculations. The wave of cAMP travels through a field of cells at a rate of about 300 μm/min (5 μm/s)

spatial cAMP gradient and the temporal cAMP gradient during cell aggregation.

The results (Fig. 5) show that the extracellular cAMP concentration profiles are approximately symmetric cAMP waves. The width at the base of the wave is about 1600 μm. Since the wave travels at a speed of about 300 μm/min, the periodicity of the wave is approximately 5–6 min. The rising flank of the wave, as well as the width at half-maximal concentration, is about 450 μm or 1.5 min. The absolute spatial gradient of the wave, $\nabla C = dC / dx$, increases during the rising flank of the wave and reaches a maximum of about 4 nM/μm at about 1 min after arrival of the cAMP wave. This absolute spatial gradient is only twofold larger than the maximal spatial gradient in the Zigmond chamber of 1.8 nM/μm (*see* Fig. 5b). The relative steepness of the wave, $\nabla C / C$, is around 0.007 (μm^{-1}) during most of the rising flank of the cAMP wave (i.e., 7 % concentration difference between front and back of a 10 μm long cell).

The cAMP waves are propagated in the field of cells with a speed $v = dx / dt$ of 300 μm/min [21]. Therefore, the temporal gradient is given by $dC / dt = v\nabla C$. Thus, the temporal gradient follows the spatial gradient and reaches a maximum value of about

17.5 nM/s at 1 min after arrival of the cAMP wave. In the Zigmond chamber the temporal gradient reaches a maximum of 3 nM/s at 2 min after application of cAMP (Fig. 1c).

5 Conclusions

The gradients formed in the modified Zigmond assay are very different from the gradients formed by a point source releasing a constant flux of cAMP. The major differences are: a temporal–spatial gradient with little distance dependency in the Zigmond assay versus a stable gradient with very strong distance dependency. It is surprising that the distance dependency of the pipette assay is often not taken into account, exemplified by the absence of information on the distance from the pipette where the chemotaxis data were obtained, a situation we also were not aware of in the past [22]. Chemotaxis in *Dictyostelium* is mainly dependent on the absolute spatial gradient resulting in only about 10 occupied receptors more at the front of the cells relative to the rear of the cell at threshold conditions, at prevailing receptor occupancies of around 1000 receptors. It will be a major effort to uncover how cells are able to deduce spatial information from a signal that is inherently very noisy due to the high average receptor occupancy. The gradient models presented here may help to design experiments to deduce the fundamental principles of gradient sensing and directed locomotion.

Acknowledgements

We thank Douwe Veltman and Ineke Keizer-Gunnink for obtaining experimental data on the Zigmond chamber (Fig. 1) and micropipettes (Figs. 3 and 4), respectively.

References

1. Szurmant H, Ordal GW (2004) Diversity in chemotaxis mechanisms among the bacteria and archaea. Microbiol Mol Biol Rev 68:301–319
2. Devreotes P, Janetopoulos C (2003) Eukaryotic chemotaxis: distinctions between directional sensing and polarization. J Biol Chem 278:20445–20448
3. Servant G, Weiner OD, Herzmark P et al (2000) Polarization of chemoattractant receptor signaling during neutrophil chemotaxis. Science 287:1037–1040
4. Affolter M, Weijer CJ (2005) Signaling to cytoskeletal dynamics during chemotaxis. Dev Cell 9:19–34
5. Parent CA, Devreotes PN (1999) A cell's sense of direction. Science 284:765–770
6. van Haastert PJ, Devreotes PN (2004) Chemotaxis: signalling the way forward. Nat Rev Mol Cell Biol 5:626–634
7. Futrelle RP (1982) *Dictyostelium* chemotactic response to spatial and temporal gradients. Theories of the limits of chemotactic sensitivity and of pseudochemotaxis. J Cell Biochem 18:197–212
8. Iijima M, Huang YE, Devreotes P (2002) Temporal and spatial regulation of chemotaxis. Dev Cell 3:469–478
9. Varnum-Finney B, Edwards KB, Voss E, Soll DR (1987) Amebae of *Dictyostelium discoi-*

deum respond to an increasing temporal gradient of the chemoattractant cAMP with a reduced frequency of turning: evidence for a temporal mechanism in ameboid chemotaxis. Cell Motil Cytoskeleton 8:7–17

10. Veltman DM, van Haastert PJ (2006) Guanylyl cyclase protein and cGMP product independently control front and back of chemotaxing *Dictyostelium* cells. Mol Biol Cell 17:3921–3929
11. Zicha D, Dunn GA, Brown AF (1991) A new direct-viewing chemotaxis chamber. J Cell Sci 99:769–775
12. Muinonen-Martin AJ, Veltman DM, Kalna G, Insall RH (2010) An improved chamber for direct visualisation of chemotaxis. PLoS One 5:e15309
13. Swanson JA, Taylor DL (1982) Local and spatially coordinated movements in *Dictyostelium discoideum* amoebae during chemotaxis. Cell 28:225–232
14. Irimia D (2010) Microfluidic technologies for temporal perturbations of chemotaxis. Annu Rev Biomed Eng 12:259–284
15. Sackmann EK, Fulton AL, Beebe DJ (2014) The present and future role of microfluidics in biomedical research. Nature 507:181–189
16. Kamholz AE, Yager P (2001) Theoretical analysis of molecular diffusion in pressure-driven laminar flow in microfluidic channels. Biophys J 80:155–160
17. Nakajima A, Ishihara S, Imoto D, Sawai S (2014) Rectified directional sensing in long-range cell migration. Nat Commun 5:5367
18. Skoge M, Yue H, Erickstad M et al (2014) Cellular memory in eukaryotic chemotaxis. Proc Natl Acad Sci U S A 111:14448–14453
19. Meier B, Zielinski A, Weber C et al (2011) Chemotactic cell trapping in controlled alternating gradient fields. Proc Natl Acad Sci U S A 108:11417–11422
20. Tomchik KJ, Devreotes PN (1981) Adenosine 3′,5′-monophosphate waves in *Dictyostelium discoideum*: a demonstration by isotope dilution-fluorography. Science 212:443–446
21. Devreotes PN, Potel MJ, MacKay SA (1983) Quantitative analysis of cyclic AMP waves mediating aggregation in *Dictyostelium discoideum*. Dev Biol 96:405–415
22. Loovers HM, Postma M, Keizer-Gunnink I et al (2006) Distinct roles of PI(3,4,5)P_3 during chemoattractant signaling in *Dictyostelium*: a quantitative in vivo analysis by inhibition of PI3-kinase. Mol Biol Cell 17:1503–1513

Chapter 27

Modeling Excitable Dynamics of Chemotactic Networks

Sayak Bhattacharya and Pablo A. Iglesias

Abstract

The study of chemotaxis has benefited greatly from computational models that describe the response of cells to chemoattractant stimuli. These models must keep track of spatially and temporally varying distributions of numerous intracellular species. Moreover, recent evidence suggests that these are not deterministic interactions, but also include the effect of stochastic variations that trigger an excitable network. In this chapter we illustrate how to create simulations of excitable networks using the Virtual Cell modeling environment.

Key words Mathematical model, Chemotaxis, Reaction-diffusion, Stochastic systems, Excitable systems

1 Introduction

The study of directed cell migration is an area in biology that has greatly benefited by the interaction of experimental and computational biologists. Numerous models have been proposed to account for various aspects of the chemotactic response, from gradient sensing [1, 2] to polarization [3, 4] to movement [5–7]. Simulations of these models recreate many of the observed behaviors of chemotactic cells, from the initial breaking of symmetry provided by the chemoattractant stimulus, to the morphological changes that are seen in migrating cells [8–11].

During the last few years, one of the most significant advances in our understanding of the mechanism driving chemotaxis in eukaryotic cells is the appreciation of the role played by excitable dynamics [12–19]. An excitable system is a dynamical system that operates near a stable equilibrium. Small-scale perturbations away from this equilibrium—be they triggered from external stimuli or stochastic fluctuations of the underlying system—are attenuated. However, if these perturbations are sufficiently strong that they cross a threshold, a large-scale, but transient response is elicited. When the excitable system incorporates diffusible elements, the system is said to be an excitable medium. In this case, the suprathreshold responses trigger traveling waves. The appearance of

Tian Jin and Dale Hereld (eds.), *Chemotaxis: Methods and Protocols*, Methods in Molecular Biology, vol. 1407,
DOI 10.1007/978-1-4939-3480-5_27,

such waves, in either cytoskeletal components or signaling molecules at the basal surface of migrating cells is a signature of excitable medium.

The presence of an excitable signaling system explains a number of observations not previously accounted by other models. For example, the triggering of a response by intrinsic, stochastic perturbations explains the random migration of unstimulated cells. Similarly, the large amplification of suprathreshold stimuli accounts for the large amplification of shallow chemoattractant gradients seen in cells. When coupled to other models, such as a gradient-sensing mechanism, realistic chemotaxis is achieved. Similarly, the inclusion of a polarity or memory module explains both persistent random migration, as well as the spatially dependent sensitivity of polarized cells that is observed in cells [11, 16, 20]. The presence of an excitable system presents challenges to developing a computational simulation. In this chapter we demonstrate how to develop such models. We begin by presenting some necessary theoretical background.

1.1 Modeling Deterministic Temporal Dynamics

When developing a model describing the temporal dynamics of a biochemical network we use ordinary differential equations (ODEs). For a network with n interacting species, labeled C_1,... ,Cn, we describe temporal changes in the concentration of species i, denoted by Ci, through the differential equation:

$$\frac{dC_i(t)}{dt} = f_i(C_1,\ldots,C_n). \tag{1}$$

The function f_i specifies the interconnections that affect the concentration of the different C_i. The specific form of this function depends on the assumptions made about the biochemical interaction. Consider the binding reaction:

$$C_1 + C_2 \underset{k_r}{\overset{k_f}{\rightleftharpoons}} C_3 \tag{2}$$

Each time the forward reaction occurs, molecules of species C_1 and C_2 bind together to create one molecule of C_3. With the reaction affinity of kf, we can use a flux term $Jf = kf \times C_1 \times C_2$ to represent the rate of consumption of C_1 and C_2, which is also the rate of production of C_3. The reverse reaction occurs with affinity kr, with one unit of C_3 decomposing into one unit of C_1 and one unit of C_2. The flux $Jr = kr \times C_3$ represents the rate the reverse reaction. The whole reaction therefore occurs at the combined rate given by the sum of the two individual rates:

$$\frac{dC_3}{dt} = -\frac{dC_1}{dt} = -\frac{dC_2}{dt} = J_f - J_r = k_f C_1 \times C_2 - k_r C_3.$$

1.2 Modeling Spatial Dynamics

The equations of the previous section are appropriate for describing reactions that take place in compartments in which the concentrations of the various species is spatially homogeneous. Of course, in developing a model of a eukaryotic chemotactic cell, it is particularly important to be able to describe the spatial distribution of intracellular markers in response to chemoattractant gradients [21]. Modeling the concentration of a biochemical species Ci that depends on time, t, and spatial dimension, x, requires the use of partial differential equations (PDE). In particular, biochemical models using reaction-diffusion equations are needed:

$$\frac{\partial C_i(x,t)}{\partial t} = f_i(C_1,\ldots,C_n) + D_i \frac{\partial^2 C_i(x,t)}{\partial x^2} \quad (3)$$

As before, the function $f_i(C_1,\ldots,Cn)$ specifies the reactions that affect the concentration of Ci—in the example of the previous section, these would be the two fluxes Jf and Jr The right-hand most term in the PDE describes diffusion of the species, where D is the diffusion coefficient of that particular species. For intracellular molecules, typical values range around 10 $\mu m^2/s$ for proteins in the cytosol, 1 $\mu m^2/s$ for lipids in the membrane, and 0.1 $\mu m^2/s$ for proteins in the membrane [22].

To solve for $Ci(x,t)$ requires that an initial distribution, $Ci(x,0)$, be known. It is also necessary to have boundary conditions. For PDEs, there are several available choices. The most common form of boundary condition is to specify the flux of the species at the boundary. For example, in an impermeable membrane, the flux would be zero.

In modeling spatial dynamics, it is sometimes necessary to consider situations were the interacting species reside on different compartments. For example: binding can occur when one reactant resides on the cell membrane (a binding site) and the other in an extracellular or intracellular volume (e.g., the cytosol). In this case, the product is on the membrane.

1.3 Modeling Stochastic Fluctuations Using the Langevin Equation

The models described above use a continuum description of chemical species. Real chemical networks, however, involve discrete molecules and reactions. These reactions occur randomly as individual molecules interact. This "biological noise" can be a significant determinant of cell behavior [23–25]. When seeking a mathematical model that accounts for this stochastic nature of these interaction, it is important to understand that it is no longer possible to describe precisely the state of the system (i.e., the number of molecules of each species) as this is now a random variable. Instead, one can only specify the probability that the system is at a particular state. Based on the various reactions, one can then describe the evolution of the system using the Chemical Master Equation

(CME), a PDE that considers all possible states of the system and the probability of transitions between states. In practice, however, this is completely impractical, as it represents an infinite system of equations. There are a number of possible approaches to work around this, from carrying out discrete simulations using Stochastic Simulation algorithm [26] or through approximations of the CME [27]. These may be appropriate for genetic regulatory elements that have small copy numbers, but impractical for the systems regulating chemotaxis in eukaryotic cells, which can number in the hundreds of thousands to millions of molecules per cell. In these cases, the Langevin approximation is more tractable [28]. In this case, the stochasticity is introduced as a noise term. As an example, the deterministic differential equation

$$\frac{dC_3}{dt} = k_f C_1 \times C_2 - k_r C_3$$

is replaced by the stochastic differential equation:

$$\frac{dC_3}{dt} = k_f C_1 \times C_2 - k_r C_3 + \sigma(C)w(t) \tag{4}$$

where $w(t)$ is a Brownian term with zero mean and unit correlation $\langle w(t), w(\tau) \rangle = \delta(t-\tau)$ The term $\sigma(C)$ gives the variance, which in general will be state-dependent [28].

An alternative approach is known as the particle method. Here, individual molecules are tracked in time and space. Using rules based on mass action dynamics, these molecules interact with neighboring molecules. An example of this approach is the Smoldyn programming environment [29, 30].

1.4 A Biased Excitable Network

As an example of a nonlinear model, we introduce a simple excitable network. The system consists of two elements: an activator (X) and an inhibitor (Y). The reaction diffusion equation for both species is given by

$$\frac{\partial X}{\partial t} = k_1 \frac{X^2}{k_m^2 + X^2} - (k_2 + k_3 Y)X + U + D_X \frac{\partial^2 X}{\partial x^2}$$

$$\frac{\partial Y}{\partial t} = k_4 X - k_5 Y + D_Y \frac{\partial^2 X}{\partial x^2}$$

These equations represent five reactions and two diffusions. The first reaction is the cooperative, autocatalytic stimulation of X and is described by the first term to the right of the equation sign in the equation for X. The form of this term is a Hill equation with coefficient 2. The second is the degradation of X, which consists of two components: $k_2X + k_3XY$. The first term represents the basal level of degradation; the second is the rate that the inhibitor speeds up this degradation. The third and fourth reaction is the rates at which the inactivator is formed (k_3X) and degraded (k_4X). The final term, U, represents the combined action of all external stimuli to the system. These can be external chemical cues—for example, the spatially dependent receptor occupancy of a spatially graded stimulus. In this case, we term the system a Biased Excitable Network (BEN). However, they could also represent stochastic fluctuations that can trigger the system. In this latter case, $U(t) = \sigma w(t)$, where $w(t)$ is Brownian noise, as described above.

We now demonstrate how to set up computational models of these systems so as to be able to simulate their behavior.

2 Materials

2.1 Compartmental Models

In general, the right-hand side of Eqs. 1, 2, or 4 are nonlinear functions of the species' concentrations. As such, analytic solutions are rarely available. It is thus necessary to solve this equation numerically, requiring a numerical simulation package. The most popular, general-purpose packages are Matlab (Mathworks, Natick, MA), Mathematica (Wolfram, Champaign, IL), and Maple (Maplesoft, Waterloo, Canada). All are relatively easy to use and provide great functionality and can handle stochastic models, though the latter requires some special toolboxes. Alternatively, a number of simulation packages specifically tailored to the biological signaling community have appeared (reviewed in Ref. [31]). In most of these, a graphical user interface allows the user to specify a biochemical interaction by selecting the type of reaction and the kinetic coefficients. The package then automatically generates and solves the necessary ODEs.

2.2 Spatial Models

For spatially varying models, the list of available simulation packages is considerably smaller. Comsol Multiphysics (Comsol, Burlington, MA), a general-purpose simulation package originally designed to work with Matlab but now independent, allows the user to specify general PDEs, or to select from one of several predefined forms, including reaction-diffusion equations such as Eq. 2. The solution is obtained using finite-element methods [32].

The Virtual Cell is one of the few simulation packages specially tailored to cell biology that can deal with spatially varying simulations. Unlike most other software packages that reside and carry out the simulations in the user's computer, the Virtual Cell software is maintained at a central server within the National Resource for Cell Analysis and Modeling (NRCAM) at the University of Connecticut Health Center [33, 34]. Using the Virtual Cell is done through a Java application over the internet. Use of the Virtual Cell requires an account, available at http://vcell.org/vcell_software/login.html. Funded through the National Center for Research Resources, a component of the National Institutes of Health, the Virtual Cell is free for users in an academic environment. General purpose tutorials are available on their web page. Importantly, recent upgrades also allow for spatially varying stochastic systems by incorporating the Smoldyn simulation framework.

3 Methods

We previously used the Virtual Cell to implement the local-excitation, global-inhibition (LEGI) mechanism, thereby demonstrating how a spatially varying, deterministic system could be simulated [35]. Here, we use the Virtual Cell to implement a biased excitable network (BEN) (*see* **Note 1**). These LEGI and BEN models could be then coupled easily to simulate the complete LEGI-BEN system [13]. The molecules involved in the BEN model, as well as their spatial distribution in the cell geometry, are specified in Table 1, while their reactions are listed in Table 2. The stochastic application reactions and molecules are specified in Table 3.

A complete model in Virtual Cell consists of three components:

1. A Biomodel component where we specify: the molecular species involved in the biological model, the physiological compartments in which they reside, and their interactions.
2. A Geometry component that allows for specifying the shapes and dimensions of each compartment.
3. A Mathematical component based on the Virtual Cell Math Description Language (VCMDL) that allows for access of mathematical formulae behind the biological model.

Table 1
List of species

Molecule	Description	Compartment	Initial condition	(Units)	Diffusion ($\mu m^2/s$)
L_source	Chemoattractant concentration at the source	EC	Variable	μM	Clamped
L_EC	Chemoattractant concentration	EC	0	μM	15
R_CM	Unoccupied receptor	CM	100	$\#/\mu m^2$	Clamped
S_CM	Occupied receptor	CM	0	$\#/\mu m^2$	0
X_CM	Activator	CM	0.14	$\#/\mu m^2$	0
Y_CM	Inhibitor	CM	2.0	$\#/\mu m^2$	0
N_CM	Noise impulse	CM	Variable	$\#/\mu m^2$	Clamped

Table 2
List of reactions

Reaction	Structure	Description	Reaction type	Reaction rate (#μm^{-2} s^{-1})	*kf* units	*kr* units
1 L_source ↔ L_EC	EC	Diffusion of cAMP from source	(*See* **Note 5**)	kf × L_source − k_r × L_EC	1 #μm^{-2} μM^{-1} s^{-1}	1 #μm^{-2} μM^{-1} s^{-1}
2 L_EC+R_CM → S_CM	CM	cAMP binding and response regulator activation	Binding	kf × L_EC × R_CM − kr × S_CM	1.66 μM^{-1} s^{-1}	0.39 s^{-1}
3 X_CM → X_CM + X_CM	CM	EN activator formation	Activator differential equation	0.04 + 0.4×N_CM + 2.0×X_CM×X_CM/ (1.0 + X_CM×X_CM) − 0.4×X_CM×Y_CM + 0.008×X_CM×S_CM	0.001 μm^2 s^{-1}/#	0.1 s^{-1}
4 Y_CM → Y_CM + Y_CM	CM	EN inhibitor formation	Inhibition differential equation	0.4×X_CM − 0.028×Y_CM	0.01 μM^{-1} s^{-1}	10 #μm^{-2} μM^{-1} s^{-1}

Table 3
Stochastic example specifications

Molecule	Description	Compartment	Initial condition	Units	Diffusion ($\mu m^2/s$)
A_CM	Reactant	CM	0.1	$\#/\mu m^2$	0.1
B_CM	Product	CM	0.0	$\#/\mu m^2$	0.1
Reaction: A_CM → B_CM					
Structure: CM		Kinetic type: mass action		Coefficients: kf = 1 s^{-1}; kr = 1 s^{-1}	

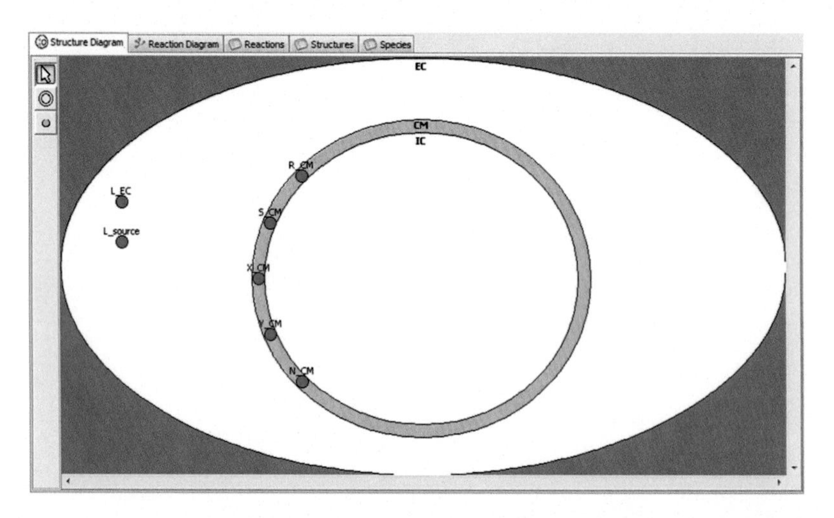

Fig. 1 Specifying the structure. The different species (*green circles*) are placed in compartments where they reside. This example shows three structures: the extracellular compartment (EC), the cell membrane (CM) and the intracellular compartment (IC). Both compartments and species are introduced using the respective tools

To create a biological model in Virtual Cell, it is necessary to define both the Biomodel component and the Geometry component. Equations in the Mathematical component are generated automatically but are also available for manual editing.

3.1 BEN Spatial Deterministic Modeling

3.1.1 Specifying the Components

1. When Virtual Cell is started, a new instance of Biomodel opens, allowing the creation of a new model.
2. Start by clicking on the unnamed compartment, and naming it "EC" (for extracellular), in the Object Properties window.
3. To add a cellular compartment, use the Compartment Tool (Fig. 1, bottom) by clicking the icon, then click the EC compartment. Now specify the name of the cell's intracellular (IC) and cell membrane (CM). A general model can include multiple compartments; however, in our example, only one compartment is needed.
4. Next, add all the molecular species to the model by using the Species Tool (Fig. 1, bottom). To start, click the Species tool

icon, followed by the EC compartment, and name the new species "L_EC" (for Ligand). Similarly, click on the Membrane and Cytosol compartments to add in molecular species that belong in each. Table 1 lists the necessary molecular species and their respective compartments.

5. When all the molecular species are added (Fig. 1, top), proceed to define interactions between them. Each reaction is defined in a specific compartment. Reactions in a volume can only involve molecules from the same volume, but reactions on the membrane can involve molecules from the membrane and neighboring compartments.

3.1.2 Specifying the Reactions

1. Start by defining the reactions for this sensing model. To do so, click on the "Reactions" icon under the Physiology heading on the left window.
2. To add a new reaction, double click on the "(add new here, e.g. $a+b\rightarrow c$)" row. In the pop-up window, choose the compartment in which the reaction is to take place, the reaction name and the expression for the reaction (Fig. 2a).
3. Once a particular reaction is entered, in the Object Properties window, change the Kinetic type (set to "Mass Action" by default) to "General [molecules/μm^2 s]" for the purpose of this example (*see* **Note 1**). Then type in the expression for the reaction rate J. An example of the ligand-binding reaction is:

$$J = kf \times R\_CM \times L\_EC - kr \times S\_CM$$

When this expression is defined, the software automatically recognizes previously undefined variables, namely the kinetic rate constants in this case. We can now specify the rate constants "kf" and "kr" of this reaction. As we are not modeling any membrane currents, the variable "I" can be ignored.

4. In this manner, proceed to define the remaining reactions (*see* **Note 2**) listed in Table 2 (Fig. 2b).
5. Save this model as BEN.

3.1.3 Specifying the Geometry

1. To define the simulation system completely, describe the geometry of the simulation: how large the cell is, where it resides in space, etc. In this example, assume that the cell is a three-dimensional sphere of radius 1.414 μm, and the chemoattractant source is positioned approximately 6.25 μm away from the cell center on the same *z*-plane (*see* **Note 3**).
2. To define this geometry, navigate to the Geometries tab under VCell DB in the lower left window. Right-click on "My Geometries" and click on New Geometry → Analytic Equations 3D. This will open up a "Geometry Editor" (Fig. 3a).

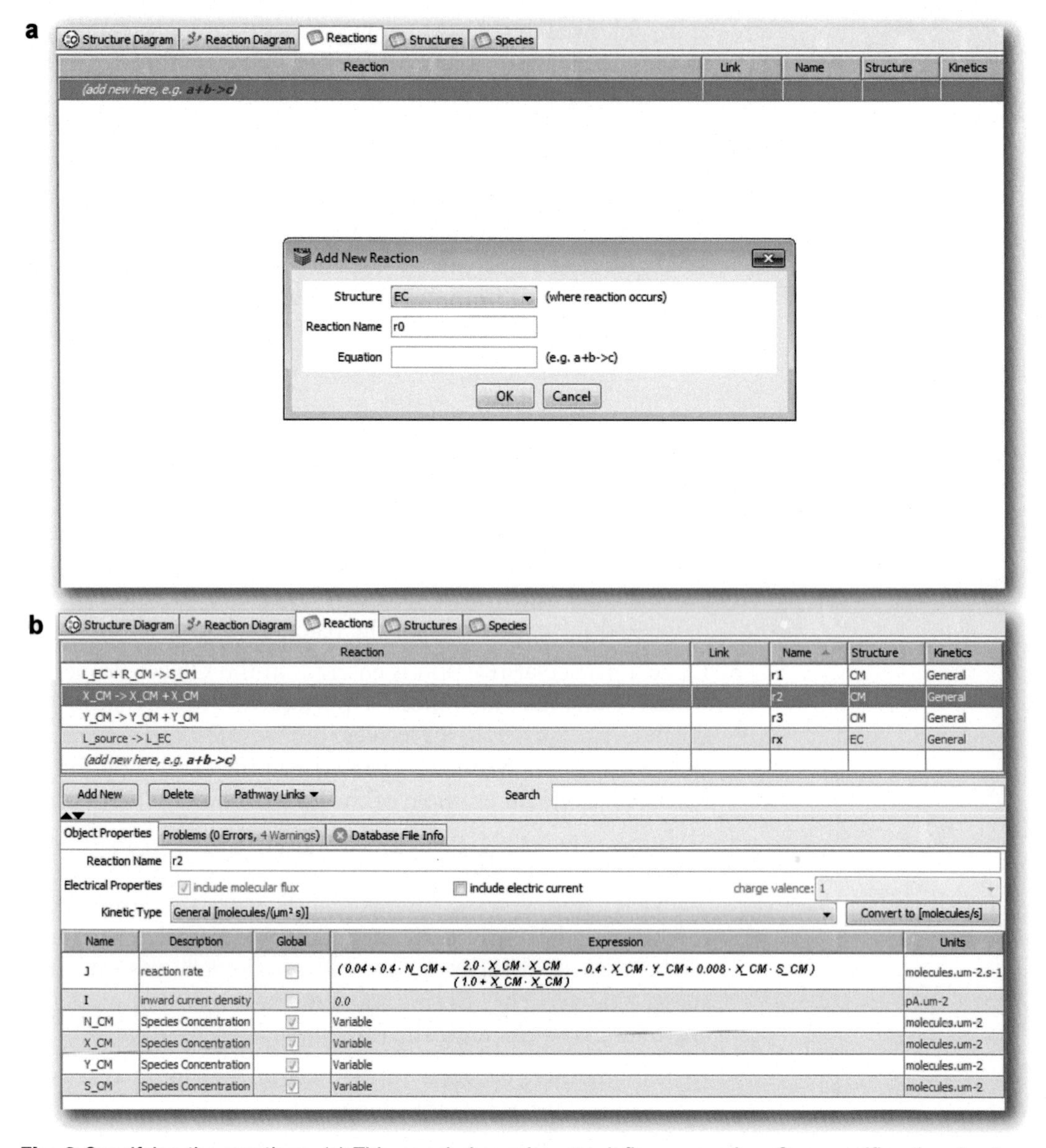

Fig. 2 Specifying the reactions. (**a**) This panel shows how to define a reaction. One specifies the structure where the reaction occurs, a name for the reaction, and an equation for it. (**b**) The complete list of reactions is shown. Pressing the "Add New" button brings up the window in panel (**a**)

3. To define an extracellular space large enough to accommodate the cell and leave room for the chemoattractant source, click on "Edit domain," and specify the domain size to be a 5 μm by 5 μm by 5 μm cube with origin at (0, 0, 0).
4. Rename the subdomain as "EC" and leave its value at "1" to represent the EC compartment.
5. Click "Add Subdomain" followed by "Analytic" to add the IC compartment. Enter the Analytic Expression: $(((-2.5+x)\times(-2.5+x)+(-2.5+y)\times(-2.5+y)+(-2.5+z)\times(-2.5+z))<2.0)$,

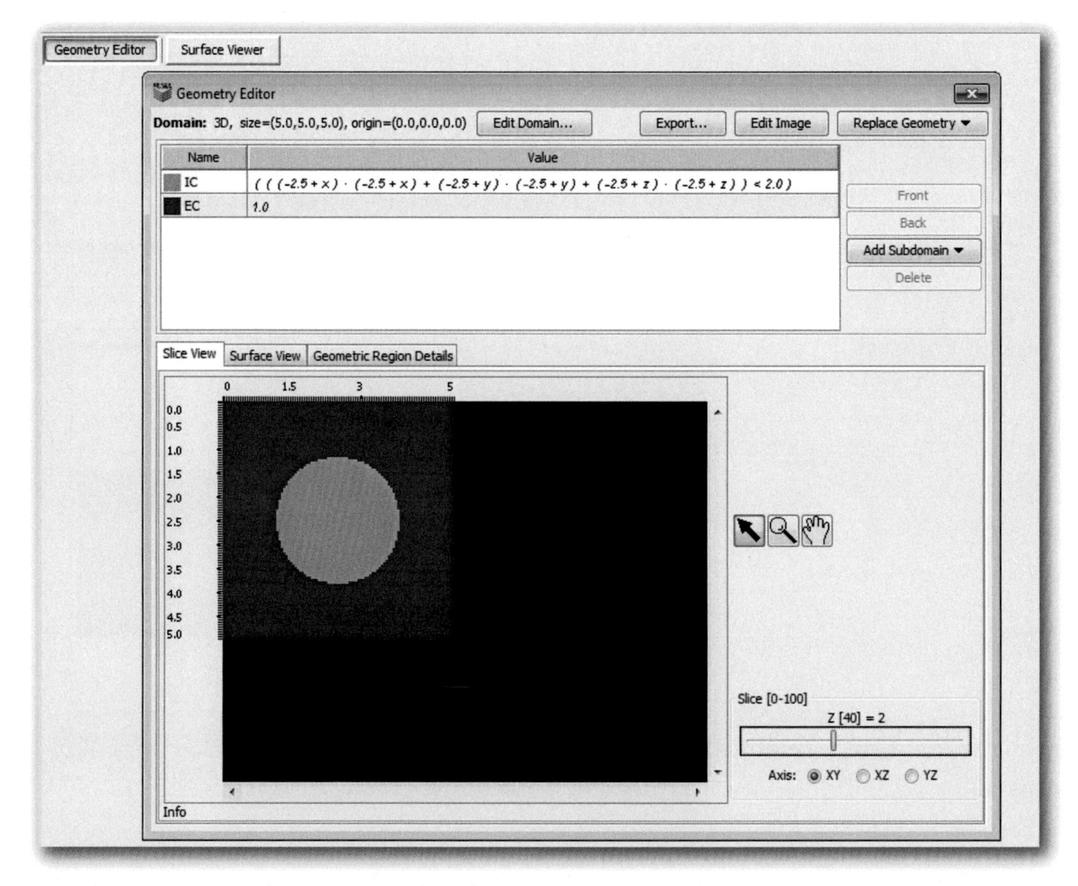

Fig. 3 Specifying the geometry. Equations are used to specify the intracellular (IC) and extracellular (EC) compartments. Note that using Smoldyn to simulate stochastic behavior requires a 3D geometry

which is the formula for a sphere with a radius 1.414 μm and centered at (2.5, 2.5, 2.5). Then rename the subvolume to "IC" (Fig. 3a).

6. Save this geometry as "3D_BEN."

3.1.4 Linking the Physiological and Geometrical Models

1. Thus far, a physiological model and a geometrical model have been created. At this point, link them to form a complete application in the Virtual Cell. Right click on the "Applications" tab on the left side of the physiological model, choose "Add New," then "Deterministic," and enter the name (e.g., "Needle") of this new application. Double clicking on "Needle" to expand it, you see the list of applications associated with this physiological model.
2. To link this application to its geometry, click on the Geometry tab under the Needle Application, and select "Geometry Definition" (Fig. 4a).
3. Click on "Add Geometry" and then Open From → Geometry Names → My Geometries → 3D_BEN.

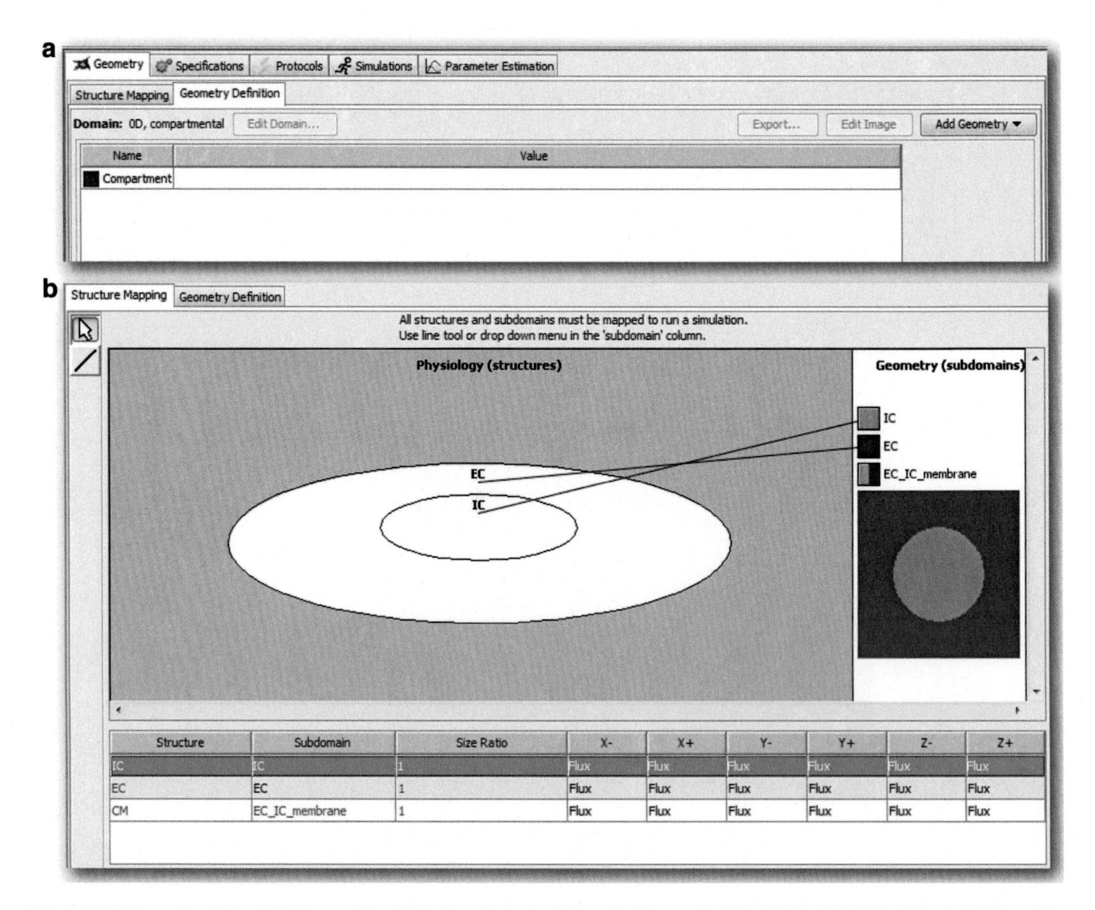

Fig. 4 Linking physiology to geometry. The two types of descriptions must be linked. (**a**) To do so, click on the Geometry tab under the Application, and select "Geometry Definition. (**b**) Under the "Structure Mapping" tab, we can link the physiology structures to their appropriate geometries. Note how the EC and IC compartments in Physiology (structures) are linked to their corresponding compartments in the Geometry subdomains

4. Now under the "Structure Mapping" tab, link the physiology structures to their appropriate geometries. Under "Subdomain" where it says "Unmapped," click the icon: link "IC" to "IC" and "EC" to "EC" (Fig. 4b).

3.1.5 Specifying the Initial Conditions

1. The initial conditions and diffusive properties of each molecular variable in the physiological model must be specified. These properties are designed to be part of each application. For each physiological model, many applications can be created; thus, many initial conditions, etc. can be tested (*see* **Note 4**).
2. In this application, it is assumed that molecules in EC are able to diffuse, while molecules in CM are not diffusible. Diffusion coefficients are found in Table 1 under the "Diffusion" column. Initial concentrations for molecules in all compartments are also found in Table 1.

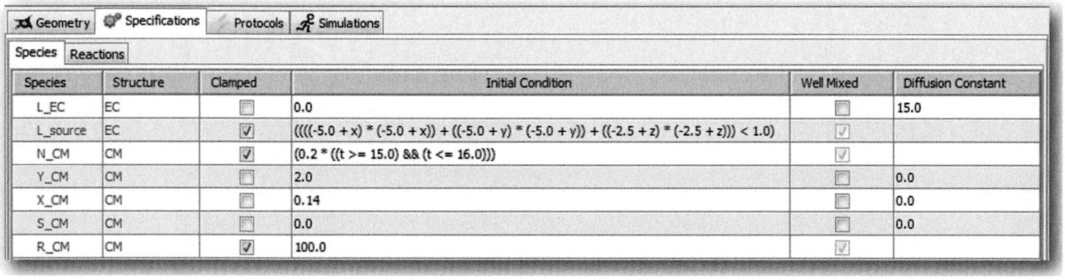

Geometry | Specifications | Protocols | Simulations

Species | Reactions

Species	Structure	Clamped	Initial Condition	Well Mixed	Diffusion Constant
L_EC	EC	☐	0.0	☐	15.0
L_source	EC	☑	((((-5.0 + x) * (-5.0 + x)) + ((-5.0 + y) * (-5.0 + y)) + ((-2.5 + z) * (-2.5 + z))) < 1.0)	☑	
N_CM	CM	☑	(0.2 * ((t >= 15.0) && (t <= 16.0)))	☑	
Y_CM	CM	☐	2.0	☐	0.0
X_CM	CM	☐	0.14	☐	0.0
S_CM	CM	☐	0.0	☐	0.0
R_CM	CM	☑	100.0	☑	

Fig. 5 Specifying initial conditions. Under "Specifications" is a list of all the species in the model. The table lists what structure they are found. "Clamped" forces the concentration to stay constant throughout the simulation. In specifying the initial condition, one can constrain the concentrations to a particular spatial domain (as done here with "L_source") or temporally (as done with "N_CM")

3. Go to the "Specifications" tab in the "Application" window. Click on each molecular species name, and modify the information on the corresponding row to match the information in Table 1 (Fig. 5).
4. The ligand source "L_source" is represented by a fixed concentration (1 μM) at a fixed location (a semicircle of radius 1 in the simulation domain). To define the fixed concentration, check the box under the "Clamped" column. To define the fixed location, use the following expression for the initial concentration:

$$(((-5.0+x)\times(-5.0+x)+(-5.0+y)\times(-5.0+y)+(-2.5+z)\times(-2.5+z))<1.0).$$

5. In this example, we want R_CM (the number of receptors) to be a constant. Therefore, we clamp that too. The initial conditions for X_CM and Y_CM are taken to be their equilibrium points in steady state.
6. Here, N_CM represents a constant impulse applied at a particular moment. In order to simulate that, in the row for the N_CM, click on "Clamped" and enter the following initial condition expression: $(0.2\times((t\geq 15.0)\ \&\&\ (t\leq 16.0)))$, which creates an impulse of concentration 0.2 μM to be applied to the system between 15 and 16 s only.
7. Save this application.

3.1.6 Running a Spatial Simulation

1. Click the "Simulation" tab in the "Application" window, and click the icon for "New Simulation" to define a new simulation. We can change the name of this simulation by double clicking on the simulation name (Fig. 6a).
2. Select this simulation, and click the icon for "Edit Simulation." A new window will open, where the simulation specifications can be modified.
3. The simulation can be started using parameters different than those previously defined, including running the simulation

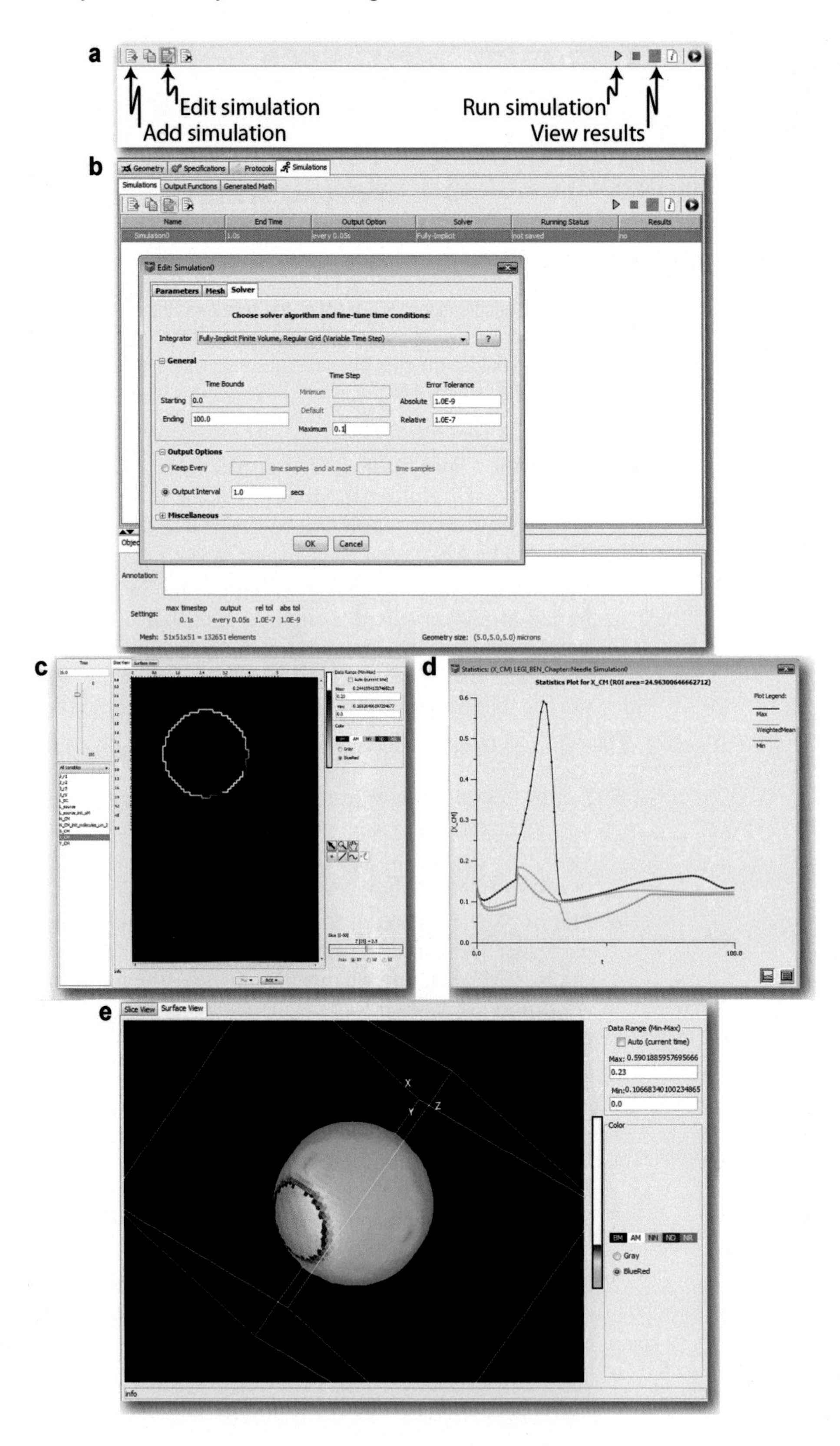
a
Edit simulation
Add simulation
Run simulation
View results
b
Geometry
Specifications
Protocols
Simulations
Output Functions
Generated Math
Name
End Time
Output Option
Solver
Running Status
Results
Simulation0
1.0s
every 0.05s
Fully-Implicit
not saved
no
Edit: Simulation0
Parameters
Mesh
Solver
Choose solver algorithm and fine-tune time conditions:
Integrator
Fully-Implicit Finite Volume, Regular Grid (Variable Time Step)
General
Time Bounds
Time Step
Error Tolerance
Starting
0.0
Ending
100.0
Minimum
Default
Maximum
0.1
Absolute
1.0E-9
Relative
1.0E-7
Output Options
Keep Every
time samples
and at most
time samples
Output Interval
1.0
secs
Miscellaneous
OK
Cancel
Annotation:
Settings:
max timestep
output
rel tol
abs tol
0.1s
every 0.05s
1.0E-7
1.0E-9
Mesh: 51x51x51 = 132651 elements
Geometry size: (5.0,5.0,5.0) microns
c
d
Statistics Plot for X_CM (ROI area=24.96300646662712)
Plot Legend:
Max
WeightedMean
Min
[X_CM]
t
0.0
100.0
e
Slice View
Surface View
Data Range (Min-Max)
Auto (current time)
Max: 0.5901885957695666
0.23
Min: 0.10668340100234865
0.0
Color
BM
AM
NN
ND
NR
Gray
BlueRed
info

with a finer spatial resolution by modifying the "Mesh" properties to contain more *X* and *Y* elements. For the purpose of this example, modify information under the "Solver" tab. Click the "Solver" tab, set maximum "Time Step" to 0.1 s, "Ending" to 100 s, and "Output Interval" to 1.0 s (Fig. 6b). The Integrator should be set to "Fully-Implicit Finite Volume, Regular Grid (Variable Time Step)." Click OK.

4. Back in the "Application" window, select this simulation and click the play icon for "Run and Save Simulation" The simulation request will be sent to the Virtual Cell server to compute, and "Running status" will indicate "completed" when the simulation is successful.
5. Click the "Results" button to view simulation results. A new window will open (Fig. 6c).
6. In this simulation, we are interested in the response of the activator X_CM to the noise impulse. Select "X_CM" from the variable list and under "Slice View" click on "ROI" → Statistics → Pre-Defined Membrane Region → OK. The time-plot for X_CM will be retrieved (Fig. 6d). We can see in the result how the max-plot (indicating the front of the cell located near the chemoattractant source) spiked larger with the noise impulse, compared to the min-plot (indicating the rear of the cell).
7. Under the Surface View, for the "X_CM," unselect the "Auto (current time)" under Data Range and enter 0.23 for Max and 0.0 for Min (Fig. 6e). Click on "Make Movie," lower the frame rate to 4 frames per second and create the movie. This movie will present a 3D visual of how the cell spiked in the front (*see* **Note 6**).

3.2 Spatial Stochastic Application Using Smoldyn

The model above is deterministic. To contrast deterministic and stochastic simulations, we consider a simple model in which a molecule is in one of two states, A and B and switches between the two.

3.2.1 Specifying the Components

1. Similar to Subheading 3.1, a new instance of Biomodel is created with similar compartments—EC, CM, and IC.
2. Two species are used here to illustrate the effect of stochastic simulations as compared to deterministic simulations. The species and their details are provided in Table 3.

Fig. 6 Running a spatial simulation. (**a**) Tabs allowing one to set up new simulations, edit old ones, and view results. After clicking "Edit Simulation" a new window opens (**b**) that allows changes to simulation specifications. (**c**) Sample simulation "Result." Here one selects the particular variable to track and time (16 s) and spatial location to plot. In this figure, we are plotting the spatial distribution of X_CM $t=16$ s of the slice with $z=2.5$ μm. The color scheme for the particular data can be specified in the *top right.* (**d**) Plot of various statistics for this simulation—shown are the maximum, minimum and weighted mean of X_CM over the 100 s of the simulation. (**e**) As an alternative to the "Slice View" of panel (**c**), one can do a Surface View

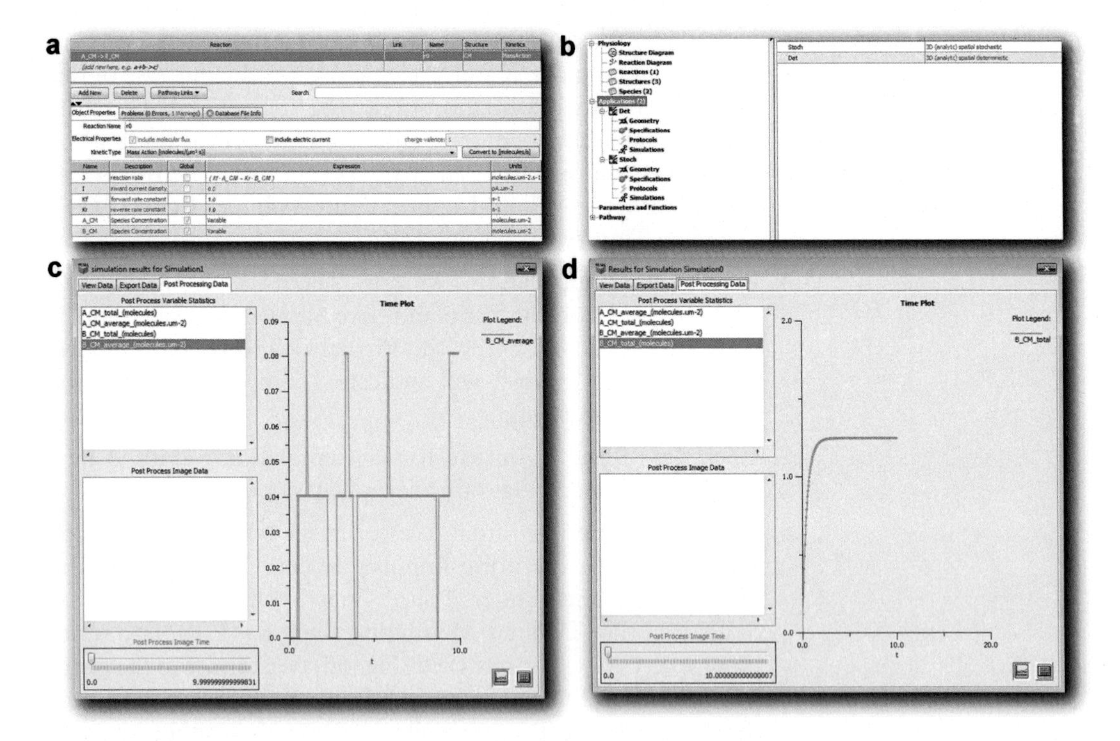

Fig. 7 Contrasting deterministic and stochastic simulations. (**a**) The model considered for the comparison is a simple system where a molecule is in one of two states (A or B) and switches between the two at rates kf and kr. (**b**) The model is used to create both a deterministic (Det) and stochastic (Stoch) application. Note that both have the same geometry and reaction description. Sample stochastic (**c**) and deterministic (**d**) simulations. In the stochastic simulation, the number of molecules in state "B" changes at discrete points in time and in specific quanta (of single molecules). In the deterministic simulation, the numbers can change in a continuum

3.2.2 Specifying the Reactions

1. For stochastic simulations, only Mass Action kinetics are permitted. Hence we cannot manipulate the reaction rates in this case and can only enter the coefficients as shown in Table 3 (Fig. 7a).
2. Save this model as "Stoch_Example."

3.2.3 Creating the Applications

1. Here we wish to perform both deterministic and stochastic simulations using the same parameters and compare the results.
2. Hence we create two applications. For the first application, right-click on "Applications" and then choose "Add New" → "Deterministic," while for the second choose "Stochastic."
3. Rename the created applications, "Det" and "Stoch" respectively (Fig. 7b).

3.2.4 Specifying the Geometry

1. In each of the created applications, we need to specify geometry. Spatial stochastic simulations require a 3D geometry.
2. Having already created a 3D geometry "3D_BEN," import this geometry and link the structures similar to Subheading 3.1. We do this for both applications.
3. Save the applications.

3.2.5 Initial Conditions

1. In both Deterministic and Stochastic applications, use the same initial conditions as listed in Table 3, so as to ensure a proper comparison base.

3.2.6 Running Simulations and Comparison

1. In the "Det" application, run the simulation similar to Subheading 3.1, using the Fully-Implicit solver with End time = 10s, Output Interval = 0.05 s and Maximum Timestep = 0.01.
2. For the "Stoch" application, run the simulation using the Spatial Stochastic Solver—Smoldyn, with the same parameters as the Deterministic case.
3. In both cases, the average time plots can be observed by clicking the "Post-processing Data" tab on viewing the results. One can then clearly (*see* Fig. 7c, d) how the stochastic element affects the output. In the Deterministic case, the product settles to an equilibrium value, whereas for the stochastic case the product oscillates randomly around the equilibrium mark.

4 Notes

1. The complete Virtual Cell implementation of the BEN model is freely available in the Virtual Cell. To access this model, go to the menu options in the main model window. Click File → Open → Biomodel. Under "Public BioModels," choose "sayak66" and you can access both the "BEN" and "Smoldyn" model. In the Virtual Cell, the default units are μM for volumetric concentrations and molecules/μm^2 for membrane-bound species. The spatial dimension is in μm.
2. The differential equations representing the activator and inhibitor cannot be written in reaction form like the ligand-binding reaction, and thus are stated in a way that directly implements the X and Y differential equations.
3. Boundary conditions for this simulation assume that there is no flux of protein across the cellular membrane. In situations where the cell and stimulus is symmetric, zero flux boundary conditions can also be used to take advantage of this symmetry. For example, in Ref. [36] a three-dimensional model of the cell used symmetry in the x, y, and z directions was developed. This allowed simulations to run on only one-eighth of the cell, saving both computer time and memory.
4. In setting the initial conditions, one would like to recreate the cell's basal levels. However, as the correct values are rarely known, it is more customary to run the simulation for some time with a stimulus, and to let the cell equilibrate at this value.
5. This reaction (rx) is not a true reaction, but rather one way of setting up diffusion of "L" from source "L_source." We can

interpret it as the source releasing the chemoattractant (L) at rate kf, with the chemoattractant diffusing through the rest of the extracellular space and degrading at a rate of kr.

6. A movie can be created using the 3D results which represent a visual of how the cell behaved in response to the stimulus. 0.23 molecules/μm^2 came out to be the average equilibrium value of the activator and thus was set to the maximum, so that white color would represent the front of the cell spiking to the stimulus. In the video, we can clearly see the white circle expanding in response to the stimulus and then setting back to equilibrium.

References

1. Levchenko A, Iglesias PA (2002) Models of eukaryotic gradient sensing: application to chemotaxis of amoebae and neutrophils. Biophys J 82:50–63
2. Levine H, Kessler DA, Rappel WJ (2006) Directional sensing in eukaryotic chemotaxis: a balanced inactivation model. Proc Natl Acad Sci U S A 103:9761–9766
3. Onsum MD, Rao CV (2009) Calling heads from tails: the role of mathematical modeling in understanding cell polarization. Curr Opin Cell Biol 21:74–81
4. Jilkine A, Edelstein-Keshet L (2011) A comparison of mathematical models for polarization of single eukaryotic cells in response to guided cues. PLoS Comput Biol 7:e1001121
5. Yang L, Effler JC, Kutscher BL et al (2008) Modeling cellular deformations using the level set formalism. BMC Syst Biol 2:68
6. Wolgemuth CW, Stajic J, Mogilner A (2011) Redundant mechanisms for stable cell locomotion revealed by minimal models. Biophys J 101:545–553
7. Vanderlei B, Feng JJ, Edelstein-Keshet L (2011) A computational model of cell polarization and motility coupling mechanics and biochemistry. Multiscale Model Simul 9:1420–1443
8. Neilson MP, Veltman DM, van Haastert PJ et al (2011) Chemotaxis: a feedback-based computational model robustly predicts multiple aspects of real cell behaviour. PLoS Biol 9:e1000618
9. Hecht I, Skoge ML, Charest PG et al (2011) Activated membrane patches guide chemotactic cell motility. PLoS Comput Biol 7:e1002044
10. Holmes WR, Edelstein-Keshet L (2012) A comparison of computational models for eukaryotic cell shape and motility. PLoS Comput Biol 8:e1002793
11. Shi C, Huang CH, Devreotes PN, Iglesias PA (2013) Interaction of motility, directional sensing, and polarity modules recreates the behaviors of chemotaxing cells. PLoS Comput Biol 9:e1003122
12. Naoki H, Sakumura Y, Ishii S (2008) Stochastic control of spontaneous signal generation for gradient sensing in chemotaxis. J Theor Biol 255:259–266
13. Xiong Y, Huang CH, Iglesias PA, Devreotes PN (2010) Cells navigate with a local-excitation, global-inhibition-biased excitable network. Proc Natl Acad Sci U S A 107: 17079–17086
14. Iglesias PA, Devreotes PN (2012) Biased excitable networks: how cells direct motion in response to gradients. Curr Opin Cell Biol 24:245–253
15. Ryan GL, Petroccia HM, Watanabe N, Vavylonis D (2012) Excitable actin dynamics in lamellipodial protrusion and retraction. Biophys J 102:1493–1502
16. Cooper RM, Wingreen NS, Cox EC (2012) An excitable cortex and memory model successfully predicts new pseudopod dynamics. PLoS One 7:e33528
17. Huang CH, Tang M, Shi C et al (2013) An excitable signal integrator couples to an idling cytoskeletal oscillator to drive cell migration. Nat Cell Biol 15:1307–1316
18. Nishikawa M, Hörning M, Ueda M, Shibata T (2014) Excitable signal transduction induces both spontaneous and directional cell asymmetries in the phosphatidylinositol lipid signaling system for eukaryotic chemotaxis. Biophys J 106:723–734
19. Tang M, Wang M, Shi C et al (2014) Evolutionarily conserved coupling of adaptive and excitable networks mediates eukaryotic chemotaxis. Nat Commun 5:5175

20. Skoge M, Yue H, Erickstad M et al (2014) Cellular memory in eukaryotic chemotaxis. Proc Natl Acad Sci U S A 111:14448–14453
21. Janetopoulos C, Ma L, Devreotes PN, Iglesias PA (2004) Chemoattractant-induced phosphatidylinositol 3,4,5-trisphosphate accumulation is spatially amplified and adapts, independent of the actin cytoskeleton. Proc Natl Acad Sci U S A 101:8951–8956
22. Postma M, Bosgraaf L, Loovers HM, Van Haastert PJ (2004) Chemotaxis: signalling modules join hands at front and tail. EMBO Rep 5:35–40
23. Balázsi G, van Oudenaarden A, Collins JJ (2011) Cellular decision making and biological noise: from microbes to mammals. Cell 144:910–925
24. Eldar A, Elowitz MB (2010) Functional roles for noise in genetic circuits. Nature 467:167–173
25. Rao CV, Wolf DM, Arkin AP (2002) Control, exploitation and tolerance of intracellular noise. Nature 420:231–237
26. Gillespie DT (2007) Stochastic simulation of chemical kinetics. Annu Rev Phys Chem 58:35–55
27. Munsky B, Khammash M (2006) The finite state projection algorithm for the solution of the chemical master equation. J Chem Phys 124:044104
28. Gillespie DT (2000) The chemical Langevin equation. J Chem Phys 113:297–306
29. Andrews SS (2012) Spatial and stochastic cellular modeling with the Smoldyn simulator. Methods Mol Biol 804:519–542
30. Andrews SS, Addy NJ, Brent R, Arkin AP (2010) Detailed simulations of cell biology with Smoldyn 2.1. PLoS Comput Biol 6:e1000705
31. Alves R, Antunes F, Salvador A (2006) Tools for kinetic modeling of biochemical networks. Nat Biotechnol 24:667–672
32. Bathe K-J (1996) Finite element procedures. Prentice Hall, Englewood Cliffs, NJ
33. Cowan AE, Moraru II, Schaff JC et al (2012) Spatial modeling of cell signaling networks. Methods Cell Biol 110:195–221
34. Resasco DC, Gao F, Morgan F et al (2012) Virtual Cell: computational tools for modeling in cell biology. Wiley Interdiscip Rev Syst Biol Med 4:129–140
35. Yang L, Iglesias PA (2009) Modeling spatial and temporal dynamics of chemotactic networks. Methods Mol Biol 571:489–505
36. Li HY, Ng WP, Wong CH et al (2007) Coordination of chromosome alignment and mitotic progression by the chromosome-based Ran signal. Cell Cycle 6:1886–1895

Index

Tian Jin and Dale Hereld (eds.), *Chemotaxis: Methods and Protocols*, Methods in Molecular Biology, vol. 1407, DOI 10.1007/978-1-4939-3480-5, © Springer Science+Business Media New York 2016

D

E

N

O

P

R

S